$$F(\omega) = \int_{-\infty}^{\infty} f(t)e^{-j\omega t}\, dt$$

$$f(t) = \int_{-\infty}^{\infty} F(\omega)e^{jt\omega}\, d\omega$$

$$Y_{zs}(\omega) = X(\omega)H(\omega) \longleftrightarrow \int_{-\infty}^{\infty} x(\tau)h(t-\tau)\, d\tau$$

$$H(\omega) = \frac{\sum_{n=0}^{M} b_n(j\omega)^n}{\sum_{n=0}^{N} a_n(j\omega)^n}$$

$$F_{dm} = \sum_{n=0}^{N-1} f_{dn} e^{-j2\pi mn/N} = F_d(m\Omega)$$

$$f_{dn} = \frac{1}{N}\sum_{m=0}^{N-1} F_{dm} e^{j2\pi nm/N} = f_d(nT_s) \qquad \frac{\Omega}{2\pi} = \frac{1}{NT_s}$$

$$F(s) = \int_{0^-}^{\infty} f(t)e^{-st}\, dt$$

$$f(t) = \frac{1}{2\pi j} \int_{\sigma-j\infty}^{\sigma+j\infty} F(s)e^{ts}\, ds$$

$$Y_{zs}(s) = X(s)H(s) \longleftrightarrow \int_{0^-}^{t} x(\tau)h(t-\tau)\, d\tau$$

$$H(s) = \frac{\sum_{n=0}^{M} b_n s^n}{\sum_{n=0}^{N} a_n s^n}$$

$$\frac{C(s)}{R(s)} = \frac{G(s)}{1 \mp G(s)H(s)}$$

Continuous and Discrete Linear Systems

HERBERT P. NEFF, JR.
The University of Tennessee

HARPER & ROW, PUBLISHERS, New York
Cambridge, Philadelphia, San Francisco,
London, Mexico City, São Paulo, Sydney

To my dad, who inspired me

Sponsoring Editor: Carl McNair
Project Editor: David Nickol
Production: Delia Tedoff
Compositor: Syntax International Pte. Ltd.
Printer and Binder: R. R. Donnelley and Sons Company
Art Studio: J&R Art Services, Inc.

Continuous and Discrete Linear Systems

Copyright © 1984 by Harper & Row, Publishers, Inc.

All rights reserved. Printed in the United States of America. No part of this book may be used or reproduced in any manner whatsoever without written permission, except in the case of brief quotations embodied in critical articles and reviews. For information address Harper & Row, Publishers, Inc., 10 East 53d Street, New York, NY 10022.

Library of Congress Cataloging in Publication Data

Neff, Herbert P., 1930–
   Continuous and discrete linear systems.

   Includes bibliographies and index.
   1. System analysis. I. Title.
QA402.N424   1984   003   83-22569
ISBN 0-06-044787-7

Contents

Preface vii

1 Systems 1

1.1 The System Concept 1
1.2 Linear Systems 4
1.3 Linear-System Classification and Description 9
1.4 Concluding Remarks 13
   References 13

2 The Continuous-Time Linear-System Differential Equations 14

2.1 Linear Differential Equations 14
2.2 The General Form 17
2.3 Conventional Methods of Solution 18
2.4 The Unit-Impulse Function 26
2.5 The Unit-Impulse Response 30
2.6 The Complete Response 44
2.7 The Second-Order System 57

2.8 Multiple Inputs and Outputs   62
2.9 Convolution   66
2.10 Stability   70
2.11 Concluding Remarks   72
   Problems   73
   References   80

3 State-Space Analysis   81

3.1 The Normal-Form Equations   82
3.2 Simplified State-Variable Analysis   87
3.3 Normal Form Block Diagrams   96
3.4 Further Consideration of Matrix Solutions   102
3.5 Numerical Methods   113
3.6 Concluding Remarks   118
   Problems   118
   References   121

4 Phasors and Fourier Series   122

4.1 The Phasor Concept   123
4.2 Trigonometric Series   128
4.3 The Fourier Trigonometric Series   132
4.4 Least Mean Square Error   136
4.5 Uses of the Fourier Series   140
4.6 Average Power   145
4.7 The Discrete Fourier Series   151
4.8 Concluding Remarks   160
   Problems   160
   References   165

5 Fourier Transform Techniques   166

5.1 Extension of the Fourier Series   167
5.2 Fourier Transform Pairs   171
5.3 The Time-Invariant Linear System (Again)   185
5.4 Periodic Excitation   189
5.5 Frequency Response   192
5.6 Additional Examples   199
5.7 Simultaneous Equations   206
5.8 The Sampling Theorem   207

5.9 The Discrete Fourier Transform   213
5.10 The Fast Fourier Transform   224
5.11 Power and Energy Spectral Densities   229
5.12 Concluding Remarks   234
     Problems   235
     References   241

6 The Laplace Transform   242

6.1 Fourier to Laplace Transform   243
6.2 Laplace Transform Pairs   250
6.3 Examples   260
6.4 A Table of Laplace Transform Pairs   268
6.6 Concluding Remarks   273
    Problems   274
    References   278

7 Applications of the Laplace Transform   279

7.1 Solution of Differential Equations   279
7.2 Simultaneous Equations   290
7.3 Appearance of the Unit Impulse   293
7.4 Other Examples   301
7.5 Stability and Feedback   307
7.6 State Variables (Again)   319
7.7 Concluding Remarks   321
    Problems   321
    References   331

8 Discrete-Time Systems   332

8.1 The Relation Between Continuous-Time and Discrete-Time Systems   334
8.2 The Unit Function and Unit Sequence   336
8.3 The Unit-Function Response   337
8.4 Linear Difference Equations   342
8.5 Examples   347
8.6 The Second-Order System   362
8.7 Frequency Response   365
8.8 State-Space Analysis   375
8.9 Simulation   387

8.10 Stability   392
8.11 Concluding Remarks   393
     Problems   394
     References   399

9 The z Transform   400

9.1 The z Transform   401
9.2 z-Transform Pairs   404
9.3 Examples   412
9.4 Linear Difference Equations (Again)   420
9.5 Frequency Response   438
9.6 Simulation   444
9.7 State-Space Analysis   456
9.8 Concluding Remarks   466
    Problems   467
    References   474

Appendices   475

Index   501

Preface

This textbook was written to present, explain, and demonstrate those techniques of linear-system analysis that can most easily be understood and used by electrical engineering students who are preparing for careers in today's rapidly changing technology. To a large extent changes have been due to the advances in integrated circuits and microprocessor techniques that have made it possible to digitally perform almost any desired operation. Thus, a modern text on the subject of linear systems must include adequate coverage of digital systems as well as the treatment of the traditional analog systems.

Most electrical engineering curricula include two years of mathematics (including calculus and some treatment of differential equations), two years of physics (or the equivalent), and a year of dc and ac circuits at the sophomore level. The first two (mathematics and physics) are necessary prerequisites for proceeding to a study of the material in this text, assuming, of course, that Kirchhoff's laws are treated in the second-year physics course. The sophomore circuits course, while not absolutely necessary for understanding the material herein, is necessary for other beginning junior electrical engineering courses (electronics, for example). It is not likely that the first two years of most electrical engineering programs will undergo drastic changes in the next several years. It is likely that more *digital* linear system analysis will be given earlier in the junior year in these programs.

Much advantageous use in the author's *Basic Electromagnetic Fields* is made of the fact that if the unit-impulse response is known (and it is often *given* by the fundamental laws), then the response for *any* source can be found by convolution. The same philosophy is followed in this text, since a linear system is involved in both cases: If the unit-impulse response (for the continuous-time case) or the unit-function response (for the discrete-time case) is known, then the response for any source can be found. This development is completely independent of any transform techniques. The method that is used for finding the unit-impulse response or the unit-function response in the most common cases is usually not employed in other texts, but is extremely simple, and only requires finding the roots of a polynomial.

Linear system analysis can be presented in several ways: discrete-time systems first, continuous-time systems first, or simultaneous presentation of discrete- and continuous-time systems. The latter was recently tried at the University of Tennessee with less than satisfactory results. Discrete-time systems are easier to analyze, so it can be argued that they should be presented first. On the other hand, beginning with the continuous-time system first is most likely a direct link to what the student has just studied in the sophomore year.

Chapter 1 is a brief introduction to linear systems. The analysis of continuous-time systems in the time domain begins in Chapter 2. State-space analysis in the time domain is the subject of Chapter 3. Chapter 4 concerns the Fourier trigonometric series, and in many cases this subject will have already been studied in the sophomore year. Frequency-domain analysis begins in Chapter 5 with the introduction of the Fourier transform. The very popular and useful Laplace transform is introduced in Chapter 6, and examples of its use are given in Chapter 7. The time-domain analysis of discrete-time systems is the subject of Chapter 8, while the z transform for discrete systems is introduced in Chapter 9. Numerical methods are used throughout where appropriate.

Emphasis is placed on the various representations of linear systems throughout the book. Among these are the differential or difference equation, the unit-impulse or unit-function response, the state-space formulation, the block diagram, and the transfer function.

The amount of time required to completely cover the material in this book is approximately nine quarter hours or six semester hours. The arrangement of the material is such that the coverage can be reduced, if desired, with no loss in continuity. Chapter 4 can be skipped, for example, if the Fourier series has already been covered. Many educators feel that state-space analysis should be reserved for more advanced courses at the senior or graduate level, so Chapter 3 (in its entirety), and parts of Chapters 7, 8, and 9 can be eliminated. In this way, much less time would be required to cover both continuous-time and discrete-time systems. Other arrangements for the presentation of the material are certainly possible.

Having successfully mastered the material in this text, the student is prepared for specialized courses in communication systems, feedback control systems, and more advanced studies in courses that may have already been begun. These would include energy conversion, electronics, electromagnetic theory, computer engineering, pattern recognition, and others.

The author would like to express his appreciation to several members of the faculty of the department of electrical engineering, University of Tennessee, Knoxville, for the helpful suggestions they made during the preparation of the manuscript for this textbook. Particular thanks go to Professors J. M. Bailey, R. C. Gonzalez, W. L. Green, and J. D. Tillman Jr. The author is also indebted to the reviewers whose labor on the manuscript is best recognized in the final product by the reviewers themselves. Finally, no project of this scope is possible without suitable staff support, and for this the author wishes to thank M. R. Bearden, R. L. Campbell, S. A. Perkins, D. R. Smiddy, and C. A. Williams.

Herbert P. Neff, Jr.

Chapter 1
Systems

In this introductory chapter we will discuss the system concept in general to determine what it is that we are to later analyze. The student has most likely already been exposed to a linear-systems course. The electrical engineering student has probably spent a year in studying network analysis, and this is one example of linear-systems analysis. We would like to broaden the linear-system concept in this chapter to include other types with which the student may not be familiar. The discrete-time system is an example. The principle of superposition will be reviewed, since it is fundamental to linear-systems analysis. Analysis will begin in Chap. 2.

1.1 THE SYSTEM CONCEPT

The electrical engineering student who is beginning the junior year has undoubtedly been exposed to the term *function* from his or her calculus courses. This term was used to include voltages and currents in the circuits courses, but it is much more general than this. In most of the material that follows, the term *function* is interchangeable with *signal*, and many of the signals that we deal with are, in fact, voltages or currents. However, it should be kept in mind that *signal* or *function* may refer to other items as well. These signals must be dependent on (or be *functions of*) some *independent variable* or *variables*, unless, of course, we are concerned with something similar to a closed circuit containing only batteries and resistors (for example). Here, we are usually concerned with only *one* independent variable, and it is usually time (*t*), but *t* can represent *any* independent variable. Consider the series *RL* circuit driven by an ideal voltage source *v*(*t*) resulting in a current *i*(*t*). The equation that describes this situation is given by Kirchhoff's voltage law:

$$v(t) = Ri(t) + L\frac{d}{dt}i(t)$$

We have also just described a *linear system*.

It should be pointed out that in some situations there may be more than one independent variable involved. The general *electric field intensity* $\mathbf{E}(x, y, z, t)$, for example, is a function of three spatial independent variables (*x*, *y*, and *z*) as well as the independent variable time (*t*). The student will encounter this type of function in a "fields" course, but it is important to be aware of the fact that many of the techniques that we will learn here (*convolution*, for example) are directly applicable to much of field theory.[1]

A *system* may be broadly described as a configuration of components or elements arranged to perform as an entity. A system has at least one *input* signal and at least one *output* signal. As indicated above, these signals are usually functions of time. Consider the single-input single-output case. If the input is called *x*(*t*), and the output called *y*(*t*), then the description above implies the *block diagram* description below in Fig. 1.1. Notice that an *instantaneous system* or *zero-memory system*, where the output at time *t* depends only on the input at time *t* (not on past or future values of the input), is implied. Thus, it is assumed, unless stated to the contrary, that the input is applied at $t = -\infty$ with no energy storage. That is, the system is at rest when the excitation or input is applied at a known instant of time, and if the behavior (output) of the system at some future time is desired, then the energy stored in all the energy storage elements must be known at the instant of application of the input.

If the system is not instantaneous it is said to be dynamic. A *dynamic system* is such that the output at some time depends on the present value of the input and some past values of the input (from energy storage, for example). A system that contains an energy-dissipating element and an energy-storage element is dynamic. The series *RC* circuit driven by an ideal voltage source is dynamic if the capacitor is already charged when the input is applied. A system that has an output at time *t* that depends only on the input in the interval $t - T$ to t ($T \geq 0$) is said to have a *memory* of length *T*. A lossless transmission line that

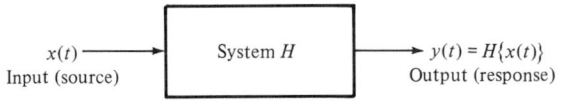

$x(t)$ → System H → $y(t) = H\{x(t)\}$
Input (source)   Output (response)

Figure 1.1. General system with one input *x*(*t*) and one output *y*(*t*).

is terminated in a matched load so that it supports an undamped traveling wave is another example of a dynamic system. An instantaneous system has a memory of zero length, hence it is also called a zero-memory system.

It is also implied in Fig. 1.1 that t is a *continuous* variable, as opposed to a *discrete* variable. The distinction between these independent variables will be discussed in more detail later on. The fact that t is a continuous variable does not alter the fact that both $x(t)$ and $y(t)$ may be discontinuous at some point or points. In spite of all the implications above we will see that the concepts developed from Fig. 1.1 can easily be extended to treat all of the cases of interest in this textbook.

Notice carefully that in Fig. 1.1 *the single input or output may consist of more than one term*. For example: $x(t) = 2t + \cos(2t) + 4\sin(3t)$, or $y(t) = t^2 + 2t + 3$. The fact remains that there is still only one input and one output. It is helpful here to remember that there is only one line entering the block, and only one line leaving the block in Fig. 1.1.

We may say from Fig. 1.1 that the system *operates* on the input signal to produce the output signal. In operational notation:

$$\boxed{y(t) = H\{x(t)\}} \tag{1.1}$$

It is also convenient to think of the block in Fig. 1.1 as *transforming* the input signal into the output signal. In more mathematical language: $H\{x(t)\}$ is a rule that assigns to each value of $x(t)$ one, and only one, value of $y(t)$. The operator may perform many tasks, such as differentiation or integration, but we will first look at an extremely simple example.

Example 1

Consider Ohm's law, $v(t) = Ri(t)$, and the circuits of Fig. 1.2. In Fig. 1.2(a), $v(t)$ is the single input signal (independent voltage source) and $i(t) = v(t)/R$ is the single output signal. The operator H is simply a multiplier, $H = 1/R$. In Fig. 1.2(b), $i(t)$ is the single input signal (independent current source) and $v(t) = Ri(t)$ is the single output signal. The operator is the multiplier $H = R$. If, in Fig. 1.2(a), $v(t) = 10 + \cos(3t)$, we still have only one input, but it consists of two parts.

If the system has M separate inputs and P separate outputs, what modifications are necessary in Fig. 1.1? It turns out that we merely replace $x(t)$ with

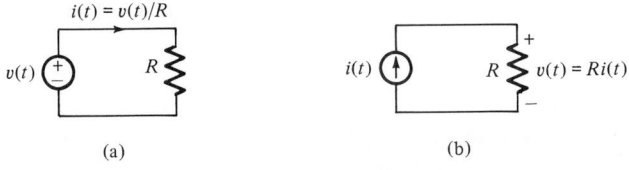

Figure 1.2. (a) Independent voltage source and resistor. (b) Independent current source and resistor.

4 SYSTEMS

$\mathbf{x}(t)$ and $\mathbf{y}(t)$ with $\mathbf{y}(t)$, where[†]

$$\mathbf{x}(t) = \begin{bmatrix} x_1(t) \\ x_2(t) \\ \vdots \\ x_M(t) \end{bmatrix} \qquad \mathbf{y}(t) = \begin{bmatrix} y_1(t) \\ y_2(t) \\ \vdots \\ y_P(t) \end{bmatrix}$$

These are column *matrices*[‡] or column *vectors* with M and P components, respectively. The use of matrices provides much simplification in the notation. Thus, the more general form of Eq. 1.1 is

$$\boxed{\mathbf{y}(t) = H\{\mathbf{x}(t)\}} \tag{1.2}$$

1.2 LINEAR SYSTEMS

We are primarily interested in linear systems in this text. The nonlinear case is treated in more advanced courses, and it is important that the linear case be understood first, since nonlinear systems are often analyzed by linearizing them (over a limited range) and then applying techniques of linear analysis.

A system is linear if it satisfies the *principle of superposition*. To be more precise, we say that a system is linear if, and only if,

$$H\{\alpha \mathbf{x}_a(t) + \beta \mathbf{x}_b(t)\} = \alpha H\{\mathbf{x}_a(t)\} + \beta H\{\mathbf{x}_b(t)\} \tag{1.3}$$

$$\mathbf{x}_a(t) = \begin{bmatrix} x_{a1}(t) \\ x_{a2}(t) \\ \vdots \\ x_{aM}(t) \end{bmatrix} \qquad \mathbf{x}_b(t) = \begin{bmatrix} x_{b1}(t) \\ x_{b2}(t) \\ \vdots \\ x_{bM}(t) \end{bmatrix}$$

Notice that there are M inputs in Eq. 1.3, and each of these input signals can consist of two terms. Notice, once again, that we are considering (here) t to be a continuous independent variable, and $\mathbf{x}_a(t)$ and $\mathbf{x}_b(t)$ are applied when the system is at rest.

If $\alpha = \beta = 1$ in Eq. 1.3,

$$H\{\mathbf{x}_a(t) + \mathbf{x}_b(t)\} = H\{\mathbf{x}_a(t)\} + H\{\mathbf{x}_b(t)\} \qquad \text{(additivity)} \tag{1.4}$$

In other words, the response of a sum input vectors is the sum of the responses. This property is called *additivity*. If $\mathbf{x}_b(t) = 0$, then Eq. 1.3 becomes

$$H\{\alpha \mathbf{x}_a(t)\} = \alpha H\{\mathbf{x}_a(t)\} \qquad \text{(homogeneity)} \tag{1.5}$$

The response of a constant multiplier times any input vector is the response of the input vector times the constant. This property is called *homogeneity*. Thus, a linear system must satisfy *both* the requirements of additivity and homogeneity, and conversely, any system that has these properties is linear. Equations 1.2–1.5 apply to the single-input single-output case when $M = P = 1$.

[†] Notice that the **boldface** type indicates a multiple input or multiple output. Notice also that there will then be M input "lines" and P output "lines" in Fig. 1.1.
[‡] Matrices and matrix operations are discussed in more detail in Appendix A.

1.2 LINEAR SYSTEMS

Consider a diode that is operated over its square-law region so that

$$y(t) = H\{x(t)\} = x^2(t)$$

Notice that this is a single-input single-output system. It is nonlinear because, if we check for additivity (Eq. 1.4 with $M = P = 1$),

$$H\{x_a(t) + x_b(t)\} = [x_a(t) + x_b(t)]^2 \neq H\{x_a(t)\} + H\{x_b(t)\} = x_a^2(t) + x_b^2(t)$$

Does this square-law device satisfy homogeneity?

$$H\{\alpha x_a(t)\} = \alpha^2 x_a^2(t) \neq \alpha H\{x_a(t)\} = \alpha x_a^2(t)$$

No!

The *balanced modulator* is simply a device that multiplies two input signals to produce its *one* output. We will encounter it again later on. Notice that here we should use $M = 2$ and $P = 1$. The system is described by

$$y(t) = H\left\{\begin{bmatrix} x_1(t) \\ x_2(t) \end{bmatrix}\right\} = x_1(t)x_2(t)$$

Checking for homogeneity first, Eq. 1.5 gives

$$H\left\{\begin{bmatrix} \alpha x_1(t) \\ \alpha x_2(t) \end{bmatrix}\right\} = \alpha x_1(t)\alpha x_2(t) \neq \alpha H\left\{\begin{bmatrix} x_1(t) \\ x_2(t) \end{bmatrix}\right\} = \alpha x_1(t)x_2(t)$$

so it is nonlinear.

We mention here that all systems that are described by linear differential equations and linear difference equations are linear.

If α in Eq. 1.5 is a *rational number*, then it can be shown that *additivity* implies *homogeneity*.

It is not necessary that the signal vectors in Eq. 1.3 contain only two terms. They may contain many:

$$H\left\{\sum_{i=1}^{n} \alpha_i \mathbf{x}_i(t)\right\} = \sum_{i=1}^{n} \alpha_i H\{\mathbf{x}_i(t)\} \tag{1.6}$$

We must be careful to remember that $\mathbf{x}_i(t)$ is still a column vector, and a double-subscripted notation is required. For example if $n = 3$ and $M = 2$, the left side of Eq. 1.6 is

$$H\left\{\begin{matrix} \alpha_1 x_{11} + \alpha_2 x_{21} + \alpha_3 x_{31} \\ \alpha_1 x_{12} + \alpha_2 x_{22} + \alpha_3 x_{32} \end{matrix}\right\}$$

We also assume (for future use) that the result in Eq. 1.6 can be extended to infinite sums and even integrals:

$$H\left\{\sum_{i=1}^{\infty} \alpha_i \mathbf{x}_i(t)\right\} = \sum_{i=1}^{\infty} \alpha_i H\{\mathbf{x}_i(t)\} \tag{1.7}$$

$$H\left\{\int_a^b \alpha(\tau)\mathbf{x}(t, \tau)\,d\tau\right\} = \int_a^b \alpha(\tau)H\{\mathbf{x}(t, \tau)\}\,d\tau \tag{1.8}$$

6 SYSTEMS

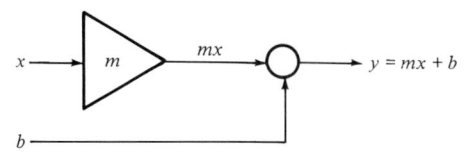

Figure 1.3. Simple system in block diagram form.

Example 2

Suppose that an instantaneous system is described by the simple block diagram in Fig. 1.3. Notice carefully that b is a *constant*, and is not considered here to be a (variable) input, although it certainly appears to be an input. Checking for additivity, we have $y = H\{x\} = mx + b$, and

$$H\{x_a + x_b\} = m(x_a + x_b) + b \neq H\{x_a\} + H\{x_b\} = mx_a + b + mx_b + b$$

Thus, even though the output is a linear function of the input, this is not a linear system. Obviously, if $b = 0$, the system is linear.

Next, suppose that b is considered to be an input signal, and $b = x_2$. Then, as seen in Fig. 1.4,

$$y = H\begin{Bmatrix} x_1 \\ x_2 \end{Bmatrix} = mx_1 + x_2$$

Checking for additivity,

$$H\begin{Bmatrix} x_{a1} + x_{b1} \\ x_{a2} + x_{b2} \end{Bmatrix} = m(x_{a1} + x_{b1}) + (x_{a2} + x_{b2}) = H\begin{Bmatrix} x_{a1} \\ x_{a2} \end{Bmatrix} + H\begin{Bmatrix} x_{b1} \\ x_{b2} \end{Bmatrix}$$

$$= mx_{a1} + x_{a2} + mx_{b1} + x_{b2}$$

Thus, this system does satisfy the additivity property. Checking for homogeneity,

$$H\begin{Bmatrix} \alpha x_1 \\ \alpha x_2 \end{Bmatrix} = \alpha m x_1 + \alpha x_2 = \alpha H\begin{Bmatrix} x_1 \\ x_2 \end{Bmatrix} = \alpha(mx_1 + x_2)$$

It also satisfies the homogeneity property. This is a linear system. The reader should carefully verify these results using Fig. 1.4 explicitly.

The preceding example demonstrates that the linear or nonlinear character of a system is sometimes nebulous. The constant b is like an initial condition,

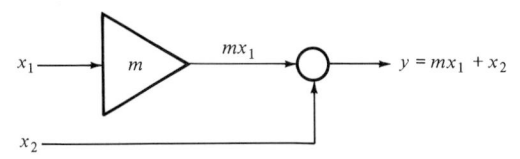

Figure 1.4. System of Fig. 1.3 with $b = x_2$ considered to be a second input.

1.2 LINEAR SYSTEMS

and as we have seen, if it is removed (set equal to zero), the system *itself* is inherently linear. This suggests that we should set all initial conditions to zero when we test a system for linearity. Another example will help clarify this.

Example 3

In the network of Fig. 1.5 $v(t)$ is the input and $i(t)$ is the output. The governing equation, obtained from Kirchhoff's current law applied at the upper node where R and L are joined is

$$i(t) = \frac{v(t)}{R} + \frac{1}{L} \int_0^t v(\tau) \, d\tau + i_L(0) \qquad t > 0$$

where $i_L(0)$ is the initial inductor current (at $t = 0$). The block labeled $s^{-1} = 1/s$ represents $\int_0^t (\ ) \, d\tau$, and s signifies differentiation.[†] Generally speaking, the existence of a nonzero value for $i_L(0)$ must be allowed, but in Fig. 1.5 an initial inductor current must arise from a source that is not shown. If we *separately* apply the signals $v_a(t)$ and $v_b(t)$ at the input, then the separate output signals are

$$i_a(t) = \frac{v_a(t)}{R} + \frac{1}{L} \int_0^t v_a(\tau) \, d\tau + i_L(0)$$

$$i_b(t) = \frac{v_b(t)}{R} + \frac{1}{L} \int_0^t v_b(\tau) \, d\tau + i_L(0)$$

The input $\alpha v_a(t) + \beta v_b(t)$ produces the output

$$i(t) = \frac{\alpha v_a(t) + \beta v_b(t)}{R} + \frac{1}{L} \int_0^t [\alpha v_a(\tau) + \beta v_b(\tau)] \, d\tau + i_L(0)$$

(a)   (b)

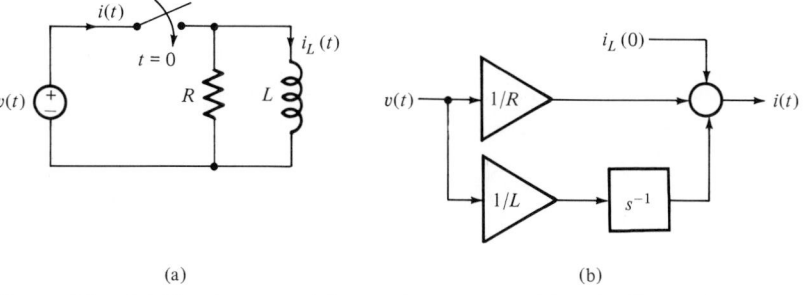

Figure 1.5. (a) Simple network for testing superposition. (b) Block diagram representing the mathematical model.

[†] Here s and s^{-1} mean nothing more than what has just been stated. We will see in Chap. 6 that s implies the same thing, plus much more, as the Laplace transform variable.

But

$$\alpha i_a(t) + \beta i_b(t) = \alpha\left[\frac{v_a(t)}{R} + \frac{1}{L}\int_0^t v_a(\tau)\,d\tau + i_L(0)\right]$$
$$+ \beta\left[\frac{v_b(t)}{R} + \frac{1}{L}\int_0^t v_b(\tau)\,d\tau + i_L(0)\right]$$

Because of $i_L(0)$ the last two equations do not agree. What we actually have, however, is a linear system with nonzero initial conditions. We conclude that all initial conditions must be removed (that is, the system must be "at rest") when it is tested for linearity.

On the other hand, it is easy to analyze a linear system with initial conditions (or initial energy storage) as a linear system by simply treating the initial conditions as inputs. This is equivalent to decomposing the response into separate parts: one resulting from the initial energy storage, and the other resulting from the system input. We will call these the *zero-input* response, $y_{zi}(t)$, and the *zero-state* response, $y_{zs}(t)$, respectively, in the next chapter. This is suggested in Fig. 1.5(b). It was also suggested in Example 2.

The examples that have been used up to this point have represented *continuous-time* systems. It would be worthwhile to consider a simple *discrete-time* system; that is, a system where the independent variable appears only at discrete instants of time $t = nT$, or $t = 0, T, 2T, \ldots$ Fig. 1.7(b) shows a discrete-time signal.

Example 4

The block diagram of a discrete system is shown in Fig. 1.6. The discrete input is $x(nT) \equiv x(n)$, the discrete output is $y(nT) = y(n)$, and the block labeled z^{-1} produces a delay in time of T (shift register).[†] This is a linear system because

$$y_a(n) = \alpha[x_a(n-1) + 2x_a(n)] \qquad \text{[produced by } \alpha x_a(n)\text{]}$$
$$y_b(n) = \beta[x_b(n-1) + 2x_b(n)] \qquad \text{[produced by } \beta x_b(n)\text{]}$$

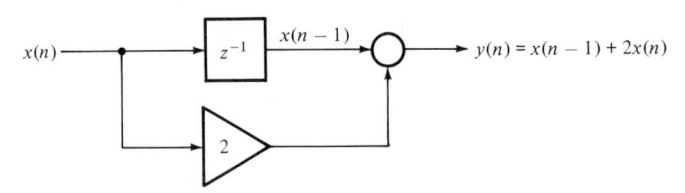

Figure 1.6. Block diagram of a discrete system.

[†] As a matter of convenience, we drop the T in the discrete independent variable $[x(nT) = x(n)]$.

and the input $\alpha x_a(n) + \beta x_b(n)$ produces

$$y(n) = \alpha x_a(n-1) + \beta x_b(n-1) + 2[\alpha x_a(n) + \beta x_b(n)]$$
$$y(n) = \alpha[x_a(n-1) + 2x_a(n)] + \beta[x_b(n-1) + 2x_b(n)]$$

which is the sum of the two separate outputs above.

Continuous and discrete systems are discussed in more detail in the next section.

1.3 LINEAR-SYSTEM CLASSIFICATION AND DESCRIPTION

We have already briefly classified systems in several ways: *instantaneous* versus *dynamic*, *continuous* versus *discrete*, and *linear* versus *nonlinear*. We would like to further classify linear systems in this section. We would also like to classify the signals themselves into several types. Remember that the system processes these signals regardless of what these signals represent: velocities, displacements, voltages, currents, the money in a savings account, the output of a digital clock, and so on. Many times we model a linear system with resistors, inductors, and capacitors, and in this case an *analog* system in *continuous* time is implied. Other times (essentially) the same operations can be performed in *discrete* time, implying a *digital* system. If we are analyzing an analog system with discrete-time techniques, or if we are simply analyzing a digital system, then a digital computer can be employed (where the signals are certainly discrete). When the term *signal* appears, it is usually obvious from the context what is meant.

We are often concerned with systems for which the input(s) and output(s) are *continuously* changing with time. A *continuous-time system* is a system (linear or nonlinear) for which the signals are capable of changing at *any* instant of time. The inputs and outputs in such a system are functions of a *continuous* variable (or variables), usually t (time). The inputs and outputs themselves may, of course, be discontinuous.

In some systems the signals will only appear at certain (discrete) instants, perhaps at regular intervals, and we may not be interested at all in the values of the signal between these discrete times. Thus, the variables are said to be discrete, and if time is the only independent variable, such systems are said to be *discrete-time systems*. A pulsed radar range-finding system is an example, since the return pulses from a target (being reflections from the pulses that are transmitted) appear only at essentially discrete instants of time.[†] Systems of this type are considered beginning with Chap. 8.

A *quantized system* is one for which the signals may assume only a finite number of values or levels, and the jumps from level to level may occur at any

---
[†] Actually, the return pulses are not discrete, but are *finite width* pulses.

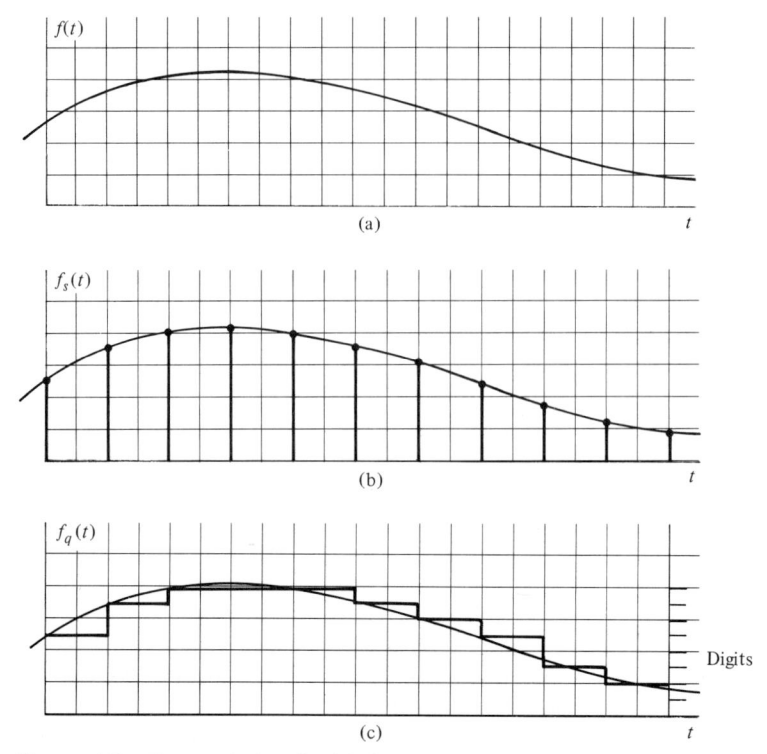

Figure 1.7. Types of signals. (a) Continuous-time signal. (b) Discrete-time signal. (c) Quantized signal.

time. The display of a digital clock is an example. Figure 1.7 shows the signals (inputs, outputs, or intermediate signals) which may occur in the systems we have classified. The continuous signal $f(t)$ appears in Fig. 1.7(a), while its *discretized* version with *uniformly* spaced samples appears in Fig. 1.7(b). *Uniform sampling* is common, but is only a special case of *nonuniform sampling*. A quantized version of $f(t)$ appears in Fig. 1.7(c). *Linear* quantization using eight equally spaced (uniform) levels was chosen for purposes of illustration. This is a special case of nonlinear quantization (unequal levels). Also, at each sample time we must choose which level is the best approximation, and here we have arbitrarily chosen to simply pick the level *closer* to $f(t)$.

Generally speaking, a *fixed*, or *stationary*, or *time-invariant*, continuous-time system is one for which the *shape* (waveform) of the output depends only on the *shape* of the input and not on the instant of application of the input. This implies that the system must be at rest when the input is applied, and if this is the case, we will obtain the same output if we apply the input at 10 AM today or 10 AM tomorrow, *except for a 24 hour time delay*. A continuous-time system described by

$$\mathbf{y}(t) = H\{\mathbf{x}(t)\}$$

is a time-invariant system if, and only if,

$$\mathbf{y}(t \pm \tau) = H\{\mathbf{x}(t \pm \tau)\} \tag{1.9}$$

for the multiple-input multiple-output case. For the single-input single-output case ($M = P = 1$) the continuous-time system

$$y(t) = H\{x(t)\}$$

is time-invariant if, and only if,

$$y(t \pm \tau) = H\{x(t \pm \tau)\} \tag{1.10}$$

A continuous-time system described by a linear differential equation with *constant* coefficients is a time-invariant system. The input-output relations for such a system are shown in Fig. 1.8 (single-input single-output).

The differential equation

$$\frac{d^2y}{dt^2} + 3\frac{dy}{dt} + 2y = x(t)$$

describes a time-invariant linear system, but Bessel's equation

$$\frac{d^2y}{dt^2} + \frac{1}{t}\frac{dy}{dt} + y = 0$$

describes a particular time-varying, but still linear, system. A system that is described by linear differential equations with time dependent coefficients is usually, but not always, time varying. A solid-state amplifier with insufficient provision for heat dissipation that allows a parameter to change value with time is a time-varying system if ambient temperature is not considered to be another input signal.

Likewise, a discrete-time system described by

$$\mathbf{y}(n) = H\{\mathbf{x}(n)\} \tag{1.11}$$

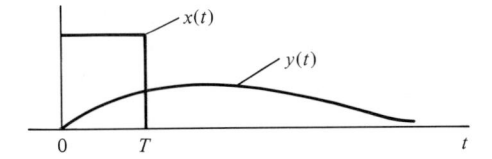

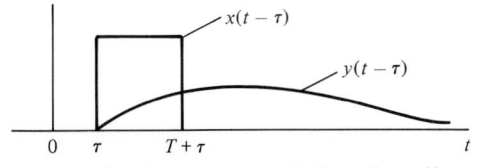

Figure 1.8. Input-output relations for a linear, time-invariant, continuous-time system.

12 SYSTEMS

is fixed or *shift-invariant*, if, and only if,

$$\mathbf{y}(n \pm m) = H\{\mathbf{x}(n \pm m)\} \tag{1.12}$$

for the multiple-input multiple-output case. For the single-input single-output case ($M = P = 1$) the discrete-time system

$$y(n) = H\{x(n)\} \tag{1.13}$$

is shift-invariant if, and only if,

$$y(n \pm m) = H\{x(n \pm m)\} \tag{1.14}$$

Again, the response does not depend on the time origin or reference, but only on the amplitudes of the discrete input signals. Linear, shift-invariant, discrete-time systems are often described by linear, constant coefficient difference equations. The input-output characteristics of a shift-invariant system are shown in Fig. 1.9. The difference equation

$$y(n + 2) + 3y(n + 1) + 2y(n) = x(n)$$

is an example of a shift-invariant system, but the difference equation

$$y(n + 2) + ny(n + 1) + 2y(n) = x(n)$$

describes a discrete system that is not shift-invariant.

A *causal*, or *physical*, or *nonanticipatory*, system is one whose output does not depend on *future values of the input* (continuous or discrete). It is a system which can be built,[†] implying that a noncausal system (where time is the independent variable) cannot be built. A *low-pass filter* for continuous-time signals will allow low-frequency signals to pass through it with very little attenuation and with nearly uniform time delays for these frequencies. The attenuation gradually becomes large near the nominal *cutoff* frequency and is very large at

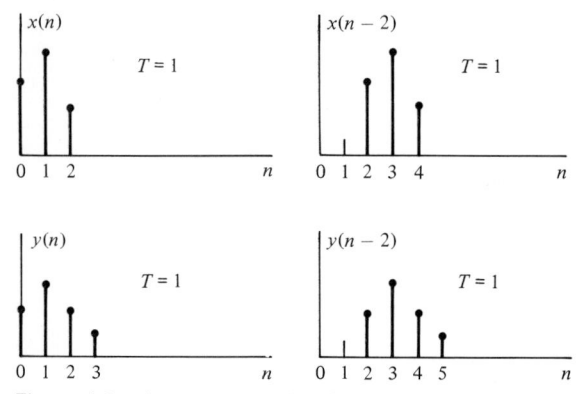

Figure 1.9. Input-output signals for a linear, shift-invariant, discrete-time system.

[†] At least in principle.

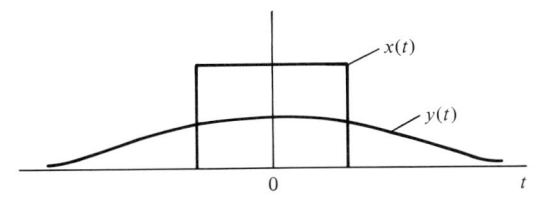

Figure 1.10. Input-output relations for a noncausal system.

higher frequencies. An *ideal* low-pass filter has infinite cutoff characteristics (where the attenuation is zero up to the cutoff frequency and discontinuously jumps to infinite attenuation thereafter) is noncausal and cannot be built. We will show later that if an impulse[†] is applied to this filter at $t = 0$, then the output started at $t = -\infty$, contradicting the first sentence of this paragraph. The characteristics of a noncausal system are shown in Fig. 1.10.

We have categorized systems variously, distinguishing those that are linear, nonlinear, causal, continuous, discrete, and so on. There are other ways in which we can describe systems in electrical engineering. There are *control* systems, *communication* systems, *electronic* systems, *electromagnetic* systems, and so on. We shall not attempt to describe all of these, and we certainly have no intention of specializing in any one of them in this textbook. We will look at the various tools and techniques available to the student who will be required to investigate some of them in future courses.

1.4 CONCLUDING REMARKS

In this chapter we introduced the concept of a system with its input and output. Multiple input and (or) output systems were mentioned with regard to superposition. They will be considered in some detail later. The principle of superposition and its relation to linearity was discussed. Several examples of linear (or other) systems were given. Systems were classified in various ways, and several of the types that were mentioned will be examined in the material to follow. The continuous-time system, which is the major concern of the first part of the remainder of this text, was described, and the discrete-time system, which is the concern of the last part of the text, was also described. Except for a brief mention where a space dimension is the independent variable (Chap. 2), we will be interested only in systems that are causal.

References

1. Neff, H. P., Jr. *Basic Electromagnetic Fields*, New York: Harper & Row, 1981.

[†] The unit-impulse function is described in Chap. 2.

Chapter 2
The Continuous-Time Linear-System Differential Equations

Now that we know what a linear system is, we will restrict ourselves in this chapter to the analysis of those continuous-time linear systems that can be described by linear ordinary differential equations. Fortunately, a large number of systems of practical importance can be described in this manner. Furthermore, many systems that are essentially nonlinear have system functions that can be linearized over a range of t or over a permissible range of input function $x(t)$ that is large enough to give useful results. Even Ohm's law has its limits, as anyone who has watched a carbon resistor go up in smoke can verify. We begin with the continuous system, rather than the discrete system, for the simple reason that the student is more likely to be familiar with the continuous system from previous courses. If we were beginning with no background (Help!), then it could be argued that the discrete system should come first because, generally speaking, it is simpler.

2.1 LINEAR DIFFERENTIAL EQUATIONS

A partial differential equation[1] is an equation which involves one or more dependent and two or more independent variables, together with partial derivatives of the dependent variables with respect to the independent variables.

For example, Poisson's equation, extremely important in "field theory," is describing a linear system. In Cartesian coordinates it is given by

$$\frac{\partial^2 \Phi}{\partial x^2} + \frac{\partial^2 \Phi}{\partial y^2} + \frac{\partial^2 \Phi}{\partial z^2} = -\rho_v/\varepsilon \quad (2.1)$$

[where $\Phi(x, y, z)$ is the scalar electric potential and is a dependent variable (the output), $\rho_v(x, y, z)$ is the electric volume charge density and is a dependent variable (the input), ε is the (constant) permittivity of the medium, and x, y, z are the three independent space variables locating the point in space where Φ is to be found] and is a partial differential equation. It is worth pointing out that the integral solution to Poisson's (linear) equation is[2]

$$\Phi(x, y, z) = \int_{-\infty}^{\infty} \int_{-\infty}^{\infty} \int_{-\infty}^{\infty} \rho_v(x', y', z')$$
$$\times \left[\frac{1}{4\pi\varepsilon \sqrt{(x-x')^2 + (y-y')^2 + (z-z')^2}} \right] dx' \, dy' \, dz' \quad (2.2)$$

It is a three-dimensional superposition integral. Fortunately, we are not primarily interested in partial differential equations in this text, although Eq. 2.2 will be mentioned again at an appropriate time. It is important, however, to point out for future reference that Eqs. 2.1 and 2.2 are describing a *linear system*.

We are more interested in the ordinary differential equation, which is an equation involving one or more dependent variables, only one independent variable, and one or more derivatives of the dependent variables with respect to the independent variable. Ohm's law can be written as an ordinary differential equation if we replace $i(t)$ with dq/dt. That is,

$$R\frac{dq}{dt} = v$$

where $q(t)$ is the electric charge in the circuit, a dependent variable (usually the output); $v(t)$ is the voltage, a dependent variable (usually the input); R is a constant; and t is the only independent variable.

Before pursuing the further study of linear differential equations, it would be wise to define a *linear* (ordinary) differential equation in such a way as to distinguish it from the *nonlinear* form. An ordinary differential equation consists of one or more terms. These terms are made up of products and quotients of explicit functions of the independent variable t (t, t^2, $\cos t$, e^t, etc.) and functions of the dependent variables and their derivatives. The terms $t(dy/dt)$ and $(1/\sin t)(d^2y/dt^2)$ are of first degree in the dependent variable $y(t)$, and the terms $x(dy/dt)$ and $xy^2(d^2y/dt^2)$ are of second and fourth degree, respectively, in the dependent variables $x(t)$ and $y(t)$. A term is said to be a *linear term* when it is of first degree in the dependent variables and their derivatives. A linear differential equation is one consisting of a sum of linear terms only. All other differential equations are nonlinear, and their treatment is beyond the scope of the material presented in this text. A series of examples will now be presented for illustrative purposes to describe various differential equations.

16 THE CONTINUOUS-TIME LINEAR-SYSTEM DIFFERENTIAL EQUATIONS

(a) Poisson's equation is

$$\frac{\partial^2 \Phi}{\partial x^2} + \frac{\partial^2 \Phi}{\partial y^2} + \frac{\partial^2 \Phi}{\partial z^2} = -\rho_v/\varepsilon$$

This is a linear partial differential equation since the terms $\partial^2 \Phi/\partial x^2$, $\partial^2 \Phi/\partial y^2$, $\partial^2 \Phi/\partial z^2$, and ρ_v are all first degree.

(b) $\dfrac{d^2 y}{dt^2} + \left(\dfrac{dy}{dt}\right)^2 = 0$

This is a nonlinear ordinary differential equation since the term $(dy/dt)^2$ is second degree.

(c) Bessel's equation is

$$\rho^2 \frac{d^2 R}{d\rho^2} + \rho \frac{dR}{d\rho} + (k^2 \rho^2 - n^2) R = 0$$

This is a linear ordinary differential equation since all terms are of first degree in R. The coefficients are not constant.

(d) $\dfrac{dy}{dt} + \cos y = x$

This is a nonlinear ordinary differential equation since $\cos y = 1 + y^2/(2!) + y^4/(4!) - \cdots$ is not of first degree. This is true of all transcendental functions.

(e) $x \dfrac{dy}{dt} + y = z$

This is a nonlinear ordinary differential equation because the term $x(dy/dt)$ is second degree in x and y.

(f) $\sin t \dfrac{dy}{dt} + ty = x$

This is a linear ordinary differential equation since all terms are first degree.

(g) $\dfrac{d^3 y}{dt^3} + 2\dfrac{d^2 y}{dt^2} + 4\dfrac{dy}{dt} = x + 2\dfrac{dx}{dt}$

This is a linear ordinary differential equation with constant coefficients, and it describes a time-invariant linear system. It is our primary concern in this chapter.

Before we proceed to the exclusive consideration of linear differential equations, it might be helpful to consider a simple nonlinear differential equation and its solution to show that superposition does not hold.

Example 1

Consider the nonlinear ordinary differential equation $y(dy/dt) = x = 1$. Its solution, easily obtained by integration, is $y = [2(t + C_1)]^{1/2}$ where C_1 is a constant. Now suppose the input is tripled so that $x = 3$. Integration once more leads to the solution $y = [6(t + C_2)]^{1/2}$ where C_2 is a constant. We have tripled the input, but the output has not tripled! It has only increased by a factor of $\sqrt{3}$ (if $C_1 = C_2$). Magnitude scaling (homogeneity) is not preserved, and this is a characteristic of nonlinear systems. Refer to Eq. 1.5.

2.2 THE GENERAL FORM

An ordinary linear differential equation can be put in the form

$$\sum_{n=0}^{N} a_n(t) \frac{d^n y}{dt^n} = \sum_{n=0}^{M} b_n(t) \frac{d^n x}{dt^n} = x_f(t) \qquad (2.3)$$

where the coefficients $a_n(t)$ and $b_n(t)$, as implied, can depend only on the independent variable t. If they do in fact depend on t, then Eq. 2.3 is a *time-variable* differential equation that describes a system that is *not* time-invariant. The equation describing the motion of a rocket must be time-variable because the mass of the rocket is decreasing with time so long as the fuel is being expended. An independent voltage source driving a series RL network where L is time-dependent is time-variable, and can be described by a differential equation. The equation is

$$Ri(t) + \frac{d}{dt}[L(t)i(t)] = v(t) = \left[R + \frac{d}{dt}L(t)\right]i(t) + L(t)\frac{d}{dt}i(t)$$

$x(t)$ is the single *source function* or *input* in Eq. 2.3, and $x_f(t)$ is the single *forcing function* as far as the differential equation is concerned. A series RC circuit driven by an ideal voltage source is described by

$$Ri(t) + \frac{1}{C}\int i(t)\,dt + C_1 = v(t)$$

or, upon differentiating both sides with respect to t

$$R\frac{d}{dt}i(t) + \frac{1}{C}i(t) = \frac{d}{dt}v(t)$$

The input function for this system is $v(t)$, but the forcing function is dv/dt. On the other hand $v(t)$ is *both* the input and forcing function for the series RL network mentioned above.

Equation 2.3 describes an Nth order linear system, and N is the highest derivative of the *single* output that appears in the Nth order linear differential equation. M, the upper index in the summation on the right side of Eq. 2.3, merely signifies the highest derivative of the *single* input that might possibly occur in the differential equation. *Notice carefully that it does not signify that there are M separate inputs even though we already* (Chap. 1) *have used, and will continue to use throughout the text, the designation,* **M-inputs** *and* **P-outputs**, *for the more general multiple-input and multiple-output case. This general case* will be treated later, but in order to make the distinction more explicit at this time, suppose that there are three separate inputs. The forcing function *vector* in a certain case might be

$$\mathbf{x}_f(t) = \begin{bmatrix} 2 & 3t^2 \dfrac{dx_1}{dt} & t\dfrac{d^2 x_1}{dt^2} & - \\ - & 5\dfrac{dx_2}{dt} & - & - \\ -3 & - & 8\dfrac{d^2 x_3}{dt^2} & 2t^2 \dfrac{d^3 x_3}{dt^3} \end{bmatrix}$$

The same remarks apply to the following paragraph and Eq. 2.4.

On the other hand, if the coefficients in Eq. 2.3 are not dependent on time, that is, if they are constants, then

$$\sum_{n=0}^{N} a_n \frac{d^n y}{dt^n} = \sum_{n=0}^{M} b_n \frac{d^n x}{dt^n} = x_f(t) \tag{2.4}$$

describes the system, and Eq. 2.4 is a constant-coefficient differential equation and the system it describes is said to be time-invariant, fixed, or stationary. Many linear systems are time-invariant, or are time-invariant on a short-time basis. This is fortunate, because in those cases the very powerful methods of the Fourier and Laplace transformations are always applicable. These will be introduced in later chapters. Notice that in Eqs. 2.3 and 2.4 we are apparently concerned with a system with a single input and a single output. Multiple inputs and outputs are no cause for concern since the principle of superposition is applicable, and this more general case will be treated later.

2.3 CONVENTIONAL METHODS OF SOLUTION

Before devising a scheme for solving Eq. 2.3 that is general enough to handle *any* input, we should consider a somewhat restricted class of inputs and a conventional method for solving the differential equations. Hopefully, this will be in the form of a review for many students since it was most likely presented in a second-year circuits or mathematics course, if not elsewhere.

Consider the simple *RL* series network of Fig. 2.1. It is analyzed with Kirchhoff's voltage law which yields

$$v(t) = Ri + L\frac{di}{dt} \tag{2.5}$$

Notice, first of all, that the equation is of *first order* since it involves only the first derivative of the dependent variable $i(t)$. This occurs because there is only *one* energy storage element present. Secondly, notice that the *source function* is $v(t)$ [the independent voltage source in Fig. 2.1(a)], and the *forcing function* is

2.3 CONVENTIONAL METHODS OF SOLUTION

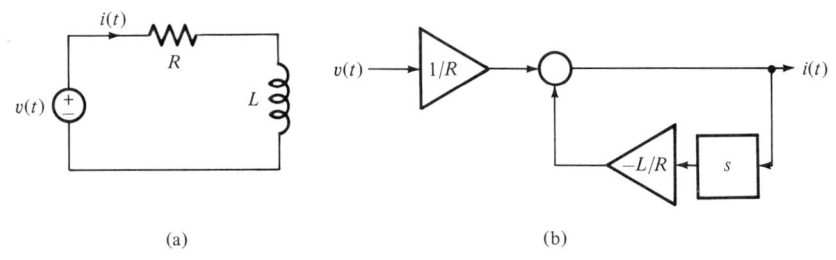

Figure 2.1. (a) Series *RL* network. (b) Block diagram representation of Eq. 2.5.

also $v(t)$ (in Eq. 2.5). This is not always the case.[†] Equation 2.4 indicates that the forcing function can be

$$b_0 x + b_1 \frac{dx}{dt} + b_2 \frac{d^2 x}{dt^2} + \cdots$$

when x is the source function. Thirdly, notice that the general solution consists of the sum of two parts.[‡] One is a solution to the homogeneous (source-free) equation

$$0 = Ri + L \frac{di}{dt} \tag{2.6}$$

It is called by many names: the *complementary function* (from classical mathematics), the *natural response*, the *transient solution* (which is not always even correct), and the *free response* among others. It is now customary, however, to call it the *zero-input response*, and it is due entirely to the stored energy in the system. The other part of the solution is the solution to the inhomogeneous equation (2.5) and it is the *particular integral* (from classical mathematics), *particular solution*, *steady-state solution* (not always correct), or *forced response*. We will call it by the current title: *zero-state response* (or *zero initial energy state response*), which indicates that the energy storage elements (L and C) are in the zero energy state. That is, all the initial energy is 0 [$i_L(0) = v_C(0) = 0$, for example], and this response is due entirely to the *forcing function*. It is important to recognize that the *form of the solution to the homogeneous equation will always be the same*. Likewise, it should also be pointed out that *the terms that describe the solution to the inhomogeneous equation* (above) *are not all the same*.

Generally speaking, the zero-state response is the more difficult of the two to obtain. A complete solution is usually not difficult to obtain, however, if the input takes on simple forms such as the constant, the polynomial, the exponential, or the sinusoid, for example. Once we have obtained a solution to the inhomogeneous equation by whatever method (including an educated guess) it is the only one. That is, it is *unique*.

[†] An example of this situation was mentioned in Sec. 2.2.
[‡] This fact should not be news to an upperclass engineering student.

20 THE CONTINUOUS-TIME LINEAR-SYSTEM DIFFERENTIAL EQUATIONS

Since the zero-input response is independent of $v(t)$ we proceed to find it first. We assign to the *zero-input* response the symbol $i_{zi}(t)$ since it is a current in this example. This notation was first mentioned in Chap. 1. Thus, we are required to find a solution to the homogeneous differential equation

$$0 = Ri_{zi} + L\frac{d}{dt}i_{zi} \tag{2.7}$$

Rearranging terms

$$\frac{di_{zi}}{i_{zi}} = -\frac{R}{L}dt$$

Integrating

$$\ln i_{zi} = -\frac{R}{L}t + C_1 = -\frac{R}{L}t + \ln C_2 \qquad C_1 \equiv \ln C_2$$

$$\ln i_{zi} - \ln C_2 = -\frac{R}{L}t = \ln\left(\frac{i_{zi}}{C_2}\right)$$

Thus, taking the antilog of both sides

$$\frac{i_{zi}}{C_2} = e^{-Rt/L}$$

or, finally

$$i_{zi}(t) = C_2 e^{-Rt/L} \tag{2.8}$$

On the other hand, with an eye on Eq. 2.7, there is only one function in all of mathematics that retains its *form* (scaled only) when differentiated or integrated: the exponential. Thus, we can simply guess the solution

$$i_{zi}(t) = e^{st}$$

and try it in Eq. 2.7

$$0 = Re^{st} + Lse^{st}$$

Since e^{st} cannot be 0 for all t, we can divide the last result by e^{st}, obtaining

$$0 = R + Ls$$

or

$$s = -R/L$$

Thus

$$i_{zi}(t) = e^{-Rt/L}$$

and since magnitude *scaling* is allowed[†] for the solution to the homogeneous equation (2.7):

$$i_{zi}(t) = C_2 e^{-Rt/L}$$

[†] This is easily verified.

which is Eq. 2.8 again. Results in this paragraph are important because, as we will see, the exponential trial $i_{zi}(t) = e^{st}$ will *always*[†] give the correct solution(s) for the zero-input response of a time-invariant linear system, *regardless of order!*

This completes the solution of the zero-input response. Notice that it contains *one* unspecified constant because it applies to a *first*-order system. Before proceeding with a complete solution to Eq. 2.5 for Fig. 2.1 we must specify an input.

Example 2

Find $i(t)$ in Fig. 2.1 if $v(t) = V_0 u(t)$, where $u(t)$ is the unit step function.[‡] We then have

$$V_0 u(t) = Ri + L\frac{di}{dt} \tag{2.9}$$

Insofar as $t > 0$ is concerned, it is certainly true that Eq. 2.9 is equivalent to

$$V_0 = Ri + L\frac{di}{dt} \qquad t > 0 \tag{2.10}$$

Notice, however, that $V_0 u(t)$ and V_0 are *not the same forcing functions!* This means that these functions *will not give the same zero-state response.* Seeking the simplest method to find the solution, we begin with Eq. 2.10. A little knowledge of the behavior of circuits is all that is necessary to recognize that the inductor eventually behaves as a short circuit to a constant source (V_0) as time increases.[§] Thus, the current is ultimately limited only by R; that is, it is given by V_0/R. We can also recognize this as the *steady-state response.* Is it the zero-state response? Substitution of

$$i_{ss}(t) = V_0/R \tag{2.11}$$

into Eq. 2.10 shows that we do indeed have a solution, but this cannot be the zero-state response for reasons that will become apparent. Notice that this response in no way accounts for the sudden jump in $v(t)$ from 0 to V_0 at $t = 0$, and this must be resolved by adding what we originally called the zero-input response, $i_{zi}(t)$. As a matter of fact, a complementary function may *always* be added in any case since it is a solution to the *homogeneous* differential equation. Thus, the complete solution must be

$$i(t) = i_{zi}(t) + i_{ss}(t)$$
$$i(t) = C_2 e^{-Rt/L} + V_0/R \tag{2.12}$$

[†] A special case, where two or more *identical* exponents occur, will be treated later.
[‡] The unit-step function is defined as

$$u(t) = \begin{cases} 0 & t < 0 \\ 1 & t > 0 \end{cases}$$

A more general definition is given in Sec. 2.4 (see Fig. 2.8).
[§] The inductor voltage is $L(di/dt)$, and with *constant* sources the current ultimately becomes constant making di/dt 0, and, hence, the inductor voltage is ultimately 0.

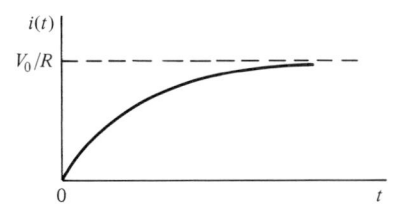

22 THE CONTINUOUS-TIME LINEAR-SYSTEM DIFFERENTIAL EQUATIONS

Figure 2.2. Current in the series RL network when $i(0) = 0$.

In order to find the constant C_2 we must know $i(t)$ at some instant of time. This time is usually that just prior to the switching action in $v(t)$; that is, $t = 0^-$. For example, if $i(0^-) = i(0^+) = 0$ (no initial energy storage in the inductor), then Eq. 2.12 gives

$$0 = C_2 + V_0/R \qquad (t = 0)$$
$$C_2 = -V_0/R$$

and the final solution is

$$i(t) = \frac{V_0}{R}(1 - e^{-Rt/L}) \qquad t > 0 \qquad (2.13)$$

This well-known result is shown in Fig. 2.2.

Now notice that the zero-input response was given by Eq. 2.8:

$$i_{zi}(t) = C_2 e^{-Rt/L} \qquad (2.8)$$

and if there is no initial energy storage, then $C_2 = 0$ and, as has already been pointed out

$$i_{zi}(t) \equiv 0$$

According to the definition of the zero-input response this is correct. If this is so, then the correct zero-state response† must be

$$i_{zs}(t) = \frac{V_0}{R}(1 - e^{-Rt/L})u(t) \qquad (2.14)$$

giving the complete solution

$$i(t) = i_{zi}(t) + i_{zs}(t)$$

$$i(t) = 0 + \frac{V_0}{R}(1 - e^{-Rt/L})u(t)$$

$$i(t) = \frac{V_0}{R}(1 - e^{-Rt/L})u(t)$$

which we already know is the correct answer for $t > 0$. We will show later how Eq. 2.14 is obtained.

---

† The subscripts zs obviously refer to $zero$-state as first noted in Chap. 1.

2.3 CONVENTIONAL METHODS OF SOLUTION

What we actually have done to obtain Eq. 2.13 is to find a particular solution to the inhomogeneous equation, which was just the steady-state solution, plus a complementary function, which was a solution to the homogeneous equation. Using this method, which is a simple one, we did not need to find the zero-state response.

If the voltage source is $v(t) = V_0 t$ in the preceding example, then the trial $i_b(t) = At + B$ will give a particular solution to be added to the complementary function $i_a(t) = Ae^{-Rt/L}$. When the voltage source is $V_0 e^{-\alpha t}$, then the trial $i_b(t) = Ae^{-\alpha t}$ gives the particular solution. Likewise, for $v(t) = V_0 \cos \omega t$, the trial $i_b(t) = A \cos \omega t + B \sin \omega t$ gives the particular solution. Generally speaking, for the types of inputs we have considered here, the suggested trials or guesses will work even for higher-order systems. We will not go into formal proofs here,[3] but a close inspection of the general equation (2.4) reveals why this is so.

Suppose the input voltage is $v(t) = J_0(t)$, where $J_0(t)$ is a Bessel function of the first kind of zero order, or perhaps the input is obtained from experimental data graphed as a function of time. What is $i(t)$? We now have no *simple* way of finding a solution. A general method does exist, however, and we shall shortly examine it. Before doing so we will examine a second-order system in which we must determine the initial conditions.

Example 3

Find the capacitor voltage, $v_C(t)$, for $-\infty < t < \infty$ in Fig. 2.3. Prior to the time that the left current source becomes active ($t = 0$), the network is in the steady-state mode, and the current from the right current source flows entirely (right to left) through the 40-Ω resistor and down through the inductor. Thus

$$v_C(t) = 40 \text{ V} \qquad t \le 0$$

This result gives us one initial condition:[†] $v_C(0^\pm) = 40$ V. Another is $i_L(0^\pm) = 1$ A.

Rather than attempting to find the differential equation that describes the circuit in Fig. 2.3, and in light of the fact we are (here) attempting to find the

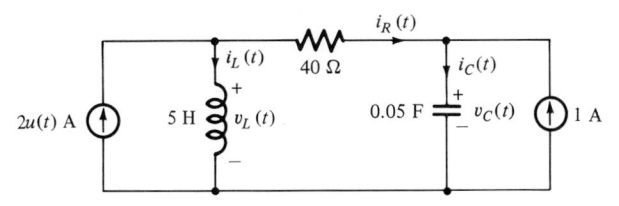

Figure 2.3. Network for Example 3.

[†] The notation $f(0^\pm)$ means $f(0)$ as approached from the left ($t < 0$) *or* from the right ($t > 0$).

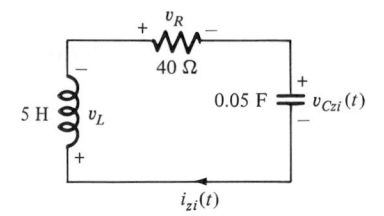

Figure 2.4. Network for the zero-input response.

solution in the simplest possible manner, let us look at the circuit for the zero-input case only (no forcing functions). Kirchhoff's voltage law for Fig. 2.4 gives

$$Ri_{zi} + L\frac{di_{zi}}{dt} + \frac{1}{C}\int_0^t i_{zi}\,dt + v_{Czi}(0) = 0$$

and differentiation gives

$$L\frac{d^2 i_{zi}}{dt^2} + R\frac{di_{zi}}{dt} + \frac{i_{zi}}{C} = 0 \tag{2.15}$$

If we can find $i_{zi}(t)$, then we can find $v_{Czi}(t)$ since

$$v_{Czi}(t) = -Ri_{zi} - L\frac{di_{zi}}{dt} \tag{2.16}$$

Equation 2.16 is not necessary at this point, however, since *every* zero-input response (natural response) in this (or any) linear system will have the *same form*!

Dividing by L, Eq. 2.15 becomes

$$\frac{d^2 i_{zi}}{dt^2} + \frac{R}{L}\frac{di_{zi}}{dt} + \frac{i_{zi}}{LC} = 0$$

or

$$\frac{d^2 i_{zi}}{dt^2} + 8\frac{di_{zi}}{dt} + 4i_{zi} = 0$$

This is a *second*-order differential equation with constant coefficients, so it must have two independent solutions (or *two* elements in its *fundamental set*). These will *always* be exponentials[†] for time-invariant linear systems, regardless of order. Thus, with much confidence, we try

$$i_{zi} = Ae^{st}$$

which gives

$$s^2 A + 8sA + 4A = 0$$

[†] The case where identical exponents (repeated roots) occurs will be treated later.

or since A is not 0
$$s^2 + 8s + 4 = 0$$
called the *characteristic equation*. Factoring
$$(s + 4 + \sqrt{12})(s + 4 - \sqrt{12}) = 0$$
So
$$\begin{aligned} s_1 &= -4 + \sqrt{12} \\ s_2 &= -4 - \sqrt{12} \end{aligned} \tag{2.17}$$

We conclude that
$$i_{zi}(t) = A_1 e^{s_1 t} + A_2 e^{s_2 t} \tag{2.18}$$
and
$$v_{Czi}(t) = B_1 e^{s_1 t} + B_2 e^{s_2 t} \tag{2.19}$$

At the present time we are looking for the solution for $t > 0$, having already found the solution for $t < 0$. In particular, we want the particular solution, and we want to find it in the simplest possible manner. Thus, since we have only constant sources (except for the jumps), we can resort to the steady state. After steady-state conditions have been reached, the left current source current flows entirely through the inductor. The current from the right current source also flows through the inductor, but it also develops 40 V across the resistor. Thus
$$v_{Css}(t) = 40 \text{ V} \qquad t > 0$$
The complete solution for $t > 0$ is
$$v_C(t) = 40 + B_1 e^{s_1 t} + B_2 e^{s_2 t} \qquad t > 0 \tag{2.20}$$
One initial condition is
$$v_C(0^+) = 40\text{V} \tag{2.21}$$
The other required initial condition involves the derivative of v_C at $t = 0^+$, and this can be found from $i_C = C(dv_C/dt)$. Kirchhoff's current law at the node at the top of the capacitor in Fig. 2.3 gives
$$i_C = C \frac{dv_C}{dt} = 1 + i_R \qquad t > 0$$
$$C \frac{dv_C}{dt} = 1 + 2 - i_L \qquad t > 0$$
$$C \frac{dv_C}{dt}\bigg|_{t=0^+} = 3 - i_L(0^+) \qquad t > 0$$
But $i_L(0^+) = 1$, so
$$\frac{dv_C}{dt}\bigg|_{t=0^+} = \frac{2}{C} = 40 \text{ V/s} \tag{2.22}$$

26 THE CONTINUOUS-TIME LINEAR-SYSTEM DIFFERENTIAL EQUATIONS

Using Eqs. 2.21 and 2.22 in 2.20 gives

$$40 = 40 + B_1 + B_2$$
$$40 = +s_1 B_1 + s_2 B_2$$

Solving for B_1 and B_2

$$B_1 = 20/\sqrt{12}$$
$$B_2 = -20/\sqrt{12}$$

Therefore

$$v_C(t) = 40 + \frac{20}{\sqrt{12}} \left[e^{-(4-\sqrt{12})t} - e^{-(4+\sqrt{12})t} \right] \qquad t > 0 \qquad (2.23)$$

The results for all time can be written

$$v_C(t) = 40u(-t) + 40u(t) + \frac{20}{n\sqrt{12}} \left[e^{-(4-\sqrt{12})t} + e^{-(4+\sqrt{12})t} \right] u(t)$$

$$v_C(t) = 40 + \frac{20}{\sqrt{12}} \left[e^{-(4-\sqrt{12})t} - e^{-(4+\sqrt{12})t} \right] u(t) \qquad (2.24)$$

The methods used to find solutions in Examples 2 and 3 were relatively simple primarily because the sources were *constant* (except for the sudden jumps). Many readers have already been exposed to this type of solution, but it should once again be emphasized that if the sources are more complicated, then a more general technique will be necessary. The definitions given earlier for the zero-input response and the zero-state response are general, clear, and concise. We will use these forms in the general technique that is to be developed later.

Finally, the definition of the steady-state response and the transient response should be given in a concise manner. They are simple. The steady-state response is that part of the *complete response* which does not go to 0 as t goes toward ∞. The transient response is that part of the *complete response* which does go to 0 as t goes toward ∞. In Eq. 2.24 the steady-state part is simply 40 V, while the transient part is

$$\frac{20}{\sqrt{12}} \left[e^{-(4-\sqrt{12})t} - e^{-(4+\sqrt{12})t} \right] \quad \text{(transient)}$$

2.4 THE UNIT-IMPULSE FUNCTION

Before attempting to present a more general method of analyzing those time-invariant linear systems described by differential equations, it would be wise to introduce the unit-impulse function. We will need to use this function extensively in the material to follow.

The unit impulse function or Dirac delta function, $\delta(t)$, is not a true function in a strict mathematical sense but a *distribution* (or a singularity function).

2.4 THE UNIT-IMPULSE FUNCTION

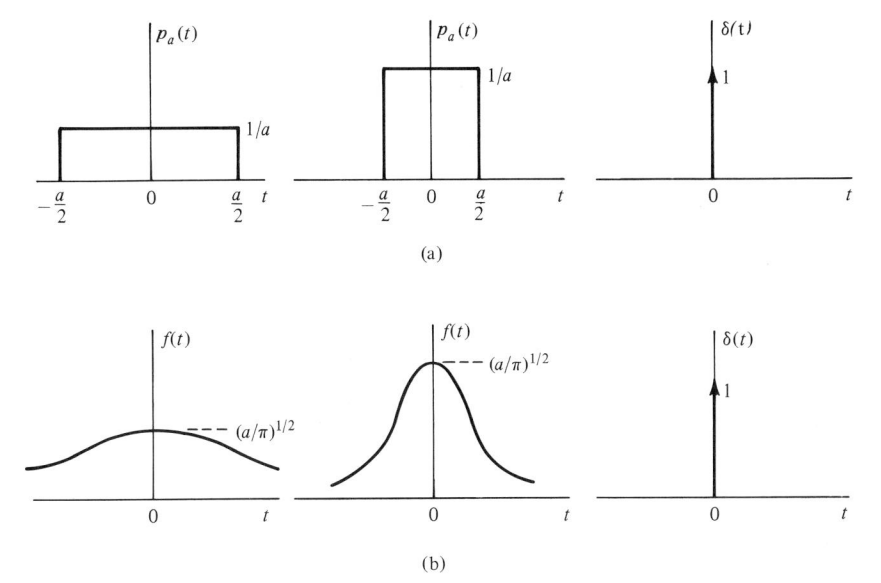

Figure 2.5. Two examples of functions which may represent unit impulses in limiting form. (a) Rectangular pulse. (b) Continuous function $f(t) = (a/\pi)^{1/2} e^{-at^2}$.

It can be considered the limit of a wide variety of true functions or other distributions if they have certain properties. Instead of becoming involved in a detailed discussion of these properties, we will examine two simple cases to observe these properties.

Consider the rectangular pulse $p_a(t)$ as shown in Fig. 2.5(a). Its amplitude is $1/a$ while its width is a so that its area is unity regardless of a. Then, the unit impulse can be regarded as

$$\delta(t) = \lim_{a \to 0} p_a(t)$$

We can visualize the height approaching ∞ while the width is simultaneously approaching 0. The increasing height, decreasing width, and unity area are all characteristics of functions which may ultimately represent unit impulses. In Fig. 2.5(b) we have

$$f(t) = (a/\pi)^{1/2} e^{-at^2}$$

$$\left.\begin{array}{l} f(0) \to \infty \quad t = 0 \\ a \to \infty \end{array}\right\} \text{ increasing height}$$

$$\left.\lim_{a \to \infty} f(t) = 0 \quad |t| > 0 \right\} \text{ decreasing width}$$

$$\int_{-\infty}^{\infty} f(t)\, dt = 1 \quad \text{unity area}$$

$$\lim_{a \to \infty} f(t) = \delta(t)$$

28 THE CONTINUOUS-TIME LINEAR-SYSTEM DIFFERENTIAL EQUATIONS

Although it is not necessary to do so, we will consider $\delta(t)$ to be an even function of t. That is, $\delta(t) = \delta(-t)$. The amplitude of the impulse function is commonly called its *moment* or *weight*, and *an impulse with zero moment is 0*. That is, $0 \cdot \delta(t) = 0$. Notice that $f(t)\delta(t - \tau) = f(\tau)\delta(t - \tau)$.

Some of the more useful properties of impulse functions are listed below. If $f(t)$ is continuous at $t = b$, then

(a) $\int_a^c f(t)\delta(t - b)\, dt = f(b) \qquad a < b < c$

$\qquad\qquad\qquad\quad\; = 0 \qquad\quad\; b < a \;\text{ or }\; b > c$ (2.25)

This is the extremely important *sampling*[†] property of the impulse function. In order to show that Eq. 2.25 is correct, we first recognize that $\delta(t - b)$ is a unit impulse which exists at $t = b$, and is 0 elsewhere. Thus, the left side of Eq. 2.25 may be written

$$\int_a^c f(b)\delta(t - b)\, dt = f(b) \int_a^c \delta(t - b)\, dt$$

and the remaining integral is that for finding the area (unity) of a unit impulse so long as $a < b < c$. Thus, Eq. 2.25 is established. This sampling property is regarded by some as defining the unit-impulse function itself (see Fig. 2.6).

In this regard, if we write an equivalent form of Eq. 2.25

$$f(t) = \int_{-\infty}^{\infty} f(\tau)\delta(t - \tau)\, d\tau \qquad (2.26)$$

we can regard it as an expression for $f(t)$ as a *superposition of impulses*. Notice that the limits have been changed to ensure that the impulse is within, and the dummy variable is now τ. To show the superposition of impulses it is only necessary to express the arbitrary, but continuous, $f(t)$ as an approximation that results from considering it to be a series of side-by-side pulses. If $p_b(t - nb)$ is a *unit amplitude pulse of width b centered at* $t = nb$, then the pulse centered at $t = nb$ with amplitude $f(nb)$ is $f(nb)p_b(t - nb)$. Thus, as seen in Fig. 2.7

$$f(t) \approx \sum_{n=-\infty}^{\infty} f(nb)p_b(t - nb)$$

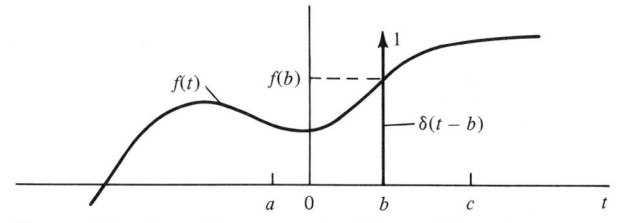

Figure 2.6. Sampling property of the unit impulse.

[†] Sometimes called the *sifting* property.

2.4 THE UNIT-IMPULSE FUNCTION

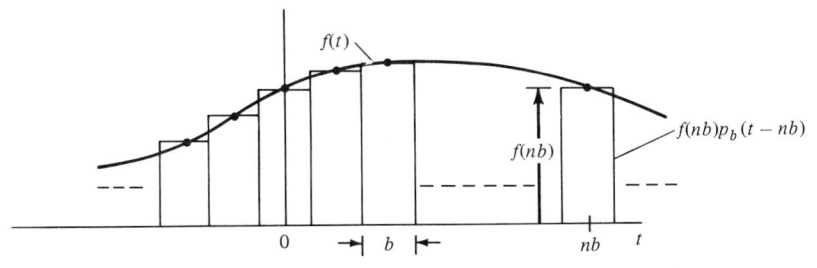

Figure 2.7. Function f(t) approximated as a superposition of pulses.

or

$$f(t) \approx \sum_{n=-\infty}^{\infty} f(nb) \left[\frac{1}{b} p_b(t-nb) \right] b \qquad (2.27)$$

Notice that the *area* of the bracketed term is unity regardless of b. In the limit as b tends to 0 the bracketed term becomes $\delta(t-\tau)$ and the sum becomes an integral in the usual way:

$$f(t) = \lim_{b \to 0} \sum_{n=-\infty}^{\infty} f(nb) \left[\frac{1}{b} p_b(t-nb) \right] b = \int_{-\infty}^{\infty} f(\tau) \delta(t-\tau) d\tau$$

thus establishing Eq. 2.26. Similarly

(b) $\int_a^c f(t) \delta'(t-b) dt = -f'(b) \quad a < b < c$
$\qquad \qquad \qquad \qquad \qquad \quad = 0 \qquad \qquad b < a \text{ or } b > c$ $\qquad (2.28)$

$$\delta'(t-b) = \frac{d}{dt}[\delta(t-b)]$$

This result is obtained from Eq. 2.25 by integrating by parts.

(c) $\int_a^c f(t) \delta^n(t-b) dt = (-1)^n f^n(b)$
$\qquad \qquad \qquad \qquad \qquad \quad = 0 \quad b < a \text{ or } b > c$ $\qquad (2.29)$

$$\delta^n(t-b) = \frac{d^n}{dt^n}[\delta(t-b)] \qquad f^n(b) = \frac{d^n}{dt^n}[f(t)]_{t=b}$$

This result is obtained from Eq. 2.28 by extension.

(d) $\delta[f(t)] = \sum_{n=-\infty}^{\infty} \frac{1}{\left|\dfrac{df}{dt}\right|} \delta(t-b_n) \qquad f(b_n) = 0 \qquad (2.30)$

Equation 2.30[4] is useful in time-domain radiation studies, as well as in other areas, but we are unlikely to encounter it in the material to follow (except perhaps in a problem at the end of the chapter).

(e) $\dfrac{d}{dt}[u(t)] = \delta(t) \qquad (2.31)$

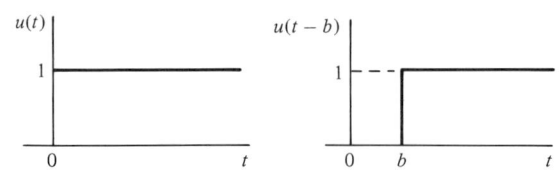

Figure 2.8. Unit-step function.

or

$$u(t) = \int_{-\infty}^{t} \delta(t') \, dt' \tag{2.32}$$

Equations 2.31 and 2.32 relate the unit-step function, shown in Fig. 2.8, to the unit-impulse function. The unit step is defined by

$$u(t) = \begin{cases} 1 & t > 0 \\ 0 & t < 0 \end{cases} \tag{2.33}$$

or more generally

$$u[f(t)] = \begin{cases} 1 & f(t) > 0 \\ 0 & f(t) < 0 \end{cases} \tag{2.34}$$

Notice finally that the unit of an impulse function will be that of the reciprocal of its argument.[†] Why?

2.5 THE UNIT-IMPULSE RESPONSE

Equation 2.26 may be written as

$$x(t) = \int_{-\infty}^{\infty} x(\tau) \delta(t - \tau) \, d\tau \tag{2.35}$$

where $x(t)$ is the *source or input* and is related to the *forcing function* by Eq. 2.3 [or 2.4 in the time-invariant case]. As we have seen, this form may be regarded as an expression for $x(t)$ in the form of a superposition of impulses. The *zero-state response* of the system is obtained from Eq. 1.1 and Fig. 1.1 if the *input* is applied when the system is at rest (no zero-input response):

$$y_{zs}(t) = H\left\{\int_{-\infty}^{\infty} x(\tau) \delta(t - \tau) \, d\tau\right\} \quad \text{(zero-state response)} \tag{2.36}$$

But τ is just a dummy variable, and t is the independent variable in the system we are describing. Thus, H operates on t, not τ, and

$$y_{zs}(t) = \int_{-\infty}^{\infty} x(\tau) H\{\delta(t - \tau)\} \, d\tau \tag{2.37}$$

[†] That is, $\delta(t)$ is measured in s^{-1} if t is time, for example.

2.5 THE UNIT-IMPULSE RESPONSE

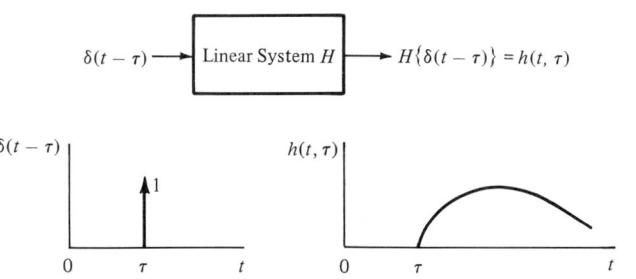

```
δ(t − τ) ──→ [ Linear System H ] ──→ H{δ(t − τ)} = h(t, τ)
```

Figure 2.9. Unit impulse applied at $t = \tau$ to the system, producing the unit-impulse response $h(t, \tau)$.

The term $H\{\delta(t - \tau)\}$ in Eq. 2.37 is the response of this system when a unit impulse is applied as the *input* at $t = \tau$. That is

$$h(t, \tau) = H\{\delta(t - \tau)\} \tag{2.38}$$

is the unit-impulse response of the system as shown in Fig. 2.9. *Notice that the unit-impulse response is a zero-state response.* Thus, Eq. 2.37 can be written as

$$\boxed{y_{zs}(t) = \int_{-\infty}^{\infty} x(\tau)h(t, \tau)\, d\tau} \quad \text{(zero-state response)} \tag{2.39a}$$

Given any reasonable system† and any reasonable input, we can find the output by means of Eq. 2.39a. This output or response is the *zero-state response* since initial energy storage was not included. The initial energy can always be included by means of the zero-input response (complementary function) which can be added later if necessary. We can now state with considerable confidence and justification that the unit-impulse response of a system completely characterizes the system. This will become even more meaningful shortly. If we can find this unit-impulse response, we should be well on our way to analyzing continuous linear systems with *any* source(s).

In terms of the forcing function $x_f(t)$ and its unit-impulse response, that we logically call $h_f(t, \tau)$, the equivalent of Eq. 2.39a is

$$\boxed{y_{zs}(t) = \int_{-\infty}^{\infty} x_f(\tau)h_f(t, \tau)\, d\tau} \quad \text{(zero-state response)} \tag{2.39b}$$

The unit-impulse response $h_f(t, \tau)$ is the response when the forcing function $x_f(t)$ (not necessarily the input) is $\delta(t - \tau)$.

We next show that the real burden is that of solving for the zero-input response. It is a solution to the homogeneous form of Eq. 2.3:

$$\sum_{n=0}^{N} a_n(t) \frac{d^n y_{zi}}{dt^n} = 0 \tag{2.40}$$

---

† Any reasonable *linear* system whose unit-impulse response can be found.

32 THE CONTINUOUS-TIME LINEAR-SYSTEM DIFFERENTIAL EQUATIONS

If we can find the zero-input response, $y_{zi}(t)$, then the unit-impulse response for the forcing function $x_f(t)$ that we called $h_f(t, \tau)$ (above Eq. 2.39b) is given by

$$\boxed{h_f(t, \tau) = \frac{A h_f^a(t, \tau) + B h_f^b(t, \tau)}{A + B}} \qquad (2.41)$$

where

$$h_f^a(t, \tau) = y_{zi}(t, \tau) u(t - \tau) \qquad (2.42)$$
$$h_f^b(t, \tau) = -y_{zi}(t, \tau) u(\tau - t) \qquad (2.43)$$

subject to the conditions

$$\left.\begin{array}{l} y_{zi}\big|_{t=\tau} = 0 \\[4pt] \dfrac{dy_{zi}}{dt}\bigg|_{t=\tau} = 0 \\[4pt] \dfrac{d^2 y_{zi}}{dt^2}\bigg|_{t=\tau} = 0 \\[4pt] \cdots \\[4pt] \dfrac{d^{N-1} y_{zi}}{dt^{N-1}}\bigg|_{t=\tau} = \dfrac{1}{a_N(\tau)} \end{array}\right\} \qquad (2.44)$$

That is, for a first-order system

$$y_{zi}\big|_{t=\tau} = \frac{1}{a_1(\tau)} \quad (N = 1)$$

while for a second-order system

$$\left.\begin{array}{l} y_{zi}\big|_{t=\tau} = 0 \\[4pt] \dfrac{dy_{zi}}{dt}\bigg|_{t=\tau} = \dfrac{1}{a_2(\tau)} \end{array}\right\} \quad (N = 2)$$

and for a third-order system

$$\left.\begin{array}{l} y_{zi}\big|_{t=\tau} = 0 \\[4pt] \dfrac{dy_{zi}}{dt}\bigg|_{t=\tau} = 0 \\[4pt] \dfrac{d^2 y_{zi}}{dt^2}\bigg|_{t=\tau} = \dfrac{1}{a_3(\tau)} \end{array}\right\} \quad (N = 3)$$

and so forth.

A and B are constants that are usually easy to determine from the nature of the system under investigation. The proof that $h_f(t, \tau)$ as given by Eqs. 2.41

2.5 THE UNIT-IMPULSE RESPONSE

through 2.44 satisfies

$$\sum_{n=0}^{N} a_n(t) \frac{d^n}{dt^n} [h_f(t, \tau)] = \delta(t - \tau) \tag{2.45}$$

is supplied in Appendix B. Notice, once again, that $h_f(t, \tau)$ is the unit impulse response for $x_f(t)$, not for $x(t)$.

It is worthwhile to demonstrate how Eq. 2.41 is derived for a simple case. Consider the second-order (time-variable) system:

$$a_2(t) \frac{d^2 y}{dt^2} + a_1(t) \frac{dy}{dt} + a_0(t) y = x_f(t)$$

We want a solution to

$$a_2(t) \frac{d^2 h_f}{dt^2} + a_1(t) \frac{dh_f}{dt} + a_0(t) h_f = \delta(t - \tau)$$

It is clear that $y_{zi}(t, \tau)$ is a solution to the preceding equation, so long as $t \neq \tau$. That is,

$$h_f(t, \tau) = y_{zi}(t, \tau) \qquad t \neq \tau$$

since the differential equation is homogeneous for $t \neq \tau$. Then, either (1) $h_f(t, \tau) = \pm y_{zi}(t, \tau) u(t - \tau)$, or (2) $h_f(t, \tau) = \pm y_{zi}(t, \tau) u(\tau - t)$, or (3) a combination of (1) and (2) exists, since the two possibilities *never overlap in time*. Trying (1):

$$a_0(t) h_f(t, \tau) = \pm a_0(t) y_{zi}(t, \tau) u(t - \tau)$$

$$a_1(t) \frac{dh_f(t, \tau)}{dt} = \pm a_1(\tau) y_{zi}(t, \tau)|_{t=\tau} \delta(t - \tau) \pm a_1(t) \frac{dy_{zi}(t, \tau)}{dt} u(t - \tau)$$

$$a_2(t) \frac{d^2 h_f(t, \tau)}{dt^2} = \pm a_2(\tau) y_{zi}(t, \tau)|_{t=\tau} \delta'(t - \tau) \pm a_2(\tau) \frac{dy_{zi}(t, \tau)}{dt}\bigg|_{t=\tau} \delta(t - \tau)$$

$$\pm a_2(t) \frac{d^2 y_{zi}(t, \tau)}{dt^2} u(t - \tau)$$

Adding, and remembering that $y_{zi}(t, \tau)$ is a solution to the homogeneous equation

$$a_2(t) \frac{d^2 h_f}{dt^2} + a_1(t) \frac{dh_f}{dt} + a_0(t) h_f = \pm a_1(\tau) y_{zi}(t, \tau)|_{t=\tau} \delta(t - \tau)$$

$$\pm a_2(\tau) y_{zi}(t, \tau)|_{t=\tau} \delta'(t - \tau)$$

$$\pm a_2(\tau) \frac{dy_{zi}(t, \tau)}{dt}\bigg|_{t=\tau} \delta(t - \tau)$$

34 THE CONTINUOUS-TIME LINEAR-SYSTEM DIFFERENTIAL EQUATIONS

Now, $\delta'(t - \tau)$ (called a *doublet*) cannot be allowed, so we require that $y_{zi}(\tau, \tau) = 0$ (2.44 with $N = 2$), giving

$$a_2(t)\frac{d^2 h_f}{dt^2} + a_1(t)\frac{dh_f}{dt} + a_0(t)h_f = \pm a_2(\tau)\frac{dy_{zi}(t, \tau)}{dt}\bigg|_{t=\tau} \delta(t - \tau)$$

The left side of the preceding equation is $\delta(t - \tau)$, therefore the $\pm$ sign must be $+$, and, in agreement with 2.44 ($N = 2$),

$$\frac{dy_{zi}(t, \tau)}{dt}\bigg|_{t=\tau} = \frac{1}{a_2(\tau)}$$

Following the same procedure for (2), we find that the $\pm$ sign is $-$, and once again,

$$\frac{dy_{zi}(t, \tau)}{dt}\bigg|_{t=\tau} = \frac{1}{a_2(\tau)}$$

Since a *unit impulse* is produced by $y_{zi}(t, \tau)u(t - \tau)$, an impulse of strength $A/(A + B)$ is produced by $y_{zi}(t, \tau)u(t - \tau)A/(A + B)$. Likewise, an impulse of strength $B/(A + B)$ is produced by $-y_{zi}(t, \tau)u(\tau - t)B/(A + B)$. Thus, a *unit impulse* is produced by

$$h_f(t, \tau) = \frac{Ay_{zi}(t, \tau)u(t - \tau) - By_{zi}(t, \tau)u(\tau - t)}{A + B}$$

This result is identical to Eq. 2.41 (with 2.42, 2.43, and 2.44).

We now want to find a solution for $h(t, \tau)$, the unit-impulse response for the input $x(t)$. Equation 2.3 can be written as two equations:

$$\sum_{n=0}^{N} a_n(t)\frac{d^n y}{dt^n} = x_f(t) \qquad (2.3a)$$

$$\sum_{n=0}^{M} b_n(t)\frac{d^n x}{dt^n} = x_f(t) \qquad (2.3b)$$

We claim that Eq. 2.39b is the zero-state solution to Eq. 2.3a. If this is the case, then in order to find the unit-impulse response for $x(t)$, we merely use Eq. 2.3b with an input that is a unit impulse applied at some *arbitrary* time t_1, $x(t) = \delta(t - t_1)$:

$$x_f(t) = \sum_{n=0}^{M} b_n(t)\frac{d^n}{dt^n}\{\delta(t - t_1)\}$$

This gives the forcing function $x_f(t)$ when the input $x(t)$ is a unit-impulse function, $\delta(t - t_1)$. Then $h(t, t_1)$ is found from Eq. 2.39b, replacing t with τ to obtain $x_f(\tau)$:

$$h(t, t_1) = \int_{-\infty}^{\infty} \sum_{n=0}^{M} b_n(\tau)\frac{d^n}{d\tau^n}\{\delta(\tau - t_1)\}h_f(t, \tau)\,d\tau$$

$$h(t, t_1) = \sum_{n=0}^{M} \int_{-\infty}^{\infty} b_n(\tau)\frac{d^n}{d\tau^n}\{\delta(\tau - t_1)\}h_f(t, \tau)\,d\tau$$

2.5 THE UNIT-IMPULSE RESPONSE

Using Eq. 2.29 (see Prob. 39):

$$h(t, t_1) = \sum_{n=0}^{M} (-1)^n \frac{d^n}{d\tau^n} \{b_n(\tau) h_f(t, \tau)\}_{\tau=t_1}$$

Since t_1 is arbitrary, we can use $t_1 = \tau$ and obtain

$$h(t, \tau) = \sum_{n=0}^{M} (-1)^n \frac{d^n}{d\tau^n} \{b_n(\tau) h_f(t, \tau)\} \qquad (2.46a)$$

In many systems the input function and the forcing function are identical, and in this case $h(t, \tau) = h_f(t, \tau)$. For the time-invariant system

$$h(t, \tau) = \sum_{n=0}^{M} b_n (-1)^n \frac{d^n}{d\tau^n} \{h_f(t, \tau)\} \qquad (2.46b)$$

For time-invariant systems it obviously makes no difference, except for a time delay, when the unit impulse is applied. That is, we will get the same solution (assuming our work is always without error) today or tomorrow, the only difference being a 24-hour shift in time base. Thus, for time-invariant systems $h(t, \tau) = h(t - \tau)$, $h_f(t, \tau) = h_f(t - \tau)$, and Eq. 2.46b becomes[†]

$$h(t - \tau) = \sum_{n=0}^{M} b_n \frac{d^n}{dt^n} \{h_f(t - \tau)\} \quad \text{(time-invariant system)} \qquad (2.46c)$$

If we can apply the unit impulse at our leisure, we might as well pick $\tau = 0$ as in Fig. 2.10(a). In this case Eq. 2.46c becomes

$$h(t) = \sum_{n=0}^{M} b_n \frac{d^n}{dt^n} \{h_f(t)\} \quad \text{(time-invariant system)} \qquad (2.46d)$$

Before looking at some examples of finding the unit-impulse response, we need to examine the implications of Eqs. 2.41–2.43:

$$h_f(t, \tau) = \frac{A h_f^a(t, \tau) + B h_f^b(t, \tau)}{A + B}$$

$$h_f^a(t, \tau) = y_{zi}(t, \tau) u(t - \tau)$$
$$h_f^b(t, \tau) = -y_{zi}(t, \tau) u(\tau - t)$$

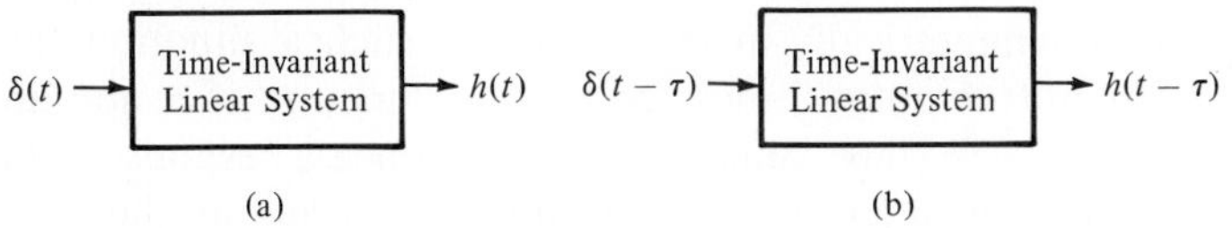

$\delta(t) \rightarrow$ [Time-Invariant Linear System] $\rightarrow h(t)$ $\quad \delta(t - \tau) \rightarrow$ [Time-Invariant Linear System] $\rightarrow h(t - \tau)$

(a)   (b)

Figure 2.10. Unit-impulse response of a time-invariant linear system. (a) Unit impulse applied at $t = 0$. (b) Unit impulse applied at $t = \tau$.

[†] Notice that $d^n/d\tau^n \{f(t - \tau)\} = (-1)^n d^n/dt^n \{f(t - \tau)\}$.

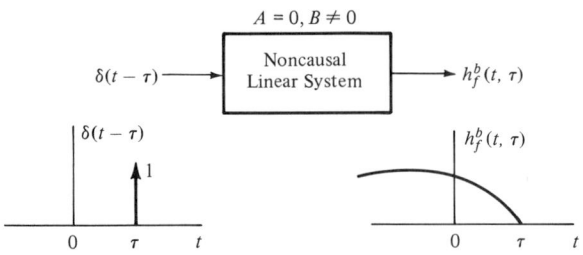

36 THE CONTINUOUS-TIME LINEAR-SYSTEM DIFFERENTIAL EQUATIONS

Figure 2.11. Noncausal part of the unit-impulse response when time is the independent variable.

If B is not 0, the system is said to be *noncausal*, or *anticipatory*, or *physically unrealizable* (if t represents time, as implied) because the unit-impulse response $h_f^b(t, \tau)$ appears only *prior* to the time $t = \tau$ when the unit impulse is applied due to the presence of $u(\tau - t)$ in Eq. 2.43:

$$h_f^b(t, \tau) = -y_{zi}(t, \tau)u(\tau - t) \tag{2.43}$$

This is shown in Fig. 2.11 and is clearly impossible realistically speaking, even though it is certainly possible mathematically speaking. Such systems cannot be built in practice.

On the other hand, if $A \neq 0$ and $B = 0$, we have a system that is always *causal* because of the presence of $u(t - \tau)$ in Eq. 2.42:

$$h_f^a(t, \tau) = y_{zi}(t, \tau)u(t - \tau) \tag{2.42}$$

or, for Eq. 2.41:

$$\boxed{h_f(t, \tau) = h_f^a(t, \tau) = y_{zi}(t, \tau)u(t - \tau)} \quad \text{(causal system)} \tag{2.47}$$

This causal system is certainly possible because the output never appears *before* the impulse is applied.

Equation 2.41 indicates that *every* linear differential equation has separate causal and noncausal unit-impulse responses. Our primary concern is with the causal unit-impulse response.

Before we jump to any false conclusions, and set $B = 0$ forever, we should remember that time is not the only possible independent variable in linear-systems analysis. The independent variable might just as well be some other physical quantity. Consider, for example, a dipole antenna in free space and oriented along the z axis in a cartesian coordinate system and excited (the input) at a small gap located symmetrically at the origin. In such a situation the magnetic vector potential (the output) must be an even function of z, and an excitation at the origin ($z = 0$) must simultaneously produce a response *both* at $z = \eta$ and $z = -\eta$. In this situation, with distance appearing as the only variable,[†] the concept of causality has no meaning, and we must use Eq. 2.41

[†] Actually, time t is also a variable in this problem, but it is usually suppressed by the use of phasor forms since we are usually interested only in the sinusoidal steady-state solution.

2.5 THE UNIT-IMPULSE RESPONSE

with $A = B$ giving

$$h_f(z, \eta) = y_{zi}(z, \eta) \frac{u(z - \eta) - u(\eta - z)}{2} \tag{2.48}$$

Example 4

What is the unit-impulse response for

$$\frac{d^2y}{dz^2} + k^2 y = x = x_f \quad [\text{Note: } h_f(z, \eta) = h(z, \eta)]$$

This equation applies in its essential details to the dipole system of the preceding paragraph. When $x = 0$, it is known as the equation of simple harmonic motion (when z is time), and its solution is well known and easily verified:

$$y_{zi} = D_1 \cos kz + D_2 \sin kz \quad (\text{zero-input response})$$

Using Eq. 2.44, with t replaced by z and τ replaced by η, we have

$$y_{zi}\big|_{z=\eta} = 0 = D_1 \cos k\eta + D_2 \sin k\eta$$

$$\frac{dy_{zi}}{dz}\bigg|_{z=\eta} = 1 = -kD_1 \sin k\eta + kD_2 \cos k\eta$$

Solving this pair of simultaneous equations gives

$$D_1 = -\frac{1}{k} \sin k\eta \qquad D_2 = \frac{1}{k} \cos k\eta$$

and therefore

$$y_{zi}(z, \eta) = \frac{1}{k} \sin k(z - \eta)$$

Using Eq. 2.48, the unit-impulse response is

$$h(z, \eta) = \frac{1}{2k} \sin k(z - \eta)[u(z - \eta) - u(\eta - z)] = h(z - \eta)$$

since this is a "space"-invariant system. More compactly

$$h(z - \eta) = \frac{1}{2k} \sin k|z - \eta|$$

The first form is preferred when differentiation is required, and it is worthwhile at this point to verify that we have found the unit-impulse response. We must show that

$$\frac{d^2h}{dz^2} + k^2 h = \delta(z - \eta)$$

Thus

$$h = \frac{1}{2k}\sin k(z-\eta)u(z-\eta) - \frac{1}{2k}\sin k(z-\eta)u(\eta-z)$$

$$\frac{dh}{dz} = \frac{1}{2k}\sin k(0)\delta(z-\eta) + \frac{1}{2k}\sin k(0)\delta(z-\eta)$$

$$+ \frac{1}{2}\cos k(z-\eta)u(z-\eta) - \frac{1}{2}\cos k(z-\eta)u(\eta-z)$$

$$\frac{d^2h}{dz^2} = \frac{1}{2}\cos k(0)\delta(z-\eta) + \frac{1}{2}\cos k(0)\delta(z-\eta)$$

$$- \frac{k}{2}\sin k(z-\eta)u(z-\eta) + \frac{k}{2}\sin k(z-\eta)u(\eta-z)$$

$$= \delta(z-\eta) - k^2 h$$

$$\frac{d^2h}{dz^2} + k^2 h = \delta(z-\eta) \qquad [\sin k0 = 0, \cos k0 = 1]$$

Example 5

Find the causal unit-impulse response for the time-variable system described by

$$t\frac{dy}{dt} + y = x = x_f \quad [\text{Note: } h_f(t, \tau) = h(t, \tau)]$$

Since we are dealing with a causal system, $B = 0$ and $h(t, \tau) = y_{zi}(t, \tau)u(t-\tau)$. We must solve

$$t\frac{dy_{zi}}{dt} + y_{zi} = 0$$

Since this is a first-order ($N = 1$) equation, it should be easy to solve by integration. Rearranging

$$\frac{dy_{zi}}{y_{zi}} = -\frac{dt}{t}$$

Integrating

$$\ln y_{zi} = -\ln t + C_1 = -\ln t + \ln C = \ln \frac{C}{t}$$

Thus

$$y_{zi} = C/t$$

There is only one equation in the set 2.44 since $N - 1 = 0$, so

$$C/\tau = 1/a_1(\tau) = 1/\tau$$

or

$$C = 1$$

and

$$y_{zi} = 1/t$$

Thus, since $x(t) = x_f(t)$, and $h(t, \tau) = h_f(t, \tau)$, Eq. 2.47 gives

$$h(t, \tau) = \frac{1}{t} u(t - \tau)$$

Check

$$\frac{dh}{dt} = \frac{1}{\tau} \delta(t - \tau) - \frac{1}{t^2} u(t - \tau)$$

$$t\frac{dh}{dt} + h = \frac{\tau}{\tau} \delta(t - \tau) - \frac{1}{t} u(t - \tau) + \frac{1}{t} u(t - \tau) = \delta(t - \tau)$$

Example 6

Find the causal unit-impulse response for the physically realizable time-invariant system shown in Fig. 2.12. Kirchhoff's voltage law gives

$$v(t) = Ri + L\frac{di}{dt} + \frac{1}{C}\int i\, dt \qquad (2.49)$$

since we have no initial energy when finding the unit-impulse response. The block diagram representation of Eq. 2.49 is shown in Fig. 2.12(b). Differentiating

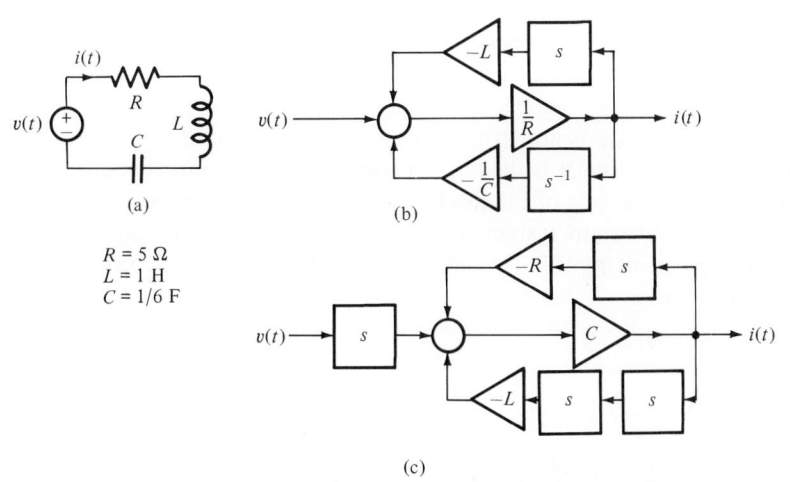

Figure 2.12. (a) Series *RLC* network. (b) Block diagram representation of Eq. 2.49. (c) Block diagram representation of Eq. 2.50.

40 THE CONTINUOUS-TIME LINEAR-SYSTEM DIFFERENTIAL EQUATIONS

Eq. 2.49 gives

$$\frac{dv}{dt} = R\frac{di}{dt} + L\frac{d^2i}{dt^2} + \frac{i}{C} \quad [\text{Note: } h_f(t, \tau) \neq h(t, \tau)] \tag{2.50}$$

Notice that the source function is $v(t)$, but the forcing function in Eq. 2.50 is dv/dt. Equation 2.50 is described in Fig. 2.12(c). The homogeneous equation is

$$L\frac{d^2i_{zi}}{dt^2} + R\frac{di_{zi}}{dt} + \frac{i_{zi}}{C} = 0 \tag{2.51}$$

In order to simplify matters, let $R = 5\,\Omega$, $L = 1\,\text{H}$, and $C = \frac{1}{6}\,\text{F}$, so that Eq. 2.51 becomes

$$\frac{d^2i_{zi}}{dt^2} + 5\frac{di_{zi}}{dt} + 6i_{zi} = 0 \tag{2.52}$$

We already know that for time-invariant systems exponential solutions *always* exist, so that if we assume

$$i_{zi} = e^{st}$$

then

$$s^2 e^{st} + 5se^{st} + 6e^{st} = 0 \qquad e^{st} \neq 0$$

or

$$s^2 + 5s + 6 = 0 \quad (characteristic\ equation)^\dagger$$

or

$$(s + 2)(s + 3) = 0$$

Thus, $s = -2, -3$, and with constants D_1 and D_2

$$i_{zi} = D_1 e^{-2t} + D_2 e^{-3t}$$

This zero-input response is constructed from the *fundamental set* (e^{-2t}, e^{-3t}), and it is easy to verify that it is the correct solution to the homogeneous equation. The *elements* or *modes* of the fundamental set are e^{-2t} and e^{-3t}.

This is a time-invariant system, and we can apply the impulse at any time, so why not apply it at $t = 0$? This makes the work somewhat easier. Using Eq. 2.44 with $\tau = 0$

$$i_{zi}|_{t=0} = 0 = D_1 + D_2$$

$$\left.\frac{di_{zi}}{dt}\right|_{t=0} = 1 = -2D_1 - 3D_2$$

---

† This polynomial represents the *characteristics* of this linear system. It is *unique* to this system. Hence: characteristic equation.

so $D_1 = 1$, $D_2 = -1$, and $i_{zi} = e^{-2t} - e^{-3t}$, and since $B = 0$ for this causal system

$$h_f(t) = [e^{-2t} - e^{-3t}]u(t)$$

Since the system is time-invariant

$$h_f(t, \tau) = h_f(t - \tau) = [e^{-2(t-\tau)} - e^{-3(t-\tau)}]u(t - \tau) \qquad (2.53)$$

We can easily find the unit-impulse response for $x(t) = v(t)$ using Eq. 2.46d. Notice that in Eq. 2.4 the only nonzero coefficient in the forcing function is $b_1 = 1$. Therefore, Eq. 2.46d gives ($n = 1$ only)

$$h(t) = \frac{dh_f}{dt} = [3e^{-3t} - 2e^{-2t}]u(t)$$

or

$$h(t - \tau) = [3e^{-3(t-\tau)} - 2e^{-2(t-\tau)}]u(t - \tau) \qquad (2.54)$$

It is worthwhile to perform a check to show that Eq. 2.54 actually is the unit-impulse response for $v(t)$ in Example 6. If it is, then according to Eq. 2.37 or 2.39a

$$i_{zs}(t) = \int_{-\infty}^{\infty} v(\tau)[3e^{-3(t-\tau)} - 2e^{-(t-\tau)}]u(t - \tau)\, d\tau$$

Now, $u(t - \tau)$ is 0 if $\tau > t$, so

$$i_{zs}(t) = \int_{-\infty}^{t} v(\tau)[3e^{-3(t-\tau)} - 2e^{-2(t-\tau)}]\, d\tau$$

Suppose $v(t) = u(t)$. Then

$$i_{zs}(t) = \int_{-\infty}^{t} u(\tau)[3e^{-3(t-\tau)} - 2e^{-2(t-\tau)}]\, d\tau$$

or[†]

$$i_{zs}(t) = \int_{0}^{t} [3e^{-3(t-\tau)} - 2e^{-2(t-\tau)}]\, d\tau\, [u(t)]$$

$$i_{zs}(t) = \left[3e^{-3t}\int_{0}^{t} e^{3\tau}\, d\tau - 2e^{-2t}\int_{0}^{t} e^{2\tau}\, d\tau\right] u(t)$$

$$i_{zs}(t) = (e^{-2t} - e^{-3t})u(t)$$

Comparing this unit-step response with Eq. 2.53 shows that a unit-step input for $v(t)$ produces the same thing as a unit-impulse input for dv/dt.

The better student may recall from a previous course in circuits that a unit-step voltage, $v(t) = u(t)$, applied at $t = 0$ for the ideal voltage source in Fig. 2.12(a) will indeed produce the current

$$i_{zs}(t) = [e^{-2t} - e^{-3t}]u(t)$$

[†] Notice the presence of $u(t)$ as a multiplier in the following equation. It indicates that t must be greater than zero.

(Eq. 2.53 with $\tau = 0$) for this *overdamped RLC* circuit. Others may have doubts, and when in doubt, the best philosophy is to "go back to the fundamentals." In this example this means go back to Kirchhoff's voltage law, which is just a restatement of *conservation of energy*. The voltage across the resistor is

$$v_{Rzs}(t) = Ri_{zs}(t) = [5e^{-2t} - 5e^{-3t}]u(t)$$

while the voltage across the inductor is

$$v_{Lzs}(t) = L\frac{di_{zs}}{dt} = [e^{-2t} - e^{-3t}]\delta(t) + [-2e^{-2t} + 3e^{-3t}]u(t)$$

$$= [1 - 1]\delta(t) + [-2e^{-2t} + 3e^{-3t}]u(t)$$

$$v_{Lzs}(t) = [-2e^{-2t} + 3e^{-3t}]u(t)$$

The capacitor voltage is given by the *indefinite integral* or *antiderivative* of the current multiplied by $1/C$:

$$v_{Czs}(t) = 6\int (e^{-2t} - e^{-3t})u(t)\,dt$$

If we integrate by parts with $u = 6u(t)$ and $dv = e^{-2t} - e^{-3t}$, so that $du = 6\delta(t)\,dt$ and $v = -\frac{1}{2}e^{-2t} + \frac{1}{3}e^{-3t}$, then

$$v_{Czs}(t) = 6(-\tfrac{1}{2}e^{-2t} + \tfrac{1}{3}e^{-3t})u(t) - 6\int(-\tfrac{1}{2}e^{-2t} + \tfrac{1}{3}e^{-3t})\delta(t)\,dt$$

$$v_{Czs}(t) = (-3e^{-2t} + 2e^{-3t})u(t) - 6(-\tfrac{1}{6})\int \delta(t)\,dt$$

But the antiderivative of $\delta(t)$ must be $u(t)$; that is

$$\int \delta(t)\,dt = u(t)$$

since

$$\frac{d}{dt}u(t) = \delta(t)$$

Thus

$$v_{Czs}(t) = (-3e^{-2t} + 2e^{-3t})u(t) + u(t)$$

This result is at least partially obvious in light of the fact that it is obvious that the unit-step input voltage must ultimately ($t \to \infty$) leave the capacitor charged with 1 V! Summing the voltages

$$v(t) = v_{Rzs}(t) + v_{Lzs}(t) + v_{Czs}(t)$$

we easily obtain $u(t) = u(t)$, verifying Kirchhoff's voltage law. Notice that both initial conditions, $i_{zs}(0^-) = 0$, and $v_{Czs}(0^-) = 0$, for the zero-state responses that we found are also satisfied for $t = 0^+$.

Finally, we should verify that a unit-impulse voltage, $v(t) = \delta(t)$, for the ideal voltage source will produce the current

$$i_{zs}(t) = [3e^{-3t} - 2e^{-2t}]u(t)$$

2.5 THE UNIT-IMPULSE RESPONSE

(Eq. 2.54 with $\tau = 0$). The voltage across the resistor is

$$v_{Rzs}(t) = Ri_{zs}(t) = [15e^{-3t} - 10e^{-2t}]u(t)$$

and the inductor voltage is

$$v_{Lzs}(t) = L\frac{di_{zs}}{dt} = [3e^{-3t} - 2e^{-2t}]\delta(t) + [-9e^{-3t} + 4e^{-2t}]u(t)$$

$$= [3 - 2]\delta(t) + [-9e^{-3t} + 4e^{-2t}]u(t)$$

$$v_{Lzs}(t) = \delta(t) + [-9e^{-3t} + 4e^{-2t}]u(t)$$

The capacitor voltage is

$$v_{Czs}(t) = \frac{1}{C}\int i_{zs}(t)\,dt = 6[-e^{-3t} + e^{-2t}]u(t)$$

The last result is obtained by integrating by parts as before. Summing the voltages

$$v(t) = v_{Rzs}(t) + v_{Lzs}(t) + v_{Czs}(t)$$

or $\delta(t) = \delta(t)$

verifying Kirchhoff's law once again.

Do the results make sense on physical grounds in this case? The voltage source is ideal (a short circuit when inactive), so after the impulse of voltage is gone ($t > 0$), the current $i_{zs}(t)$ is trapped in a sourceless RLC series circuit. This current *is not* 0 at $t = 0^+$! In spite of the fact that it is the current in the inductor, this current cannot be 0 at $t = 0^+$ if energy is to be conserved. The inductor voltage thus contains an impulse because of the discontinuous jump in its current at $t = 0$. This (unit) impulse of voltage is exactly that required to account for the unit impulse at the source! *All* voltages, as well as the current, die out for $t \to \infty$, thus the steady-state responses are all 0. Recall that for the unit-step voltage input there was one response (the capacitor voltage) that did not die out for $t \to \infty$.

We have described a *general* method for finding the unit-impulse response for all linear systems that are described by linear differential equations. Linear systems that are not time-invariant or are noncausal are included. The method required that we first find the zero-input response (or complementary function), and for systems that are not time-invariant it may be very difficult to find the zero-input response. On the other hand for time-invariant systems (which is our primary concern) we have seen that the problem is merely that of solving for the roots of the characteristic equation (a polynomial in s). This is a relatively simple matter. Since the zero-input response leads to the unit-impulse response, and this, in turn, leads to the zero-state response, we can make a very positive statement. *For a time-invariant linear system it is only necessary to solve for the roots of a polynomial to find the complete solution for the response function.*

44 THE CONTINUOUS-TIME LINEAR-SYSTEM DIFFERENTIAL EQUATIONS

There are other methods for finding the unit-impulse response. We can first find the unit-step response using the conventional methods of Sec. 2.3. This is usually not difficult. The unit-impulse response is then obtained as the time derivative of the unit-step response. This method has already been mentioned with regard to Example 6. In Example 2 we found the zero-state response (the current) for a series RL circuit when driven by the voltage source $V_0 u(t)$. It was given by Eq. 2.13. There are two modifications that we need to make in this equation before using it to find the unit-impulse response for the current in the RL circuit. First, we obviously set V_0 equal to 1. Secondly, we formally and explicitly recognize that Eq. 2.13 *only applies for* $t > 0$. We do this by simply multiplying by $u(t)$. The unit-step response then is

$$\frac{1}{R}(1 - e^{-Rt/L})u(t)$$

and its time derivative is the unit-impulse response

$$h(t) = \frac{1}{L} e^{-Rt/L} u(t)$$

It is often possible to obtain the unit-impulse response for simple systems by applying a unit impulse at the input of the block diagram representation, and then performing the indicated operations. Since this method is of limited usefulness, we will give no example.

Finally, as we will discover later, the unit-impulse response for a time-invariant linear system can be found as the *inverse Fourier or Laplace transform of the transfer function of the system*. This is usually very easy to do.

2.6 THE COMPLETE RESPONSE

The zero-input response of a differential equation is the solution of the homogeneous form of the differential equation, that is, when the input $x(t)$ is 0. It is comprised of the fundamental set, and a fundamental set is a set of N linearly independent solutions. There is, however, no unique fundamental set. The fundamental set we obtained in Example 4 was $\cos kz$, $\sin kz$, but another fundamental set is e^{jkz}, e^{-jkz}. Instead of becoming involved in a detailed discussion at this point, we will simply mention that a sufficient condition for a set to be linearly independent, and therefore a fundamental set, is that the determinant

$$\begin{vmatrix} f_1 & f_2 & \cdots & f_N \\ \dfrac{df_1}{dt} & \dfrac{df_2}{dt} & \cdots & \dfrac{df_N}{dt} \\ \cdots & \cdots & \cdots & \cdots \\ \dfrac{d^{N-1}f_1}{dt^{N-1}} & \dfrac{d^{N-1}f_2}{dt^{N-1}} & \cdots & \dfrac{d^{N-1}f_N}{dt^{N-1}} \end{vmatrix}$$

(for the case where time is the independent variable) be nonzero. The determinant above is called the *Wronskian* of the functions. For example,

$$\begin{vmatrix} \cos kz & \sin kz \\ -k \sin kz & k \cos kz \end{vmatrix} = k(\cos^2 kz + \sin^2 kz) = k \neq 0$$

so the set $\cos kz$, $\sin kz$ is linearly independent. Also

$$\begin{vmatrix} e^{jkz} & e^{-jkz} \\ jke^{jkz} & -jke^{-jkz} \end{vmatrix} = -jk - jk = -j2k \neq 0$$

so the set e^{jkz}, e^{-jkz} is also linearly independent.

Unfortunately, there are no standard techniques for finding the zero-input response for systems that are not time-invariant. For first-, or perhaps second-, order systems integration will sometimes work (as in Example 5). Sometimes a power series will lead to a solution (as in Bessel's equation). In other cases an educated guess will give a solution.

If the system is time-invariant, assuming a solution of the form e^{st} in the homogeneous equation

$$\sum_{n=0}^{N} a_n \frac{d^n y_{zi}}{dt^n} = 0 \tag{2.55}$$

will *always* lead to a polynomial in s:

$$a_N s^N + a_{N-1} s^{N-1} + \cdots + a_1 s + a_0 = 0$$

or

$$s^N + \frac{a_{N-1}}{a_N} s^{N-1} + \cdots + \frac{a_1}{a_N} s + \frac{a_0}{a_N} = 0 \tag{2.56}$$

Redefining the constants gives a simpler looking form:

$$\boxed{s^N + C_{N-1} s^{N-1} + \cdots + C_1 s + C_0 = 0} \tag{2.57}$$

Equation 2.57 is called the *characteristic equation* of the system for obvious reasons. It should be mentioned that the roots of the characteristic equation may be real, imaginary, or complex. They are called *eigenvalues* or characteristic values.

Sometimes a system will have repeated roots. If, in an Nth order system, the ith root of the characteristic equation occurs n times, the elements of the fundamental set will be (without proof)

$$e^{s_1 t}, e^{s_2 t}, \ldots, e^{s_i t}, te^{s_i t}, t^2 e^{s_i t}, \ldots, t^{n-1} e^{s_i t}, \ldots, e^{s_N t} \tag{2.58}$$

Example 7

For the system with the characteristic equation

$$(s + 2)(s^2 + 2s + 1) = 0$$

or

$$(s+2)(s+1)^2 = 0$$

the fundamental set is e^{-2t}, e^{-t}, te^{-t}, and the zero-input response is

$$y_{zi} = D_1 e^{-2t} + D_2 e^{-t} + D_3 t e^{-t}$$

Notice that an Nth order linear homogeneous differential equation must have N independent elements in its fundamental set.

The *general form* for the zero-state response of a linear system has been found to be Eq. 2.39a:

$$y_{zs}(t) = \int_{-\infty}^{\infty} x(\tau) h(t, \tau) \, d\tau \quad \text{(zero-state response, general case)} \quad (2.39)$$

For a time-invariant linear system $h(t, \tau) = h(t - \tau)$, so

$$\boxed{y_{zs}(t) = \int_{-\infty}^{\infty} x(\tau) h(t - \tau) \, d\tau} \quad \text{(zero-state response, time-invariant system)}$$
(2.59)

Equation 2.59 and its more general multidimensional forms, Eq. 2.2, for example, have been variously described as *convolution integrals*, *Helmholtz integrals* (from field theory), and *superposition integrals*. The term *convolution* appears most frequently, but the term *superposition* is more descriptive, not only because of the nature of the integral itself, but because it is a reminder that it only applies to linear systems. Convolution will be examined in detail later.

It is worthwhile to digress at this point and reconsider Eq. 2.2 as a more general form of Eq. 2.59. In electrostatics, a *point charge* is a finite charge occupying zero volume at a point in space (x', y', z'). It therefore has infinite charge density, and its density can be described as a spatial unit-impulse function for 1 C of charge. As a consequence of Coulomb's law, it can be shown that the potential (Φ) at x, y, z due to a unit (1 C) point charge at x', y', z' is

$$\frac{1}{4\pi\varepsilon R} = \frac{1}{4\pi\varepsilon[(x-x')^2 + (y-y')^2 + (z-z')^2]^{1/2}}$$

where $R = [\underline{\qquad}]^{1/2}$ is the *interval* or distance from the point charge to the point where Φ is measured. Now look at Eq. 2.2:

$$\Phi(x, y, z) = \int_{-\infty}^{\infty} \int_{-\infty}^{\infty} \int_{-\infty}^{\infty} \rho_v(x', y', z') \frac{1}{4\pi\varepsilon R} \, dx' \, dy' \, dz' \quad (2.2)$$

The potential Φ is the response, ρ_v is a *general* volume charge density source, and $1/(4\pi\varepsilon R)$ is the unit-impulse response for this linear system. Equation 2.2 is therefore a three-dimensional (space) form of Eq. 2.59!

We now return to the single-variable (t) case. For a causal system the response must not occur before the source is applied. This means that the unit-impulse response will always contain the multiplier $u(t - \tau)$, as is demonstrated by Eq. 2.47. This, in turn, means that the integrand in Eq. 2.39a is 0 if $\tau > t$. Therefore, for a causal system

$$y_{zs}(t) = \int_{-\infty}^{t} x(\tau)h(t, \tau)\, d\tau \quad \text{(zero-state response, causal system)} \quad (2.60)$$

For a system for which the input is applied at some finite time, for example $t = 0$, $x(t)$ will always appear as being multiplied by $u(t)$, explicitly, or otherwise. This type problem is called the *initial value problem*, and realistically speaking we *always* have this type problem. Thus, because of $u(t)$ [or $u(\tau)$ in the integrand]

$$y_{zs}(t) = \int_{0}^{\infty} x(\tau)h(t, \tau)\, d\tau \quad \text{(zero-state response, system starting at } t = 0\text{)} \quad (2.61)$$

Any combination of the system characteristics given by Eqs. 2.60 and 2.61 is possible. Fortunately, we are most often interested in the time-invariant, causal linear system starting at $t = 0$, and in this case

$$y_{zs}(t) = u(t) \int_{0}^{t} x(\tau)h(t - \tau)\, d\tau \quad \begin{array}{l}\text{(zero-state response, system time-}\\ \text{invariant, causal, starting at } t = 0\text{)}\end{array} \quad (2.62)$$

Notice that the integral is 0 for $t < 0$, so the multiplier $u(t)$ is inserted in Eq. 2.62.

The *complete* response of a general linear system is the sum of the zero-input response and the zero-state response:

$$y(t) = y_{zi}(t) + y_{zs}(t) \quad (2.63)$$

or

$$y(t) = \sum_{n=1}^{N} E_n e^{s_n t} + \int_{-\infty}^{\infty} x(\tau)h(t, \tau)\, d\tau \quad (2.64)$$

for an Nth order system. Equation 2.64 assumes that no roots are repeated.

Recall that in Sec. 2.3 the complete response was the sum of a particular solution and the complementary function. The initial conditions *were applied to the complete response*. The zero-state response, however, is the same, whether there is initial energy storage or not. It is due to the forcing function alone, and this forcing function may be a linear combination of the source function and its derivatives (see Eq. 2.3 or Example 6). If several sources are present, the forcing function may be a linear combination of several of the sources and their derivatives. Example 9 will demonstrate this. If there is initial energy storage when a source is suddenly applied, then not all of the initial conditions are 0. These initial conditions must be accounted for in the complete solution, and since this cannot be done by the zero-state response, it must be done by the addition of a zero-input response, $y_{zi}(t)$. The safest course to follow, if we are to avoid pitfalls, is simply that of adding a zero-input response (with N unknown constants for an Nth order system) to the zero-state response (Eq. 2.63) to form the complete solution. The initial conditions can then be applied to the complete solution to determine the unknown constants. Notice that in Eq. 2.63 (and 2.64) *both* $y_{zi}(t)$ and $y_{zs}(t)$ can contain transient terms.

48 THE CONTINUOUS-TIME LINEAR-SYSTEM DIFFERENTIAL EQUATIONS

Next we will present a summary of the method given in this chapter for solving ordinary linear differential equations, whether for a time-invariant (constant coefficients) or otherwise (variable coefficients) system.

1. Find the zero-input response. This is a solution to the homogeneous equation, and for an Nth order equation it will contain N independent solutions or elements of the fundamental set. If the system is time-invariant, the problem simply becomes that of finding the roots of an Nth order polynomial, the characteristic equation of the system, Eq. 2.57. The elements of the fundamental set will be $e^{s_1 t}, e^{s_2 t}, e^{s_3 t}, \ldots$. If some of the roots are repeated, then the elements of the fundamental set are given by Eq. 2.58. If the system is not time-invariant, then finding the zero-input response may be difficult if the order of the system is higher than $N = 1$. We may be forced to resort to special techniques not treated explicitly in this material.
2. Find the unit-impulse response from the zero-input response by using Eqs. 2.41–2.44 and 2.46b.
3. Find the zero-state response, $y_{zs}(t)$, using the superposition integral, Eq. 2.39a, 2.59, 2.60, 2.61, or 2.62.
4. Form the complete solution by adding a new zero-input response to the zero-state response. Notice that this zero-input response contains N unknown constants.
5. Apply the N given (or determined) initial conditions to the complete solution to evaluate the N constants.

Notice that steps 4 and 5 can be skipped if there is no initial energy storage.

We investigated two simple circuits in Sec. 2.3 (Examples 2 and 3) using what we called "conventional methods." We would now like to analyze these circuits using the general method of this section.

Example 8

The series RL circuit of Example 2 and Fig. 2.1 is described by Eq. 2.9:

$$V_0 u(t) = Ri + L \frac{di}{dt} \quad [\text{Note: } h_f(t) = h(t)]$$

The zero-input response is easily obtained and was given by Eq. 2.8.

$$i_{zi}(t) = D_1 e^{-Rt/L} \quad (D_1 = A)$$

Using Eq. 2.44 ($N - 1 = 0$)

$$\frac{1}{L} = D_1$$

Therefore, with Eq. 2.42

$$h(t - \tau) = \frac{1}{L} e^{-R(t-\tau)/L} u(t - \tau) \tag{2.65}$$

2.6 THE COMPLETE RESPONSE

and with Eq. 2.62 for this time-invariant causal system that is starting at $t = 0$

$$i_{zs}(t) = u(t) \int_0^t \frac{V_0}{L} e^{-R(t-\tau)/L} \, d\tau$$

or

$$i_{zs}(t) = \frac{V_0}{R}(1 - e^{-Rt/L})u(t)$$

The complete response is

$$i(t) = E_1 e^{-Rt/L} + \frac{V_0}{R}(1 - e^{-Rt/L})u(t)$$

subject to the initial condition $i(0) = 0$ (Example 2). Thus

$$i(0) = 0 = E_1$$

and

$$i(t) = \frac{V_0}{R}(1 - e^{-Rt/L})u(t)$$

which agrees with Eq. 2.13 and Fig. 2.2 for $t > 0$.
Next, suppose that (somehow) $i(0) = I_0$; that is, there is an initial current. In this case

$$i(0) = I_0 = E_1$$

Therefore, the complete solution is

$$i(t) = I_0 e^{-Rt/L} + \frac{V_0}{R}(1 - e^{-Rt/L})u(t) \qquad t > 0$$

This result satisfies both the differential equation and the initial condition, and is therefore the correct solution.

As a reminder, consider the fact that for an *arbitrary* source voltage, $v(t)$, Eqs. 2.60 and 2.65 give

$$i_{zs}(t) = \int_{-\infty}^{t} v(\tau) \frac{1}{L} e^{-R(t-\tau)/L} \, d\tau$$

for this *RL* circuit. Working with the methods of Sec. 2.3 we would probably be forced to start over again.

Example 9

We would now like to rework Example 3. Let us call the left current source $i_1(t)$ and the right source $i_2(t)$. Proceeding in this manner complicates matters very little and allows a more general approach. Our immediate problem (which we were essentially able to avoid in Example 3) then is to analyze the circuit of Fig. 2.3. We choose nodal analysis to accomplish our goal. Kirchhoff's current law gives two equations for the two node voltages, $v_L(t)$ and $v_C(t)$:

$$i_1(t) = \frac{1}{L}\int v_L\, dt + \frac{1}{R}[v_L(t) - v_C(t)] \tag{2.66}$$

$$i_2(t) = C\frac{dv_C}{dt} + \frac{1}{R}[v_C(t) - v_L(t)] \tag{2.67}$$

Differentiating the last two equations yields

$$\frac{di_1}{dt} = \frac{1}{L}v_L(t) + \frac{1}{R}\frac{dv_L}{dt} - \frac{1}{R}\frac{dv_C}{dt} \tag{2.68}$$

$$\frac{di_2}{dt} = C\frac{d^2v_C}{dt^2} + \frac{1}{R}\frac{dv_C}{dt} - \frac{1}{R}\frac{dv_L}{dt} \tag{2.69}$$

Equation 2.69 can be written

$$\frac{1}{R}\frac{dv_L}{dt} - \frac{1}{R}\frac{dv_C}{dt} = C\frac{d^2v_C}{dt^2} - \frac{di_2}{dt} \tag{2.70}$$

From Eq. 2.67

$$v_L(t) = RC\frac{dv_C}{dt} + v_C(t) - Ri_2(t) \tag{2.71}$$

Substituting Eqs. 2.70 and 2.71 into 2.68 gives

$$\frac{di_1}{dt} = \frac{RC}{L}\frac{dv_C}{dt} + \frac{1}{L}v_C(t) - \frac{R}{L}i_2(t) + C\frac{d^2v_C}{dt^2} - \frac{di_2}{dt}$$

and rearranging

$$\frac{d^2v_C}{dt^2} + \frac{R}{L}\frac{dv_C}{dt} + \frac{1}{LC}v_C(t) = \frac{1}{C}\frac{di_1}{dt} + \frac{1}{C}\frac{di_2}{dt} + \frac{R}{LC}i_2(t) \tag{2.72}$$

Notice that there are two source functions in this circuit (i_1 and i_2), but the *forcing function* in Eq. 2.72 can be considered to be the right side of Eq. 2.72. This makes it unnecessary to write *two* different equations: one that describes $v_C(t)$ in terms of $i_1(t)$, and another that describes $v_C(t)$ in terms of $i_2(t)$. Thus, we use the forcing function

$$\frac{1}{C}\frac{di_1}{dt} + \frac{1}{C}\frac{di_2}{dt} + \frac{R}{LC}i_2(t) = x_f(t)$$

When we insert the numerical values for R, L, and C from Fig. 2.3, Eq. 2.72 becomes

$$\frac{d^2v_C}{dt^2} + 8\frac{dv_C}{dt} + 4v_C(t) = 20\frac{di_1}{dt} + 20\frac{di_2}{dt} + 160i_2(t) \tag{2.73}$$

2.6 THE COMPLETE RESPONSE

Notice that the homogeneous equation is characteristic of the system and (for Eq. 2.73) agrees with Eq. 2.15. The zero-input response can be found here, but from Eq. 2.19 it is

$$v_{Czi}(t) = D_1 e^{s_1 t} + D_2 e^{s_2 t}$$

where

$$\left.\begin{matrix} s_1 = -4 + \sqrt{12} \\ s_2 = -4 - \sqrt{12} \end{matrix}\right\} \quad (2.17)$$

Using Eq. 2.44

$$0 = D_1 + D_2$$
$$1 = s_1 D_1 + s_2 D_2$$

so that

$$D_1 = \frac{1}{s_1 - s_2} \qquad D_2 = -\frac{1}{s_1 - s_2}$$

Equation 2.42 gives the unit-impulse response for $x_f(t)$:

$$h_f(t - \tau) = \frac{1}{s_1 - s_2} [e^{s_1(t-\tau)} - e^{s_2(t-\tau)}] u(t - \tau) \quad (2.74)$$

Substituting the *forcing function* and Eq. 2.74 into 2.39b:

$$v_{Czs}(t) = \int_{-\infty}^{t} \left[20 \frac{di_1}{d\tau} + 20 \frac{di_2}{d\tau} + 160 i_2(\tau) \right] \frac{1}{s_1 - s_2} [e^{s_1(t-\tau)} - e^{s_2(t-\tau)}] d\tau \quad (2.75)$$

Equation 2.75 applies for *any* inputs i_1 and i_2. In Fig. 2.3 and Example 3 we had $i_1(t) = 2u(t)$ and $i_2(t) = 1$. For these values Eq. 2.75 becomes

$$v_{Czs}(t) = \frac{1}{s_1 - s_2} \int_{-\infty}^{t} [40\delta(\tau) + 160][e^{s_1(t-\tau)} - e^{s_2(t-\tau)}] d\tau$$

The integration is relatively simple, but we do need to remember that t must be greater than 0 if the impulse is to be within the limits.

$$v_{Czs}(t) = \frac{40}{s_1 - s_2} [e^{s_1} - e^{s_2 t}] u(t) - \frac{160}{s_1(s_1 - s_2)} + \frac{160}{s_2(s_1 - s_2)}$$

Using Eq. 2.17, this becomes

$$v_{Czs}(t) = 40 + \frac{20}{\sqrt{12}} [e^{-(4-\sqrt{12})t} - e^{-(4+\sqrt{12})t}] u(t) \quad (2.76)$$

for all t. This result is identical to that given by Eq. 2.24.

One of the difficulties that was encountered in the preceding example is that of uncoupling the pair of equations 2.66 and 2.67, and this is typical of multiple-input and(or) multiple-output systems. It is relatively easy to remove this difficulty by using an *operator* notation.† Let the operator p be defined by $p \equiv d/dt$, so that $pi = di/dt$. In the same way $p^{-1}i = i/p = \int i\,dt$. Using these, we can write Eqs. 2.66 and 2.67 as

$$i_1 = \frac{1}{L}\left(\frac{v_L}{p}\right) + \frac{v_L}{R} - \frac{v_C}{R}$$

and

$$i_2 = C(pv_C) + \frac{v_C}{R} - \frac{v_L}{R}$$

The advantage of the operator forms in the last two equations is that they can be treated *algebraically*, and the equations can easily be uncoupled. Multiplying the first of these equations by p, and then rearranging both

$$pi_1 = \left(\frac{1}{L} + \frac{p}{R}\right)v_L - \left(\frac{p}{R}\right)v_C$$

$$i_2 = \left(-\frac{1}{R}\right)v_L + \left(pC + \frac{1}{R}\right)v_C$$

Solving for v_C using Cramer's rule

$$v_C = \frac{[(1/L) + (p/R)]i_2 + (p/R)i_1}{(Cp^2/R) + (Cp/L) + (1/LR)}$$

or

$$\left(\frac{C}{R}\right)p^2 v_C + \left(\frac{C}{L}\right)pv_C + \left(\frac{1}{LR}\right)v_C = \left(\frac{1}{R}\right)pi_1 + \left(\frac{1}{R}\right)pi_2 + \left(\frac{1}{L}\right)i_2$$

or

$$p^2 v_C + \left(\frac{R}{L}\right)pv_C + \left(\frac{1}{LC}\right)v_C = \left(\frac{1}{C}\right)pi_1 + \left(\frac{1}{C}\right)pi_2 + \left(\frac{R}{LC}\right)i_2$$

Since $p = d/dt$ and $p^2 = d^2/dt^2$

$$\frac{d^2 v_C}{dt^2} + \frac{R}{L}\frac{dv_C}{dt} + \frac{1}{LC}v_C = \frac{1}{C}\frac{di_1}{dt} + \frac{1}{C}\frac{di_2}{dt} + \frac{R}{LC}i_2$$

which is identical to Eq. 2.72. The perceptive reader will recognize that what was just accomplished is simply a "streamlined" version of what we had already done in Example 9.

We used no initial conditions nor any zero-input response to obtain Eq. 2.76. These were accounted for when we used the source functions $i_1(t)$ and $i_2(t)$, and used values for them that were good for *all time*. Suppose, however, that we care nothing about the response for $t < 0$, but only want to know what

---

† See DelToro in the list of references at the end of the chapter. We originally used s to signify differentiation and s^{-1} to signify integration when drawing block diagrams (Chap. 1). Here, we are using p and p^{-1} to signify the same operations in the integro-differential equations in order to avoid confusion with the Laplace transform variable s to be introduced in Chap. 6.

2.6 THE COMPLETE RESPONSE

it is for $t > 0$. Problems of this type, where there is some sort of switching action at $t = 0$ (or some other reference time), are common. When there is initial energy storage, we can add a zero-input response to accommodate the initial conditions. Alternatively, we can add a source for $t < 0$ that produces the given initial conditions and then becomes inactive at $t = 0$. The latter method, although always possible, may be difficult to implement in many cases because finding the correct source may not be a simple matter.

In order to use the zero-input response, let us reconsider Example 9, assuming that we are only interested in finding $v_C(t)$ for $t > 0$ as in Example 3. The initial conditions for the problem that we want to solve are those that were originally used in Example 3:

$$\left. \begin{array}{l} v_C(0^+) = 40 \text{ V} \\ i_L(0^+) = 1 \text{ A} \\ \dfrac{dv_C}{dt}\bigg|_{t=0^+} = 40 \text{ V/s} \end{array} \right\} \quad (2.77)$$

Our first problem is that of finding the zero-state response. We want to find this response for

$$i_1(t) = 2u(t)$$
$$i_2(t) = u(t)$$

Since this response is a zero-state response, it will automatically provide

$$\left. \begin{array}{l} v_{C_{zs}}(0^+) = 0 \\ i_{L_{zs}}(0^+) = 0 \end{array} \right\} \quad (2.78)$$

Notice particularly that it will not give

$$\frac{dv_{C_{zs}}}{dt}\bigg|_{t=0^+} = 0$$

but, in fact, will give

$$\frac{dv_{C_{zs}}}{dt}\bigg|_{t=0^+} = 60 \text{ V/s}$$

This can be seen by going back to the equations below Eq. 2.21, where we find that

$$C\frac{dv_{C_{zs}}}{dt}\bigg|_{t=0^+} = 3 - i_L(0^+) \qquad i_L(0^+) = 0$$

or

$$\frac{dv_{C_{zs}}}{dt}\bigg|_{t=0^+} = \frac{3}{C} = 60 \text{ V/s} \quad (2.79)$$

The zero-state response is found from the general form, Eq. 2.75:

$$v_{C_{zs}}(t) = \int_{-\infty}^{t} [40\delta(\tau) + 20\delta(\tau) + 160u(\tau)] \frac{1}{s_1 - s_2} [e^{s_1(t-\tau)} - e^{s_2(t-\tau)}] d\tau$$

or

$$v_{Czs}(t) = \frac{60}{s_1 - s_2} \int_{-\infty}^{t} \delta(\tau)[e^{s_1(t-\tau)} - e^{s_2(t-\tau)}] d\tau$$

$$+ \frac{160}{s_1 - s_2} \int_{-\infty}^{t} u(\tau)[e^{s_1(t-\tau)} - e^{s_2(t-\tau)}] d\tau$$

$$v_{Czs}(t) = \frac{60}{s_1 - s_2}[e^{s_1 t} - e^{s_2 t}]u(t)$$

$$- \frac{160 \, u(t)}{s_1(s_1 - s_2)}[1 - e^{s_1 t}] + \frac{160 \, u(t)}{s_2(s_1 - s_2)}[1 - e^{s_2 t}]$$

$$v_{Czs}(t) = 40u(t) - 34.434 e^{s_1 t} u(t) - 5.566 e^{s_2 t} \ (t) \tag{2.80}$$

The general form for the zero-input response is

$$v_{Czi}(t) = A_1 e^{s_1 t} + A_2 e^{s_2 t}$$

The complete response is

$$v_C(t) = v_{Czi}(t) + v_{Czs}(t)$$
$$v_C(t) = A_1 e^{s_1 t} + A_2 e^{s_2 t} + 40u(t) - [34.434 e^{s_1 t} + 5.566 e^{s_2 t}]u(t) \tag{2.81}$$

Applying the first and third elements of Eq. 2.77 to 2.81

$$40 = A_1 + A_2$$
$$40 = s_1 A_1 + s_2 A_2 + 60$$

Solving for A_1 and A_2 gives

$$A_1 = 40.207 \qquad A_2 = -0.207$$

Therefore

$$v_C(t) = 40.207 e^{s_1 t} - 0.207 e^{s_2 t}$$
$$+ 40u(t) - [34.434 e^{s_1 t} + 5.566 e^{s_2 t}]u(t) \qquad t > 0 \tag{2.82}$$

or, in a simpler form

$$v_C(t) = 40 + 5.773[e^{s_1 t} - e^{s_2 t}] \qquad t > 0 \tag{2.83}$$

This result agrees with Eqs. 2.24 and 2.76 for $t > 0$.

The preceding work has shown that we must be very careful when adding a zero-input response to accommodate given initial conditions.

Example 10

The source voltage $v(t) = 10 \cos(2t)$ was applied to the RLC circuit of Fig. 2.12(a) at some time in the distant past. Find $i(t)$. Here, we are seeking the zero-state response alone. The differential equation relating $v(t)$ and $i(t) = i_{zs}(t)$ is Eq. 2.50:

$$\frac{1}{L} \frac{dv}{dt} = \frac{d^2 i_{zs}}{dt} + \frac{R}{L} \frac{di_{zs}}{dt} + \frac{1}{LC} i_{zs} \tag{2.50}$$

2.6 THE COMPLETE RESPONSE

or

$$\frac{dv}{dt} = \frac{d^2 i_{zs}}{dt} + 5\frac{di_{zs}}{dt} + 6i_{zs}$$

The unit-impulse response for the forcing function $x_f(t) = dv/dt$ was given by Eq. 2.53:

$$h_f(t - \tau) = [e^{-2(t-\tau)} - e^{-3(t-\tau)}]u(t - \tau) \qquad (2.53)$$

Equation 2.39b [with $h_f(t, \tau) = h_f(t - \tau)$] gives the zero-state response:

$$i_{zs}(t) = \int_{-\infty}^{t} \frac{dv}{d\tau}[e^{-2(t-\tau)} - e^{-3(t-\tau)}] d\tau$$

$$= -20e^{-2t} \int_{-\infty}^{t} \sin(2\tau)e^{2\tau} d\tau + 20e^{-3t} \int_{-\infty}^{t} \sin(2\tau)e^{3\tau} d\tau \qquad (2.84)$$

The integrals should cause no difficulty. They can be found in most math handbooks[1] or in App. C. The result is

$$i_{zs}(t) = -\tfrac{5}{2}[2\sin(2t) - 2\cos(2t)] + \tfrac{20}{13}[3\sin(2t) - 2\cos(2t)]$$
$$i_{zs}(t) = 1.96\cos(2t + 11.31°)$$

The reader who has been exposed to a first course in network analysis should recognize that the use of *phasors* makes this *steady-state* problem trivial. We will mention phasors again in a later chapter.

What is the response if we create a transient by *suddenly* applying the same source? That is, we now use $v(t) = 10u(t)\cos(2t)$, and assume no initial energy storage. The only difference between this example and the last is the presence of $u(t)$, but now

$$\frac{dv}{dt} = -20u(t)\sin(2t) + 10\delta(t)$$

Thus

$$i_{zs}(t) = -20\int_0^t \sin(2\tau)[e^{-2(t-\tau)} - e^{-3(t-\tau)}] d\tau$$
$$+ 10\int_{-\infty}^t \delta(\tau)[e^{-2(t-\tau)} - e^{-3(t-\tau)}] d\tau$$
$$= -20e^{-2t}u(t)\int_0^t \sin(2\tau)e^{2\tau} d\tau + 20e^{-3t}\int_0^t \sin(2\tau)e^{3\tau} d\tau$$
$$+ 10u(t)[e^{-2t} - e^{-3t}]$$
$$= -\tfrac{20}{8}[2\sin(2t) - 2\cos(2t) + 2e^{-2t}]u(t)$$
$$+ \tfrac{20}{13}[3\sin(2t) - 2\cos(2t) + 2e^{-3t}]u(t) + 10[e^{-2t} - e^{-3t}]u(t)$$
$$= [\tfrac{65}{13}e^{-2t} - \tfrac{90}{13}e^{-3t} - \tfrac{5}{13}\sin(2t) + \tfrac{25}{13}\cos(2t)]u(t)$$
$$i_{zs}(t) = 1.96\cos(2t + 11.31°)u(t) + \tfrac{1}{13}(65e^{-2t} - 90e^{-3t})u(t)$$

There are several things to notice in comparing this result to the previous one. The steady-state responses, as expected, are the same. The transient term, created by the switching action of the source, dies out. In the second case, the same result could have been obtained by the method of Sec. 2.3: that is, assume a steady-state response of the form $A \cos(2t + \theta)$, and add a complementary function, $A_1 e^{-2t} + A_2 e^{-3t}$. Notice also that in the second case it is easy to show that $i(0^+) = 0$,

$$\left.\frac{di}{dt}\right|_{t=0^+} = 10$$

Thus

$$v_C(0^+) = v(0^+) - Ri(0^+) = L \left.\frac{di}{dt}\right|_{t=0^+}$$

$$v_C(0^+) = 10 - 0 - 10 = 0$$

and the initial energy is indeed 0.

It may be more convenient (analytically and experimentally) to characterize the input-output relationship in terms of the *unit-step response* rather than the unit-impulse response. The unit-step response is the integral of the unit-impulse response, and the unit-impulse response is the derivative of the unit-step response. The unit-step (zero-state) response for a time-invariant system is $[x(t) = u(t)]$

$$y_{uzs}(t) = \int_{-\infty}^{\infty} u(\tau) h(t - \tau) d\tau$$

$$y_{uzs}(t) = \int_{0}^{\infty} h(t - \tau) d\tau$$

An obvious change of variable gives

$$\boxed{y_{uzs}(t) = \int_{-\infty}^{t} h(\tau) d\tau} \quad \text{(unit-step response, zero state)} \quad (2.85)$$

Differentiating gives

$$h(t) = \frac{dy_{uzs}}{dt} \quad (2.86)$$

For a causal time-invariant system with arbitrary input

$$y_{zs}(t) = \int_{-\infty}^{t} x(\tau) h(t - \tau) d\tau$$

Integrating by parts with $u = x$, $dv = h(t - \tau) d\tau$ we have

$$v = -\int_{-\infty}^{t-\tau} h(\eta) d\eta = -y_{uzs}(t - \tau)$$

from Eq. 2.85. Thus

$$y_{zs}(t) = \left. -x(\tau) y_{uzs}(t - \tau) \right|_{-\infty}^{t} + \int_{-\infty}^{t} \frac{dx}{d\tau} y_{uzs}(t - \tau) d\tau$$

If $x(-\infty)$ and $y_{uzs}(0)$ are 0

$$\boxed{y_{zs}(t) = \int_{-\infty}^{t} \frac{dx}{d\tau} y_{uzs}(t-\tau)\, d\tau}\quad \begin{array}{l}\text{(zero-state response in terms of}\\ \text{the unit-step response)}\end{array} \quad (2.87)$$

Sometimes $x(t)$ begins discontinuously at some instant of time, say $t = 0$. We can replace $x(t)$ with $x(t)u(t)$, whose derivative is $x(0)\delta(t) + (dx/dt)u(t)$. Using this result in Eq. 2.87 gives

$$y_{zs}(t) = \int_{-\infty}^{t}\left[x(0)\delta(\tau) + \frac{dx}{d\tau}u(\tau)\right] y_{uzs}(t-\tau)\, d\tau$$

$$y_{zs}(t) = x(0)y_{uzs}(t) + \int_{0}^{t}\frac{dx}{d\tau} y_{uzs}(t-\tau)\, d\tau \qquad (2.88)$$

2.7 THE SECOND-ORDER SYSTEM

Second-order systems have already appeared in the examples. They occur frequently in engineering analysis, not only initially but often as approximations for higher-order systems. Because of this, a generic form has evolved and is commonly used in all areas of engineering. Figure 2.13(a) shows a parallel *RLC* network being driven by an independent current source. The integro-differential equation is obtained by nodal analysis and Kirchhoff's current law:

$$i(t) = \frac{1}{R}v(t) + C\frac{dv}{dt} + \frac{1}{L}\int v\, dt$$

for no initial energy storage. Recalling the differential equation that was found for the *series RLC* circuit, we see that it is the same as the above equation except for symbols. Mathematically, we say that they are *dual networks*, and the equations describing them are *dual equations*. Differentiating the last equation, we obtain

$$\frac{di}{dt} = C\frac{d^2v}{dt^2} + \frac{1}{R}\frac{dv}{dt} + \frac{1}{L}v(t)$$

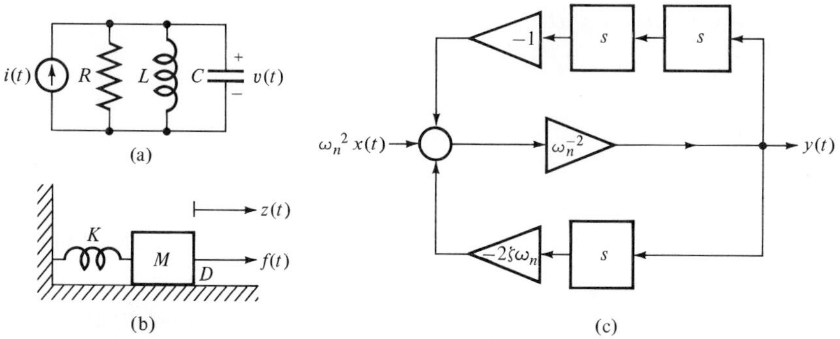

Figure 2.13. (a) Parallel *RLC* network. (b) Friction, mass, spring, mechanical system. (c) Generic block diagram for the second-order system.

Another simple second-order system is shown in Fig. 2.13(b). It is a translational mechanical system, but the electrical engineer should have little trouble using Newton's law (and Hooke's law) to obtain

$$f(t) = M\frac{d^2z}{dt^2} + D\frac{dz}{dt} + Kz(t)$$

where $f(t)$ is the applied force, z is the displacement, M is the mass, K is the linear spring constant, and D is the viscous friction.[†] The generic block diagram for a second-order system, of which we have just given two examples, is shown in Fig. 2.13(c). Its differential equation is dual to those above:

$$\frac{d^2y}{dt^2} + 2\zeta\omega_n\frac{dy}{dt} + \omega_n^2 y(t) = \omega_n^2 x(t) = x_f(t) \qquad (2.89)$$

where

ω_n = undamped natural frequency
ζ = damping ratio
$\omega_d = \omega_n\sqrt{1 - \zeta^2}$ = damped natural frequency ($\zeta < 1$)
$\alpha_d = \omega_n\sqrt{\zeta^2 - 1}$ ($\zeta > 1$)
$\zeta\omega_n \equiv \alpha$ = damping coefficient ($1/\alpha$ = time constant)

The characteristic equation is

$$s^2 + 2\zeta\omega_n s + \omega_n^2 = 0 \qquad (2.90)$$

The unit-impulse response for this time-invariant system is obtained from the zero-input response:

$$y_{zi}(t) = D_1 e^{s_1 t} + D_2 e^{s_2 t}$$

where

$$y_{zi}(0) = 0 = D_1 + D_2$$

$$\left.\frac{dy_{zi}}{dt}\right|_{t=0} = 1 = s_1 D_1 + s_2 D_2$$

or

$$D_1 = \frac{1}{s_1 - s_2} \qquad D_2 = -\frac{1}{s_1 - s_2}$$

Thus

$$y_{zi}(t) = \frac{1}{s_1 - s_2}(e^{s_1 t} - e^{s_2 t})$$

[†] See Appendix E for a brief treatment of analogous systems.

2.7 THE SECOND-ORDER SYSTEM

and

$$h_f(t-\tau) = \frac{1}{s_1 - s_2}[e^{s_1(t-\tau)} - e^{s_2(t-\tau)}]u(t-\tau) \tag{2.91}$$

for the forcing function $x_f(t) = \omega_n^2 x(t) = \delta(t)$. The zero-state response is

$$y_{zs}(t) = \int_{-\infty}^{t} \omega_n^2 x(\tau) h_f(t-\tau)\,d\tau$$

$$y_{zs}(t) = \frac{\omega_n^2}{s_1 - s_2}\int_{-\infty}^{t} x(\tau)[e^{s_1(t-\tau)} - e^{s_2(t-\tau)}]\,d\tau$$

$$y_{zs}(t) = \frac{\omega_n^2}{s_1 - s_2}\left[e^{s_1 t}\int_{-\infty}^{t} x(\tau)e^{-s_1\tau}\,d\tau - e^{s_2 t}\int_{-\infty}^{t} x(\tau)e^{-s_2\tau}\,d\tau\right] \tag{2.92}$$

The unit-step response is $y_{uzs}(t)$. In terms of the forcing function $\omega_n^2 x(t)$ it is given by Eq. 2.92 with $x_f(t) = \omega_n^2 x(t) = u(t)$:

$$y_{uzs}(t) = \frac{u(t)e^{s_1 t}}{s_1 - s_2}\int_0^t e^{-s_1\tau}\,d\tau - \frac{u(t)e^{s_2 t}}{s_1 - s_2}\int_0^t e^{-s_2\tau}\,d\tau$$

$$y_{uzs}(t) = \frac{1 - e^{s_2 t}}{s_2(s_1 - s_2)}u(t) - \frac{1 - e^{s_1 t}}{s_1(s_1 - s_2)}u(t) \tag{2.93}$$

There are four cases to be considered.
(1) *Overdamped*, $\zeta > 1$

$$\begin{aligned}s_1 &= -\zeta\omega_n - \omega_n\sqrt{\zeta^2 - 1} = -(\alpha + \alpha_d)\\ s_2 &= -\zeta\omega_n + \omega_n\sqrt{\zeta^2 - 1} = -(\alpha - \alpha_d)\end{aligned} \tag{2.94}$$

Using Eq. 2.91 with $\tau = 0$

$$h_f(t) = \frac{1}{2\alpha_d}[e^{-(\alpha - \alpha_d)t} - e^{-(\alpha + \alpha_d)t}]u(t) \tag{2.95}$$

Using Eq. 2.93

$$y_{uzs}(t) = \frac{1}{2\alpha_d}\left[\frac{1 - e^{-(\alpha - \alpha_d)t}}{\alpha - \alpha_d} - \frac{1 - e^{-(\alpha + \alpha_d)t}}{\alpha + \alpha_d}\right]u(t) \tag{2.96}$$

(2) *Critically Damped*, $\zeta = 1$
In this case $\alpha_d = 0$, and we have a pair of repeated roots. The zero-input response is

$$y_{zi}(t) = D_1 e^{s_1 t} + D_2 t e^{s_1 t}$$

We can follow the usual procedure to find $h(t)$, but it is easier to simply use L'Hospital's rule for $\alpha_d \to 0$ in Eqs. 2.95 and 2.96. The result is

$$h_f(t) = t e^{-\alpha t} u(t) \tag{2.97}$$

and

$$y_{uzs}(t) = \frac{1}{\alpha^2}[1 - e^{-\alpha t}(1 + \alpha t)]u(t) \tag{2.98}$$

(3) *Underdamped,* $\zeta < 1$

$$s_1 = -\zeta\omega_n - j\omega_n\sqrt{1-\zeta^2} = -\alpha - j\omega_d$$
$$s_2 = -\zeta\omega_n + j\omega_n\sqrt{1-\zeta^2} = -\alpha + j\omega_d$$

Equation 2.91 with $\tau = 0$ gives

$$h_f(t) = \frac{e^{-\alpha t}}{\omega_d}\sin(\omega_d t)u(t) \tag{2.99}$$

while Eq. 2.92 with $\omega_n^2 x(t) = u(t)$ gives

$$y_{uzs}(t) = \frac{1}{\omega_n^2}\left\{1 - \frac{e^{-\alpha t}}{\omega_d}[\alpha\sin(\omega_d t) + \omega_d\cos(\omega_d t)]\right\}u(t) \tag{2.100}$$

(4) *No Damping,* $\zeta = 0$
From Eq. 2.99 ($\alpha = 0$, $\omega_d = \omega_n$)

$$h_f(t) = \frac{\sin(\omega_n t)}{\omega_n}u(t) \tag{2.101}$$

while from Eq. 2.100

$$y_{uzs}(t) = \frac{1}{\omega_n^2}[1 - \cos(\omega_n t)]u(t) \tag{2.102}$$

The results of these four cases are shown in Fig. 2.14 for $\zeta = 0, 0.5, 1, 2$. Notice that the decreasing loss (decreasing ζ) results in less and less damping

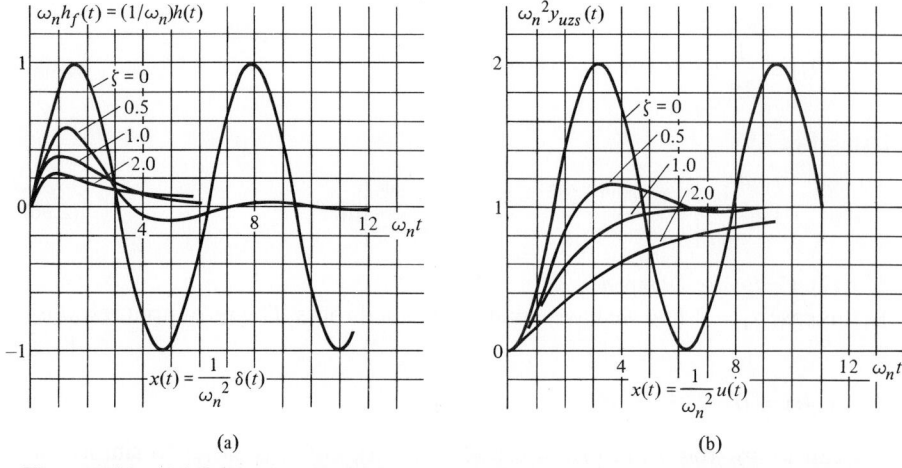

Figure 2.14. (a) Second-order system normalized impulse response for various amounts of damping. (b) Second-order system normalized step response for various amounts of damping.

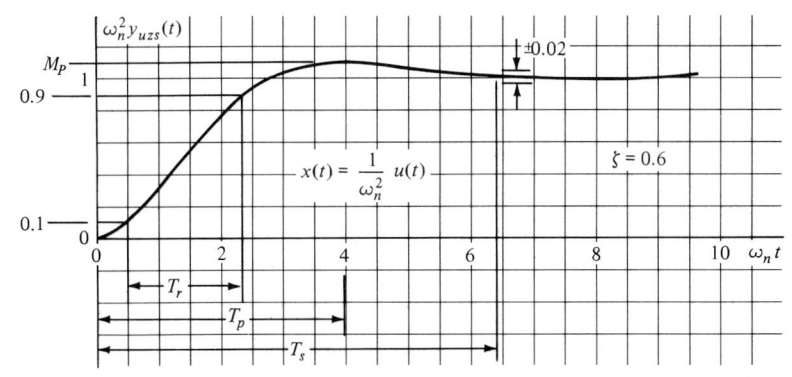

Figure 2.15. Performance measures for an underdamped second-order system ($\zeta = 0.6$).

and, finally, no damping. These are universal curves in that the abscissa values are for $\omega_n t$, while the ordinate values in the case of the impulse response are $\omega_n h_f(t) = (1/\omega_n) h(t)$ $[x(t) = (1/\omega_n^2)\delta(t)]$, and are $\omega_n^2 y_{uzs}(t)$ $[x(t) = (1/\omega_n^2)u(t)]$ for the step response.

Performance measures are defined in various ways, but are usually given in terms of the step response. One way to measure the speed of response is through the 10 to 90 percent *rise time*,[5,6] T_r, as seen in Fig. 2.15. The peak time,[†] T_p, is easily found to be

$$T_p = \pi/\omega_d = \frac{\pi}{\omega_n\sqrt{1-\zeta^2}} \qquad (\zeta < 1) \qquad (2.103)$$

and the *peak response* $M_p = \omega_n^2 y_{uzs}(T_p)$ is

$$\omega_n^2 y_{uzs}(T_p) = 1 + e^{-\zeta\pi/(1-\zeta^2)^{1/2}} \qquad (\zeta < 1) \qquad (2.104)$$

The *percent overshoot* is

$$\text{P.O.} = 100 e^{-\zeta\pi/(1-\zeta^2)^{1/2}} \text{ (percent)} \qquad (\zeta < 1) \qquad (2.105)$$

The *settling time* is the time required for response to settle within ± 2 percent of its final value. This is four time constants, so

$$T_s = \frac{4}{\alpha} = \frac{4}{\zeta\omega_n} \qquad (2.106)$$

A compromise must be made in choosing ζ if one is interested in *both* a fast response and a small overshoot.

[†] The time derivative of $\omega_n^2 y_{uzs}(t)$ is $\omega_n^2 h_f(t)$, and (for $\zeta < 1$) this is first 0 for $\omega_d t = \pi$.

Example 11

What is the maximum displacement of the mass in Fig. 2.13(b) if $f(t) = 10u(t)$ N, and the differential equation

$$f(t) = 10u(t) = 2\frac{d^2z}{dt^2} + 4.8\frac{dz}{dt} + 8z(t)$$

describes the system? Our first task is that of putting the last equation into the form of Eq. 2.89. The coefficient of the second derivative should be unity. Thus

$$5u(t) = \frac{d^2z}{dt^2} + 2.4\frac{dz}{dt} + 4z(t) \tag{2.107}$$

and $\omega_n^2 = 4$, or $\omega_n = 2$. Also, $2\zeta\omega_n = 2.4$, or $\zeta = 2.4/(2\omega_n) = 2.4/4 = 0.6$. The forcing function in Eq. 2.107 is $5u(t)$, but the result given by Eq. 2.100 for this underdamped case was derived for a forcing function given by $\omega_n^2 x(t) = u(t)$. Thus, Eq. 2.100 must be multiplied by the scale factor 5. It then gives the zero-state response

$$z_{zs}(t) = \frac{5}{4}\left\{1 - \frac{e^{-\alpha t}}{\omega_d}[\alpha \sin(\omega_d t) + \omega_d \cos(\omega_d t)]\right\}u(t)$$

Now

$$\omega_d = \omega_n\sqrt{1-\zeta^2} = 2\sqrt{1 - 0.36} = 1.6$$
$$\alpha = \zeta\omega_n = 0.6(2) = 1.2$$

Therefore

$$z_{zs}(t) = 1.25\{1 - 0.625e^{-1.2t}[1.2\sin(1.6t) + 1.6\cos(1.6t)]\}u(t) \tag{2.108}$$

Notice that the steady-state response is $z_{ss}(t) = 1.25$ m, and this agrees with Eq. 2.107.

The peak time is

$$T_p = \pi/\omega_d = 1.96 \text{ s}$$

while the peak response (not forgetting the scale factor 1.25) is

$$z_{max} = [1 + e^{-\zeta\pi/(1-\zeta^2)^{1/2}}]1.25 = 1.37 \text{ m}$$

Notice that Fig. 2.15 applies for this example when the scale factor is used.

2.8 MULTIPLE INPUTS AND OUTPUTS

If a linear system has M inputs and P outputs, then *each* zero-state output may possibly depend on every input. Suppose that we first calculate the pth zero-state output due to the mth input. Then

$$y_{pzs}(t) = H\{x_m(t)\}$$

2.8 MULTIPLE INPUTS AND OUTPUTS

But, according to the sampling property of the unit impulse function

$$x_m(t) = \int_{-\infty}^{\infty} \delta_m(t - \tau) x_m(\tau) \, d\tau$$

where $\delta_m(t - \tau)$ is a unit impulse applied at the mth input at time τ. Thus

$$y_{pzs}(t) = H\left\{\int_{-\infty}^{\infty} \delta_m(t - \tau) x_m(\tau) \, d\tau\right\}$$

$$= \int_{-\infty}^{\infty} H\{\delta_m(t - \tau)\} x_m(\tau) \, d\tau$$

The response of the system at the pth output due to this unit impulse applied to the mth input at time τ is $h_{pm}(t, \tau)$. That is:

$$H\{\delta_m(t - \tau)\} \equiv h_{pm}(t, \tau)$$

Therefore

$$y_{pzs}(t) = \int_{-\infty}^{\infty} h_{pm}(t, \tau) x_m(\tau) \, d\tau$$

The pth output due to all M inputs acting together is given by superposition as

$$y_{pzs}(t) = \sum_{m=1}^{M} \int_{-\infty}^{\infty} h_{pm}(t, \tau) x_m(\tau) d\tau \tag{2.109}$$

An equation of the form of Eq. 2.109 appears for each output (each p), and the resulting set of equations can best be described in matrix form:

$$\boxed{\mathbf{y}_{zs}(t) = \int_{-\infty}^{\infty} \mathbf{h}(t, \tau) \mathbf{x}(\tau) \, d\tau} \tag{2.110}$$

where $\mathbf{y}_{zs}(t)$ is the column *matrix* or *vector* representing the outputs, $\mathbf{x}(t)$ is the column vector of inputs and $\mathbf{h}(t, \tau)$ is the P by M (P rows and M columns) matrix of unit-impulse responses. Matrix operations are covered in Appendix A.

Equation 1.2 in Chap. 1 gave the zero-state response in matrix form for M inputs and P outputs. This was not originally called the zero-state response, but, since no initial energy storage was allowed, it is the zero-state response. Thus

$$\mathbf{y}_{zs}(t) = H\{\mathbf{x}(t)\} \tag{1.2}$$

and we now know that the operation on the right side gives the right side of Eq. 2.110.

Equation 2.110 is certainly concise, but before deciding that this is a method we want to use extensively, it would be wise to try it on an example.

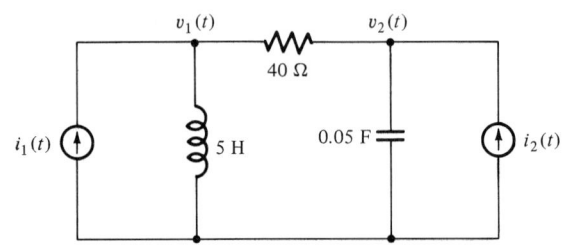

64 THE CONTINUOUS-TIME LINEAR-SYSTEM DIFFERENTIAL EQUATIONS

Figure 2.16. Circuit for Example 11.

Example 12

Find **h** for the circuit shown below in Fig. 2.16. Notice that this circuit is the same as that in Fig. 2.3 and Examples 3 and 9. We want to find the zero-state response. Consider i_1 and i_2 to be the two inputs and v_1 and v_2 to be the two outputs. Nodal analysis gives the *coupled* equations

$$i_1 = \frac{1}{L}\int v_1\, dt + \frac{v_1}{R} - \frac{v_2}{R} \tag{2.111}$$

$$i_2 = C\frac{dv_2}{dt} + \frac{v_2}{R} - \frac{v_1}{R} \tag{2.112}$$

Our first task is that of uncoupling these equations, and we have already seen below Example 9 that the best way of accomplishing this is to use the operator forms. We found that Eqs. 2.111 and 2.112 become

$$p i_1 = \left(\frac{1}{L} + \frac{p}{R}\right)v_1 - \left(\frac{p}{R}\right)v_2 \tag{2.113}$$

$$i_2 = \left(-\frac{1}{R}\right)v_1 + \left(pC + \frac{1}{R}\right)v_2 \tag{2.114}$$

since $v_1 = v_L$ and $v_2 = v_C$. Using Cramer's rule on this pair of algebraic equations gives the *uncoupled* equations

$$\frac{d^2 v_1}{dt^2} + \frac{R}{L}\frac{dv_1}{dt} + \frac{1}{LC}v_1 = R\frac{d^2 i_1}{dt^2} + \frac{1}{C}\frac{di_1}{dt} + \frac{1}{C}\frac{di_2}{dt}$$

$$\frac{d^2 v_2}{dt^2} + \frac{R}{L}\frac{dv_2}{dt} + \frac{1}{LC}v_1 = \frac{1}{C}\frac{di_1}{dt} + \frac{1}{C}\frac{di_2}{dt} + \frac{R}{LC}i_2$$

or

$$\frac{d^2 v_1}{dt^2} + 8\frac{dv_1}{dt} + 4v_1 = 40\frac{d^2 i_1}{dt^2} + 20\frac{di_1}{dt} + 20\frac{di_2}{dt} = x_{f1}(t)$$

$$\frac{d^2 v_2}{dt^2} + 8\frac{dv_2}{dt} + 4v_2 = 20\frac{di_1}{dt} + 20\frac{di_2}{dt} + 160 i_2 = x_{f2}(t)$$

2.8 MULTIPLE INPUTS AND OUTPUTS

Now, the unit-impulse responses *for the forcing function* will be the same for the last two equations since the left sides are *dual*. Thus

$$y_{zi}(t) = D_1 e^{s_1 t} + D_2 e^{s_2 t}$$

where

$$\begin{aligned} s_1 &= -4 + \sqrt{12} \\ s_2 &= -4 - \sqrt{12} \end{aligned} \quad (2.115)$$

Solving for $h_f(t)$ in the usual way, we obtain

$$h_f(t) = \frac{1}{s_1 - s_2}(e^{s_1 t} - e^{s_2 t}) u(t) \quad (2.116)$$

In order to put the results in the form of Eq. 2.110 we must find $h_{pm}(t - \tau)$. The response for v_1 when i_1 is a unit impulse applied at $t = 0$ and $i_2 = 0$ is $h_{11}(t)$, and is found by using Eq. 2.46d

$$h_{11}(t) = 40\frac{d^2 h_f}{dt^2} + 20\frac{dh_f}{dt}$$

$$h_{11}(t) = 40\delta(t) + \frac{40}{s_1 - s_2}(s_1^2 e^{s_1 t} - s_2^2 e^{s_2 t}) u(t)$$

$$+ \frac{20}{s_1 - s_2}(s_1 e^{s_1 t} - s_2 e^{s_2 t}) u(t) \quad (2.117)$$

The response for v_1 when i_2 is a unit impulse applied at $t = 0$ and $i_1 = 0$ is $h_{12}(t)$, and

$$h_{12}(t) = 20\frac{dh_f}{dt}$$

$$h_{12}(t) = \frac{20}{s_1 - s_2}(s_1 e^{s_1 t} - s_2 e^{s_2 t}) u(t) \quad (2.118)$$

The response for v_2 when i_1 is a unit impulse applied at $t = 0$ and $i_2 = 0$ is $h_{21}(t)$, and

$$h_{21}(t) = 20\frac{dh_f}{dt} = h_{12}(t) \quad (2.119)$$

This last result should not be surprising in light of the *reciprocity theorem*.[7] Finally, h_{22} is the response for v_2 when i_2 is a unit impulse applied at $t = 0$ and $i_1 = 0$:

$$h_{22}(t) = 20\frac{dh_f}{dt} + 160 h_f$$

$$h_{22}(t) = \frac{20}{s_1 - s_2}(s_1 e^{s_1 t} - s_2 e^{s_2 t}) u(t) + \frac{160}{s_1 - s_2}(e^{s_1 t} - e^{s_2 t}) u(t) \quad (2.120)$$

Equation 2.110 can be written as

$$\begin{bmatrix} v_{1zs}(t) \\ v_{2zs}(t) \end{bmatrix} = \begin{bmatrix} \int_{-\infty}^{t} h_{11}(t-\tau)i_1(\tau)\,d\tau + \int_{-\infty}^{t} h_{12}(t-\tau)i_2(\tau)\,d\tau \\ \int_{-\infty}^{t} h_{21}(t-\tau)i_1(\tau)\,d\tau + \int_{-\infty}^{t} h_{22}(t-\tau)i_2(\tau)\,d\tau \end{bmatrix}$$

We found the voltage across the capacitor $v_2(t) = v_C(t)$ when $i_1(t) = 2u(t)$ and $i_2(t) = 1$ in Example 9. Here, we have

$$v_{2zs}(t) = \int_{-\infty}^{t} 2u(\tau)h_{21}(t-\tau)\,d\tau + \int_{-\infty}^{t} h_{22}(t-\tau)\,d\tau$$

$$= 2u(t) \int_{0}^{t} \frac{20}{s_1 - s_2}[s_1 e^{s_1(t-\tau)} - s_2 e^{s_2(t-\tau)}]\,d\tau$$

$$+ \int_{-\infty}^{t} \frac{20}{s_1 - s_2}[(s_1 + 8)e^{s_1(t-\tau)} - (s_2 + 8)e^{s_2(t-\tau)}]\,d\tau$$

The integration gives

$$v_{2zs}(t) = 40 + \frac{20}{\sqrt{12}}(e^{s_1 t} - e^{s_2 t})u(t)$$

which is the same result as that obtained in Example 8.

Even though the preceding example was relatively simple, the procedure became somewhat tedious, and would become more so for higher-order systems. The method to be introduced in the next chapter is helpful in this regard.

2.9 CONVOLUTION

Equation 2.59:

$$y_{zs}(t) = \int_{-\infty}^{t} x(\tau)h(t-\tau)\,d\tau \tag{2.59}$$

was called a *superposition* or *convolution* type integral earlier. We are primarily interested in a causal time-invariant system starting at some finite time, say $t = 0$, in which case we have Eq. 2.62:

$$y_{zs}(t) = \int_{0}^{t} x(\tau)h(t-\tau)\,d\tau \tag{2.121}$$

Notice that the multiplier $u(t)$ that appeared in Eq. 2.62 has been dropped here for simplicity. The convolution of two continuous-time signals is the integral

$$f(t) = \int_{0}^{t} f_1(\tau)f_2(t-\tau)\,d\tau \tag{2.122}$$

and is denoted by

$$f(t) = f_1(t) * f_2(t) \tag{2.123}$$

2.9 CONVOLUTION

A simple change of variable $(t - \tau) = \eta$ in Eq. 2.122 gives

$$f(t) = \int_0^t f_1(t - \tau) f_2(\tau) \, d\tau \tag{2.124}$$

or

$$f(t) = f_2(t) * f_1(t) \tag{2.125}$$

which shows that this operation is *commutative*. It is also *associative*:

$$[f_1(t) * f_2(t)] * f_3(t) = f_1(t) * [f_2(t) * f_3(t)] \tag{2.126}$$

Notice carefully that when t changes (Eq. 2.122) both the product $f_1(\tau)f_2(t - \tau)$ and the upper limit of the integral change.

It is helpful in visualizing convolution to look at a graphical example.

Example 13

Find $f(t) = f_1(t) * f_2(t)$ for $f_1(t)$ and $f_2(t)$ shown in Fig. 2.17. We will use Eq. 2.122. The function $f_2(-\tau)$ is simply $f_2(\tau)$ *folded* about the ordinate (vertical axis) as shown in Fig. 2.18(a). The function $f_2(t_1 - \tau)$ is the result of *shifting* $f_2(-\tau)$ to the right by an amount equal to t_1. This is shown in Fig. 2.18(b).

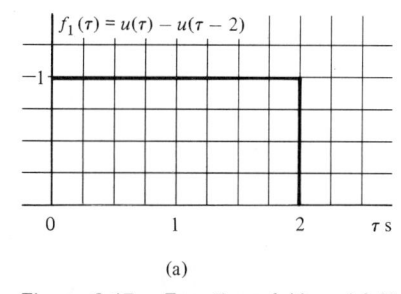

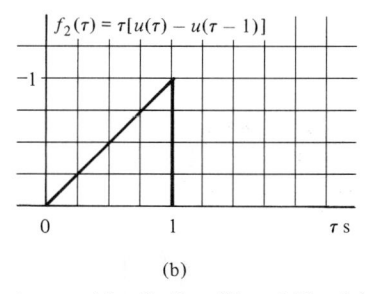

Figure 2.17. Functions $f_1(t)$ and $f_2(t)$ to be used for finding $f(t) = f_1(t) * f_2(t)$.

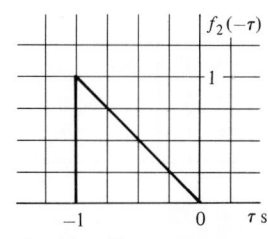

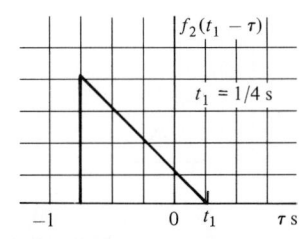

Figure 2.18. (a) Function $f_2(-\tau)$. (b) Function $f_2(t_1 - \tau)$ for $t_1 = 0.25$ s.

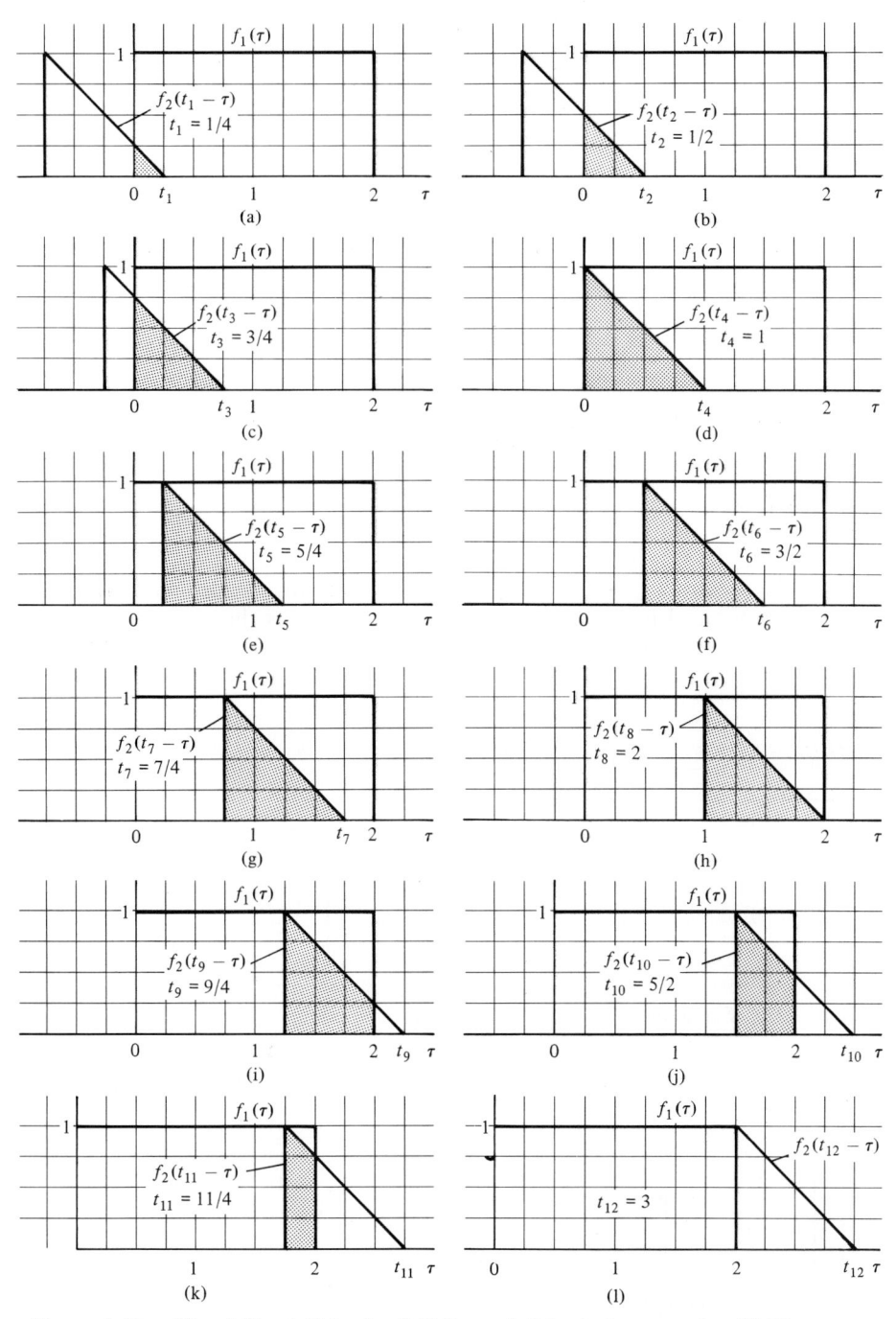

68 THE CONTINUOUS-TIME LINEAR-SYSTEM DIFFERENTIAL EQUATIONS

Figure 2.19. $f(t) = f_1(t) * f_2(t)$ for $t = 0.25$ through 3.0 s in increments of 0.25 s.

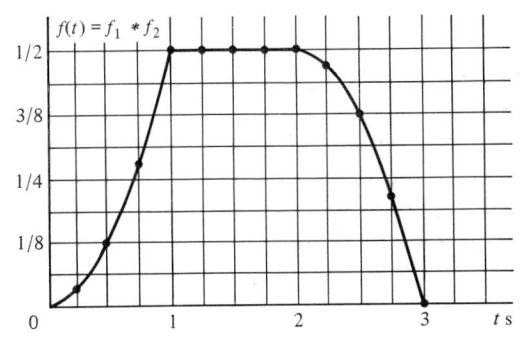

Figure 2.20. Function $f(t) = f_1(t) * f_2(t)$.

Since

$$f(t_1) = \int_0^{t_1} f_1(\tau) f_2(t_1 - \tau)\, d\tau$$

the function $f(t_1)$ is the area (*integral*) under the curve given by the product of $f_1(\tau)$ and $f_2(t_1 - \tau)$.

In Fig. 2.19(a) the area is $\frac{1}{2}(\frac{1}{4})(\frac{1}{4}) = \frac{1}{32}$. For Fig. 2.19(b) the area is $\frac{1}{2}(\frac{1}{2})(\frac{1}{2}) = \frac{4}{32}$. For Fig. 2.19(c) it is $\frac{1}{2}(\frac{3}{4})(\frac{3}{4}) = \frac{9}{32}$. This process continues until $t_{12} = 3$ where the area is 0 and $f(t_{12}) = 0$. Also, $f(t) = 0$, $t > 3$.

The function $f(t) = f_1(t) * f_2(t)$ is shown in Fig. 2.20. We have, of course, found $f(t)$ geometrically at the points $t_1, t_2, \ldots, t_{12}$, and interpolated to make the continuous plot in Fig. 2.20. Analytically the function $f(t)$ is

$$f(t) = \begin{cases} t^2/2 & 0 \leq t \leq 1 \\ \frac{1}{2} & 1 \leq t \leq 2 \\ -(t-2)^2/2 + \frac{1}{2} & 2 \leq t \leq 3 \\ 0 & t \leq 3 \end{cases}$$

The reader is well advised to actually perform the analytic calculation of $f(t)$, for this shows that even for this relatively simple example the work becomes tedious. It is not trivial! The reader is also encouraged to demonstrate the commutative property by making sketches similar to Fig. 2.19 for $f_1(t - \tau)$ and $f_2(\tau)$.

The preceding example has demonstrated convolution from a geometrical point of view. The idea is basically simple. We *fold, shift, multiply,* and *integrate.* Even though the convolution integral represents a *formal* solution for the zero-state response of a time-invariant linear system, the integral itself may be difficult to evaluate in some cases. Do not forget that numerical integration can always be performed, and *must* be performed in those cases where the forcing function and(or) the unit-impulse response (or the unit-step response) are available only in numerical form.

2.10 STABILITY

System stability is best discussed with the aid of the Laplace transform, and this will be done when this powerful technique is introduced in Chap. 6. On the other hand there are some conclusions that can be reached based on our work up to this point. For the sake of simplicity we restrict the present discussion to those linear systems that are causal and, therefore, physically realizable (can be built). The unit-impulse response for the forcing function $x_f(t)$ is given by Eq. 2.47:

$$h_f(t, \tau) = y_{zi}(t, \tau)u(t - \tau) \tag{2.47}$$

for an applied unit impulse $\delta(t - \tau)$. All *real* systems have loss (resistance in the case of a lumped electrical network), and since an applied impulse does not continually supply energy to the system, we say that

$$\left. \begin{array}{l} h_f(t, \tau) \to 0 \quad \text{for} \quad t \to \infty \\ \text{or} \\ y_{zi}(t, \tau) \to 0 \quad \text{for} \quad t \to \infty \end{array} \right\} \tag{2.127}$$

for *stable* passive physical systems. That is, the unit-impulse response and the zero-input response must die out to 0 as t approaches ∞. Example 5 is a case in point. For systems that are time-invariant, as well as causal, Eq. 2.47 becomes (for $\tau = 0$)

$$h_f(t) = y_{zi}(t)u(t) \tag{2.128}$$

If $h_f(t)$ and $h(t)$ die out[†] for $t \to \infty$, then $y_{zi}(t)$ must die out for $t \to \infty$. Thus

$$\boxed{\begin{array}{c} y_{zi}(t) \to 0 \\ t \to \infty \end{array}} \quad \text{(stable, time-invariant, causal linear system)} \tag{2.129}$$

Since $y_{zi}(t)$ is the zero-input response, and it has the general form[‡]

$$y_{zi}(t) = D_1 e^{s_1 t} + D_2 e^{s_2 t} + \cdots + D_N e^{s_N t}$$

we conclude that *if such a system is to be stable then the real parts of all the roots $s_1, s_2, \ldots, s_N$ of the characteristic equation*

$$s^N + C_{N-1} s^{N-1} + \cdots + C_1 s + C_0 = 0$$

must be negative. Each element in the fundamental set will then go to 0, because if $s_n = \sigma + j\omega$

$$e^{s_n t} = e^{\sigma t} e^{j\omega t} \to 0 \quad \sigma < 0$$
$$t \to \infty$$

[†] See Eq. 2.46d.
[‡] Repeated roots cause no trouble because the elements of the fundamental set have the form $t^{n-1} e^{s_i t}$ (Eq. 2.58) which dies out with increasing t so long as $\text{Re}\{s_i\} < 0$.

2.10 STABILITY

A statement that is equivalent to Eq. 2.129 is that the *roots of the characteristic equation must lie in the left half of the complex* $(s = \sigma + j\omega)$ *plane*. There are two simple techniques available for determining the *number* of roots of the characteristic equation that lie in the *right half* of the s plane without actually finding these roots. Thus, if one of these tests is applied and shows that there are no roots in the right half of the s plane, then the system is stable. They are called the Routh test and the Hurwitz test. The Routh test is presented in App. D. It is equivalent to the Hurwitz test.

An equivalent statement regarding the stability of a system is that if the system is stable then *every bounded input produces a bounded output*. A sufficient condition[8] for this to occur is that $h(t)$ be absolutely integrable. That is

$$\int_0^\infty |h(t)|\,dt \le K_2 < \infty \quad (K_2 > 0 \text{ is a constant}) \tag{2.130}$$

To show this let $x(t)$, the input, be bounded

$$|x(t)| < K_1 < \infty \quad (K_1 > 0 \text{ is a constant}) \tag{2.131}$$

Now, for a causal system

$$y_{zs}(t) = \int_{-\infty}^{t} x(\tau)h(t-\tau)\,d\tau$$

and upon changing the variable: $t - \tau = \eta$

$$y_{zs}(t) = \int_0^\infty h(\eta)x(t-\eta)\,d\eta \tag{2.132}$$

The absolute value of an integral is less than or equal to the integral of the absolute value of the integrand, so

$$|y_{zs}(t)| \le \int_0^\infty |h(\eta)x(t-\eta)|\,d\eta$$

$$\le \int_0^\infty |h(\eta)|\,|x(t-\eta)|\,d\eta \le K_1 \int_0^\infty |h(\eta)|\,d\eta$$

Using inequality 2.130

$$|y_{zs}(t)| \le K_1 K_2$$

so the output is bounded if 2.130 holds.

Figure 2.14 for the second-order system is a good example of the philosophy of the preceding paragraphs. The only case where the normalized unit-impulse response does not die out is for $\zeta = 0$ which is the lossless (and unrealizable) case (the system is said to be *marginally stable* in that case). The pair of conjugate roots lie *on* the imaginary axis as in Fig. 2.21 ($\zeta = 0$). The normalized unit-impulse responses for the four cases in Fig. 2.21 are shown in Fig. 2.14.

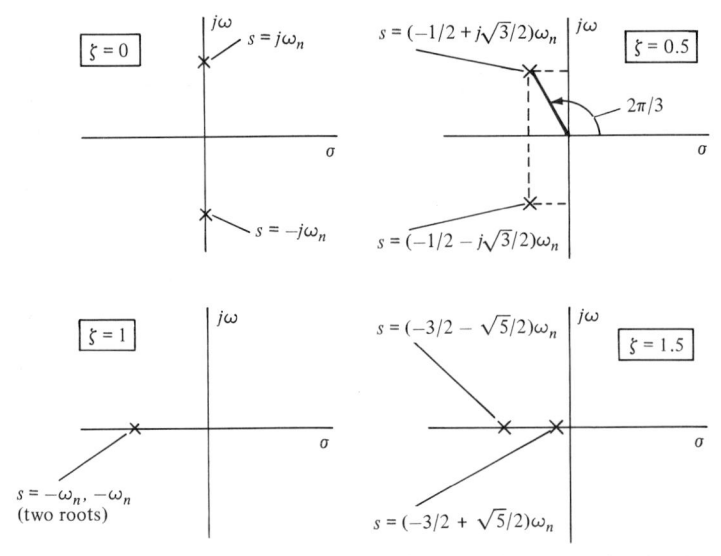

Figure 2.21. Loci of the roots of the characteristic equation for the second-order system.

2.11 CONCLUDING REMARKS

The method presented in this chapter for obtaining the complete response for linear systems that can be described by linear ordinary differential equations is a general one. It departs somewhat from classical methods but is relatively straightforward. It requires that we first find the zero-input response, and this response gives us the unit-impulse response, which in turn *completely characterizes the system*. This statement will have more meaning for us as we proceed to the material in later chapters. The complete response is given by Eq. 2.64, and it is important to recognize that if the zero-input response is known, then $h(t, \tau)$ is known, and the zero-state response can be found regardless of the form of the input, even if we are forced to use numerical integration.

In fact, the input may not even be known in terms of analytic functions, but may be given by an experimental graph, perhaps obtained in the laboratory. We can still perform the required integration numerically. A good approximation to the unit-impulse response can be obtained if, in the laboratory, we apply an intense, but short-duration, repetitive rectangular pulse of unit area (approximating a unit impulse) to the system. The pulse duration must be small compared to times over which significant changes in the response are expected, and the period (time between successive pulses) must be large enough so that any transients (short-term effects) have essentially died out. This last sentence will become more obvious in the material to follow. On the other hand it may be much easier in practice to apply (the approximation to) a unit step to the system and consider this to be the generic input. We also have found superposition integrals for finding the response of a linear system to arbitrary excitation in terms of the unit-step response (see Probs. 18 and 19, Chap. 4).

It would be wise for us to summarize the method developed in this chapter for treating the frequently encountered *causal* and *time-invariant linear system*. We have found a general solution for its differential equation description:

$$a_N \frac{d^N y}{dt^N} + \cdots + a_2 \frac{d^2 y}{dt^2} + a_1 \frac{dy}{dt} + a_0 y$$

$$= b_0 x + b_1 \frac{dx}{dt} + \cdots + b_M \frac{d^M x}{dt^M} = x_f(t) \qquad (2.4)$$

The solution is (no repeated roots)

$$y(t) = \int_{-\infty}^{t} x_f(\tau) h_f(t - \tau) \, d\tau + E_1 e^{s_1 t} + E_2 e^{s_2 t} + \cdots + E_N e^{s_N t} \qquad (2.133)$$

where

$$h_f(t) = [D_1 e^{s_1 t} + D_2 e^{s_2 t} + \cdots + D_N e^{s_N t}] u(t) = y_{zi}(t) u(t) \qquad (2.134)$$

subject to the *special* initial conditions

$$y_{zi}(0) = 0, \ y'_{zi}(0) = 0, \ldots, \ y_{zi}^{N-1}(0) = 1/a_N \qquad (2.135)$$

and where $s_1, s_2, \ldots, s_N$ are the N solutions to the Nth order polynomial (characteristic equation)

$$a_N s^N + \cdots + a_2 s^2 + a_1 s + a_0 = 0 \qquad (2.136)$$

The N constants $E_1, E_2, \ldots, E_N$ are determined from N initial conditions to form the zero-input response.

It was pointed out in the material following Example 9 that the use of the p operator greatly facilitates the uncoupling of simultaneous integro-differential equations that arise when multiple-input, multiple-output systems are being analyzed. This occurs because the integro-differential equations are converted into algebraic equations that are easy to uncouple with standard methods such as Cramer's rule.

In the next chapter we will present another method with which multiple-input, multiple-output systems can be treated. The method of *state-variables* converts an Nth order system description into N first-order equations. It is also ideal for numerical solutions, and so we have postponed consideration of these until the next chapter.

There are other (continuous) linear-system characteristics that need to be investigated. Frequency response, for example, is best treated by Fourier methods, and this will be done in Chaps. 4 and 5.

PROBLEMS

1. Determine A such that for $\tau \to 0 \ (\tau > 0)$, $f(t) \to \delta(t)$. Use a table of integrals for finding the area.

    (a) $f(t) = A \dfrac{\sin(t/\tau)}{t/\tau}$

THE CONTINUOUS-TIME LINEAR-SYSTEM DIFFERENTIAL EQUATIONS

(b) $f(t) = A\left[\dfrac{\sin(t/\tau)}{t/\tau}\right]^2$

(c) $f(t) = Ae^{-|t|/\tau}$

(d) $f(t) = A(1 - |t|/\tau)[u(t + \tau) - u(t - \tau)]$

2. Use the sampling property of the unit-impulse function to evaluate the following:

(a) $\displaystyle\int_{-\infty}^{\infty} \cos(6t)\delta(t - 2)\, dt$

(b) $\displaystyle\int_{-\infty}^{\infty} \delta(t - \tau)\delta(\tau - 1)\, d\tau$

(c) $\displaystyle\int_{-\infty}^{t} \delta(\tau - 1)\, d\tau$

(d) $\displaystyle\int_{0}^{3} f(\tau)\delta(\tau + 2)\, d\tau$

(e) $\displaystyle\int_{0}^{\pi} \delta(\cos t)\, dt$

3. What does the function $u(\cos t) - \tfrac{1}{2}$ represent?

4. Sketch $u[f(t)]$ for the $f(t)$ shown in Fig. 2.22.

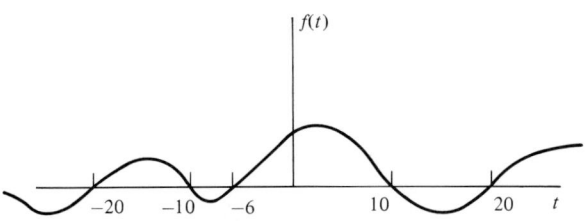

Figure 2.22. Arbitrary function f(t).

5. Evaluate the following:

(a) $\dfrac{d}{dt}\{u[f(t)]\}$

(b) $\dfrac{d}{dr}[f(r)u(r)]$

(c) $\dfrac{d}{dt}[\sin(t)u(t)]$

(d) $\dfrac{d}{dt}[\cos(t)u(t)]$

(e) $\dfrac{d}{dt}[t\delta(t)]$

(f) $\dfrac{d^2}{dt^2}[t^2 u(t)]$

6. If $h(t, \tau) = e^{-|t-\tau|}$ and $x(t) = t$, find $y(t)$.

7. Repeat Prob. 6 if the system is unchanged, but the input is changed to $x(t) = u(t)$.

8. Describe the following differential equations as to whether they are ordinary or partial.

(a) $\dfrac{d^2x}{dt^2} + \dfrac{dy}{dt} + y = 0 \qquad x = x(t), \quad y = y(t)$

(b) $\dfrac{\partial \Phi}{\partial x} + \dfrac{\partial \Phi}{\partial y} + x = 0 \qquad \Phi = \Phi(x, y)$

(c) $\dfrac{d}{dx}\left(\dfrac{\partial \Phi}{\partial t}\right) = 0 \qquad \Phi = t^2 + \dfrac{dt}{dx}$

(d) $\dfrac{\partial \alpha}{\partial \rho} + \dfrac{1}{\rho}\dfrac{\partial \alpha}{\partial \phi} + \dfrac{\partial \alpha}{\partial z} = 0 \qquad \alpha = \alpha(\rho, \phi, z)$

9. Describe the systems that are listed below as to whether they are time-invariant or time-variable.

(a) $\dfrac{d^2y}{dt^2} + k^2 y = x(t)$

(b) $\dfrac{d}{dt}(yt^3) + y = 0$

(c) $\dfrac{1}{t}\dfrac{dy}{dt} + \dfrac{1}{t}y = 0$

(d) $\dfrac{dy}{dt} + \sin(t)y = 0$

(e) $\dfrac{dy}{dt} + \cos(t)y = 0 \qquad 0 \le t \le 0.01$

10. Describe the following systems in terms of whether or not they are causal.

(a) $h(t, \tau) = \dfrac{t^2}{\tau} u(\tau - t)$ 　　(b) $h(t, \tau) = 2(t - \tau)u(\tau - t) + u(t - \tau)$

(c) $h(t, \tau) = u(t - \tau)$ 　　(d) $h(t, \tau) = (t - \tau)u(t - \tau)$

(e) $h(t, \tau) = \sin(t - \tau)u(t - \tau)$

11. Describe the causal nonlinear system,

$$\dfrac{d^2y}{dt^2} + 3y + f(y) = x(t)$$

where $f(y)$ is shown in Fig. 2.23, as a *piecewise-linear* system.

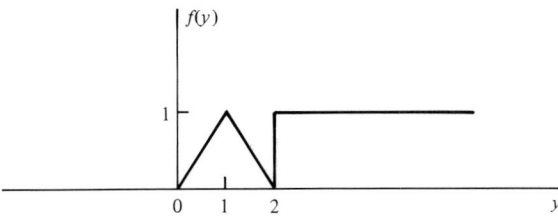

Figure 2.23. f(y) versus y.

76 THE CONTINUOUS-TIME LINEAR-SYSTEM DIFFERENTIAL EQUATIONS

12. Find $y_{zs}(t)$ for $0 < y < 1$ in Prob. 11 if $x(t) = u(t) - u(t-1)$.
13. If the Taylor series expansion for $\cos(y)$ about the point $y = 0$ is $\cos(y) = 1 - y^2/2! + \cdots$, linearize the differential equation

$$\frac{d^2y}{dt^2} + 4y\cos(y) = x$$

about this point, and state what constraints exist. Notice that $y = 0$ when $x = 0$.

14. Show that the causal unit-impulse response

$$h(t) = [e^{-t} - \tfrac{3}{2}e^{-2t} + \tfrac{1}{2}e^{-4t}]u(t)/6$$

describes the time-invariant linear system

$$2\frac{d^3y}{dt^3} + 14\frac{d^2y}{dt^2} + 28\frac{dy}{dt} + 16y = x(t)$$

15. Show that the causal unit-impulse response

$$h(t, \tau) = \frac{1}{t}u(t - \tau)$$

describes the time-variable linear system

$$t\frac{dy}{dt} + y = x$$

16. Show that

$$y_{zi}(t) = D_1 e^{-t} + D_2 e^{-3t}$$

is a zero-input response for the time-invariant system

$$\frac{d^2y}{dt^2} + 4\frac{dy}{dt} + 3y = x(t)$$

17. Show that the general unit-impulse response for Prob. 16 is

$$\frac{\tfrac{1}{2}e^{-(t-\tau)} - \tfrac{1}{2}e^{-3(t-\tau)}}{A + B}[Au(t-\tau) - Bu(\tau - t)]$$

18. Show that the causal unit-impulse response for Prob. 16 is

$$[\tfrac{1}{2}e^{-(t-\tau)} - \tfrac{1}{2}e^{-3(t-\tau)}]u(t-\tau)$$

19. The unit-impulse response for a time-invariant, causal system is

$$h(t) = [\tfrac{1}{2}e^{-2t} - e^{-3t} + \tfrac{1}{2}e^{-4t}]u(t)$$

Find the differential equation that describes the system.

20. What is the fundamental set for the time-invariant linear system whose characteristic equation is

$$s^3 + 7s^2 + 16s + 12 = 0$$

It has a pair of repeated roots.

21. What is the fundamental set for the linear system whose differential equation is

$$\frac{d^3y}{dt^3} + 4\frac{d^2y}{dt^2} + 6\frac{dy}{dt} + 4y = x$$

One element of the set is $e^{(-1-j)t}$.

22. Find the zero-input response for the system in Prob. 21 if

$$y(0) = 1 \qquad y'(0) = 0 \qquad y''(0) = 0$$

23. Find the causal zero-state response for the causal system

$$\frac{d^2y}{dt^2} + 4\frac{dy}{dt} + 4y = 3\frac{dx}{dt} + 2x = x_f(t)$$

where $x(t) = e^{-3t}u(t)$.

24. Find the zero-state response for the system in Prob. 23 if $x_f(t) = tu(t)$. What are the steady-state and transient parts? Is the system stable?

25. Show that the *unit-ramp response* (zero state) for a causal linear system is

$$y_{rzs}(t) = u(t)\int_0^t \tau h(t, \tau)\, d\tau \qquad x(t) = tu(t)$$

26. Find the complete causal response for the causal system

(a) $\dfrac{d^2y}{dt^2} + 4\dfrac{dy}{dt} + 4y = \sin(\omega_0 t) \qquad y(0) = 1 \quad y'(0) = 0$

(b) $\dfrac{d^2y}{dt^2} + 4\dfrac{dy}{dt} + 4y = \sin(\omega_0 t)u(t), \qquad y(0) = 1 \quad y'(0) = 0$

27. What kind of second-order system is given by

$$\frac{d^2y}{dt^2} + 2\frac{dy}{dt} + 4y = x$$

Find α, ω_n, ζ, and ω_d. When a unit step is applied, at what value of time is the peak value of y attained? $y(0) = y'(0) = 0$.

28. Find the causal unit-impulse response for the time-variable system

$$t^2\frac{d^2y}{dt^2} - 2t\frac{dy}{dt} + 2y = x$$

if the fundamental set is t, t^2.

29. State whether the following systems are linear or nonlinear, indicating the nonlinear terms

(a) $t\dfrac{dy}{dt} + y = 0$ \qquad (d) $\sin(t)\dfrac{dy}{dt} + y = 0$

(b) $\dfrac{1}{y}\dfrac{dy}{dt} + 1 = 0$ \qquad (e) $\sin(y)\dfrac{dy}{dt} + y = 0$

(c) $\dfrac{dy}{dt} + y^2 = 0$ \qquad (f) $\dfrac{d^2y}{dt^2} + y\dfrac{dy}{dt} + y^2 = 0$

78 THE CONTINUOUS-TIME LINEAR-SYSTEM DIFFERENTIAL EQUATIONS

(g) $y = mx + b$ $x =$ input   (h) $e^t \dfrac{dy}{dt} + y = 0$

30. Show that

$$C\frac{d^2v}{dt^2} + \frac{1}{R}\frac{dv}{dt} + \frac{1}{L}v = 0$$

describes the circuit in Fig. 2.24. Find α, ω_n, ζ, and ω_d. Find the zero-input response if $R = 25\,\Omega$, $L = 50$ mH, $C = 5\,\mu$F, $v(0) = 50$ V, and $i_L(0) = 0$. Notice that $i_L = -v/R - C\,dv/dt$, so that

$$\left.\frac{dv}{dt}\right|_{t=0} = -\frac{i_L(0)}{C} - \frac{v(0)}{CR}$$

Can a steady-state response exist?

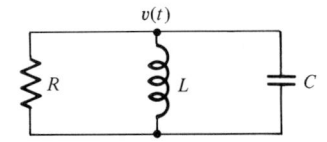

Figure 2.24. Parallel RLC circuit, source free.

31. The circuit of Fig. 2.24 has no energy storage when the current source $i_s(t)$ is placed across the resistor at $t = -\infty$.
(a) Find $v(t)$ if $i_s(t) = 10u(t)$ A.
(b) Repeat if $i_s(t) = 10\sin(2000t)$ A.
Identify the transient and steady-state parts in (a) and (b).

32. Does the circuit of Fig. 2.25 represent a time-invariant linear system? Treat the problem of finding $v(t)$ as an initial value problem.
(a) Find $v(t)$, $t \geq 0$, if $v_1(t) = 60$ V.
(b) Repeat if $v_1(t) = 60\cos(5t)$ V.

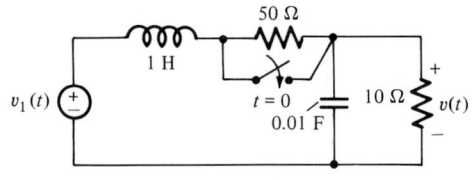

Figure 2.25. Circuit.

33. Find $y_1(t)$ and $y_2(t)$ for the system in Fig. 2.26 if $x_1(t) = \sin(t)$, $x_2(t) = 2\cos(3t)$, and $x_3(t) = t$.

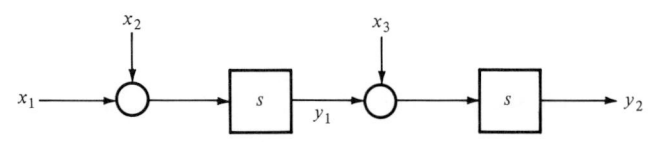

Figure 2.26. Block diagram of a time-invariant linear system.

34. Find the fundamental set for the time-variable linear system

$$t^2 \frac{d^2 y}{dt^2} + 2t \frac{dy}{dt} - 2y = 0$$

Hint: try $y = t^a$

35. Find the differential equation that describes the system of Fig. 2.27.

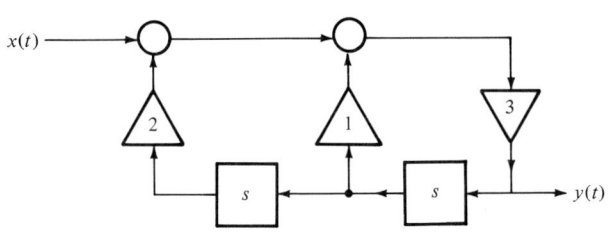

Figure 2.27. Time-invariant linear system.

36. Find the differential equation that describes the system of Fig. 2.28.

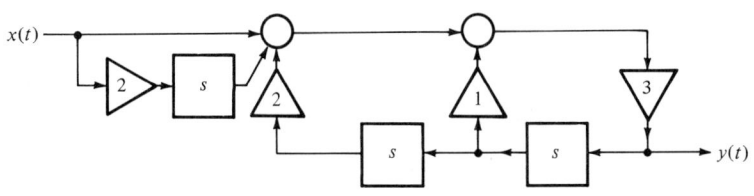

Figure 2.28. Time-invariant linear system.

37. Find the differential equation that describes the linear system of Fig. 2.29.

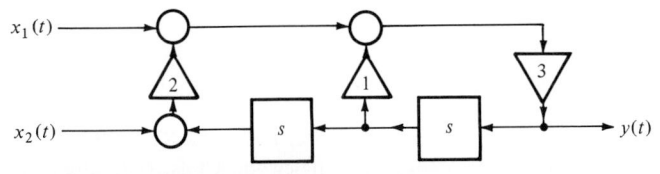

Figure 2.29. Multiple-input linear system.

38. The block diagram of a *zero-order hold* system is shown in Fig. 2.30. It is used to smooth a discrete input sequence. Find its output if $x(t) = A[u(t) - u(t - T_1)]$, $T_1 < T$.

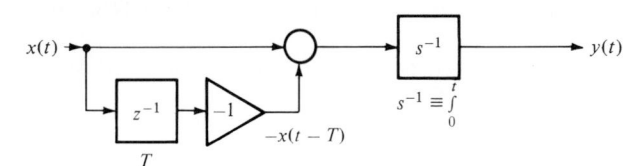

Figure 2.30. Zero-order hold system.

39. Establish Eq. 2.46a using Eq. 2.3b and integrating by parts.

80 THE CONTINUOUS-TIME LINEAR-SYSTEM DIFFERENTIAL EQUATIONS

40. Use the p operator to uncouple the simultaneous differential equations below.
(a) $y_1' - y_1 + 2y_2 = x_1$
$y_1 - y_2' - y_2 = x_2$
(b) $y_1' + y_1 + y_2' + 3y_2 = x_1$
$y_1 + y_2' + y_2 = x_2$
(c) $y_1' + y_1 + y_2' = x_1$
$y_1' + 3y_1 - y_2' - 2y_2 = x_2$
(d) Find the zero-state response (integral form) for y_1 and y_2 in (a). Also, see Probs. 1, 2, and 3, Chap. 3.

41. Use the p operator to find the differential equation for u for the loudspeaker of Fig. 2.31 and Eqs. E.26 and E.27 (App. E) repeated below.

$$L\frac{di}{dt} + Ri + (2\pi r B N)u = v_s(t) \qquad \text{(E.26)}$$

$$M\frac{du}{dt} + Du + \frac{1}{K}\int u\,dt - (2\pi r B N)i = 0 \qquad \text{(E.27)}$$

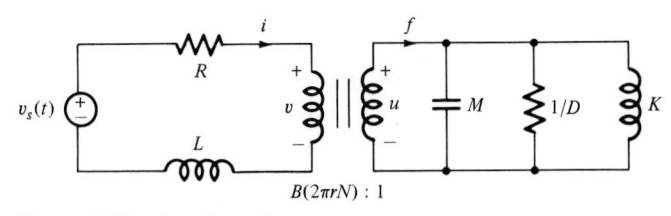

Figure 2.31. Loudspeaker.

References

1. Spiegel, M. R. *Mathematical Handbook*, Schaum Outline Series. New York: McGraw-Hill, 1968.
2. Neff, H. P., Jr. *Basic Electromagnetic Fields*. New York: Harper & Row, 1981.
3. Del Toro, V. *Electrical Engineering Fundamentals*. Englewood Cliffs, N.J.: Prentice-Hall, 1972.
4. Jackson, J. D. *Classical Electrodynamics*. New York: John Wiley & Sons, 1962.
5. Dorf, R. C. *Modern Control Systems*, 3rd ed. Reading, Mass.: Addison-Wesley, 1980.
6. Peebles, P. Z., Jr. *Communication System Principles*. Reading, Mass.: Addison-Wesley, 1976.
7. Hayt, W. H., Jr., and Kemmerly, J. E. *Engineering Circuit Analysis*, 3rd ed. New York: McGraw-Hill, 1978.
8. McGillem, C. D., and Cooper, G. R. *Continuous and Discrete Signal and System Analysis*. New York: Holt, Rinehart and Winston, 1974.

Chapter 3
State-Space Analysis

The time-domain analysis presented up to this point has been primarily concerned with linear systems containing one input and one output and has, for the most part, utilized the unit-impulse response for obtaining the response due to the actual input. If we have a linear system with more than one input and more than one output, we can invoke the very powerful principle of superposition and consider each output separately. In other words each output consists of separate parts, each of which is due to each input acting alone. We considered examples[†] of this type problem and used the p operator to uncouple the simultaneous integro-differential equations that arose using only algebra. This is a simple procedure. The emphasis was on finding the output or outputs, and we apparently were not very concerned over the internal workings of the system. This is not, however, always the case in practice. Many of the difficulties encountered in Chap. 2 are removed by what we call *state-space analysis*.

The state-space method allows us to *organize* the complicated equations resulting from higher-order systems and (or) systems with multiple inputs and outputs into *first-order* forms that are easier to treat and are compatible with programs for use in a digital computer (or a programmable calculator). This

[†] Following Example 9 or Example 12 (Chap. 2).

method allows us to analyze systems whose complexity would perhaps overwhelm us if we took the direct approach of Chap. 2. The key word is organization. The state-space method requires that the description of an Nth order system be recast into the form of N first-order descriptions called the *normal form*. The dependent variables in this description are called the *state variables*, because they provide at any time a description of the state (usually the *energy state*) of the system. The outputs may, or may not, be state variables, but, in any case, are available as linear combinations of the state variables. The advantages of this scheme are numerous.

1. It provides an *organized* approach to the analysis of linear systems in any area (engineering, economics, ecology, and so on).
2. It provides information about what is going on *inside* a system, assuming that *all* of the system variables are ultimately found. This may be important in modeling a system first to determine if any component will be subject to higher than rated voltage, current, force, and so on, that would destroy it, or perhaps cause it to behave in a nonlinear manner. Of course any complete analysis ultimately provides the same information but may not be as easy to implement.
3. It treats systems with multiple inputs and (or) multiple outputs.
4. It provides a simpler approach to the treatment of time-variable systems; however, as in the conventional approach it will not allow the analysis of *all* time-variable systems but only special cases.
5. The *normal-form* equations can be applied to many nonlinear systems (not assuring us of a solution, of course).
6. Numerical techniques are easily applied when the normal-form equations are used.
7. The normal-form equations can be cast into *matrix form*, greatly simplifying the notation for systems of *any* order.

3.1 THE NORMAL-FORM EQUATIONS

In state-space analysis we obtain a set of equations whose dependent variables can describe the *energy* state of the system, and so they are logically called the state variables. Insofar as electric circuits are concerned, then, we are interested in the inductor currents and capacitor voltages because these tell us at any instant of time the energy stored in the inductors and capacitors, respectively. It should be pointed out that the use of inductor currents and capacitor voltages is not *necessary* for the state-variable formulation, but it is *sufficient*. That is, other sets of voltages and currents can be used. Since we are not all experts in circuit analysis, we will refrain from involving ourselves in a detailed discussion of how we obtain the *normal-form* equations for state-variable analysis in a completely general case.[1] Instead, we simply note that when the time derivative of each state variable is expressed as a linear combination of *all* of the state variables and forcing functions, the equations are said to be in the normal

3.1 THE NORMAL-FORM EQUATIONS

form. Thus, a system with N energy storage elements will have N state-variable equations in normal form. For the time being we will concern ourselves only with zero-state responses.

Example 1

Consider the series RLC network in Fig. 3.1. We want to find the normal-form equations. Kirchhoff's voltage law gives

$$v(t) = Ri_L + L\frac{di_L}{dt} + \frac{1}{C}\int i_L\,dt + K$$

and it is obvious that there will be two state variables: the inductor current i_L and the capacitor voltage v_C. Thus, we have

$$v = Ri_L + L\frac{di_L}{dt} + v_C \tag{3.1}$$

and

$$i_L = C\frac{dv_C}{dt} \tag{3.2}$$

Since the ordering of the terms in the state-variable equation is important, especially for systems with many elements, let us arbitrarily consider i_L first and v_C second. Then, rewriting Eqs. 3.1 and 3.2

$$\frac{di_L}{dt} = -\frac{R}{L}i_L - \frac{1}{L}v_C - \frac{1}{L}v$$

$$\frac{dv_C}{dt} = \frac{1}{C}i_L$$

or

$$i'_L = -\frac{R}{L}i_L - \frac{1}{L}v_C - \frac{1}{L}v \tag{3.3}$$

$$v'_C = +\frac{1}{C}i_L + 0\cdot v_C + 0\cdot v \tag{3.4}$$

Notice that these are *first-order coupled* equations, and they are systematically ordered in normal form.

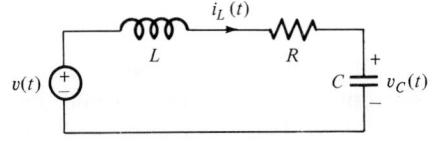

Figure 3.1. Series RLC network.

If we are to organize this method in an efficient manner we should utilize a general notation. Let the state variables be called $q_1, q_2, \ldots, q_N$, and let the *source* functions be $x_1, x_2, \ldots, x_N$. In this regard notice especially that these are just the *inputs* in a more general multiple-input system. Even in a multiple-input system where *inductor currents* and *capacitor voltages* are used as the state variables, the forcing function will involve linear combinations of the inputs and *not linear combinations of the inputs and their derivatives*. The reason[1] for this can be seen in Example 1. We obtained the normal-form equations directly from Kirchhoff's voltage and current laws that did not involve derivatives of the source. On the other hand, if we had specified v_L and i_C as the state variables in Example 1, then

$$v = Ri_C + v_L + \frac{1}{C}\int i_C \, dt$$

$$v' = Ri_C' + v_L' + \frac{1}{C}i_C$$

But, since

$$v_L = Li_C'$$

we have

$$v' = \frac{R}{L}v_L + v_L' + \frac{1}{C}i_C$$

The last two equations can then be put in normal form (v_L, i_C):

$$v_L' = -\frac{R}{L}v_L - \frac{1}{C}i_C + v'$$

$$i_C' = \frac{1}{L}v_L + 0 \cdot i_C + 0$$

where the first contains the derivative of the input voltage. The state variables can be advantageously chosen in *any linear system*.

Now, let the coefficients multiplying the source functions be b_{ij}. In the same way let the coefficients multiplying the state variables on the right side of Eqs. 3.3 and 3.4 be a_{ij}. Thus, we could write Eqs. 3.3 and 3.4 as

$$q_1' = a_{11}q_1 + a_{12}q_2 + b_{11}x_1 \tag{3.5}$$
$$q_2' = a_{21}q_1 + 0 \cdot q_2 + 0 \cdot x_1 \tag{3.6}$$

Those of us who are even slightly acquainted with matrix notation[†] and manipulation should quickly recognize the opportunity to write Eqs. 3.5 and 3.6

[†] A brief discussion of matrix manipulations is supplied in App. A.

as the *single* matrix equation

$$\boxed{\mathbf{q}'(t) = \mathbf{A}\mathbf{q}(t) + \mathbf{B}\mathbf{x}(t)} \tag{3.7}$$

which is general and applies regardless of how many state variables or inputs there are in the system. This, then, is an immediate advantage of the scheme: organization. The solution of Eq. 3.7 allows us to solve for *all* of the dependent variables in a system. That is, in the case of a network, it allows us to find all of the inductor currents and capacitor voltages; and from these any other currents and voltages.

In expanded matrix form for Eqs. 3.5 and 3.6, Eq. 3.7 becomes

$$\begin{bmatrix} q'_1 \\ q'_2 \end{bmatrix} = \begin{bmatrix} a_{11} & a_{12} \\ a_{21} & 0 \end{bmatrix} \begin{bmatrix} q_1 \\ q_2 \end{bmatrix} + \begin{bmatrix} b_{11} \\ 0 \end{bmatrix} [x_1] \tag{3.8}$$

or for Eqs. 3.3 and 3.4

$$\begin{bmatrix} i'_L \\ v'_C \end{bmatrix} = \begin{bmatrix} -\dfrac{R}{L} & -\dfrac{1}{L} \\ \dfrac{1}{C} & 0 \end{bmatrix} \begin{bmatrix} i_L \\ v_C \end{bmatrix} + \begin{bmatrix} -\dfrac{1}{L} \\ 0 \end{bmatrix} [v] \tag{3.9}$$

The column matrix or *vector* **q**

$$\mathbf{q} = \begin{bmatrix} q_1 \\ q_2 \end{bmatrix} \tag{3.10}$$

is called the *state vector*. Here it has two rows and one column (2×1), but in general it has N rows and 1 column ($N \times 1$). The column matrix **q**′

$$\mathbf{q}' = \begin{bmatrix} q'_1 \\ q'_2 \end{bmatrix} \tag{3.11}$$

is its derivative, and is also $N \times 1$ in dimension. The square matrix **A**

$$\mathbf{A} = \begin{bmatrix} a_{11} & a_{12} \\ a_{21} & a_{22} \end{bmatrix} \tag{3.12}$$

is $N \times N$ and is logically called the *system matrix*. The input coefficient matrix **B**

$$\mathbf{B} = \begin{bmatrix} b_{11} & b_{12} & \cdots & b_{1M} \\ b_{21} & b_{22} & \cdots & b_{2M} \end{bmatrix} \tag{3.13}$$

has N rows and M columns (for M inputs) in general. In the present example there is only one input. Finally

$$\mathbf{x} = \begin{bmatrix} x_1 \\ x_2 \\ \vdots \\ x_M \end{bmatrix} \tag{3.14}$$

a column matrix, is called the *source function vector*, and has M rows and one column for M inputs ($M \times 1$). We can also define the product $\mathbf{Bx}$ as the *forcing function vector* $\mathbf{f}$

$$\mathbf{Bx} \equiv \mathbf{f} \tag{3.15}$$

Notice that $\mathbf{B}$ and $\mathbf{x}$ (or $\mathbf{A}$ and $\mathbf{q}$) must be *conformable*; that is, the number of columns in the first ($\mathbf{B}$) must equal the number of rows in the second ($\mathbf{x}$) in order for the matrix multiplication $\mathbf{Bx}$ to exist.

If the system is *not* time-invariant, then the matrices $\mathbf{A}$ and $\mathbf{B}$ are time-dependent; that is the elements are $a_{ij}(t)$ and $b_{ij}(t)$, respectively, and Eq. 3.7 is written

$$\mathbf{q}'(t) = \mathbf{A}(t)\mathbf{q}(t) + \mathbf{B}(t)\mathbf{x}(t) \tag{3.16}$$

Suppose there are P output signals. Then there will be P equations describing this, and each requires N coefficients multiplying the state variables. Thus, we define a $\mathbf{C}$ matrix with elements c_{ij}. It will have P rows and N columns. In addition, each output equation may possibly contain terms involving each of the M inputs. Thus, we need a $P \times M$ matrix $\mathbf{D}$ with elements d_{ij} to multiply the $M \times 1$ input matrix. Therefore, the output equation in matrix form is

$$\boxed{\mathbf{y}(t) = \mathbf{C}\mathbf{q}(t) + \mathbf{D}\mathbf{x}(t)} \tag{3.17}$$

The matrices $\mathbf{C}$ and $\mathbf{D}$ are written $\mathbf{C}(t)$ and $\mathbf{D}(t)$ for systems that are time-variable. Before attempting to work with these complicated forms we will look at another simple example of finding the normal-form equations.

Example 2

Find the normal-form equations for the network in Fig. 3.2. The state variables are v_C and i_L (choosing that order also). For the node labeled v_C Kirchhoff's current law gives for the current leaving the node

$$-i_s + \frac{v_C}{R_1} + C\frac{dv_C}{dt} + i_L = 0$$

For the mesh containing R_2, L, and C Kirchhoff's voltage law gives

$$v_C - L\frac{di_L}{dt} - R_2 i_L = 0$$

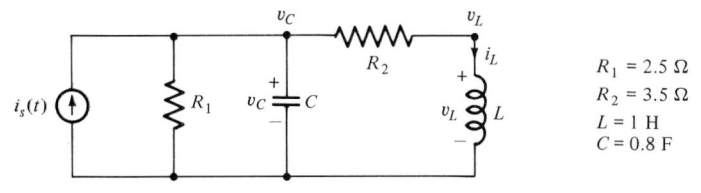

$R_1 = 2.5\ \Omega$
$R_2 = 3.5\ \Omega$
$L = 1\ \text{H}$
$C = 0.8\ \text{F}$

Figure 3.2. Network for Example 2.

Inserting the numerical values for the parameters, and arranging into normal form gives[†]

$$\left.\begin{array}{l}v'_C = -0.5v_C - 1.25i_L + 1.25i_s\\ i'_L = v_C - 3.5i_L\end{array}\right\} \quad (3.18)$$

or

$$\left.\begin{array}{l}q'_1 = -0.5q_1 - 1.25q_2 + 1.25x_1\\ q'_2 = 1q_1 - 3.5q_2\end{array}\right\} \quad (3.19)$$

where

$$q_1 = v_C \qquad q_2 = i_L \qquad x_1 = i_s$$

$$\mathbf{A} = \begin{bmatrix} a_{11} & a_{12} \\ a_{21} & a_{22} \end{bmatrix} = \begin{bmatrix} -0.5 & -1.25 \\ 1 & -3.5 \end{bmatrix}$$

$$\mathbf{B} = \begin{bmatrix} 1.25 \\ 0 \end{bmatrix}$$

If the single output is $v_L(t) = Li'_L$, for example, we have

$$v_L = Lv_C - 3.5Li_L$$

or

$$y_1 = c_{11}q_1 + c_{12}q_2$$

In the next section we will solve for v_C and i_L.

3.2 SIMPLIFIED STATE-VARIABLE ANALYSIS

Equation 3.7 is the (matrix) normal form for state-space analysis. In order to solve for **q** we must either explicitly or implicitly uncouple the set of coupled simultaneous equations represented by Eq. 3.7. Eventually we must find the *fundamental set* which represents the system, and this means, in turn, that we can find the *unit-impulse response for the system*. Recall from Chap. 2 that these last two items are the same for *any* dependent variable in the system. If we can also find the forcing function associated with each state variable (that is, for the differential equation containing only *one* state variable), then convolution will allow us to solve for the state variable. It is this procedure we will follow in this section. It has the advantage of being directly related to the material in the first two chapters, while not being obscured by matrix manipulations. We will *indirectly* uncouple Eq. 3.7.

Our first task, as mentioned above, is that of finding the fundamental set. This must be done no matter what scheme we employ because the fundamental set is at the heart of linear-system analysis. It is *fundamental*. As we have seen

---

[†] Notice once more that derivatives of the source do not appear.

before, finding the fundamental set requires that we find the characteristic equation and its roots, the eigenvalues. This is the same situation we faced earlier in Chap. 2, with one exception. There, we were usually dealing with only one dependent variable, and here we are dealing with perhaps several:

$$\mathbf{q} = \begin{bmatrix} q_1 \\ q_2 \\ \vdots \\ q_N \end{bmatrix}$$

How do we find the characteristic equation? This question is answered by addressing it to the theory of linear algebra. Instead of writing the answer down, let us see if we can sneak up on it in terms of something we already know is correct. Consider the homogeneous form of Eq. 3.7 in *first order* (only one state variable):

$$q' = a_{11}q$$

or

$$a_{11}q = q'$$

and assume that

$$q' = \lambda q \tag{3.20}$$

As mentioned above, we know that this is correct because we know that the solution to Eq. 3.20 is an exponential, which repeats its form upon differentiation. Thus, we are asking for a solution to

$$a_{11}q = \lambda q \tag{3.21}$$

or

$$a_{11}q - \lambda q = 0$$

or

$$(a_{11} - \lambda)q = 0 \tag{3.22}$$

Since q is not 0, we have the characteristic equation

$$a_{11} - \lambda = 0 \tag{3.23}$$

and its solution for the eigenvalue λ

$$\lambda = a_{11} \tag{3.24}$$

The *eigenvector* is q.

In the same way, for an Nth order case (N dimensional space), and looking at Eq. 3.21 for support, we are asking for a solution to the matrix equation

$$\mathbf{Aq} = \lambda \mathbf{q} = \lambda \mathbf{Iq} \tag{3.25}$$

where

$$I = \begin{bmatrix} 1 & 0 & 0 & \\ 0 & 1 & 0 & \\ 0 & 0 & 1 & \ddots \\ \vdots & \vdots & \vdots & & 1 \end{bmatrix} \quad (3.26)$$

is the *identity matrix*. Thus, we have[†]

$$\mathbf{Aq} - \lambda \mathbf{Iq} = \mathbf{0}$$

or

$$(\mathbf{A} - \lambda \mathbf{I})\mathbf{q} = \mathbf{0} \quad (3.27)$$

For $N = 2$, Eq. 3.27 is

$$\begin{bmatrix} a_{11} - \lambda & a_{12} \\ a_{21} & a_{22} - \lambda \end{bmatrix} \begin{bmatrix} q_1 \\ q_2 \end{bmatrix} = \begin{bmatrix} 0 \\ 0 \end{bmatrix} \quad (3.28)$$

Since q_1 and q_2 cannot both be 0 (if we are to avoid a trivial solution), the determinant of the matrix multiplying $\mathbf{q}$ must be 0. That is

$$\begin{vmatrix} a_{11} - \lambda & a_{12} \\ a_{21} & a_{22} - \lambda \end{vmatrix} \equiv 0 \quad (3.29)$$

or

$$(a_{11} - \lambda)(a_{22} - \lambda) - a_{12}a_{21} = 0$$

or

$$\lambda^2 - (a_{11} + a_{22})\lambda + a_{11}a_{22} - a_{12}a_{21} = 0 \quad (3.30)$$

a quadratic in λ, forming the characteristic equation. In general then, the eigenvalues are found from the characteristic equation

$$\boxed{\text{determinant of } (\mathbf{A} - \lambda \mathbf{I}) = |\mathbf{A} - \lambda \mathbf{I}| = 0} \quad (3.31)$$

The $\mathbf{q}$ in Eq. 3.27 gives the eigenvectors, about which we will say more later on. Notice that *what* we are doing is not really new, but some of the words are.

The general characteristic equation can be put in the polynomial form

$$\lambda^N + C_{N-1}\lambda^{N-1} + \cdots + C_1\lambda + C_0 = 0 \quad (3.32)$$

and the eigenvalues $\lambda_1, \lambda_2, \ldots, \lambda_N$ can be found from this polynomial form. These give the elements (exponentials) of the fundamental set and the zero-input response

$$q_{zi}(t) = D_1 e^{\lambda_1 t} + D_2 e^{\lambda_2 t} + \cdots + D_N e^{\lambda_N t} \quad (3.33)$$

[†] Notice that $\mathbf{O}$ is the *zero vector* (see Eq. 3.28).

assuming no repeated roots. Notice, once again, that this form is the same for *any* dependent variable, *including the state variables.*

We use the special set of initial conditions given by the set 2.44 to find a unit-impulse response:

$$q_{zi}(0) = 0, \; q'_{zi}(0) = 0, \; q''_{zi}(0) = 0, \ldots, q_{zi}^{N-1}(0) = 1 \quad (3.34)$$

Application of Eq. 3.34 to 3.33 leads to N equations in N unknowns for the D_n. These can be solved by any method, including the use of matrices, but it is worthwhile pointing out that since all the initial conditions in Eq. 3.34 are homogeneous, except the last one, $\mathbf{D}^N$, as a column vector[†]

$$\mathbf{D}^N = \begin{bmatrix} D_1 \\ D_2 \\ \vdots \\ D_N \end{bmatrix}$$

is given by the *last column* of the *inverse* of the square matrix

$$\boldsymbol{\lambda}^N = \begin{bmatrix} 1 & 1 & 1 & \cdots & \\ \lambda_1 & \lambda_2 & \lambda_3 & & \\ \lambda_1^2 & \lambda_2^2 & \lambda_3^2 & & \\ \vdots & \vdots & \vdots & & \\ \lambda_1^{N-1} & \lambda_2^{N-1} & \lambda_3^{N-1} & \cdots & \lambda_N^{N-1} \end{bmatrix} \quad (3.35)$$

That is, for a general term (for distinct roots)

$$\boxed{D_n = \frac{1}{|\boldsymbol{\lambda}^N|} \times \text{cofactor of } (\lambda_n)^{N-1}} \quad (3.36)$$

Then, the unit-impulse response is

$$h_{fs}(t) = q_{zi}(t)u(t)$$

where $q_{zi}(t)$ is given by Eqs. 3.33 and 3.34. Notice carefully that this is not the unit-impulse response due to unit impulses applied for $x_1, x_2, \ldots$, but, in fact, is the unit-impulse response for a *forcing function* we have yet to find.

Each state variable can then be represented by an Nth order inhomogeneous differential equation whose constant coefficients are those of the characteristic equation. That is, in matrix form

$$\mathbf{q}^N + C_{N-1}\mathbf{q}^{N-1} + C_{N-2}\mathbf{q}^{N-2} + \cdots + C_1\mathbf{q}' + C_0\mathbf{q} = \mathbf{f}_s \quad (3.37)$$

where

$$\mathbf{q}^N \equiv \frac{d^N \mathbf{q}}{dt^N}$$

[†] Do not confuse $\mathbf{D}^N$ with $\mathbf{D}$, the matrix multiplying $\mathbf{x}$ (Eq. 3.17).

The "forcing function" $\mathbf{f}_s$ will depend only on $\mathbf{f}, \mathbf{f}', \mathbf{f}'', \ldots$, and not on $\mathbf{q}$. $\mathbf{f}$ is given by Eq. 3.15:

$$\mathbf{f} = \mathbf{Bx} \tag{3.15}$$

and $\mathbf{f}_s$ is given by Eq. 3.37 when

$$\mathbf{q}' = \mathbf{Aq} + \mathbf{Bx} = \mathbf{Aq} + \mathbf{f}$$
$$\mathbf{q}'' = \mathbf{Aq}' + \mathbf{f}'$$
$$\mathbf{q}''' = \mathbf{Aq}'' + \mathbf{f}''$$
$$\vdots$$

are substituted for $\mathbf{q}', \mathbf{q}'', \mathbf{q}''', \ldots$, and all derivatives of $\mathbf{q}$ are eliminated. Using a superscript on $\mathbf{f}_s$ to indicate the order, this process gives

$$\mathbf{f}_s^1 = \mathbf{f}_1 \quad (N = 1) \tag{3.38}$$
$$\mathbf{f}_s^2 = \mathbf{Af} + \mathbf{f}' + C_1\mathbf{f} \quad (N = 2) \tag{3.39}$$
$$\mathbf{f}_s^3 = \mathbf{A}(\mathbf{Af}) + \mathbf{Af}' + \mathbf{f}'' + C_2(\mathbf{Af}) + C_2\mathbf{f}' + C_1\mathbf{f} \quad (N = 3) \tag{3.40}$$
$$\mathbf{f}_s^4 = \mathbf{A}(\mathbf{A}(\mathbf{Af})) + \mathbf{A}(\mathbf{Af}') + \mathbf{Af}'' + \mathbf{f}'''$$
$$+ C_3\mathbf{A}(\mathbf{Af}) + C_3\mathbf{Af}' + C_3\mathbf{f}''$$
$$+ C_2\mathbf{Af} + C_2\mathbf{f}' + C_1\mathbf{f} \quad (N = 4)$$

or, in general

$$\mathbf{f}_s^N = \mathbf{A}^{N-1}\mathbf{f} + \mathbf{A}^{N-2}\mathbf{f}' + \cdots + \mathbf{Af}^{N-2} + \mathbf{f}^{N-1}$$
$$+ C_{N-1}\mathbf{A}^{N-2}\mathbf{f} + C_{N-1}\mathbf{A}^{N-3}\mathbf{f}' + \cdots + C_{N-1}\mathbf{Af}^{N-3} + C_{N-1}\mathbf{f}^{N-2}$$
$$+ C_{N-2}\mathbf{A}^{N-3}\mathbf{f} + C_{N-2}\mathbf{A}^{N-4}\mathbf{f}' + \cdots + C_{N-2}\mathbf{f}^{N-3}$$
$$\vdots$$
$$+ C_2\mathbf{Af} + C_2\mathbf{f}' + C_1\mathbf{f} \tag{3.41}$$

where

$$\mathbf{f}^N = \frac{d^N\mathbf{f}}{dt^N}$$

and

$$\mathbf{A}^N\mathbf{f} \equiv \mathbf{A}(\mathbf{A}(\mathbf{A}\cdots\mathbf{f}))\cdots \tag{3.42}$$

N matrix multiplications.

When $\mathbf{f}_s$ has been found, then $\mathbf{q}$ is simply given by convolution:

$$\mathbf{q} = \int_{-\infty}^{\infty} \mathbf{f}_s(\tau) h_{fs}(t - \tau)\, d\tau$$

$$\mathbf{q} = \int_{-\infty}^{\infty} \mathbf{f}_s(\tau) q_{zi}(t - \tau) u(t - \tau)\, d\tau$$

$$\boxed{\mathbf{q} = \int_{-\infty}^{t} \mathbf{f}_s(\tau) q_{zi}(t - \tau)\, d\tau} \tag{3.43}$$

Remember that $\mathbf{q}$ and $\mathbf{f}_s$ are column vectors. Also, it should be pointed out as a reminder that the results of this section give the zero-state response for a time-invariant linear system.

92 STATE-SPACE ANALYSIS

The mathematics has been on the heavy side, and we need to look at a simple example to find out what is really going on.

Example 3

Solve for the state variables in Example 2. We have

$$\mathbf{A} = \begin{bmatrix} -0.5 & -1.25 \\ 1 & -3.5 \end{bmatrix}$$

so using Eq. 3.29, the characteristic equation is

$$\begin{vmatrix} -0.5 - \lambda & -1.25 \\ 1 & -3.5 - \lambda \end{vmatrix} = 0$$

or

$$(-0.5 - \lambda)(-3.5 - \lambda) + 1.25 = 0$$

or

$$\lambda^2 + 4\lambda + 3 = 0 \qquad C_1 = 4,\ C_0 = 3$$

Solving this quadratic we have

$$\lambda_1 = -1 \qquad \lambda_2 = -3$$

and

$$q_{zi}(t) = D_1 e^{-t} + D_2 e^{-3t}$$

subject to

$$q_{zi}(0) = 0 \qquad q'_{zi}(0) = 1$$

Thus

$$0 = D_1 + D_2$$
$$1 = -D_1 - 3D_2$$

This set of equations is easy to solve by any method, but let us use Eqs. 3.35 and 3.36 as a check:

$$\lambda^2 = \begin{bmatrix} 1 & 1 \\ -1 & -3 \end{bmatrix}$$

$$D_1 = \frac{(-1)^3}{-2} \times 1 = +\frac{1}{2}$$

$$D_2 = \frac{(-1)^4}{-2} \times 1 = -\frac{1}{2}$$

Therefore

$$h_{fs}(t) = \tfrac{1}{2}(e^{-t} - e^{-3t})u(t)$$

Using Eq. 3.39

$$\mathbf{f}_s^2 = \mathbf{Af} + 4\mathbf{f} + \mathbf{f}' = (\mathbf{A} + 4\mathbf{I})\mathbf{f} + \mathbf{f}'$$

or

$$\begin{bmatrix} f_{s1} \\ f_{s2} \end{bmatrix} = \begin{bmatrix} -0.5+4 & -1.25 \\ 1 & -3.5+4 \end{bmatrix} \begin{bmatrix} f_1 \\ f_2 \end{bmatrix} + \begin{bmatrix} f'_1 \\ f'_2 \end{bmatrix}$$

or

$$\begin{bmatrix} f_{s1} \\ f_{s2} \end{bmatrix} = \begin{bmatrix} 3.5 & -1.25 \\ 1 & 0.5 \end{bmatrix} \begin{bmatrix} f_1 \\ f_2 \end{bmatrix} + \begin{bmatrix} f'_1 \\ f'_2 \end{bmatrix}$$

or

$$f_{s1} = 3.5f_1 - 1.25f_2 + f'_1$$
$$f_{s2} = f_1 + 0.5f_2 + f'_2$$

In the present example $f_1 = 1.25i_s = 1.25x_1$, $f_2 = 0$, so

$$f_{s1} = 3.5f_1 + f'_1$$
$$f_{s2} = f_1$$

Equation 3.43 gives the state variables

$$q_1(t) = \tfrac{1}{2} \int_{-\infty}^{t} [3.5f_1(\tau) + f'_1(\tau)][e^{-(t-\tau)} - e^{-3(t-\tau)}] d\tau$$

$$q_2(t) = \tfrac{1}{2} \int_{-\infty}^{t} [f_1(\tau)][e^{-(t-\tau)} - e^{-3(t-\tau)}] d\tau$$

or

$$v_C(t) = \int_{-\infty}^{t} [2.1875 i_s(\tau) + 0.625 i''_s(\tau)][e^{-(t-\tau)} - e^{-3(t-\tau)}] d\tau$$

$$i_L(t) = \int_{-\infty}^{t} [0.625 i_s(\tau)][e^{-(t-\tau)} - e^{-3(t-\tau)}] d\tau$$

Now, given i_s, the problem of finding the zero-state response for the state variables is formally solved. The capacitor voltage and inductor current are given by the last two equations, and any other voltage or current in Fig. 3.2 can be found from v_C and i_L. Notice that superposition applies in Eq. 3.43, and the state variables due to each state-form input $(x_1, x_2, \ldots)$ *acting alone* can be found, if desired. Notice also that a zero-input response, made up of the fundamental set, can always be added to accommodate any initial conditions that might be present. The initial conditions will normally be known or given for $y_1(0), y_2(0), y_3(0), \ldots, y_N(0)$, and an Nth order system will have N elements in its fundamental set, so that the zero-input response will have N coefficients that must be determined. In order to find these coefficients, we must use the equations that describe the system to evaluate the derivatives at $t = 0$. The normal-form equations will suffice for this. The complete solution once again is the sum of the zero-state response and a zero-input response (with its N coefficients). The initial conditions are applied to this complete solution to evaluate the coefficients and give the final solution for the state variables.

It is worthwhile to use some numbers in the last example. This will give us confidence in the method and will also serve for comparison purposes with another method to be introduced in the next section.

Example 4

Let $i_s(t) = 4.8$ A, $-\infty < t < \infty$, and find $v_C(t)$ and $i_L(t)$ in Example 3. With this (dc) input we have only a steady-state response. Reference to Fig. 3.2 shows that since the inductor acts as a short circuit and the capacitor acts as an open circuit, we have

$$v_C = i_s \frac{R_1 R_2}{R_1 + R_2} = 4.8 \frac{2.5(3.5)}{2.5 + 3.5} = 7 \text{ V}$$

$$i_L = i_s \frac{R_1}{R_1 + R_2} = 4.8 \frac{2.5}{2.5 + 3.5} = 2 \text{ A}$$

On the other hand, results of Example 3 give

$$v_C(t) = \int_{-\infty}^{t} 10.5[e^{-(t-\tau)} - e^{-3(t-\tau)}] d\tau$$

$$= 10.5 e^{-t} e^{\tau} \Big|_{-\infty}^{t} - 10.5 e^{-3t}(e^{3\tau}/3) \Big|_{-\infty}^{t}$$

$$= 10.5 - 3.5 = 7 \text{ V}$$

$$i_L(t) = \int_{-\infty}^{t} 3[e^{-(t-\tau)} - e^{-3(t-\tau)}] d\tau$$

$$= 3 e^{-t} e^{\tau} \Big|_{-\infty}^{t} - 3 e^{-3t}(e^{3\tau}/3) \Big|_{-\infty}^{t}$$

$$= 3 - 1 = 2 \text{ A}$$

For an input $i_s(t) = 9.6 e^{-4t} u(t)$, we have

$$i'_s(t) = 9.6 \delta(t) - 38.4 e^{-4t} u(t)$$

$$v_C(t) = 21 u(t) \int_0^t e^{-4\tau}[e^{-(t-\tau)} - e^{-3(t-\tau)}] d\tau$$

$$+ 6 u(t) \int_{-\infty}^{t} \delta(\tau)[e^{-(t-\tau)} - e^{-3(t-\tau)}] d\tau$$

$$- 24 u(t) \int_0^t e^{-4\tau}[e^{-(t-\tau)} - e^{-3(t-\tau)}] d\tau$$

and

$$i_L(t) = 6 u(t) \int_0^t e^{-4\tau}[e^{-(t-\tau)} - e^{-3(t-\tau)}] d\tau$$

giving

$$v_C(t) = (5 e^{-t} - 3 e^{-3t} - 2 e^{-4t}) u(t)$$
$$i_L(t) = (2 e^{-t} - 6 e^{-3t} + 4 e^{-4t}) u(t)$$

These are the zero-state responses for $v_C(t)$ and $i_L(t)$, respectively, and zero-input responses, having the form $A_1 e^{-t} + A_2 e^{-3t}$, can be added in case $i_L(0) \neq 0$ or $v_C(0) \neq 0$. Notice that the steady-state responses are both 0, and the zero-state response is entirely transient.

3.2 SIMPLIFIED STATE-VARIABLE ANALYSIS

We now briefly return to the unit-impulse response. It is now possible for us to find the unit-impulse response for any state variable q_n due to any input x_n. It is perhaps best to show this for the simple case $N = 2$, and rather than using general results, we will simply use the system in Example 3. We had found

$$q_1(t) = \int_{-\infty}^{t} [2.1875 i_s(\tau) + 0.625 i'_s(\tau)][e^{-(t-\tau)} - e^{-3(t-\tau)}] d\tau$$

and

$$q_2(t) = \int_{-\infty}^{t} [0.625 i_s(\tau)][e^{-(t-\tau)} - e^{-3(t-\tau)}] d\tau$$

The unit-impulse response for $q_1(t)$ with $i_s(t) = \delta(t)$ is called $h_{11}(t)$, and the unit-impulse response for $q_2(t)$ due to $i_s(t) = \delta(t)$ is called $h_{21}(t)$. Remember that $x_2(t) = 0$, so two other unit-impulse responses, $h_{12}(t)$ and $h_{22}(t)$, due to x_2 are not present in this example.

$$h_{11}(t) = \int_{-\infty}^{t} [2.1875 \delta(\tau) + 0.625 \delta'(\tau)][e^{-(t-\tau)} - e^{-3(t-\tau)}] d\tau$$

$$h_{21}(t) = \int_{-\infty}^{t} [0.625 \delta(\tau)][e^{-(t-\tau)} - e^{-3(t-\tau)}] d\tau$$

Using Eqs. 2.25 and 2.28 we obtain

$$h_{11}(t) = 2.1875(e^{-t} - e^{-3t})u(t) - 0.625(e^{-t} - 3e^{-3t})u(t)$$
$$h_{21}(t) = 0.625(e^{-t} - e^{-3t})u(t)$$

or

$$h_{11}(t) = (1.5625 e^{-t} - 0.3125 e^{-3t})u(t)$$
$$h_{21}(t) = (0.625 e^{-t} - 0.625 e^{-3t})u(t)$$

Is this a stable system?

The examples in this section have been relatively simple because it was a second-order system. For higher-order systems we will probably need to use a digital computer, but we do have a method for programming the computer.

We close this section with a brief summary of the method that has been presented for finding the state variables in an Nth order linear system.

1. Arrange the equations describing the system in normal form (Eq. 3.7).
2. Find the eigenvalues λ_i from the characteristic equation (Eq. 3.31). Notice that for a high-order system this will be done with a programmable calculator or computer.
3. Solve for the coefficients (D_n) in the zero-state response using Eq. 3.36 (or 3.47 for repeated roots). This gives $h_{fs}(t) = q_{zi}(t)u(t)$, the unit-impulse response.
4. Find the forcing function $\mathbf{f}_s(t)$ using the appropriate N in Eq. 3.41. Notice, once more, that this will be done with a computer using the matrix forms for a high-order system. For a second-order system the equation is

$$\mathbf{f}_s^2 = (\mathbf{A} + C_1 \mathbf{I})(\mathbf{Bx}) + \mathbf{Bx}'$$

96 STATE-SPACE ANALYSIS

5. Calculate the state variables with Eq. 3.43.
6. Find the outputs with Eq. 3.17.
7. Add a zero-input response (if necessary).

Notice that the entire process can be treated very nicely with matrix forms.

3.3 NORMAL-FORM BLOCK DIAGRAMS

It is advantageous to be able to draw block diagram representations where the state variables and "outputs," which may be linear combinations of the state variables, appear. This is true even if some of the blocks in such a diagram do not correspond to elements in the system that is being mathematically modeled.

Consider the feedback system of Fig. 3.3. For this system

$$y = \int (x + ay)\, dt$$

or

$$y' = ay + x \tag{3.44}$$

and we recognize this as a first-order normal-form equation. Thus, Fig. 3.3 gives us an easy way to draw the block diagram for an Nth order system that is modeled in normal form.

Consider Fig. 3.2. The normal-form equations were the set 3.19:

$$q_1' = -0.5q_1 - 1.25q_2 + 1.25x_1$$
$$q_2' = 1q_1 - 3.5q_2 \tag{3.19}$$

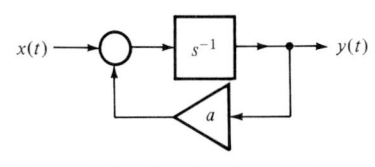

Figure 3.3. Feedback system.

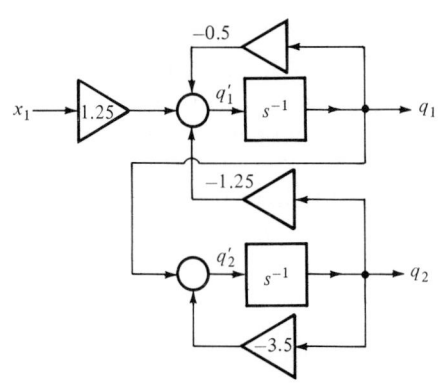

Figure 3.4. Normal-form block diagram for Fig. 3.2.

The block diagram representation of this ($N = 2$) pair of normal-form equations is shown in Fig. 3.4.

Example 5

Draw the block diagram for the series RLC network using the normal-form equations when the output is the voltage $v_0(t)$ in Fig. 3.5. The normal-form equations were 3.3 and 3.4:

$$i'_L = -\frac{R}{L}i_L - \frac{1}{L}v_C - \frac{1}{L}v$$

$$v'_C = +\frac{1}{C}i_L$$

and now

$$v_0 = v_L + v_C = Li'_L + v_C$$

The block diagram of Fig. 3.5 is shown in Fig. 3.6 with $v_0(t)$ as the output.

There are two things that should be pointed out with regard to the output $v_0(t)$. First, we could save an integrator by noting that it is also true that $v_0 = v - Ri_L$, and both v and i_L are already available on the block diagram. Secondly, as has already been mentioned, the equation for the output can also be put in matrix form. In the present example this is certainly not necessary, but in a higher-order system, the matrix notation may be very advantageous.

Before proceeding it would be wise to consider a slightly more complicated example.

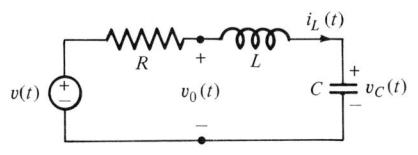

Figure 3.5. Series RLC network.

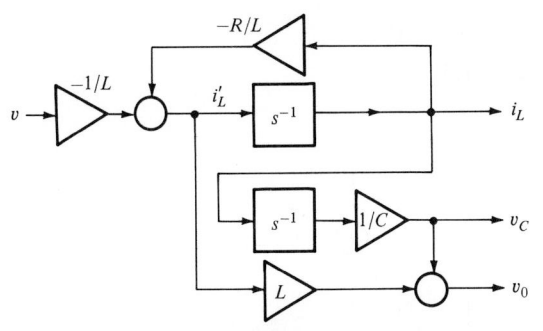

Figure 3.6. Block diagram for the system shown in Fig. 3.5 showing the state variables and the output.

Example 6

Find the normal-form equations, block diagram, and "formal" (zero-state) solution for the network in Fig. 3.7. In a simple manner it can be thought of as a filter with the low frequencies from both sources appearing in the (output) voltage $v_2(t)$, while the high frequencies from both sources appear in $v_1(t)$. It is not too difficult to obtain the normal-form equations if it is remembered that $v'_C = i_C/C$ and $i'_L = v_L/L$.

$$v'_C(t) = -\frac{1}{C}\frac{1}{R_1+R'_p}v_C(t) - \frac{1}{C}\frac{R'_p}{R_1+R'_p}i_L(t) + \frac{1}{C}\frac{R'_p}{R_1+R'_p}\left[\frac{v_s(t)}{R_s}+i_s(t)\right]$$

$$i'_L(t) = +\frac{1}{L}\frac{R'_p}{R_1+R'_p}v_C(t) - \frac{1}{L}\left(R_2 + \frac{R_1 R'_p}{R_1+R'_p}\right)i_L(t) + \frac{1}{L}\frac{R_1 R'_p}{R_1+R'_p}\left[\frac{v_s(t)}{R_s}+i_s(t)\right]$$

where

$$R'_p = \frac{R_s R_p}{R_s + R_p}$$

We choose

$$R_1 = R_2 = 10^3 \ \Omega$$
$$L = 10^3 \ H$$
$$C = 10^{-9} \ F$$
$$R_s = R_p = 2 \times 10^3 \ \Omega$$

Thus, $R'_p = 10^3 \ \Omega$, and

$$v'_C = -(0.5 \times 10^6)v_C - (0.5 \times 10^9)i_L + 0.25 \times 10^6 v_s + 0.5 \times 10^9 i_s$$
$$i'_L = +(0.5 \times 10^3)v_C - (1.5 \times 10^6)i_L + 0.25 \times 10^3 v_s + 0.5 \times 10^6 i_s$$

or

$$\left.\begin{array}{l} q'_1 = a_{11}q_1 + a_{12}q_2 + b_{11}x_1 + b_{12}x_2 \\ q'_2 = a_{21}q_1 + a_{22}q_2 + b_{21}x_1 + b_{22}x_2 \end{array}\right\} \quad (3.45)$$

It is obvious that

$$a_{11} = -0.5 \times 10^6 \qquad a_{12} = -0.5 \times 10^9$$
$$a_{21} = 0.5 \times 10^3 \qquad a_{22} = -1.5 \times 10^6$$

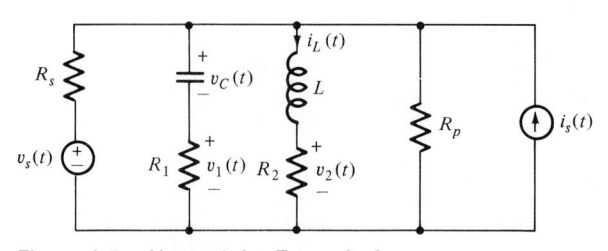

Figure 3.7. Network for Example 6.

and
$$b_{11} = 0.25 \times 10^6 \qquad b_{12} = 0.5 \times 10^9$$
$$b_{21} = 0.25 \times 10^3 \qquad b_{22} = 0.5 \times 10^6$$

The output equations are simply
$$v_1 = R_1 C v'_C = 10^{-6} v'_C$$
$$v_2 = R_2 i_L = 10^3 i_L$$

or
$$v_1 = -0.5 v_C - 0.5 \times 10^3 i_L + 0.25 v_s + 0.5 \times 10^3 i_s$$
$$v_2 = 10^3 i_L$$

In generalized notation (Eq. 3.17) these are
$$y_1 = c_{11} q_1 + c_{12} q_2 + d_{11} x_1 + d_{12} x_2$$
$$y_2 = 0 \cdot q_1 + c_{22} q_2 + 0 \cdot x_1 + 0 \cdot x_2 \tag{3.46}$$

Here, it is obvious that
$$c_{11} = -0.5 \qquad c_{12} = -0.5 \times 10^3$$
$$c_{21} = 0 \qquad c_{22} = 10^3$$
$$d_{11} = 0.25 \qquad d_{12} = 0.5 \times 10^3$$
$$d_{21} = 0 \qquad d_{22} = 0$$

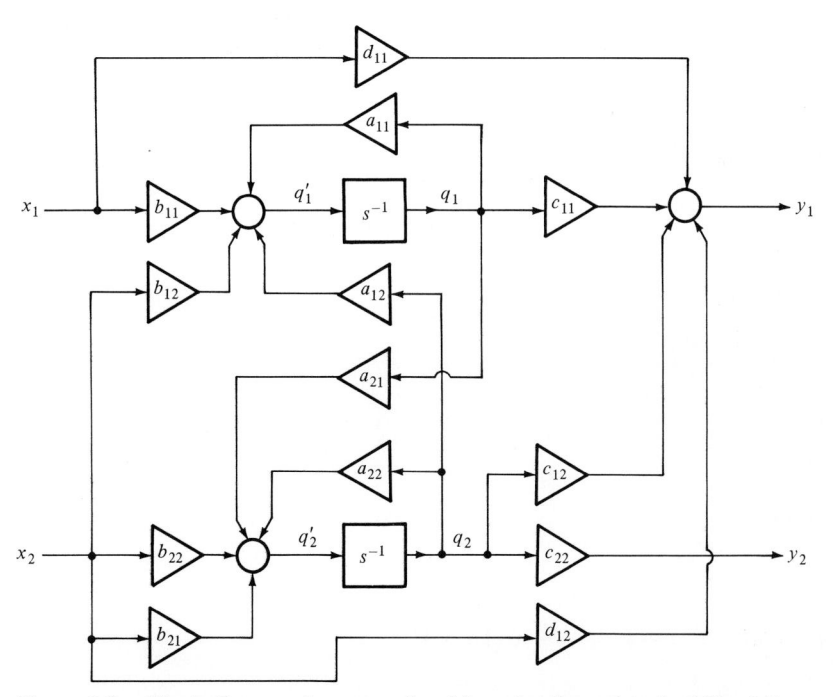

Figure 3.8. Block diagram in generalized form for the network of Fig. 3.7.

The block diagram can be drawn in several forms, but utilizing the generalized notation of Eqs. 3.45 and 3.46 gives Fig. 3.8. Notice that even for this second-order system the block diagram is relatively complex.

The system matrix is

$$\mathbf{A} = \begin{bmatrix} a_{11} & a_{12} \\ a_{21} & a_{22} \end{bmatrix}$$

and the characteristic equation is

$$\begin{vmatrix} a_{11} - \lambda & a_{12} \\ a_{21} & a_{22} - \lambda \end{vmatrix} = 0$$

or

$$\lambda^2 - (a_{11} + a_{22})\lambda + a_{11}a_{22} - a_{12}a_{21} = 0 \tag{3.30}$$

Numerically, the last equation is

$$\lambda^2 + 2 \times 10^6 \lambda + 10^{12} = 0 \qquad C_1 = 2 \times 10^6 \qquad C_0 = 10^{12}$$
$$(\lambda + 10^6)(\lambda + 10^6) = 0$$

Thus, we have repeated roots, and proceed cautiously. A zero-input response is given by Eq. 2.58:

$$q_{zi}(t) = D_1 e^{\lambda_1 t} + D_2 t e^{\lambda_1 t} \qquad \lambda_1 = -10^6$$

It is still true, even for repeated roots, that $\mathbf{D}^N$ as a column vector is given by the last column of the inverse of the square matrix λ^N, and hence a general term is given by

$$D_n = \frac{1}{|\lambda^N|} \times \begin{array}{l} \text{cofactor of the element in the } N\text{th} \\ \text{(last) row and } n\text{th column of } \lambda^N \end{array} \tag{3.47}$$

The difference is that, because of the repeated roots, the fundamental set is given in 2.58, and the elements of λ^N are *not* those of Eq. 3.35. In the present case, for example

$$\lambda^N = \begin{bmatrix} 1 & 0 \\ \lambda_1 & 1 \end{bmatrix}$$

$$D_1 = \tfrac{1}{1} \times (-0) = 0$$
$$D_2 = \tfrac{1}{1} \times (+1) = 1$$

Therefore

$$h_{fs}(t) = t e^{\lambda_1 t} u(t)$$

and

$$\mathbf{f}_s^2 = \mathbf{A}\mathbf{f} + C_1 \mathbf{f} + \mathbf{f}' = (\mathbf{A} + C_1 \mathbf{I})\mathbf{f} + \mathbf{f}' \tag{3.48}$$

or

$$\begin{bmatrix} f_{s1} \\ f_{s2} \end{bmatrix} = \begin{bmatrix} 1.5 \times 10^6 & -0.5 \times 10^9 \\ 0.5 \times 10^3 & 0.5 \times 10^6 \end{bmatrix} \begin{bmatrix} f_1 \\ f_2 \end{bmatrix} + \begin{bmatrix} f'_1 \\ f'_2 \end{bmatrix}$$

3.3 NORMAL-FORM BLOCK DIAGRAMS

So

$$f_{s1} = 1.5 \times 10^6 f_1 - 0.5 \times 10^9 f_2 + f'_1 \\ f_{s2} = 0.5 \times 10^3 f_1 + 0.5 \times 10^6 f_2 + f'_2 \tag{3.49}$$

But

$$\mathbf{f} = \mathbf{Bx}$$

or

$$f_1 = 0.25 \times 10^6 x_1 + 0.5 \times 10^9 x_2 \\ f_2 = 0.25 \times 10^3 x_1 + 0.5 \times 10^6 x_2 \tag{3.50}$$

Combining Eqs. 3.48 and 3.15 gives the matrix form

$$\mathbf{f}_s^2 = (\mathbf{A} + C_1 \mathbf{I})(\mathbf{Bx} + \mathbf{Bx}') \tag{3.51}$$

which holds for any second-order system. Alternatively, we can simply combine Eqs. 3.49 and 3.50 directly:

$$f_{s1} = 0.25 \times 10^{12} x_1 + 0.5 \times 10^{15} x_2 + 0.25 \times 10^6 x'_1 + 0.5 \times 10^9 x'_2 \tag{3.52}$$

$$f_{s2} = 0.25 \times 10^9 x_1 + 0.5 \times 10^{12} x_2 + 0.25 \times 10^3 x'_1 + 0.5 \times 10^6 x'_2 \tag{3.53}$$

The formal zero-state solutions for the state variables are

$$q_1(t) = \int_{-\infty}^{t} f_{s1}(\tau)(t - \tau) e^{-10^6(t - \tau)} d\tau \tag{3.54}$$

$$q_2(t) = \int_{-\infty}^{t} f_{s2}(\tau)(t - \tau) e^{-10^6(t - \tau)} d\tau \tag{3.55}$$

The outputs are given by Eq. 3.46.

Example 7

Using the system matrix below, find the unit-impulse response $h_{fs}(t)$. We have

$$\mathbf{A} = \begin{bmatrix} 1 & 0 & 2 \\ 1 & -1 & 1 \\ -1 & 0 & -4 \end{bmatrix}$$

The eigenvalues are $\lambda_1 = -1, \lambda_2 = -1, \lambda_3 = -2$ for the characteristic equation

$$\lambda^3 + 4\lambda^2 + 5\lambda + 2 = 0 \quad \begin{cases} C_0 = 2 \\ C_1 = 5 \\ C_2 = 4 \end{cases}$$

or

$$(\lambda + 1)^2(\lambda + 2) = 0$$

Therefore

$$y_{zi}(t) = D_1 e^{-t} + D_2 t e^{-t} + D_3 e^{-2t}$$

and

$$0 = D_1 + 0 + D_3$$
$$0 = -D_1 + D_2 - 2D_3$$
$$1 = D_1 - 2D_2 + 4D_3$$

or

$$\begin{bmatrix} 0 \\ 0 \\ 1 \end{bmatrix} = \begin{bmatrix} 1 & 0 & 1 \\ -1 & 1 & -2 \\ 1 & -2 & 4 \end{bmatrix} \begin{bmatrix} D_1 \\ D_2 \\ D_3 \end{bmatrix}$$

Thus

$$\lambda^3 = \begin{bmatrix} 1 & 0 & 1 \\ -1 & 1 & -2 \\ 1 & -2 & 4 \end{bmatrix}$$

and

$$D_n = \frac{1}{|\lambda^3|} \times \begin{array}{l} \text{cofactor of the element in the} \\ \text{3rd row and } n\text{th column of } \lambda^3 \end{array}$$

Solving

$$D_1 = \frac{(-1)^4}{-1} \begin{vmatrix} 0 & 1 \\ 1 & -2 \end{vmatrix} = 1$$

$$D_2 = \frac{(-1)^5}{-1} \begin{vmatrix} 1 & 1 \\ -1 & -2 \end{vmatrix} = 1$$

$$D_3 = \frac{(-1)^6}{-1} \begin{vmatrix} 1 & 0 \\ -1 & 1 \end{vmatrix} = -1$$

Therefore

$$h_{fs}(t) = (e^{-t} + te^{-t} - e^{-3t})u(t)$$

3.4 FURTHER CONSIDERATION OF MATRIX SOLUTIONS

We now return to the normal-form equation (3.7) with $f = Bx$:

$$q' = Aq + f \tag{3.7}$$

and consider the simple case $N = 1$ from the classical viewpoint.

$$q' = a_{11}q + f \tag{3.56}$$

or

$$q' - a_{11}q = f$$

Multiplying by $e^{-ta_{11}}$ gives
$$e^{-ta_{11}}q' + (-a_{11}e^{-ta_{11}})q = e^{-ta_{11}}f$$
The left side of this equation can be expressed as
$$\frac{d}{dt}(e^{-ta_{11}}q) = e^{-ta_{11}}f$$
or
$$d(e^{-ta_{11}}q) = e^{-ta_{11}}f\,dt$$
Integrating (and assuming that $e^{-ta_{11}}q \to 0$ for $t \to \infty$)
$$\int_{-\infty}^{t} d(e^{-ta_{11}}q) = \int_{-\infty}^{t} e^{-\tau a_{11}}f\,d\tau$$
or
$$e^{-ta_{11}}q = \int_{-\infty}^{t} e^{-\tau a_{11}}f\,d\tau$$
$$q(t) = e^{ta_{11}} \int_{-\infty}^{t} e^{-\tau a_{11}}f\,d\tau$$
At $t = 0$
$$q(0) = \int_{-\infty}^{0} e^{-\tau a_{11}}f\,d\tau$$
Thus
$$q(t) = e^{ta_{11}} \int_{-\infty}^{0} e^{-\tau a_{11}}f\,d\tau + e^{ta_{11}} \int_{0}^{t} e^{-\tau a_{11}}f\,d\tau$$
or
$$q(t) = e^{ta_{11}}q(0) + e^{ta_{11}} \int_{0}^{t} e^{-\tau a_{11}}f\,d\tau \qquad (3.57)$$

Notice that this is the sum of the zero-input response and the zero-state response exactly as we would have found it using other methods.

We now wish to consider the matrix form, Eq. 3.7, and do exactly the same thing (in proper matrix order, of course). We have

$$\mathbf{q}' - \mathbf{A}\mathbf{q} = \mathbf{f} \qquad (3.58)$$

Proceeding as before, we need to investigate $e^{-t\mathbf{A}}$. As a *power series* it is

$$e^{-t\mathbf{A}} = \mathbf{I} - t\mathbf{A} + \frac{t^2}{2!}\mathbf{A}\mathbf{A} - \cdots \qquad (3.59)$$

$e^{t\mathbf{A}}$, which is the inverse of $e^{-t\mathbf{A}}$, is called the *state-transition matrix*. It is an $N \times N$ matrix whose elements are generally functions of time. Thus, we premultiply Eq. 3.58 by $e^{-t\mathbf{A}}$ (as we did in the first-order case):

$$e^{-t\mathbf{A}}(\mathbf{q}' - \mathbf{A}\mathbf{q}) = e^{-t\mathbf{A}}\mathbf{f}$$

or

$$\frac{d}{dt}(e^{-t\mathbf{A}}\mathbf{q}) = e^{-t\mathbf{A}}\mathbf{f}$$

Integrating (as before)

$$e^{-t\mathbf{A}}\mathbf{q} = \int_{-\infty}^{t} e^{-\tau\mathbf{A}}\mathbf{f}\, d\tau \tag{3.60}$$

Premultiplying Eq. 3.60 by $e^{+t\mathbf{A}}$ gives

$$\mathbf{q} = e^{t\mathbf{A}} \int_{-\infty}^{t} e^{-\tau\mathbf{A}}\mathbf{f}\, d\tau \tag{3.61}$$

since

$$(e^{t\mathbf{A}})(e^{-t\mathbf{A}}) = \mathbf{I} \quad \text{or} \quad (e^{t\mathbf{A}}) = (e^{-t\mathbf{A}})^{-1}$$

Separating Eq. 3.61 into the zero-input and zero-state response gives

$$\boxed{\mathbf{q} = e^{t\mathbf{A}}\mathbf{q}(0) + e^{t\mathbf{A}} \int_{0}^{t} e^{-\tau\mathbf{A}}\mathbf{f}\, d\tau} \tag{3.62}$$

The output is given by Eq. 3.17.

Equation 3.62 constitutes a complete formal solution to the problem of finding the state variables in an Nth order system. Before rejoicing too much, however, we should take note of the fact that if we can only express the state-transition matrix, $e^{t\mathbf{A}}$ and its inverse $e^{-t\mathbf{A}}$ (both of which are needed) as infinite series, we would have each $q_n(t)$ expressed as infinite power series in t. This is just too much for hand calculations. A computer, however, could handle all this numerical work without much difficulty. Thus, without much regret we leave the infinite series idea until that time when we are forced to solve a high-order system numerically. Fortunately, there is a simpler way.

We found in Eq. 3.31 that

$$|\mathbf{A} - \lambda\mathbf{I}| = 0 \tag{3.31}$$

is the characteristic equation of the system, and it can be put in the polynomial form of Eq. 3.32:

$$\sum_{m=0}^{N} C_m \lambda^m = 0 \tag{3.32}$$

where $C_N \equiv 1$. The Caley-Hamilton theorem[2] states that every $N \times N$ matrix satisfies its own characteristic equation. This means that

$$\sum_{m=0}^{N} C_m \mathbf{A}^m = \mathbf{0} \tag{3.63}$$

where $\mathbf{A}^0 \equiv \mathbf{I}$ (the identity matrix) and $\mathbf{0}$ is the null matrix with all zero elements. Expanding Eq. 3.63

$$\mathbf{A}^N + C_{N-1}\mathbf{A}^{N-1} + \cdots + C_1\mathbf{A} + C_0\mathbf{I} = \mathbf{0}$$

or

$$\mathbf{A}^N = -C_{N-1}\mathbf{A}^{N-1} - \cdots - C_1\mathbf{A} - C_0\mathbf{I} \tag{3.64}$$

Now multiply Eq. 3.64 by $\mathbf{A}$:

$$\mathbf{A}^{N+1} = -C_{N-1}\mathbf{A}^N - \cdots - C_1\mathbf{A}^2 - C_0\mathbf{A}$$

Substitute Eq. 3.64:

$$\mathbf{A}^{N+1} = -C_{N-1}(-C_{N-1}\mathbf{A}^{N-1} - \cdots - C_1\mathbf{A} - C_0\mathbf{I}) - \cdots - C_1\mathbf{A}^2 - C_0\mathbf{A}$$
$$\mathbf{A}^{N+1} = (C_{N-1}^2 - C_{N-2})\mathbf{A}^{N-1} + (C_{N-1}C_{N-2} - C_{N-3})\mathbf{A}^{N-2} + \cdots + C_{N-1}C_0\mathbf{I}$$

This process can be repeated so that *any power of* $\mathbf{A}$ can be expressed as a series in powers of $\mathbf{A}$ with the *highest power* in the series being only $N - 1$, not infinity (Eq. 3.59). Thus, any function of $\mathbf{A}$, even if it calls for $\mathbf{A}^m$, $m \to \infty$, can be put in the form

$$\boxed{\mathbf{f}(\mathbf{A}) = \sum_{m=0}^{N-1} \beta_m \mathbf{A}^m} \tag{3.65}$$

which only has N terms. For example

$$\mathbf{f}(\mathbf{A}) = e^{t\mathbf{A}} = \mathbf{I} + t\mathbf{A} + \frac{t^2}{2!}\mathbf{A}^2 + \cdots$$

has an infinite number of terms, but each one can be put in the form of Eq. 3.65 with coefficients $a_m, b_m, c_m, \ldots$, and the coefficients of like powers of $\mathbf{A}$ can be combined to give Eq. 3.65. Thus

$$e^{t\mathbf{A}} = \sum_{m=0}^{N-1} \beta_m \mathbf{A}^m \tag{3.66}$$

In order to use this result we must be able to find β_m. If we go back to Eq. 3.32

$$\lambda^N + C_{N-1}\lambda^{N-1} + \cdots + C_1\lambda + C_0 = 0$$
$$\lambda^N = -C_{N-1}\lambda^{N-1} - \cdots - C_1\lambda - C_0 \tag{3.67}$$
$$\lambda^{N+1} = -C_{N-1}\lambda^N - \cdots - C_1\lambda^2 - C_0\lambda$$

Substituting Eq. 3.67

$$\lambda^{N+1} = (C_{N-1}^2 - C_{N-2})\lambda^{N-1} + (C_{N-1}C_{N-2} - C_{N-3})\lambda^{N-2} + \cdots + C_{N-1}C_0$$

Thus, if this process is repeated, we see that any power of λ can be expressed as a series in powers of λ with the highest power in the series being only $N - 1$. Thus, any function of λ can be put in the form

$$f(\lambda) = \sum_{m=0}^{N-1} \beta_m \lambda^m$$

and, in particular

$$e^{\lambda t} = \sum_{m=0}^{N-1} \beta_m \lambda^m \qquad (3.68)$$

where the β in Eq. 3.68 are identically the same as the β in Eq. 3.66. Equation 3.68 represents N equations (one for each eigenvalue, assuming that none are repeated) in the N unknown β_m. Thus, we have a relatively easy way to find the β_m and, hence, $e^{t\mathbf{A}}$ (Eq. 3.66).

Example 8

Find $e^{t\mathbf{A}}$ for Example 3, where $\lambda_1 = -1$, $\lambda_2 = -3$, and

$$\mathbf{A} = \begin{bmatrix} -0.5 & -1.25 \\ 1 & -3.5 \end{bmatrix}$$

According to Eq. 3.66

$$e^{t\mathbf{A}} = \sum_{m=0}^{1} \beta_m \mathbf{A}^m = \beta_0 \mathbf{I} + \beta_1 \mathbf{A}$$

and according to Eq. 3.68

$$\left. \begin{array}{l} e^{-t} = \beta_0 - \beta_1 \\ e^{-3t} = \beta_0 - 3\beta_1 \end{array} \right\} \quad \begin{array}{l} \beta_0 = \tfrac{1}{2}(3e^{-t} - e^{-3t}) \\ \beta_1 = \tfrac{1}{2}(e^{-t} - e^{-3t}) \end{array}$$

Thus, using Eq. 3.66

$$e^{t\mathbf{A}} = \begin{bmatrix} \beta_0 & 0 \\ 0 & \beta_0 \end{bmatrix} + \begin{bmatrix} -0.5\beta_1 & -1.25\beta_1 \\ \beta_1 & -3.5\beta_1 \end{bmatrix}$$

$$e^{t\mathbf{A}} = \begin{bmatrix} \beta_0 - 0.5\beta_1 & -1.25\beta_1 \\ \beta_1 & \beta_0 - 3.5\beta_1 \end{bmatrix}$$

$$e^{t\mathbf{A}} = \begin{bmatrix} 1.25e^{-t} - 0.25e^{-3t} & -0.625e^{-t} + 0.625e^{-3t} \\ 0.5e^{-t} - 0.5e^{-3t} & -0.25e^{-t} + 1.25e^{-3t} \end{bmatrix} \qquad (3.69)$$

For an input $i_s(t) = x_1(t) = 9.6e^{-4t}u(t)$, as used before, we have $x_2(t) \equiv 0$, $f_1(t) = 12e^{-4t}u(t)$, $f_2(t) \equiv 0$. Equation 3.61 is

$$\mathbf{q} = \int_{-\infty}^{t} e^{(t-\tau)\mathbf{A}} \mathbf{f}(\tau) \, d\tau$$

When Eq. 3.69 is substituted, and the integration is performed, we obtain

$$q_1(t) = (5e^{-t} - 3e^{-3t} - 2e^{-4t})u(t)$$
$$q_2(t) = (2e^{-t} - 6e^{-3t} + 4e^{-4t})u(t)$$

as before.

3.4 FURTHER CONSIDERATION OF MATRIX SOLUTIONS

When the ith root in an Nth order system is repeated n times, the fundamental set is given by Eq. 2.58 ($s = \lambda$):

$$e^{\lambda_1 t}, e^{\lambda_2 t}, \ldots, e^{\lambda_i t}, t e^{\lambda_i t}, t^2 e^{\lambda_i t}, \ldots, t^{n-1} e^{\lambda_i t}, \ldots, e^{\lambda_N t}$$

Instead of trying to show how the general case is treated we will demonstrate how repeated roots of the characteristic equation are handled for the second-order case. For this case Eq. 3.68 becomes

$$\left.\begin{aligned} e^{\lambda_1 t} &= \beta_0 + \lambda_1 \beta_1 \\ e^{\lambda_2 t} &= \beta_0 + \lambda_2 \beta_1 \end{aligned}\right\} \quad \beta_0 = \frac{\lambda_2 e^{\lambda_1 t} - \lambda_1 e^{\lambda_2 t}}{\lambda_2 - \lambda_1}$$

$$\beta_1 = \frac{e^{\lambda_2 t} - e^{\lambda_1 t}}{\lambda_2 - \lambda_1}$$

and if $\lambda_2 = \lambda_1$ both β_0 and β_1 are indeterminate. L'Hospital's rule gives

$$\beta_0 = (1 - \lambda_1 t) e^{\lambda_1 t} \tag{3.70}$$
$$\beta_1 = t e^{\lambda_1 t} \tag{3.71}$$

This is the same result that is obtained if we use the first of the equations above Eq. 3.70 unchanged, but differentiate the second with respect to λ_2, and then set $\lambda_2 = \lambda_1$:

$$e^{\lambda_1 t} = \beta_0 + \lambda_1 \beta_1$$
$$t e^{\lambda_1 t} = \beta_1$$

which give Eqs. 3.70 and 3.71 again. We then proceed to find $e^{t\mathbf{A}}$ and substitute it into Eq. 3.61 or 3.17 to find $\mathbf{q}$ or $\mathbf{y}$, respectively.

There may be several reasons why it might be desirable to alter the internal structure of a system without changing the description of the system externally. One form of internal structure might be more economical than another, or one form might have better noise characteristics. In any case the state-variable formulation with its matrix forms allows us to easily apply transformations to the state vector. There is one transformation that is particularly useful from the analysis standpoint in that it essentially converts $\mathbf{A}$ into a diagonal matrix and *uncouples* the state-variable equations. This is of much value, and is the topic of the next few paragraphs.

We found in Eq. 3.31 that

$$|\mathbf{A} - \lambda \mathbf{I}| = 0 \tag{3.31}$$

is the characteristic equation of the system, and in Eq. 3.27

$$(\mathbf{A} - \lambda \mathbf{I})\mathbf{q} = \mathbf{0} \tag{3.27}$$

$\mathbf{q}$ represents the eigenvectors. Take the case $N = 2$, for example. There are two eigenvalues, λ_1 and λ_2. For λ_1, Eq. 3.27 gives

$$\begin{bmatrix} a_{11} - \lambda_1 & a_{12} \\ a_{21} & a_{22} - \lambda_1 \end{bmatrix} \begin{bmatrix} q_1 \\ q_2 \end{bmatrix} = \begin{bmatrix} 0 \\ 0 \end{bmatrix}$$

or
$$(a_{11} - \lambda_1)q_1 + a_{12}q_2 = 0 \qquad (3.72)$$
and
$$a_{21}q_1 + (a_{22} - \lambda_1)q_2 = 0 \qquad (3.73)$$
From Eq. 3.72
$$q_2 = -q_1 \frac{a_{11} - \lambda_1}{a_{12}}$$
so the eigenvector associated with λ_1, called $\mathbf{q}^1$ is

$$\mathbf{q}^1 = \begin{bmatrix} q_1 \\ -q_1 \dfrac{a_{11} - \lambda_1}{a_{12}} \end{bmatrix} = \begin{bmatrix} q_1^1 \\ q_2^1 \end{bmatrix} \qquad (3.74)$$

Notice that we have not determined both q_1^1 and q_2^1, but, rather, we have only determined q_2^1 in terms of q_1^1. That is, we have not determined the *length* of $\mathbf{q}^1$, but we have determined the relative values of its two ($N = 2$) components and, therefore, its direction. This is sufficient. In exactly the same way, the eigenvector associated with λ_2 is

$$\mathbf{q}^2 = \begin{bmatrix} q_1 \\ -q_1 \dfrac{a_{11} - \lambda_2}{a_{12}} \end{bmatrix} = \begin{bmatrix} q_1^2 \\ q_2^2 \end{bmatrix} \qquad (3.75)$$

Next, form the square matrix of eigenvectors, called the *modal matrix* $\mathbf{T}$

$$\mathbf{T} = \begin{bmatrix} q_1^1 & q_1^2 \\ q_2^1 & q_2^2 \end{bmatrix} \qquad (3.76)$$

Using $\mathbf{T}$ we can write Eq. 3.25 (for any N) as

$$\mathbf{AT} = \mathbf{T}\lambda \qquad (3.77)$$

where λ is the diagonal matrix

$$\lambda = \begin{bmatrix} \lambda_1 & 0 & 0 & \cdots & 0 \\ 0 & \lambda_2 & 0 & \cdots & 0 \\ 0 & 0 & \lambda_3 & \cdots & 0 \\ \vdots & \vdots & \vdots & \ddots & 0 \\ 0 & 0 & \cdots & & \lambda_N \end{bmatrix} \qquad (3.78)$$

It is not difficult to verify Eq. 3.77 for $N = 2$. We obtain λ by premultiplying both sides of Eq. 3.77 by $\mathbf{T}^{-1}$

$$\mathbf{T}^{-1}\mathbf{AT} = \mathbf{T}^{-1}\mathbf{T}\lambda = \lambda \qquad (3.79)$$

The left side of Eq. 3.79 is called a *similarity transformation*.

Example 9

Find **T** and $\mathbf{T}^{-1}$ for Example 3, where $\lambda_1 = -1$, $\lambda_2 = -3$, and

$$\mathbf{A} = \begin{bmatrix} -0.5 & -1.25 \\ 1 & -3.5 \end{bmatrix}$$

We have from Eqs. 3.74 and 3.75

$$\mathbf{q}^1 = \begin{bmatrix} q_1 \\ -q_1 \times \dfrac{-0.5 + 1}{-1.25} \end{bmatrix} = \begin{bmatrix} q_1 \\ 0.4 q_1 \end{bmatrix}$$

$$\mathbf{q}^2 = \begin{bmatrix} q_1 \\ -q_1 \times \dfrac{-0.5 + 3}{-1.25} \end{bmatrix} = \begin{bmatrix} q_1 \\ 2 q_1 \end{bmatrix}$$

The usual procedure is that of normalizing the eigenvectors so that the sum of the squares is unity. Thus

$$\mathbf{q}^1 = \begin{bmatrix} \dfrac{1}{\sqrt{1.16}} \\ \dfrac{0.4}{\sqrt{1.16}} \end{bmatrix} \qquad \mathbf{q}^2 = \begin{bmatrix} \dfrac{1}{\sqrt{5}} \\ \dfrac{2}{\sqrt{5}} \end{bmatrix}$$

Therefore

$$\mathbf{T} = \begin{bmatrix} \dfrac{1}{\sqrt{1.16}} & \dfrac{1}{\sqrt{5}} \\ \dfrac{0.4}{\sqrt{1.16}} & \dfrac{2}{\sqrt{5}} \end{bmatrix}$$

and

$$\mathbf{T}^{-1} = \dfrac{\sqrt{5.8}}{1.6} \begin{bmatrix} \dfrac{2}{\sqrt{5}} & -\dfrac{1}{\sqrt{5}} \\ -\dfrac{0.4}{\sqrt{1.16}} & \dfrac{1}{\sqrt{1.16}} \end{bmatrix} = \begin{bmatrix} 1.25\sqrt{1.16} & -\dfrac{\sqrt{1.16}}{1.6} \\ -0.25\sqrt{5} & \dfrac{\sqrt{5}}{1.6} \end{bmatrix}$$

Our next task is that of uncoupling the state-variable equations. This can be accomplished if we first make a change of variable. Let

$$\mathbf{r} = \mathbf{T}^{-1}\mathbf{q} \tag{3.80}$$

so
$$\mathbf{r}' = \mathbf{T}^{-1}\mathbf{q}' \tag{3.81}$$
and
$$\mathbf{q} = \mathbf{Tr} \tag{3.82}$$
$$\mathbf{q}' = \mathbf{Tr}' \tag{3.83}$$

When the last two equations are substituted into the normal-form equation (3.7), we obtain
$$\mathbf{Tr}' = \mathbf{ATr} + \mathbf{f} \qquad \mathbf{f} = \mathbf{Bx}$$
and premultiplying by $\mathbf{T}^{-1}$ gives
$$\mathbf{T}^{-1}\mathbf{Tr}' = \mathbf{r}' = \mathbf{T}^{-1}\mathbf{ATr} + \mathbf{T}^{-1}\mathbf{f}$$
Using Eq. 3.79
$$\mathbf{r}' = \boldsymbol{\lambda}\mathbf{r} + \mathbf{T}^{-1}\mathbf{f} \tag{3.84}$$
$\mathbf{T}^{-1}\mathbf{f}$ is a new forcing function that we designate by $\mathbf{f}_r$. Thus
$$\mathbf{f}_r = \mathbf{T}^{-1}\mathbf{f} \tag{3.85}$$
and
$$\mathbf{r}' = \boldsymbol{\lambda}\mathbf{r} + \mathbf{f}_r \tag{3.86}$$

Since $\boldsymbol{\lambda}$ is diagonal, Eq. 3.86 represents N *first-order, uncoupled differential equations*! That is, the nth equation is simply
$$r'_n = \lambda_n r_n + f_{rn}$$
or
$$r'_n - \lambda_n r_n = f_{rn}$$
whose unit-impulse response is
$$h_{f_r}(t) = e^{\lambda_n t} u(t)$$
Therefore
$$r_n(t) = e^{\lambda_n t} \int_{-\infty}^{t} f_{rn}(\tau) e^{-\lambda_n \tau} d\tau$$
or, in matrix form[†]
$$\mathbf{r}(t) = \int_{-\infty}^{t} e^{\boldsymbol{\lambda}(t-\tau)} \mathbf{f}_r(\tau) d\tau \tag{3.87}$$

[†] $e^{\boldsymbol{\lambda} t}$ is the diagonal matrix with elements given by $e^{\lambda_1 t}, e^{\lambda_2 t}, \ldots, e^{\lambda_N t}$

3.4 FURTHER CONSIDERATION OF MATRIX SOLUTIONS

We can separate Eq. 3.87 into the zero-input response and the zero-state response:

$$\boxed{\mathbf{r}(t) = e^{\lambda t}\mathbf{r}(0) + \int_0^t e^{\lambda(t-\tau)}\mathbf{f}_r(\tau)d\tau} \qquad (3.88)$$

When $\mathbf{r}(t)$ has been found, $\mathbf{q}(t)$ is found by Eq. 3.82

$$\boxed{\mathbf{q} = \mathbf{T}\mathbf{r}} \qquad (3.89)$$

and, once again, we have formally found the state variables.

Again, it would be wise to try this new technique on a simple example.

Example 10

Repeat Example 4 using the techniques of this section. We found $\mathbf{T}$ and $\mathbf{T}^{-1}$ in Example 9. Equation 3.85 gives ($x_1 = i_s$, $x_2 = 0$):

$$\mathbf{f}_r = \begin{bmatrix} 1.25\sqrt{1.16} & -\dfrac{\sqrt{1.16}}{1.6} \\ -0.25\sqrt{5} & \dfrac{\sqrt{5}}{1.6} \end{bmatrix} \begin{bmatrix} f_1 \\ 0 \end{bmatrix}$$

$$f_{r1} = 1.25\sqrt{1.16}\, f_1$$
$$f_{r2} = -0.25\sqrt{5}\, f_1$$

Notice that these are not the same forcing functions as in Example 3 (same problem, first method). Equation 3.87 gives

$$r_1(t) = e^{-t}\int_{-\infty}^t 1.25\sqrt{1.16}\, f_1(\tau) e^{\tau}\, d\tau$$

$$r_2(t) = -e^{-3t}\int_{-\infty}^t 0.25\sqrt{5}\, f_1(\tau) e^{3\tau}\, d\tau$$

with $i_s(t) = 4.8$ A, we have

$$r_1(t) = 7.5\sqrt{1.16}\, e^{-t}\int_{-\infty}^t e^{\tau}\, d\tau = 7.5\sqrt{1.16}$$

$$r_2(t) = 1.5\sqrt{5}\, e^{-3t}\int_{-\infty}^t e^{3\tau}\, d\tau = -0.5\sqrt{5}$$

With Eq. 3.82

$$\mathbf{q} = \begin{bmatrix} \dfrac{1}{\sqrt{1.16}} & \dfrac{1}{\sqrt{5}} \\ \dfrac{0.4}{\sqrt{1.16}} & \dfrac{2}{\sqrt{5}} \end{bmatrix} \begin{bmatrix} 7.5\sqrt{1.16} \\ -0.5\sqrt{5} \end{bmatrix}$$

or

$$q_1 = 7.5 - 0.5 = 7$$
$$q_2 = 3 - 1 = 2$$

as before! Notice that the normalization of **T** had no effect on the result.
For an input $i_s(t) = 9.6\,e^{-4t}u(t)$, as used before

$$r_1(t) = 15\sqrt{1.16}\,e^{-t}u(t)\int_0^t e^{-4\tau}e^{\tau}\,d\tau$$

$$r_2(t) = -3\sqrt{5}\,e^{-3t}u(t)\int_0^t e^{-4\tau}e^{3\tau}\,d\tau$$

or

$$r_1(t) = -5\sqrt{1.16}\,e^{-t}u(t)e^{-3\tau}\Big|_0^t = -5\sqrt{1.16}(e^{-4t} - e^{-t})u(t)$$

$$r_2(t) = 3\sqrt{5}\,e^{-3t}u(t)e^{-\tau}\Big|_0^t = 3\sqrt{5}(e^{-4t} - e^{-3t})u(t)$$

Then

$$\mathbf{q} = \begin{bmatrix} \dfrac{1}{\sqrt{1.16}} & \dfrac{1}{\sqrt{5}} \\ \dfrac{0.4}{\sqrt{1.16}} & \dfrac{2}{\sqrt{5}} \end{bmatrix} \begin{bmatrix} -5\sqrt{1.16}(e^{-4t} - e^{-t})u(t) \\ 3\sqrt{5}(e^{-4t} - e^{-3t})u(t) \end{bmatrix}$$

or

$$q_1(t) = (5e^{-t} - 3e^{-3t} - 2e^{-4t})u(t)$$
$$q_2(t) = (2e^{-t} - 6e^{-3t} + 4e^{-4t})u(t)$$

as before.

The methods we described for calculating the state variables are similar, but different. The amount of effort required for each is about the same (conservation of difficulty), at least for the simple example we used. The first method (Sec. 3.2) required that we only partially invert a matrix (Eq. 3.35), but the calculation of the forcing function (Eq. 3.41) required the use of a lengthy formula (for large N). The third method required that we calculate the eigenfunctions to form the modal matrix and then find its inverse.

The special case of repeated roots of the characteristic equation needs to be investigated. This was mentioned earlier where it was stated that if the ith root of an Nth order differential equation occurs n times, the fundamental set is

$$e^{s_1 t}, e^{s_2 t}, \ldots, e^{s_i t}, te^{s_i t}, \ldots, t^{n-1}e^{s_i t}, \ldots, e^{s_N t} \qquad (2.58)$$

In this case the matrix method just discussed will not work without modification. This unfortunate fact results because at least one pair (for one repeated root) of eigenvectors, q^n and q^{n+1} for example, will be identical, and thus at least two columns of **T** will be identical. Therefore, the determinant of **T** is 0, and $(T)^{-1}$ does not exist.

Example 11

Given

$$A = \begin{bmatrix} 1 & 0 & 2 \\ 1 & -1 & 1 \\ -3 & 0 & -4 \end{bmatrix}$$

find **T**. Solving for the eigenvalues in the usual way we find $\lambda_1 = -1, \lambda_2 = -1$, and $\lambda_3 = -2$. Finding the eigenvectors by using

$$\begin{bmatrix} 1-\lambda & 0 & 2 \\ 1 & -1-\lambda & 1 \\ -3 & 0 & -4-\lambda \end{bmatrix} \begin{bmatrix} q_1 \\ q_2 \\ q_3 \end{bmatrix} = 0$$

with $\lambda_1, \lambda_2, \lambda_3$ gives

$$q^1 = q^2 = \begin{bmatrix} q_1 \\ \tfrac{1}{5}q_1 \\ -\tfrac{3}{5}q_1 \end{bmatrix} \quad q^3 = \begin{bmatrix} q_1 \\ \tfrac{1}{6}q_1 \\ -\tfrac{1}{2}q_1 \end{bmatrix}$$

So normalizing

$$T = \begin{bmatrix} \dfrac{5}{\sqrt{35}} & \dfrac{5}{\sqrt{35}} & \dfrac{6}{\sqrt{46}} \\ \dfrac{1}{\sqrt{35}} & \dfrac{1}{\sqrt{35}} & \dfrac{1}{\sqrt{46}} \\ \dfrac{0.6}{\sqrt{35}} & \dfrac{0.6}{\sqrt{35}} & \dfrac{0.6}{\sqrt{46}} \end{bmatrix}$$

and $|T| \equiv 0$.

The method introduced at the beginning of Sec. 3.4 that used the state-transition matrix e^{tA} will work with repeated roots, and we looked at one simple example of this. It is primarily an advanced numerical technique.

3.5 NUMERICAL METHODS[3]

The other sections of this chapter were devoted to developing analytical methods for systems analysis with state variables. On the other hand, the method we briefly examined that used the state-transition matrix is primarily adapted to numerical methods using a digital computer. The simpler impulse-response method leads to a superposition or convolution type integral, Eq. 3.43. It is relatively simple to perform the integration numerically, and so we will simply leave it at that.

The normal form, Eq. 3.7, is ideal for numerical work when we approximate the derivative by its prelimit-difference form. That is, for one dependent variable

$$\frac{dq}{dt} = \lim_{h \to 0} \frac{q(t+h) - q(t)}{h} \approx \frac{q(t+h) - q(t)}{h}$$

or

$$q(t+h) \approx q(t) + h\frac{dq}{dt} = q(t) + hq'(t)$$

Using matrix notation, anticipating the normal form, we have (approximately)

$$\mathbf{q}(t+h) = \mathbf{q}(t) + h\mathbf{q}'(t) \tag{3.90}$$

Substituting Eq. 3.7 into 3.90 gives

$$\mathbf{q}(t+h) = \mathbf{q}(t) + h\mathbf{A}\mathbf{q}(t) + h\mathbf{f}(t) \tag{3.91}$$

or

$$\boxed{\mathbf{q}(t+h) = (\mathbf{I} + h\mathbf{A})\mathbf{q}(t) + h\mathbf{f}(t)} \tag{3.92}$$

It is now easy to start at $t = t_0$ ($t_0 = 0$, usually) and step forward (step-by-step) in time by an amount h, finding $q(t_0)$, $q(t_0 + h)$, $q(t_0 + 2h)$, ... It should be obvious that the amount of error incurred will generally depend on h, and this error will be negligible if we make h small enough (and therefore increase the time required to calculate the state variable at some time in the future).

Example 12

Find an approximate numerical solution for $\mathbf{q}(t)$ in Example 4. We had $x_1 = i_s$, $i_s = 4.8$ A, $x_2 = 0$, and $f_1 = 1.25x_1$:

$$\mathbf{A} = \begin{bmatrix} -0.5 & -1.25 \\ 1 & -3.5 \end{bmatrix} \quad \mathbf{f} = \begin{bmatrix} 6 \\ 0 \end{bmatrix}$$

Thus

$$\mathbf{I} + h\mathbf{A} = \begin{bmatrix} 1 & 0 \\ 0 & 1 \end{bmatrix} + h\begin{bmatrix} -0.5 & -1.25 \\ 1 & -3.5 \end{bmatrix}$$

We should choose h so that it is small compared to any expected time constant. We may have smaller time constants in the zero-state response, but they are

1 s and $\frac{1}{3}$ s in the zero-input response, so we begin by trying $h = 0.1$ s:

$$\mathbf{q}(t+h) = \begin{bmatrix} 0.95 & -0.125 \\ 0.1 & 0.65 \end{bmatrix} \begin{bmatrix} q_1(t) \\ q_2(t) \end{bmatrix} + \begin{bmatrix} 0.6 \\ 0 \end{bmatrix}$$

or

$$\begin{bmatrix} v_C(t+h) \\ i_L(t+h) \end{bmatrix} = \begin{bmatrix} 0.95 & -0.125 \\ 0.1 & 0.65 \end{bmatrix} \begin{bmatrix} v_C(t) \\ i_L(t) \end{bmatrix} + \begin{bmatrix} 0.6 \\ 0 \end{bmatrix}$$

Next, we need to find a place to start. With this in mind, and recognizing from Fig. 3.2 that if we specify $v_C(0) = 7$ V and $i_L(0) = 2$ A (the steady-state values), we should obtain $v_C(t) = 7$ V and $i_L(t) = 2$ A for all time. We specify initial conditions

$$v_C(0) = 7 \text{ V}$$
$$i_L(0) = 2 \text{ A}$$
$$v_C(t+h) = 0.95v_C(t) - 0.125i_L(t) + 0.6$$
$$i_L(t+h) = 0.10v_C(t) + 0.650i_L(t)$$

so

$$v_C(h) = 0.95v_C(0) - 0.125i_L(0) + 0.6$$
$$i_L(h) = 0.10v_C(0) + 0.650i_L(0)$$
$$v_C(h) = 0.95(7) - 0.125(2) + 0.6 = 7 \text{ V}$$
$$i_L(h) = 0.10(7) + 0.650(2) = 2 \text{ A}$$

In the same way

$$v_C(2h) = 7 \text{ V}$$
$$i_L(2h) = 2 \text{ A}$$

and so on, forever. In this case the answer is exact ($dy = 0$).

Next, let us try $x_1 = 1.25i_s$, $i_s = 9.6e^{-4t}u(t)$, $x_2 = 0$, and start at $t = 0$ with no initial energy: $v_C(0) = i_L(0) = 0$, as in Example 4. For $h = 0.05$

$$v_C(t+h) = 0.975v_C(t) - 0.0625i_L(t) + 0.6e^{-4t}$$
$$i_L(t+h) = 0.050v_C(t) + 0.8250i_L(t)$$
(3.93)

while for $h = 0.025$

$$v_C(t+h) = 0.9875v_C(t) - 0.03125i_L(t) + 0.3e^{-4t}$$
$$i_L(t+h) = 0.0250v_C(t) + 0.9125i_L(t)$$
(3.94)

Results for these two cases, using a hand-held programmable calculator, along with the exact solutions

$$\left.\begin{array}{l} v_C(t) = (5e^{-t} - 3e^{-3t} - 2e^{-4t})u(t) \\ i_L(t) = (2e^{-t} - 6e^{-3t} + 4e^{-4t})u(t) \end{array}\right\}$$
(3.95)

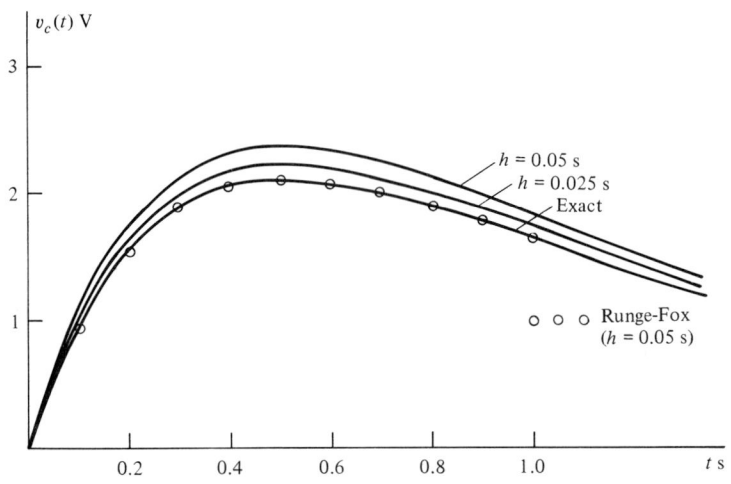

Figure 3.9. $v_C(t)$ versus t, approximate and exact solutions.

found in Example 4 are shown in Figs. 3.9 and 3.10. Notice that halving h essentially halves the error.

The numerical method just presented is extremely simple, its accuracy primarily determined by how much time the analyst is willing to spend on the calculation. There are many other higher-order numerical methods that are more accurate and more complicated. The Runge-Fox method is one. We have

$$\mathbf{q}'(t) = \mathbf{A}\mathbf{q}(t) + \mathbf{f}(t)$$

and

$$\mathbf{q}'(t + h) = \mathbf{A}\mathbf{q}(t + h) + \mathbf{f}(t + h)$$

Adding the last two equations gives

$$\frac{\mathbf{q}'(t + h) + \mathbf{q}'(t)}{2} = \frac{1}{2}\mathbf{A}\mathbf{q}(t + h) + \frac{1}{2}\mathbf{A}\mathbf{q}(t) + \frac{1}{2}\mathbf{f}(t + h) + \frac{1}{2}\mathbf{f}(t)$$

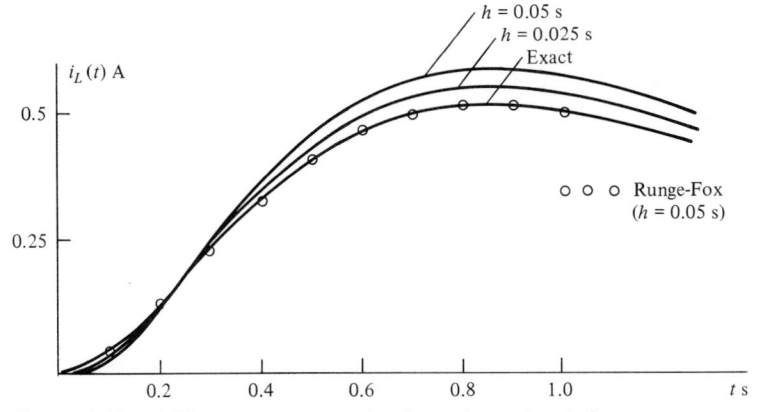

Figure 3.10. $i_L(t)$ versus t, approximate and exact solutions.

But, for small h, $\mathbf{q}'(t+h) \approx \mathbf{q}'(t)$, so
$$\frac{\mathbf{q}'(t+h) + \mathbf{q}'(t)}{2} \approx \frac{\mathbf{q}(t+h) - \mathbf{q}(t)}{h}$$

So
$$\mathbf{q}(t+h) - \mathbf{q}(t) = \frac{h}{2}\mathbf{A}\mathbf{q}(t+h) + \frac{h}{2}\mathbf{A}\mathbf{q}(t) + \frac{h}{2}\mathbf{f}(t+h) + \frac{h}{2}\mathbf{f}(t)$$

The last equation is, of course, only an approximation. Transposing, and using $\mathbf{I}$ for proper matrix manipulation
$$\left(\mathbf{I} - \frac{h}{2}\mathbf{A}\right)\mathbf{q}(t+h) = \left(\mathbf{I} + \frac{h}{2}\mathbf{A}\right)\mathbf{q}(t) + \frac{h}{2}\mathbf{f}(t+h) + \frac{h}{2}\mathbf{f}(t)$$

Therefore

$$\boxed{\mathbf{q}(t+h) = \left(\mathbf{I} - \frac{h}{2}\mathbf{A}\right)^{-1}\left(\mathbf{I} + \frac{h}{2}\mathbf{A}\right)\mathbf{q}(t) + \left(\mathbf{I} - \frac{h}{2}\mathbf{A}\right)^{-1}\left[\frac{h}{2}\mathbf{f}(t+h) + \frac{h}{2}\mathbf{f}(t)\right]}$$

(3.96)

It is obvious now why this method is more complicated: it requires a matrix inversion.

Example 13

Repeat the second part of Example 12, using the Runge-Fox method with $h = 0.05$.

$$\left(\mathbf{I} - \frac{h}{2}\mathbf{A}\right) = \begin{bmatrix} 1 & 0 \\ 0 & 1 \end{bmatrix} - 0.025 \begin{bmatrix} -0.5 & -1.25 \\ 1 & -3.5 \end{bmatrix}$$

$$\left(\mathbf{I} - \frac{h}{2}\mathbf{A}\right) = \begin{bmatrix} 1.0125 & 0.03125 \\ -0.025 & 1.0875 \end{bmatrix}$$

$$\left(\mathbf{I} + \frac{h}{2}\mathbf{A}\right) = \begin{bmatrix} 0.9875 & -0.03125 \\ 0.025 & 0.9125 \end{bmatrix}$$

$$\left(\mathbf{I} - \frac{h}{2}\mathbf{A}\right)^{-1} = \frac{1}{1.101875}\begin{bmatrix} 1.0875 & -0.03125 \\ 0.025 & 1.0125 \end{bmatrix} = \begin{bmatrix} 0.9870 & -0.0284 \\ 0.0227 & 0.9189 \end{bmatrix}$$

Notice the round-off in the last calculation.

$$\left(\mathbf{I} - \frac{h}{2}\mathbf{A}\right)^{-1}\left(\mathbf{I} + \frac{h}{2}\mathbf{A}\right) = \begin{bmatrix} 0.9740 & -0.0568 \\ 0.0454 & 0.8378 \end{bmatrix}$$

Combining the terms required in Eq. 3.96 gives

$$\left.\begin{aligned} v_C(t+h) &= 0.974 v_C(t) - 0.0568 i_L(t) + 0.5385 e^{-4t} \\ i_L(t+h) &= 0.0454 v_C(t) + 0.8378 i_L(t) + 0.0125 e^{-4t} \end{aligned}\right\}$$

(3.97)

118   STATE-SPACE ANALYSIS

Equation 3.97 should be compared to Eq. 3.93. Again, $v_C(t)$ and $i_L(t)$ were calculated with a hand-held programmable calculator. They are plotted in Figs. 3.9 and 3.10. The error for this method is about 0.5 percent for most of the range of t.

3.6   CONCLUDING REMARKS

In this chapter the concepts of time-domain state-variable analysis were presented. This method essentially organizes a higher-order system with N energy storage elements into a system of N *first-order* differential equations in N unknowns. We considered only the time-invariant linear system (constant coefficients), and three methods for solving these equations were examined. The first method is based on the unit-impulse response and is an extension of the concepts of the first two chapters. The second is more complicated but was pursued in some detail. The third is simple and elegant but must be modified for repeated roots of the characteristic equation. We chose to leave that course for move advanced study.

Finally, some numerical solutions were obtained for a simple system. It is important to remember that *any* time-invariant linear system, regardless of its order, can be analyzed by these methods, and they apply to any of the time-invariant systems that we have studied to this point. This assumes, of course, that, in the case of a network, we have the network given to us so that the normal-form equations can be obtained.

PROBLEMS

1. (a) Arrange the coupled equations (below) into normal form.

$$y_1' - y_1 + 2y_2 = z_1 \qquad z_1 = z_1(t)$$
$$y_1 - y_2' - y_2 = z_2 \qquad z_2 = z_2(t)$$

   (b) Find **A**.
   (c) Find $h_{fs}(t)$.
   (d) Find $\mathbf{f}^N$.
   (e) Find the formal zero-state response (integrals) for **y**.

2. Repeat Prob. 1 for

$$y_1' + y_1 + y_2' + 3y_2 = z_1$$
$$y_1 + y_2' + y_2 = z_2$$

3. Repeat Prob. 1 for

$$y_1' + y_1 + y_2' = z_1$$
$$y_1' + 3y_1 - y_2' - 2y_2 = z_2$$

4. Repeat Prob. 1 for

$$y_1' = y_2$$
$$y_2' = y_3$$
$$y_3' = y_1$$

5. Repeat Prob. 1 for

$$y_1' - 3y_2 = z_1$$
$$y_1 - y_2' - y_3 = z_2$$
$$y_2 + y_3' = z_3$$

6. Solve Prob. 1 for $z_1(t) = 4u(t)$, $z_2(t) = -3u(t)$.
7. Repeat Prob. 6 if the initial conditions $y_1(0) = 1$ and $y_2(0) = 0$ are imposed.
8. Solve Prob. 2 for $z_1(t) = tu(t)$, $z_2(t) = 0$.
9. Solve Prob. 3 for $z_1(t) = 2\sin(2t)$, $z_2(t) = \cos(2t)$.
10. Find a zero-input response for Prob. 4 if $y_1(0) = 1$, $y_2(0) = 0$, and $y_3(0) = 0$.
11. Solve Prob. 5 for $z_1(t) = 5u(t)$, $z_2(t) = tu(t)$, $z_3(t) = 0$.
12. Find the zero-state response for

$$ty_1' + y_2 = z_1$$
$$y_1 + ty_2' = z_2$$

by any method.

13. Show that t, $1/t^5$ is a fundamental set for

$$ty_1' - 3ty_2' + 5y_2 = z_1$$
$$y_1 + ty_2' + y_2 = z_2$$

14. Obtain the normal-form equations for the mechanical system that is shown in Fig. E.4 (App. E). Use the velocity of the masses and the spring force as the (three) state variables.
15. Solve Example 12, Chap. 2, using state variables. Compare the result for $v_2 (=v_C)$ in Fig. 3.11 to that given earlier. Compare the methods.

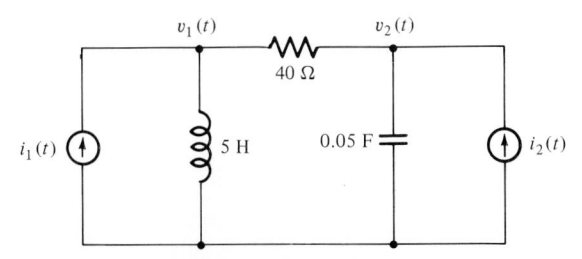

Figure 3.11. Circuit for Prob. 15.

16. Find $v_C(t)$ in Fig. 3.12 using the state variables i_L and v_C.

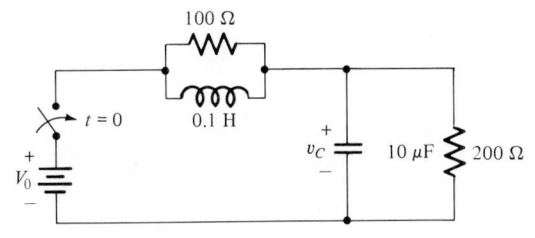

Figure 3.12. Circuit for Prob. 16.

120   STATE-SPACE ANALYSIS

17. Find $i_L(t)$ in Fig. 3.13 using the state variables i_L and v_C.

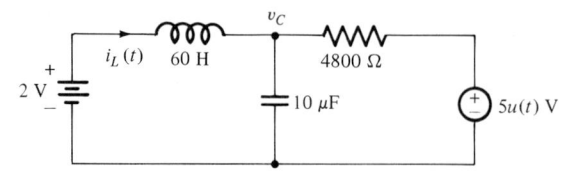

Figure 3.13. Circuit for Prob. 17.

18. Repeat Prob. 8 with the method that uses the modal matrix (Eq. 3.86).
19. Repeat Prob. 1 with the method that uses the modal matrix.
20. If

$$y'_1 = y_1 + 2y_3 + x_1 \qquad x_1(t) = 5u(t)$$
$$y'_2 = y_1 - y_2 + y_3 + x_2 \qquad x_2(t) = -2u(t)$$
$$y'_3 = -3y_1 - 4y_3$$

find y_1, y_2, and y_3 (zero state) in integral form.

21. Repeat Prob. 3 numerically using $h = 0.025$ s. Use the method of Example 12. Plot $y_1(t)$ and $y_2(t)$ versus t. Let $z_1 = z_2 = u(t)$.
22. Repeat Prob. 3 numerically using the Runge-Fox method. Plot $y_1(t)$ and $y_2(t)$ versus t. Let $z_1 = z_2 = u(t)$.
23. The second-order differential equation describing the mechanical system in Fig. 3.14

$$M\frac{d^2z}{dt^2} + D\frac{dz}{dt} + Kz = u(t)$$

Using state variables $y_1(t) = z(t)$ and $y_2(t) = dz/dt$, find the normal-form equations.

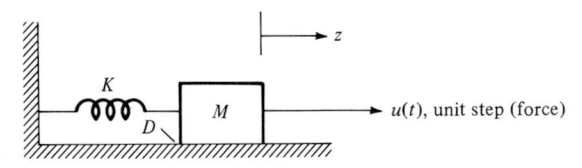

Figure 3.14. Translational mechanical system.

24. Find e^{tA} in closed form for

(a) $A = \begin{bmatrix} 1 & 0 \\ \frac{1}{2} & \frac{1}{2} \end{bmatrix}$ (b) $A = \begin{bmatrix} 1 & -1 \\ 1 & 1 \end{bmatrix}$

(c) $A = \begin{bmatrix} 3 & 2 \\ 0 & -1 \end{bmatrix}$ (d) $A = \begin{bmatrix} 1 & 3 \\ 2 & 1 \end{bmatrix}$

25. Find $q_1(t)$ and $q_2(t)$ for Prob. 24 if $f_1(t) = u(t)$ and $f_2(t) = 0$.

26. Find the normal-form equations in matrix form for the system of Fig. 3.15.

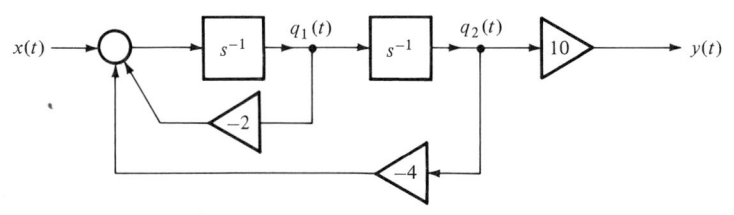

Figure 3.15. Linear system.

27. What value of K makes the system in Fig. 3.16 stable?

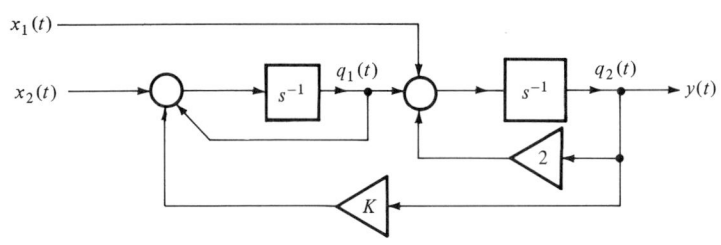

Figure 3.16. Linear system.

References

1. Hayt, W. H., Jr., and Kemmerly, J. E. *Engineering Circuit Analysis*, 3rd ed. New York: McGraw-Hill, 1978.
2. Gabel, R. A., and Roberts, R. A. *Signals and Linear Systems*. New York: John Wiley & Sons, 1980.
3. James, M. L., Smith, G. M., and Wolford, J. C. *Applied Numerical Methods for Digital Computation*, 2nd ed. New York: Harper & Row, 1977.

Chapter 4
Phasors and Fourier Series

Many of us are familiar with the use of phasors in circuit theory when the excitation is *sinusoidal* and we are interested only in the *steady-state* response. This situation covers many applications of a practical nature. Because we are so familiar with the ease of use of phasors, we sometimes forget that in most of circuit theory what the phasor concept really does for us is to convert differential equations into algebraic equations. This is a tremendous advantage, because algebraic equations are usually easy to solve, and once the phasor solution is obtained, it is relatively easy to obtain the steady-state time-domain solution from it. In this chapter we will review the use of phasors and show how to obtain phasor forms from the differential equations of the preceding chapters.

   Next, we will investigate the use of a special trigonometric series, called a Fourier series, to extend the phasor concept so that it can treat *periodic* functions, or systems with periodic excitation. This may be a new topic for many of us, so we will be very careful about the manner in which we treat it. Once we master the concepts of the series itself, the extension of the phasor concept is relatively simple. We will be able to do such things as examine the ripple at the output of a low-pass filter when excited by a rectified version of a sine wave.

4.1 THE PHASOR CONCEPT

Consider a time-invariant, causal linear system starting at $t = 0$. Assume that it has one input and one output and can be described by an Nth order ordinary differential equation. The response is[†] (Eq. 2.64)

$$y(t) = \int_0^t x(\tau)h(t - \tau)\,d\tau + \sum_{n=1}^{N} E_n e^{s_n t} \quad \text{(complete response)} \tag{4.1}$$

The first term is the zero-state response, and the second term is the zero-input response. $x(t)$ is the source function. The unit-impulse response is

$$h(t) = y_{zi}(t)u(t)$$

$$y_{zi}(t) = \sum_{n=1}^{N} D_n e^{s_n t} \quad \text{(subject to the set 2.44)} \tag{4.2}$$

We now assume that the forcing function is sinusoidal; that is

$$x(t) = A \cos(\omega t + \theta)u(t) \tag{4.3}$$

Instead of using $x(t)$ as given by Eq. 4.3, however, let us use the *complex* form

$$x_c(t) = A e^{j(\omega t + \theta)} u(t)$$
$$= [A \cos(\omega t + \theta) + jA \sin(\omega t + \theta)]u(t) \tag{4.4}$$

since it is simple to manipulate. Notice that

$$x(t) = \text{Re}\{x_c(t)\} \tag{4.5}$$

that is, $x(t)$ is the real part of $x_c(t)$. Substituting Eqs. 4.2 and 4.4 into 4.1 gives a *complex* response that we call $y_c(t)$.

$$y_c(t) = \int_0^t A e^{j(\omega\tau + \theta)} \sum_{n=1}^{N} D_n e^{s_n(t-\tau)}\,d\tau + \sum_{n=1}^{N} E_n e^{s_n t}$$

$$= A e^{j\theta} \sum_{n=1}^{N} D_n e^{s_n t} \int_0^t e^{-(s_n - j\omega)\tau}\,d\tau + \sum_{n=1}^{N} E_n e^{s_n t}$$

$$= A e^{j\theta} \sum_{n=1}^{N} D_n e^{s_n t} \left[\frac{e^{-(s_n - j\omega)\tau}}{-s_n + j\omega}\right]_0^t + \sum_{n=1}^{N} E_n e^{s_n t}$$

$$= A e^{j\theta} \sum_{n=1}^{N} D_n e^{s_n t} \frac{e^{-(s_n - j\omega)t} - 1}{-s_n + j\omega} + \sum_{n=1}^{N} E_n e^{s_n t}$$

$$y_c(t) = A e^{j\theta} \sum_{n=1}^{N} \frac{D_n}{s_n - j\omega}[e^{s_n t} - e^{j\omega t}]$$

$$+ \sum_{n=1}^{N} E_n e^{s_n t} \quad \text{(complete complex response)} \tag{4.6}$$

[†] Assuming that there are no repeated roots of the characteristic equation.

We next assume that we are only interested in the steady state, which was defined in Chap. 2 to be the response as t approaches ∞. If the system is stable, then $h(t)$ tends to 0 as t approaches ∞ (also defined in Chap. 2). Examination of Eq. 4.2 reveals that the real part of s_n must be negative if the system we are discussing is to be stable, because in that case every term in Eq. 4.2 contains a decreasing exponential. Thus, Eq. 4.6 becomes for the steady state only

$$y_{css}(t) = Ae^{j(\omega t + \theta)} \sum_{n=1}^{N} \frac{D_n}{-s_n + j\omega}$$

or

$$y_{css}(t) = x_c(t) \left[\sum_{n=1}^{N} \frac{D_n}{-s_n + j\omega} \right] \tag{4.7}$$

Notice that the zero-input response (entirely transient) has completely disappeared, and that part of the zero-state response contributing to the transient response has also disappeared! The term in brackets is in general a complex quantity called the *transfer function* of the system. Notice in Eq. 4.7 that the complex input, $Ae^{j(\omega t + \theta)}$, has apparently gone through the whole process appearing intact as a multiplier in the output. This suggests some interesting possibilities. First of all, there is no need to carry the $e^{j\omega t}$ term all the way through. We can suppress it initially without loss of generality! With $e^{j\omega t}$ suppressed, the quantity $Ae^{j\theta}$ is called the *phasor input*, or phasor excitation, and the resulting output is called the *phasor output*. Thus, in phasor form[†]

$$X(\omega) \equiv Ae^{j\theta} = A\underline{|\theta} \quad \text{(phasor input)} \tag{4.8}$$

$$H(\omega) \equiv \sum_{n=1}^{N} \frac{D_n}{-s_n + j\omega} \quad \text{(transfer function)} \tag{4.9}$$

and the phasor output is (from Eq. 4.7)

$$Y(\omega) = X(\omega)H(\omega) \tag{4.10}$$

Thus, as demonstrated by Eq. 4.10 and Fig. 4.1, the phasor output is the product of the phasor input and the transfer function of the system. Equation 4.10 is an *algebraic equation, not a differential equation*! The relation between Fig. 4.1 and Fig. 1.1 will be examined in the next chapter. The transfer function concept of Eq. 4.10 and Fig. 4.1 is *not* limited to systems described by differential equations. Notice that there is no subscript used on $Y(\omega)$, but this phasor

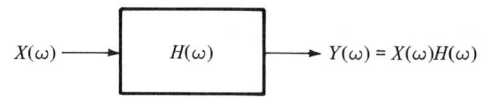

$X(\omega) \longrightarrow \boxed{H(\omega)} \longrightarrow Y(\omega) = X(\omega)H(\omega)$

Figure 4.1. Transfer function concept.

[†] Strictly speaking we should use $X(j\omega)$, but for the sake of simplicity we use $X(\omega)$. Also, it should be mentioned that in power computations with phasors it is common to define the phasor as $(A/\sqrt{2})\underline{|\theta}$. The *magnitude* in this case is an *rms* rather than a *peak* value. See Sec. 4.6.

(*complex*) quantity will shortly lead us to the *real* zero-state response, $y_{zs}(t)$, that is also the steady-state response, $y_{ss}(t)$, for this case (where the sinusoidal excitation has been applied forever). The *complex* zero-state response, $y_{czs}(t)$, is the same as the steady-state response due to a complex exponential forcing function $x_c(t) = Ae^{j(\omega t + \theta)}$ that has been applied forever with no zero-input response. Using this forcing function and Eq. 4.2 in Eq. 2.59 gives

$$y_{czs}(t) = \int_{-\infty}^{t} Ae^{j(\omega\tau + \theta)} \sum_{n=1}^{N} D_n e^{s_n(t-\tau)} d\tau$$

that easily reduces to (real part of $s_n < 0$)

$$y_{czs}(t) = Ae^{j(\omega t + \theta)} \sum_{n=1}^{N} \frac{D_n}{-s_n + j\omega}$$

which is identical to Eq. 4.7.

We still want a true time-domain solution. That is, we still want $y_{ss}(t)$ for $x_{ss}(t) = A\cos(\omega t + \theta)$. This is a linear system, so superposition applies, and *the real part of* $x_{css}(t) = Ae^{j(\omega t + \theta)}$ [that is, $A\cos(\omega t + \theta)$] *produces the real part of* $y_{css}(t)$. Likewise, *the imaginary part of* $x_{css}(t)$ [$A\sin(\omega t + \theta)$] *produces the imaginary part of* $y_{css}(t)$. Therefore, $y_{ss}(t)$ for $x_{ss}(t) = A\cos(\omega t + \theta)$ is given by

$$y_{zs}(t) = y_{ss}(t) = \text{Re}\left\{Ae^{j(\omega t + \theta)} \sum_{n=1}^{N} \frac{D_n}{-s_n + j\omega}\right\} \tag{4.11}$$

If we are working with the phasor forms where $e^{j\omega t}$ has been suppressed, we must reinsert it. Thus, for phasor forms

$$y_{zs}(t) = y_{ss}(t) = \text{Re}\{Y(\omega)e^{j\omega t}\} = \text{Re}\{X(\omega)H(\omega)e^{j\omega t}\} \tag{4.12}$$

Equation 4.7 clearly shows that every term in the complex steady-state output has the same time dependence ($e^{j\omega t}$) for the complex input $x_c(t)$. Thus, differentiating any term in y_{css} simply produces the original term *multiplied by* $j\omega$. Likewise, integration produces the original term *divided by* $j\omega$. This is mentioned because working along these lines gives us a method of finding the transfer function which is simpler than using Eq. 4.9 directly. An example is in order.

Example 1

Consider the *RLC* series circuit excited by $v(t) = V_m\cos(\omega t + \theta)$. The circuit is shown in Fig. 2.12(a). We would like to find the *steady-state* current. The integro-differential equation for this case is

$$V_m\cos(\omega t + \theta) = L\frac{di}{dt} + Ri + \frac{1}{C}\int i\,dt$$

The phasor form of this equation is

$$V(\omega) = V_m\underline{|\theta} = j\omega LI(\omega) + RI(\omega) + \frac{1}{j\omega C}I(\omega)$$

or

$$V_m\lfloor\theta = \left[R + j\omega L + \frac{1}{j\omega C}\right]I(\omega)$$

or

$$I(\omega) = \frac{V_m\lfloor\theta}{R + j\omega L + 1/j\omega C}$$

The transfer function is $Y(\omega)/X(\omega)$ from Eq. 4.10, so $H(\omega) = I(\omega)/V(\omega)$:

$$H(\omega) = \frac{1}{R + j(\omega L - 1/\omega C)} = \frac{1}{Z(\omega)}$$

where $Z(\omega) = V(\omega)/I(\omega)$ is the *ratio of phasor voltage to current*, called the *impedance*, and here is given by

$$Z(\omega) = R + j(\omega L - 1/\omega C) = 1/H(\omega)$$

The phasor form of the circuit is shown in Fig. 4.2. For the phasor forms we are only using *complex algebra*! Now, $I(\omega)$ can also be written as

$$I(\omega) = \frac{V_m\lfloor\theta}{[R^2 + (\omega L - 1/\omega C)^2]^{1/2}\lfloor\theta_z}$$

where

$$\theta_z = \tan^{-1}\left(\frac{\omega L - 1/\omega C}{R}\right)$$

so that

$$I(\omega) = \frac{V_m}{|Z|}\lfloor\theta - \theta_z \quad \textit{(polar form)}$$

The time-domain current is obtained by multiplying $I(\omega)$ by $e^{j\omega t}$ and taking the real part of the resultant:

$$i(t) = \text{Re}\left\{\frac{V_m e^{j(\omega t + \theta - \theta_z)}}{|Z|}\right\}$$

$$i(t) = \frac{V_m}{|Z|}\cos(\omega t + \theta - \theta_z) = i_{zs}(t)$$

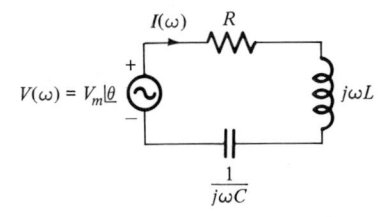

Figure 4.2. Phasor form of the *RLC* series circuit.

The results of this example should be quite familiar to those who have been exposed to a sophomore level course in circuit theory.

Suppose that $R = 1\,\Omega, L = 1\,\text{H}, C = \frac{1}{2}\,\text{F}, V_M = 10\,\text{V}, \theta = 0°$, and $\omega = 2\,\text{rad/s}$ in Example 1. These numbers give

$$\theta_z = \tan^{-1}(1) = 45° \qquad Z_1 = \sqrt{2}\underline{|45°}$$

and

$I(2) = 7.07\underline{|-45°}$ (phasor)
$i(t) = 7.07\cos(2t - 45°)$ (time domain)

Next, suppose that $\omega = 4$ and $V_m = 5$ with all other parameters unchanged. This gives

$$\theta_z = \tan^{-1}(3.5) = 74.05° \qquad Z_2 = 3.64\underline{|74.05°}$$

and

$I(4) = 1.37\underline{|-74.05°}$ (phasor)
$i(t) = 1.37\cos(4t - 74.05°)$ (time domain)

Thus, for *any* ω, it is easy to find the time-domain current using the phasor concept.

If $v(t) = 10\cos(2t) + 5\cos(4t)$, then both of the cosinusoidal terms in the preceding paragraph are being applied to the series RLC circuit, and they are being applied simultaneously to form the input. Notice that (coincidentally) $v(t)$ is *periodic*: that is, $v(t) = v(t \pm nT), n = 0, 1, 2,\ldots$. What is the current now? This is a linear system, so the principle of superposition applies, and the response due to several terms in the input or sources acting simultaneously is the *sum* of the responses due to each acting alone. Therefore

$$i(t) = 7.07\cos(2t - 45°) + 1.37\cos(4t - 74.05°)$$

In general, then, if the source voltage for Example 1 is

$$v(t) = \sum_{n=0}^{N} V_n \cos(n\omega_T t) \quad \text{(periodic function)}$$

then the resulting current is given by

$$i(t) = \sum_{n=0}^{N} \frac{V_n}{Z_n} \cos(n\omega_T t - \theta_n) \quad \text{(periodic function)}$$

where

$$Z_n = \sqrt{R^2 + (n\omega_T L - 1/n\omega_T C)^2}$$

$$\theta_n = \tan^{-1}\left(\frac{n\omega_T L - 1/n\omega_T C}{R}\right)$$

and

$$\omega_T = 2\pi/T \quad \text{(radian) fundamental frequency}$$

We have chosen $v(t)$ here so that it is periodic, but it should be pointed out that the principle of superposition applies in this linear system whether the terms making up $v(t)$ are harmonically related (or periodic, $n = 0, 1, 2, \ldots$) or not. Thus if the source voltage is

$$v(t) = \sum_{n=0}^{N} V_n \cos \omega_n t \quad \text{(nonperiodic function)}$$

where ω_n is completely arbitrary, then the resulting current is

$$i(t) = \sum_{n=0}^{N} \frac{V_n}{Z_n} \cos(\omega_n t - \theta_n) \quad \text{(nonperiodic function)}$$

where

$$Z_n = \sqrt{R^2 + (\omega_n L - 1/\omega_n C)^2}$$

and

$$\theta_n = \tan^{-1}\left(\frac{\omega_n L - 1/\omega_n C}{R}\right)$$

We will lean heavily on these concepts in the following material.

4.2 TRIGONOMETRIC SERIES

Consider the finite, periodic $[f(t) = f(t \pm nT)]$ trigonometric series of cosine terms which are even functions $[f(t) = f(-t)]$, and together represent an even function:

$$f_c(t) = \sum_{n=0}^{N} A_n \cos(n\omega_T t) = \sum_{n=0}^{N} A_n \cos\left(n\frac{2\pi}{T}t\right) \qquad (4.13)$$

where

$$A_0 = \text{average value term}$$
$$A_1 \cos(\omega_T t) = \text{fundamental term}$$
$$A_2 \cos(2\omega_T t) = \text{second harmonic term}, \ldots$$
$$\omega_T = \text{(radian) fundamental frequency}$$
$$T = \frac{2\pi}{\omega_T} = \text{period}$$

We now pose some questions of considerable practical importance. Can the series of Eq. 4.13 be used to approximate an *arbitrary* even function of t, and if so, how accurate will the series representation be? Is the series easy to obtain,

4.2 TRIGONOMETRIC SERIES

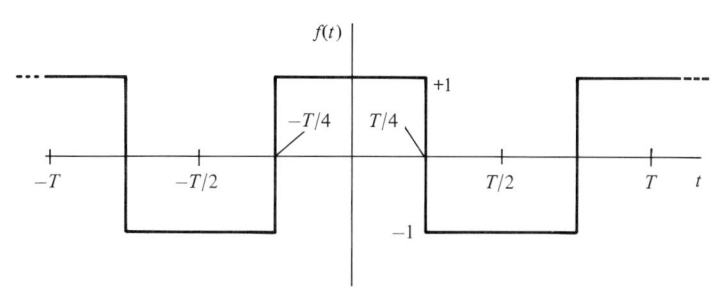

Figure 4.3. Symmetrical square wave, period T.

or, put another way, are the coefficients easy to calculate? Perhaps we can try our ideas on an example. Suppose we want to find a series of the form of Eq. 4.13 for the symmetrical square wave of Fig. 4.3. Perhaps the most obvious idea to occur to a novice investigator is that of forcing Eq. 4.13 to hold at N evenly spaced (or unevenly spaced, in a more general method) "matching points" in $-T/2 \leq t \leq T/2$. Because of the symmetry in this particular example, the matching points can be chosen in $0 \leq t \leq T/2$, and, also, $A_0 = 0$ since $f(t)$ in Fig. 4.3 has zero average value. Thus, we are led to consider that if the matching points are t_m, then

$$f_c(t_m) = \sum_{n=1}^{N} A_n \cos\left(\frac{n2\pi t_m}{T}\right) \qquad m = 1, 2, \ldots, N \qquad (4.14)$$

must hold at N points in $0 \leq t \leq T/2$. So this scheme generates N simultaneous equations (one for each t_m) for the N unknown coefficients A_n. This looks like an attractive scheme. As a matter of fact, the scheme may be formalized in a more general way.[1] This consists of beginning with Eq. 4.13 and multiplying both sides by some "test function" and then integrating over the range of validity. For the matching point scheme the testing function must be $\delta(t - t_m)$. Thus, ($A_0 = 0$)

$$\int_0^{T/2} f_c(t)\delta(t - t_m) \, dt = \sum_{n=1}^{N} A_n \int_0^{T/2} \cos\left(\frac{n2\pi t}{T}\right) \delta(t - t_m) \, dt$$

or, for each $m = 1, 2, \ldots, N$

$$f_c(t_m) = \sum_{n=1}^{N} A_n \cos\left(\frac{n2\pi t_m}{T}\right)$$

which is Eq. 4.14 again. Thus, it seems that the commonsense matching-point idea is only a special case of a general method.

Suppose we choose

$$t_m = (m - \tfrac{1}{2})T/(2N) \qquad m = 1, 2, \ldots, N \qquad (4.15)$$

which gives N evenly spaced matching points in $0 \leq t \leq T/2$. Then, Eq. 4.14 represents N simultaneous equations (one for each m) in the N unknown A_n.

Some finite-order approximations for N even are

$$f_c(t) \approx 1.4142 \cos\left(\frac{2\pi t}{T}\right) + 0 \quad (N = 2)$$

$$f_c(t) \approx 1.3066 \cos\left(\frac{2\pi t}{T}\right) + 0 - 0.5412 \cos\left(\frac{6\pi t}{T}\right) + 0 \quad (N = 4)$$

$$f_c(t) \approx 1.2879 \cos\left(\frac{2\pi t}{T}\right) + 0 - 0.4714 \cos\left(\frac{6\pi t}{T}\right) + 0$$

$$+ 0.3451 \cos\left(\frac{10\pi t}{T}\right) + 0 \quad (N = 6)$$

$$f_c(t) \approx 1.2733 \cos\left(\frac{2\pi t}{T}\right) + 0 - 0.4245 \cos\left(\frac{6\pi t}{T}\right) + 0$$

$$+ 0.2548 \cos\left(\frac{10\pi t}{T}\right) + 0 + \cdots \quad (N = 160)$$

Figure 4.4 shows the results for $N = 2, 4, 6$, and the scheme seems to be working since the approximation becomes better as N increases. On the other hand, there are some obvious features which are disconcerting to say the least. Among these are:

1. A_n depends on N. That is, the A_n must be recalculated each time N is changed.
2. The method becomes unwieldy, requiring the use of a computer, for large N.
3. There is apparently no way of knowing how accurate an approximation we have for a given N.
4. There is no guarantee, in general, that the resulting series will converge, although the series does converge for the example we have chosen.

There is a better way to represent $f_c(t)$.

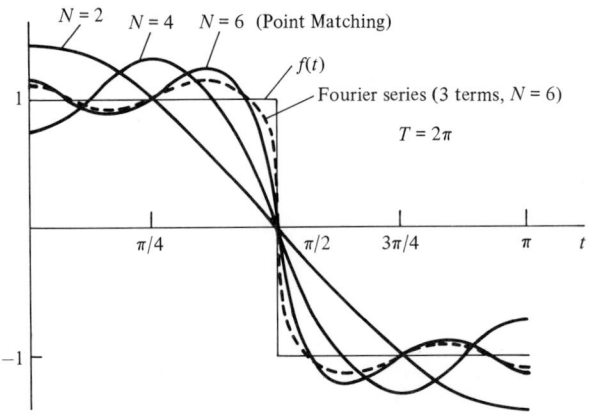

Figure 4.4. Low-order trigonometric series representation of a symmetrical square wave ($T = 2\pi$).

Instead of using point matching where the testing functions are $\delta(t - t_m)$, let us use testing functions which are $\cos(m2\pi t/T)$, $m = 1, 2, \ldots, N$, and integrate over the full range of validity, $-T/2 \le t \le T/2$. In this case

$$\int_{-T/2}^{T/2} f_c(t)\cos\left(\frac{m2\pi t}{T}\right) dt = \sum_{n=0}^{N} A_n \int_{-T/2}^{T/2} \cos\left(\frac{m2\pi t}{T}\right)\cos\left(\frac{n2\pi t}{T}\right) dt$$

The simple change of variable $2\pi t/T = \theta$, $dt = (T/2\pi)\,d\theta$, on the right side gives

$$\int_{-T/2}^{T/2} f_c(t)\cos\left(\frac{m2\pi t}{T}\right) dt = \sum_{n=0}^{N} \frac{A_n T}{2\pi} \int_{-\pi}^{\pi} \cos(m\theta)\cos(n\theta)\,d\theta$$

It is easy to show that the integral on the right is (a) 2π if $m = n = 0$, (b) π if $m = n \ne 0$, and (c) 0 if $m \ne n$. Functions with properties like the cosine just mentioned are said to be *orthogonal over the interval*. There are other functions one may encounter besides the sine and cosine functions which are orthogonal: the Bessel functions, Walsh functions, and the Legendre functions, for example. The last equation can now be written in parts:

$$\int_{-T/2}^{T/2} f_c(t)\,dt = A_0 T \qquad m = n = 0$$

$$\int_{-T/2}^{T/2} f_c(t)\cos\left(\frac{n2\pi t}{T}\right) dt = A_n T/2 \qquad m = n \ne 0$$

$$\int_{-T/2}^{T/2} f_c(t)\cos\left(\frac{m2\pi t}{T}\right) dt = 0 \qquad m \ne n$$

or

$$A_n = \begin{cases} \dfrac{1}{T}\int_{-T/2}^{T/2} f_c(t)\,dt & m = n = 0 \\[1em] \dfrac{2}{T}\int_{-T/2}^{T/2} f_c(t)\cos\left(\dfrac{n2\pi t}{T}\right) dt & m = n \ne 0 \end{cases} \tag{4.16}$$

When Eq. 4.16 is applied to the square wave of Fig. 4.3, we obtain

$$A_n = \begin{cases} 0 & m = 0, 2, 4, 6, \ldots \\ 2\dfrac{\sin(n\pi/2)}{n\pi/2} & m = 1, 3, 5, \ldots \end{cases}$$

The resulting series is

$$f_c(t) = 1.2732 \cos\left(\frac{2\pi t}{T}\right) + 0 - 0.4244 \cos\left(\frac{6\pi t}{T}\right) + 0$$

$$+ 0.2546 \cos\left(\frac{10\pi t}{T}\right) + 0 + \cdots$$

for *all N*.

If we now compare our second attempt to the first, we find that:

1. A_n does not depend on N. The coefficients never change when N is changed. Thinking in terms of simultaneous equations which this method (like the first) generates, we have to invert a matrix which *has only a main diagonal* as a result of the orthogonality of the cosine functions! This type matrix can *always be inverted*, even in infinite order!
2. The method is extremely simple to implement. Equation 4.16 is the formula for A_n for any n.
3. We will show later that this method guarantees that $f_c(t)$ "fits" $f(t)$ with the *least mean square error*.
4. Convergence can be tested (since we know all A_n) by conventional means.

We note in passing that the coefficients, as calculated by the first method, are apparently converging ($N = 160$) to the same values as those given by the second method.

The series as calculated by the second method is known as a *Fourier trigonometric series*. Results are shown in Fig. 4.4 for $N = 6$.

4.3 THE FOURIER TRIGONOMETRIC SERIES

The infinite trigonometric series

$$f(t) = \frac{a_0}{2} + \sum_{n=1}^{\infty} \left[a_n \cos\left(\frac{n2\pi t}{T}\right) + b_n \sin\left(\frac{n2\pi t}{T}\right) \right] \tag{4.17}$$

is called a *Fourier trigonometric series if the coefficients are calculated by the Euler formulas*

$$a_n = \frac{2}{T} \int_{-T/2}^{T/2} f(t) \cos\left(\frac{n2\pi t}{T}\right) dt \tag{4.18}$$

and

$$b_n = \frac{2}{T} \int_{-T/2}^{T/2} f(t) \sin\left(\frac{n2\pi t}{T}\right) dt \tag{4.19}$$

The formula for a_n, as indicated in the previous section, is obtained from Eq. 4.17 by multiplying both sides by $\cos(m2\pi t/T)$ and integrating from $-T/2$ to $T/2$ (or $C - T/2$ to $C + T/2$) on t. The formula for b_n is obtained by multiplying both sides of Eq. 4.17 by $\sin(m2\pi t/T)$ and also integrating over one period (T). Notice that if we set $n = 0$ in Eq. 4.18 then

$$\frac{a_0}{2} = \frac{1}{T} \int_{-T/2}^{T/2} f(t) \, dt$$

which by definition is the average value of $f(t)$.

The Fourier trigonometric series is periodic and is therefore useful for representing an arbitrary periodic function. If we are interested in a function

4.3 THE FOURIER TRIGONOMETRIC SERIES

only within a certain interval, then that interval can be defined to be one period, or less, and the Fourier series can be made to represent the function in that interval.

Example 2

Expand the function $f(t) = 10$ in the interval $0 \le t < 5$ s using cosines only. The function $f(t)$ is shown in Fig. 4.5(a), and a periodic representation of $f(t)$ (in $0 \le t < 5$) is shown in Fig. 4.5(b). This periodic symmetrical square wave has an average value of 0 and a period $T = 20$ s. The average value of any periodic function will always be given by $a_0/2$. This function was examined in the preceding section. Using Eqs. 4.18 and 4.19 we find

$$a_0 = 0$$

$$a_n = 20 \frac{\sin(n\pi/2)}{n\pi/2}$$

$$b_n = 0$$

so

$$f(t) = \sum_{n=1}^{\infty} 20 \frac{\sin(n\pi/2)}{n\pi/2} \cos\left(\frac{n\pi t}{10}\right)$$

or

$$f(t) = \sum_{\substack{n=1 \\ (n \text{ odd})}}^{\infty} \frac{40}{n\pi} (-1)^{(n-1)/2} \cos\left(\frac{n\pi t}{10}\right)$$

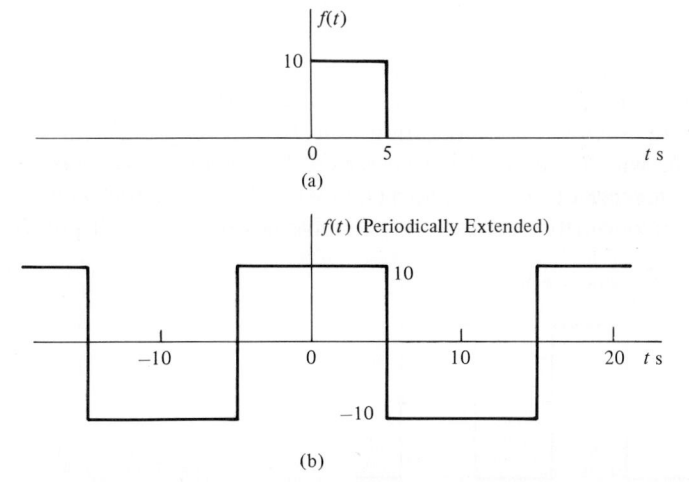

Figure 4.5. (a) $f(t) = 10$. (b) Periodic representation of $f(t)$ with $T = 20$ s, containing cosine terms only ($a_0 = 0$).

Example 3

Expand the function $f(t) = 10$ in the interval $0 < t < 5$ s using sines only. The periodic representation of $f(t)$ is shown in Fig. 4.6. Notice that since the sine functions are odd functions $[f(t) = -f(-t)]$, we must use an odd function for the periodic extension of $f(t)$. Using Eqs. 4.18 and 4.19 we find

$$a_0 = 0, \qquad a_n = 0$$

$$b_n = \frac{20}{n\pi}[1 - \cos(n\pi)]$$

so

$$f(t) = \sum_{n=1}^{\infty} \frac{20}{n\pi}[1 - \cos(n\pi)]\sin\left(\frac{n\pi t}{5}\right)$$

or

$$f(t) = \sum_{\substack{n=1 \\ (n \text{ odd})}}^{\infty} \frac{40}{n\pi} \sin\left(\frac{n\pi t}{5}\right)$$

In leaving Examples 2 and 3, we take note of the fact that in Example 2 we could have chosen *any* period T so long as $T \geq 20$ s, and in Example 3, we could choose any T so long as $T \geq 10$ s. Different values of T will give different series, so neither of these two examples has a unique Fourier series representation.

We saw in the last two examples that a periodic function, or a nonperiodic function that can be periodically extended, can be expressed as an infinite trigonometric series in an infinite variety of ways, but because of the orthogonal properties of the sine and cosine function, the easy way to calculate the coefficients is by way of Eqs. 4.18 and 4.19, giving the Fourier trigonometric series. We now ask the question, Can all periodic functions be expressed as Fourier trigonometric series? The answer is yes for almost all periodic functions of engineering interest. The conditions on $f(t)$ are expressed in the Dirichlet conditions which state that if $f(t)$ (period T) is *bounded* and has no more than a *finite number* of *discontinuities* (jumps) and *maxima* and *minima*, then the Fourier trigonometric series for $f(t)$ converges to $f(t)$ at all points

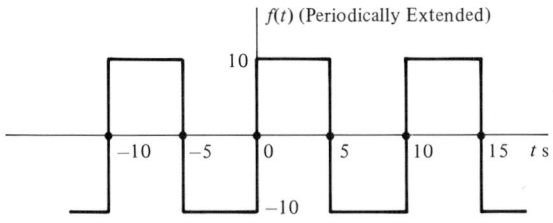

Figure 4.6. Periodic representation of $f(t)$ with $T = 10$ s, containing sine terms only.

where $f(t)$ is continuous. At points where $f(t)$ is discontinuous, the series converges to the *average* of the left- and right-hand limits.

Example 4

Calculate $f(0)$ for the Fourier series in Example 2. Since we found

$$f(t) = \sum_{\substack{n=1 \\ (n \text{ odd})}}^{\infty} \frac{40}{n\pi}(-1)^{(n-1)/2} \cos\left(\frac{n\pi t}{10}\right)$$

for $t = 0$

$$f(0) = \sum_{\substack{n=1 \\ (n \text{ odd})}}^{\infty} \frac{40}{n\pi}(-1)^{(n-1)/2}$$

or

$$f(0) = 10\left(\frac{4}{\pi}\right)\left[1 - \tfrac{1}{3} + \tfrac{1}{5} - \tfrac{1}{7} + \cdots\right]$$

The bracketed term is a well-known series that converges to $\pi/4$. Therefore, $f(0) = 10$, as expected. Now calculate $f(5)$, that is, $f(t)$, at the discontinuity at $t = 5$ s.

$$f(t) = \sum_{\substack{n=1 \\ (n \text{ odd})}}^{\infty} \frac{40}{n\pi}(-1)^{(n-1)/2} \cos\left(\frac{n\pi}{2}\right) = 0$$

which is the average of the left- and right-hand limits at $t = 5$. That is

$$f(5) = \tfrac{1}{2}\left\{\lim_{t \to 5} f(t)\bigg|_{t<5} + \lim_{t \to 5} f(t)\bigg|_{t>5}\right\}$$

$$f(5) = \tfrac{1}{2}[10 - 10] = 0$$

The Fourier trigonometric series can also be written in the form

$$\boxed{f(t) = a_0/2 + \sum_{n=1}^{\infty} d_n \cos\left(\frac{n2\pi t}{T} + \phi_n\right)} \qquad (4.20)$$

where

$$\boxed{d_n = \sqrt{a_n^2 + b_n^2}} \qquad (4.21)$$

and

$$\boxed{\phi_n = -\tan^{-1}(b_n/a_n)} \qquad (4.22)$$

This form is easy to derive from Eqs. 4.17, 4.18, and 4.19 and is particularly well suited for engineering problems where the use of phasors is permissible. This will be demonstrated once more in a short while.

The complex Fourier trigonometric series is perhaps the most general form and is also easy to derive. It is

$$f(t) = \sum_{n=-\infty}^{\infty} C_n e^{jn\omega_T t} \qquad \omega_T = \frac{2\pi}{T} \tag{4.23}$$

where

$$C_n = \frac{1}{T} \int_{-T/2}^{T/2} f(t) e^{-jn\omega_T t} \, dt \tag{4.24}$$

and

$$\begin{aligned} C_n &= \tfrac{1}{2}(a_n - jb_n) \\ C_{-n} &= \tfrac{1}{2}(a_n + jb_n) \end{aligned} \tag{4.25}$$

It should be pointed out that if $f(t)$ is a *real quantity*, as is usually the case, then Eq. 4.23 gives a *real* quantity. We will have need of this complex Fourier series shortly.

4.4 LEAST MEAN SQUARE ERROR

It was mentioned earlier that the Fourier trigonometric series not only has the advantage of ease of coefficient calculation over other possible trigonometric series, but also, with a finite number of terms present, it gives a "best approximation" to the actual function if the criterion for best approximation is *least mean square error*. Consider the complex or exponential form of the Fourier series as given by Eqs. 4.23 and 4.24. Next consider, as an approximation to $f(t)$, a complex series of the form of Eq. 4.23 where the coefficients, which we choose to call D_n, can be determined by any method whatsoever. For example, they can be determined by point matching, mentioned earlier. That is

$$f(t) \approx s_M(t) = \sum_{n=-M}^{M} D_n e^{jn\omega_T t} \tag{4.26}$$

Then, the error is

$$\varepsilon(t) = f(t) - s_M(t) \tag{4.27}$$

and the square of the error is

$$\begin{aligned} [\varepsilon(t)]^2 &= [f(t) - s_M(t)]^2 \\ &= [f(t)]^2 - 2f(t)s_M(t) + [s_M(t)]^2 \\ &= [f(t)]^2 - 2f(t) \sum_{-M}^{M} D_n e^{jn\omega_T t} + \sum_{n=-M}^{M} \sum_{m=-M}^{M} D_n D_m e^{j(m+n)\omega_T t} \end{aligned} \tag{4.28}$$

The integral of the square of the error over a period is

$$\int_{-T/2}^{T/2} [\varepsilon(t)]^2 \, dt = \int_{-T/2}^{T/2} [f(t)]^2 \, dt - 2 \sum_{-M}^{M} D_n \int_{-T/2}^{T/2} f(t) e^{jn\omega Tt} \, dt$$

$$+ \sum_{n=-M}^{M} \sum_{m=-M}^{M} D_n D_m \int_{-T/2}^{T/2} e^{j(m+n)\omega Tt} \, dt$$

It is easy to show that, because of the orthogonality property of sines and cosines, the last integral is 0 for $m \neq n$ and T for $m = -n$, so

$$\int_{-T/2}^{T/2} [\varepsilon(t)]^2 \, dt = \int_{-T/2}^{T/2} [f(t)]^2 \, dt - 2T \sum_{-M}^{M} D_n C_{-n} + T \sum_{-M}^{M} D_n D_{-n} \quad (4.29)$$

where we have also used Eq. 4.24 for C_{-n}.

Equation 4.29 can be written

$$\int_{-T/2}^{T/2} [\varepsilon(t)]^2 \, dt = \int_{-T/2}^{T/2} [f(t)]^2 \, dt$$

$$+ T \sum_{-M}^{M} [D_n D_{-n} - 2 D_n C_{-n} + C_n C_{-n} - C_n C_{-n}]$$

but since

$$\sum_{-M}^{M} C_n D_{-n} = \sum_{-M}^{M} C_{-n} D_n$$

$$\int_{-T/2}^{T/2} [\varepsilon(t)]^2 \, dt = \int_{-T/2}^{T/2} [f(t)]^2 \, dt + T \sum_{-M}^{M} [(D_n - C_n)(D_{-n} - C_{-n}) - C_n C_{-n}]$$

The *root mean square* error (rms error) is

$$\left[\frac{1}{T} \int_{-T/2}^{T/2} [\varepsilon(t)]^2 \, dt \right]^{1/2} = \left[\frac{1}{T} \int_{-T/2}^{T/2} [f(t)]^2 \, dt \right.$$

$$\left. + \sum_{-M}^{M} (D_n - C_n)(D_{-n} - C_{-n}) - \sum_{-M}^{M} C_n C_{-n} \right]^{1/2}$$

(4.30)

and this *is certainly minimum when* $D_n = C_n$ or $D_{-n} = C_{-n}$, that is, *when the* D_n *are the Fourier coefficients given by the Euler formula* (Eq. 4.24). Thus, the least rms error is

$$\varepsilon_{\text{rms}} = \left[\frac{1}{T} \int_{-T/2}^{T/2} [\varepsilon(t)]^2 \, dt \right]^{1/2} = \left[\frac{1}{T} \int_{-T/2}^{T/2} [f(t)]^2 \, dt - \sum_{-M}^{M} C_n C_{-n} \right]^{1/2} \quad (4.31)$$

where, from Eq. 4.25

$$C_n C_{-n} = \tfrac{1}{4}(a_n^2 + b_n^2) \quad (4.32)$$

Example 5

Find the complex Fourier trigonometric series for the triangular waveshape in Fig. 4.7 and then find the rms error for several values of M. Show that this error becomes 0 as $M \to \infty$. In the interval $-T/2 < t < T/2$ $f(t)$ is the straight line $f(t) = 2At/T$, so Eq. 4.24 gives

$$C_n = \frac{1}{T}\int_{-T/2}^{T/2}(2At/T)e^{-jn\omega_T t}\,dt = \frac{2A}{T^2}\int_{-T/2}^{T/2} t e^{-jn\omega_T t}\,dt$$

Expanding the exponential

$$C_n = \frac{2A}{T^2}\int_{-T/2}^{T/2}[t\cos(n\omega_T t) - jt\sin(n\omega_T t)]\,dt$$

but within the limits of integration t is an *odd* function, $\cos(n\omega_T t)$ is an *even* function, and $\sin(n\omega_T t)$ is an *odd* function. Thus, $t\cos(n\omega_T t)$ is an odd function, and $t\sin(n\omega_T t)$ is an *even* function. The integral of an *odd* function over symmetric limits, like $-T/2$, $T/2$ (that is, the area), is 0. In the same way, the integral of an *even* function over symmetric limits is exactly twice the integral over either half of the range ($-T/2$ to 0, or 0 to $T/2$). Quite often symmetry conditions like this exist and should be used to advantage. Thus

$$C_n = -\frac{j4A}{T^2}\int_0^{T/2} t\sin(n\omega_T t)\,dt$$

Integrating by parts, or simply using results in App. C

$$C_n = -\frac{j4A}{T^2}\left[\frac{1}{n^2\omega_T^2}\sin(n\omega_T t) - \frac{t}{n\omega_T}\cos(n\omega_T t)\right]_0^{T/2}$$

Simplifying ($\omega_T = 2\pi/T$)

$$C_n = \frac{-jA}{n\pi}\left[\frac{\sin(n\pi)}{n\pi} - \cos(n\pi)\right] \qquad C_0 = 0$$

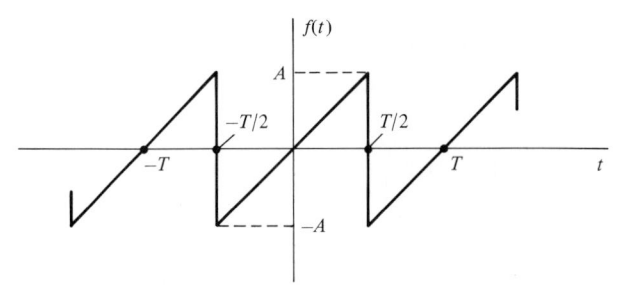

Figure 4.7. Triangular periodic wave for Example 5.

or

$$C_n = \frac{jA}{n\pi}(-1)^n \qquad n = 1, 2, 3, \ldots$$

Thus

$$f(t) = \sum_{-\infty}^{\infty} \frac{jA}{n\pi}(-1)^n e^{jn\omega_T t} \qquad n \neq 0$$

Now

$$C_n C_{-n} = \left[\frac{jA}{n\pi}(-1)^n\right]\left[\frac{jA}{-n\pi}(-1)^{-n}\right] = \left(\frac{A}{n\pi}\right)^2$$

so the rms error is (from Eq. 4.31)

$$\varepsilon_{rms} = \left[\frac{1}{T}\int_{-T/2}^{T/2}[f(t)]^2\,dt - \sum_{-M}^{M}\left(\frac{A}{n\pi}\right)^2\right]^{1/2}$$

$$= \left[\frac{1}{T}\int_{-T/2}^{T/2}(2At/T)^2\,dt - \sum_{-M}^{M}(A/n\pi)^2\right]^{1/2}$$

$$\varepsilon_{rms} = A\left[\frac{1}{3} - \frac{1}{\pi^2}\sum_{-M}^{M}\frac{1}{n^2}\right]^{1/2} = A\left[\frac{1}{3} - \frac{2}{\pi^2}\sum_{n=1}^{M}\frac{1}{n^2}\right]^{1/2}$$

Results are shown in Fig. 4.8. Notice that for $M \to \infty$

$$\sum_{n=1}^{\infty}\frac{1}{n^2} = 1 + \frac{1}{4} + \frac{1}{9} + \cdots = \frac{\pi^2}{6}$$

so

$$\varepsilon_{rms}\big|_{M\to\infty} = 0$$

Notice also that the rms error approaches 0 rather slowly, as do the C_n, and this is simply due to the fact the function we are expanding (straight-line

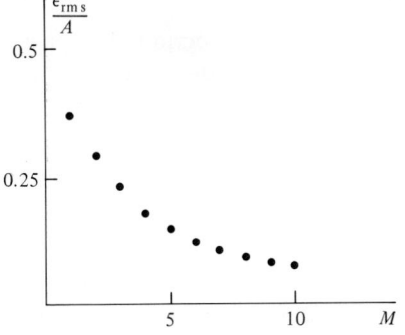

Figure 4.8. Root-mean-square error versus M, Example 5.

segments) is difficult to approximate with sine functions. We might ponder the the results of a Maclaurin series in $-T/2 < t < T/2$ for this function.

4.5 USES OF THE FOURIER SERIES

Whenever the need to expand a function of time, frequency,[†] distance, angle, and so on, arises, a Fourier series should be considered if possible; this is true even if the function is not periodic, so long as it can be periodically extended outside the range of interest. In this material we are primarily interested only in those problems where time is the only independent variable and shall proceed in this direction. We begin by considering the Fourier trigonometric series given by Eq. 4.20 as

$$f(t) = a_0/2 + \sum_{n=1}^{\infty} d_n \cos(n\omega_T t + \phi_n) \tag{4.33}$$

where

$$d_n = \sqrt{a_n^2 + b_n^2} \tag{4.34}$$

and

$$\phi_n = -\tan^{-1}(b_n/a_n) \tag{4.35}$$

We now restrict our attention to those time-invariant linear systems where the excitation is sinusoidal and we are only interested in the steady-state response. These two conditions, as pointed out earlier, are those necessary to be able to use the *phasor* concept.

If the excitation is not sinusoidal, but is periodic, then Eq. 4.33 can be used to represent the input. Inspection of this equation reveals that it is a *sum of sinusoids*, and since we are dealing with a linear system, the principle of *superposition* applies. The average value, or dc, term $a_0/2$ is a cosine with zero frequency. The output, due to an input in the form of Eq. 4.33, is the *sum* of the outputs due to each term in Eq. 4.33 *acting alone*!

The general term in Eq. 4.33 is a sinusoid $(x_0(t) = a_0/2, n = 0)$

$$x_n(t) = d_n \cos(n\omega_T t + \phi_n) \tag{4.36}$$

at a frequency $n\omega_T = n2\pi/T$, where, once again, T is the period. The phasor form of this function of time (considered to be one part of the input) is

$$X(n\omega_T) = d_n e^{j\phi_n} = d_n \lfloor \phi_n \tag{4.37}$$

The phasor output of this system is related to the phasor input through the transfer function, and is given by Eq. 4.10:

$$Y(n\omega_T) = X(n\omega_T)H(n\omega_T) \qquad (\omega = n\omega_T) \tag{4.38}$$

[†] The need for a periodic function of *frequency* arises in connection with the *discrete Fourier transform* (DFT) in the next chapter.

4.5 USES OF THE FOURIER SERIES

Since we are using the cosine function as the "reference" (so to speak) in Eq. 4.36 (and originally in Eq. 4.3), in order to recover the time-domain form $y_{zs}(t)$, we must reinsert the suppressed $e^{jn\omega_T t}$ and take the real part, as in Eq. 4.12. Thus

$$y_{nzs}(t) = \text{Re}\{Y(n\omega_T)e^{jn\omega_T t}\} = \text{Re}\{X(n\omega_T)H(n\omega_T)e^{jn\omega_T t}\}$$

or

$$y_{nzs}(t) = \text{Re}\{d_n \lfloor \phi_n \, H(n\omega_T)e^{jn\omega_T t}\}$$

or

$$y_{nzs}(t) = \text{Re}\{d_n H(n\omega_T)e^{j(n\omega_T t + \phi_n)}\} \tag{4.39}$$

The response due to *all* of the sinusoidal terms in the input is, by superposition

$$y_{zs}(t) = \frac{a_0}{2}H(0) + \sum_{n=1}^{\infty} \text{Re}\{d_n |H(n\omega_T t)| e^{j(n\omega_T t + \phi_n + \theta_n)}\}$$

or, since d_n will be real if $x(t)$ is real

$$\boxed{y_{zs}(t) = \frac{a_0}{2}H(0) + \sum_{n=1}^{\infty} d_n |H(n\omega_T)| \cos(n\omega_T t + \phi_n + \theta_n)} \tag{4.40}$$

where

$$H(n\omega_T) = |H(n\omega_T)| \lfloor \theta_n$$

Thus, we see that the only differences between this problem and that for single-frequency (ω) sinusoidal excitation are (1) the general frequency is $n\omega_T$ instead of ω and (2) we are dealing with a sum of terms rather than one. Superposition has really aided us. A formal procedure for analyzing systems of this type consists of the following:

1. Given $x(t)$, find a_n, b_n, d_n, and ϕ_n with Eqs. 4.18, 4.19, 4.34, and 4.35, respectively.
2. Find the differential equation (if not given) governing the system. Replace derivatives of $y(t)$ with $j\omega Y(\omega)$, and replace integrals of $y(t)$ with $1/(j\omega)Y(\omega)$. Analyze the system in phasor form.
3. Find the transfer function $H(\omega) = Y(\omega)/X(\omega)$.
4. Replace ω with $n\omega_T$.
5. Substitute into Eq. 4.40.

Example 6

After passing through a full-wave rectifier, the negative part of a sinusoid becomes positive. That is, the effect on $V_m \cos(\omega_T t)$ is to convert it to $V_m |\cos(\omega_T t)|$. The rectified voltage is applied to a low-pass filter to produce a predominantly

142 PHASORS AND FOURIER SERIES

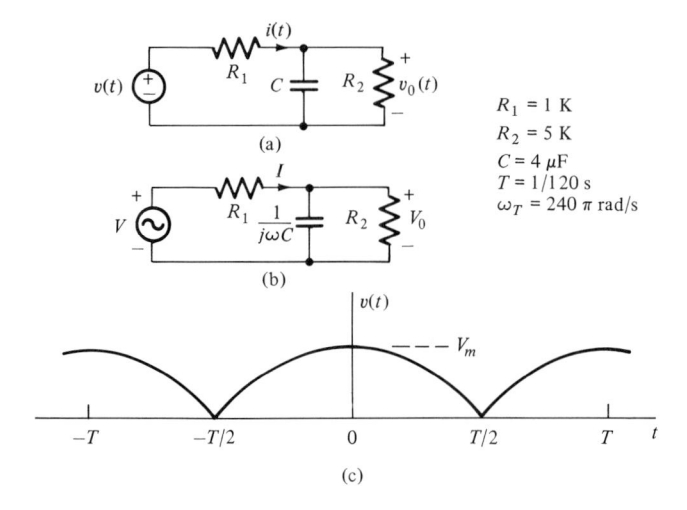

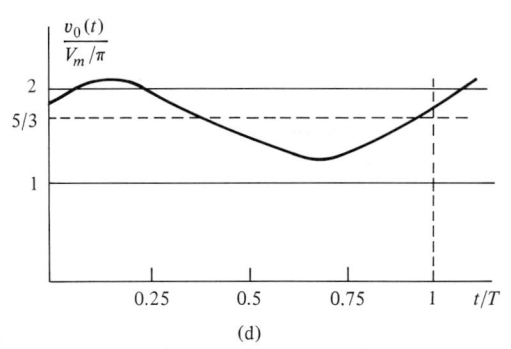

Figure 4.9. (a) Circuit to be analyzed for Example 6. (b) Phasor form of the circuit. (c) $v(t) = V_m|\cos(\omega_T t)|$, full-wave-rectified voltage input. (d) Normalized output voltage.

$R_1 = 1$ K
$R_2 = 5$ K
$C = 4$ μF
$T = 1/120$ s
$\omega_T = 240\pi$ rad/s

dc signal (with ripple) as shown in Fig. 4.9. These are the elements of a simple ac to dc power supply. Find $v_0(t)$.

(1) Since $v(t)$ is an even function, $b_n = 0$, $\phi_n = 0$, $d_n = a_n$, and

$$a_n = \frac{4V_m}{T}\int_0^{T/2} \cos(\omega_T t/2)\cos(n\omega_T t)\, dt$$

Notice carefully that $v(t) = V_m\cos(\omega_T t/2)$ in $-T/2 \le t \le T/2$. The fundamental frequency of the *periodic* rectified voltage is 120 Hz, but before being rectified, the fundamental frequency was 60 Hz, the line frequency. Proceeding:

$$a_n = \frac{2V_m}{T}\int_0^{T/2}\cos[(n+1/2)\omega_T t]\, dt + \frac{2V_m}{T}\int_0^{T/2}\cos[(n-1/2)\omega_T t]\, dt$$

This simplifies to

$$a_n = \frac{V_m}{\pi} \frac{(-1)^{n+1}}{n^2 - 1/4} = \frac{4V_m}{\pi} \frac{(-1)^n}{1 - 4n^2}$$

(2) $I = \dfrac{V}{R_1 + \dfrac{R_2 1/j\omega C}{R_2 + 1/j\omega C}}$ $V_0 = I \dfrac{R_2 1/j\omega C}{R_2 + 1/j\omega C}$

Solving algebraically

$$V_0 = V \frac{R_2}{R_1 + R_2 + j\omega C R_1 R_2}$$

(3) $H(\omega) = \dfrac{V_0}{V} = \dfrac{R_2}{R_1 + R_2 + j\omega C R_1 R_2}$

or, in polar form

$$H(\omega) = \frac{R_2}{\sqrt{(R_1 + R_2)^2 + (\omega C R_1 R_2)^2}} \exp\left(-j \tan^{-1} \frac{\omega C R_1 R_2}{R_1 + R_2}\right)$$

(4) $H(n\omega_T) = \dfrac{R_2}{\sqrt{(R_1 + R_2)^2 + (n\omega_T C R_1 R_2)^2}} \exp\left(-j \tan^{-1} \dfrac{n\omega_T C R_1 R_2}{R_1 + R_2}\right)$

(5) $v_0(t) = \dfrac{2V_m}{\pi} \dfrac{R_2}{R_1 + R_2}$

$$+ \frac{4V_m R_2}{\pi} \sum_{n=1}^{\infty} \frac{(-1)^n}{1 - 4n^2} \operatorname{Re}\left\{\frac{\exp\left[j\left(n\omega_T t - \tan^{-1}\dfrac{n\omega_T C R_1 R_2}{R_1 + R_2}\right)\right]}{\sqrt{(R_1 + R_2)^2 + (n\omega_T C R_1 R_2)^2}}\right\}$$

or

$$v_0(t) = \frac{2V_m}{\pi} \frac{R_2}{R_1 + R_2}$$

$$+ \frac{4V_m R_2}{\pi} \sum_{n=1}^{\infty} \frac{(-1)^n}{1 - 4n^2} \frac{\cos\left(n\omega_T t - \tan^{-1}\dfrac{n\omega_T C R_1 R_2}{R_1 + R_2}\right)}{\sqrt{(R_1 + R_2)^2 + (n\omega_T C R_1 R_2)^2}}$$

or

$$\frac{v_0(t)}{V_m/\pi} = 1.667 + 3.333 \sum_{n=1}^{\infty} \frac{(-1)^n}{1 - 4n^2} \frac{\cos(n\omega_T t - \tan^{-1} 0.8n\pi)}{\sqrt{1 + (0.8n\pi)^2}}$$

The average value of $v_0(t)/(V_m/\pi)$ is $1.667 = \tfrac{5}{3}$. A plot of $(v_0/V_m/\pi)$ versus t is shown in Fig. 4.9(d). Notice that the ripple is large, and this would be a poor dc power supply. We need to improve the filter by designing one (more complex)

144 PHASORS AND FOURIER SERIES

which does a better job of enhancing the dc term and rejecting all the ac terms[†] (the fundamental and all harmonics). It should be mentioned in passing that (in Chap. 5) we can find a closed-form solution to this problem. It is much easier to use with numbers:

$$\frac{v_0(t)}{V_m/\pi} = \begin{cases} 0.796e^{-2.5t/T} + 1.015\cos(\pi t/T) + 1.276\sin(\pi t/T) & 0 \le t \le T/2 \\ 9.700e^{-2.5t/T} - 1.015\cos(\pi t/T) - 1.276\sin(\pi t/T) & T/2 \le t \le T \end{cases}$$

Example 7

A class C amplifier is biased so that the output current flowing into a parallel RLC network is just the tip of a cosine waveform as shown in Fig. 4.10. If the network is tuned (i.e., resonant) to the fundamental frequency, find the output voltage. We begin by finding an equation for $i(t)$ in $-10^{-5}/8 \le t \le 10^{-5}/8$.

$$i(t) = A\cos(\omega_T t) - A_0 \qquad -10^{-5}/8 \le t \le 10^{-5}/8$$
$$i(t) = A\cos(2\pi \times 10^5 t) - A_0$$
$$i(10^{-5}/8) = 0 = A\cos(\pi/4) - A_0 = A/\sqrt{2} - A_0$$
$$i(0) = 1 = A - A_0$$

Solving the last two equations for A and A_0

$$A = \frac{\sqrt{2}}{\sqrt{2}-1} \qquad A_0 = \frac{1}{\sqrt{2}-1}$$

Therefore

$$i(t) = \frac{\sqrt{2}}{\sqrt{2}-1}\cos(\omega_T t) - \frac{1}{\sqrt{2}-1} \qquad -T/8 \le t \le T/8$$

(1) Solving for a_n ($b_n = 0$), we obtain (after some labor)

$$a_n = \frac{A}{\pi\sqrt{2}} \frac{2n\sin(n\pi/4) - 2\cos(n\pi/4)}{n^2 - 1} - \frac{A_0}{2} \frac{\sin(n\pi/4)}{n\pi/4}$$

(a)

(b)

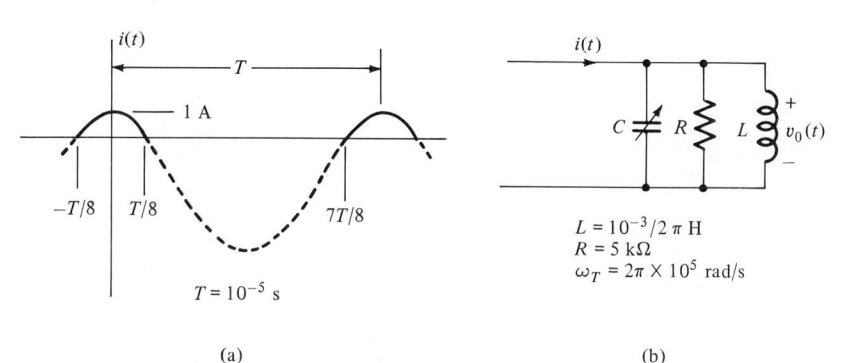

$L = 10^{-3}/2\pi$ H
$R = 5\ \text{k}\Omega$
$\omega_T = 2\pi \times 10^5$ rad/s

Figure 4.10. (a) Tip of cosine pulse applied to an *RLC* parallel network. (b) *RLC* network.

[†] See Sec. 5.5.

(2) $V_0(\omega) = \dfrac{I(\omega)}{\dfrac{1}{R} + \dfrac{1}{j\omega L} + j\omega C}$

(3) $H(\omega) = Z(\omega) = \dfrac{1}{\dfrac{1}{R} + \dfrac{1}{j\omega L} + j\omega C}$

(4) $H(n\omega_T) = \dfrac{1}{\dfrac{1}{R} + \dfrac{1}{jn\omega_T L} + jn\omega_T C}$

If the resonant frequency (tuned frequency) is ω_T, then

$$\omega_T = \dfrac{1}{\sqrt{LC}}$$

or

$$C = \dfrac{1}{L\omega_T^2} = \dfrac{10^{-6}}{20\pi} \text{ F} \qquad (Q = R/\omega L = \omega RC = 50 \text{ at } \omega = \omega_T)$$

(5) Substitute into Eq. 4.40 and use the given numbers for R, L, C, A, and A_0. This gives

$$v_0(t) = 1551 \cos(\omega_T t) + 17.1 \cos(2\omega_T t - 89.24°)$$
$$+ 6.8 \cos(3\omega_T t - 89.57°) + 2.7 \cos(4\omega_T t - 89.69°) + \cdots \text{V}$$

Since we are trying to recover only the fundamental term in this application, we have accomplished our goal with some small harmonic distortion. The distortion can be reduced by decreasing the bandwidth or increasing the Q (increasing R). We will mention this again in Sec. 5.5.

Suppose we want to make a *frequency doubler* by changing C so that resonance occurs at $2\omega_T$. In this case $C = 10^{-6}/80\pi$, and the Q is halved. Repeating steps (1) through (5), we have

$$v_0(t) = 41.4 \cos(\omega_T t + 88.5°) + 1281 \cos(2\omega_T t)$$
$$+ 43.4 \cos(3\omega_T t - 87.25°) + 13.7 \cos(4\omega_T t - 88.5°) + \cdots \text{V}$$

and, indeed, the second harmonic is the largest term.

In order to improve this system we must either decrease the bandwidth (increase Q) or completely redesign the bandpass filter (see Example 6, Chap. 5).

4.6 AVERAGE POWER

It is convenient at this point to review what is meant by average power. Consider a current $i(t)$ entering the positive terminal of a block which represents a combination of passive elements R, L, and C with the voltage $v(t)$ as labeled

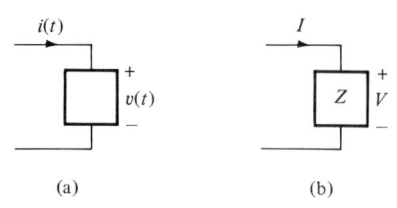

Figure 4.11. (a) Group of elements represented by a block. (b) Block represented by an impedance for the phasor form, $I = V/Z$.

in Fig. 4.11. For single-frequency sinusoidal excitation in the steady state (for which the phasor concept applies), we may use classical methods *or* phasor methods to solve for the current. If

$$v(t) = V_m \cos(\omega t + \phi) \tag{4.41}$$

then

$$i(t) = \frac{V_m}{|Z|} \cos(\omega t + \phi - \alpha) \tag{4.42}$$

where

$$Z = |Z|e^{j\alpha} = |Z|\underline{/\alpha} \tag{4.43}$$

The instantaneous power is

$$p(t) = v(t)i(t) = \frac{V_m^2}{|Z|} \cos(\omega t + \phi)\cos(\omega t + \phi - \alpha) \tag{4.44}$$

When we speak of the average value of a quantity, we must specify an *interval*. In this case we mean the *time average value* of power over an interval of *time*. Since sinusoids are periodic, we mean the time average over one period (T). It is obvious that the time interval can also be any integer number of periods, and the average power will be unchanged. Thus, using $\langle p \rangle$ to indicate the average value

$$\langle p \rangle = \frac{1}{T} \int_{-T/2}^{T/2} p(t)\, dt = \frac{1}{T} \int_{-T/2}^{T/2} \frac{V_m^2}{|Z|} \cos(\omega t + \phi)\cos(\omega t + \phi - \alpha)\, dt \tag{4.45}$$

Since $\cos(A)\cos(B) = \tfrac{1}{2}\cos(A+B) + \tfrac{1}{2}\cos(A-B)$

$$\langle p \rangle = \frac{1}{T} \frac{V_m^2}{2|Z|} \left\{ \int_{-T/2}^{T/2} \cos(2\omega t + 2\phi - \alpha)\, dt + \int_{-T/2}^{T/2} \cos(\alpha)\, dt \right\}$$

The first integral is 0, while the second is $T \cos(\alpha)$. Thus

$$\langle p \rangle = \frac{V_m^2}{2|Z|} \cos(\alpha) \tag{4.46}$$

4.6 AVERAGE POWER

On the other hand, we can easily calculate the time average power from the phasors:

$$V = V_m \underline{|\phi}$$

$$I = \frac{V}{Z} = \frac{V_m \underline{|\phi}}{|Z|\underline{|\alpha}} = \frac{V_m}{|Z|} \underline{|\phi - \alpha}$$

and

$$\langle p \rangle = \tfrac{1}{2} \operatorname{Re}\{VI^*\} \tag{4.47}$$

where I^* means *complex conjugate of I*. Thus

$$\langle p \rangle = \frac{1}{2} \operatorname{Re}\left\{ V_m \underline{|\phi} \, \frac{V_m}{|Z|} \underline{|-\phi + \alpha} \right\}$$

$$= \frac{1}{2} \operatorname{Re}\left\{ \frac{V_m^2}{|Z|} \underline{|\alpha} \right\} = \frac{V_m^2}{2|Z|} \operatorname{Re}\{\underline{|\alpha}\} = \frac{V_m^2}{2|Z|} \operatorname{Re}\{e^{j\alpha}\}$$

$$\langle p \rangle = \frac{V_m^2}{2|Z|} \cos(\alpha)$$

as before.†

What is the time average power for a situation like that in Fig. 4.11(a) if the voltage is periodic but not sinusoidal: that is, where a Fourier trigonometric series can be employed?

We have

$$v(t) = a_0/2 + \sum_{n=1}^{\infty} d_n \cos(n\omega_T t + \phi_n)$$

and

$$i(t) = \frac{a_0}{2} H(0) + \sum_{n=1}^{\infty} d_n |H(n\omega_T)| \cos(n\omega_T t + \phi_n + \theta_n)$$

from Eqs. 4.20 and 4.40, respectively. In the present problem $H(\omega) = 1/Z(\omega)$, so

$$i(t) = \frac{a_0}{2R(0)} + \sum_{n=1}^{\infty} \frac{d_n}{|Z(n\omega_T)|} \cos(n\omega_T t + \phi_n - \alpha_n)$$

where Eq. 4.43 has been used. Notice also that $Z(0) = R(0)$ since we are in the steady-state condition where (for $\omega = 0$) all inductors behave as short circuits, and all capacitors behave as open circuits, leaving only combinations of resistors. Since we intend to multiply $v(t)$ and $i(t)$ to find $p(t)$, we can avoid much confusion if $i(t)$ is written as

$$i(t) = \frac{a_0}{2R(0)} + \sum_{m=1}^{\infty} \frac{d_m}{|Z(m\omega_T)|} \cos(m\omega_T t + \phi_m - \alpha_m)$$

---

† Had we been using rms (or effective) values instead of peak values for the phasors, then the factor $\tfrac{1}{2}$ would not appear here or in Eq. 4.47. See the equation preceding 4.51.

Thus

$$p(t) = \frac{a_0^2}{4R(0)} + \frac{a_0}{2} \sum_{m=1}^{\infty} \frac{d_m}{|Z(m\omega_T)|} \cos(m\omega_T t + \phi_m - \alpha_m)$$

$$+ \frac{a_0}{2R(0)} \sum_{n=1}^{\infty} d_n \cos(n\omega_T t + \phi_n)$$

$$+ \sum_{n=1}^{\infty} \sum_{m=1}^{\infty} d_n \cos(n\omega_T T + \phi_n) \frac{d_m}{|Z(m\omega_T)|} \cos(m\omega_T t + \phi_m - \alpha_m)$$

Now, the average value of the first term is $a_0^2/4R(0)$, while the average values of the second and third terms, taken over an integer multiple of periods, are clearly 0. Thus

$$\langle p \rangle = \frac{a_0^2}{4R(0)}$$

$$+ \frac{1}{T} \int_{-T/2}^{T/2} \sum_{n=1}^{\infty} \sum_{m=1}^{\infty} \frac{d_n d_m}{|Z(m\omega_T)|} \cos(n\omega_T t + \phi_n) \cos(m\omega_T t + \phi_m - \alpha_m) \, dt$$

and

$$\cos(n\omega_T t + \phi_n)\cos(m\omega_T t + \phi_m - \alpha_m) = \tfrac{1}{2}\cos(n\omega_T t + \phi_n + m\omega_T t + \phi_m - \alpha_m)$$
$$+ \tfrac{1}{2}\cos(n\omega_T t + \phi_n - m\omega_T t - \phi_m + \alpha_m)$$

These cosine terms integrate to 0 also, except for the case when $m = n$. In this case the product of the cosines becomes

$$\tfrac{1}{2}\cos(2n\omega_T t + 2\phi_n - \alpha_n) + \tfrac{1}{2}\cos(\alpha_n)$$

the first of which still integrates to 0, while the second integrates to $(T/2)\cos(\alpha_n)$. Thus, the double infinite sum becomes a single infinite sum, and

$$\langle p \rangle = \frac{a_0^2}{4R(0)} + \sum_{n=1}^{\infty} \frac{d_n^2}{2|Z(n\omega_T)|} \cos(\alpha_n) \qquad (4.48)$$

The first term in Eq. 4.48 is the average power in an arrangement of resistances with terminal value $R(0)$, and with a voltage $a_0/2$ and current $a_0/2R(0)$, while the remaining terms (in the infinite series) are each like that in Eq. 4.46. *In this special case superposition is applicable to power*! This is a rather amazing result, and gives us another formula (other than the definition) for $\langle p \rangle$.

As a matter of fact, the average power absorbed in a resistor R due to a voltage of the form[†]

$$v(t) = \sum_{n=1}^{\infty} a_n \cos(\omega_n t)$$

when defined according to

$$\langle p \rangle = \lim_{\tau \to \infty} \frac{1}{\tau} \int_{-\tau/2}^{\tau/2} \frac{v^2(t)}{R}$$

[†] Not necessarily a Fourier series.

can easily be shown to be

$$\langle p \rangle = \frac{1}{2R} \sum_{n=1}^{\infty} a_n^2 \qquad (4.49)$$

so long as $\omega_n \neq \omega_{n+1}$ for any n; that is, so long as the sinusoids have *different periods* (or frequencies). They are not necessarily harmonically related. The *effective value* of the voltage is that value, which, when squared and divided by R, gives the same average power as Eq. 4.49. Thus

$$\langle p \rangle = \frac{V_{\text{eff}}^2}{R} \qquad (4.50)$$

Equating results in Eqs. 4.49 and 4.50

$$V_{\text{eff}}^2 = \tfrac{1}{2} \sum_{n=1}^{\infty} a_n^2 = \sum_{n=1}^{\infty} a_n^2/2$$

or

$$V_{\text{eff}} = \left(\sum_{n=1}^{\infty} a_n^2/2 \right)^{1/2}$$

For a *single* sinusoid, Eqs. 4.49 and 4.50 give

$$V_{\text{eff}} = a_n/\sqrt{2}$$

Therefore

$$V_{\text{eff}} = \sqrt{V_{1\text{eff}}^2 + V_{2\text{eff}}^2 + V_{3\text{eff}}^2 + \cdots} \qquad (4.51)$$

Results in this paragraph (Eqs. 4.49 and 4.51) apply even if the sinusoids have arbitrary phase angles.

Example 8

The waveform of Example 2 is applied as the voltage to the series combination of a 1-Ω resistor and a 1-H inductor. Find $\langle p \rangle$ using Eq. 4.48. We begin by finding $i(t)$ from

$$v(t) = \sum_{\substack{n=1 \\ (n \text{ odd})}}^{\infty} \frac{40}{n\pi} (-1)^{(n-1)/2} \cos\left(\frac{n\pi t}{10}\right) \qquad \omega_T = \frac{\pi}{10}$$

and

$$Z = R + j\omega L = \sqrt{R^2 + \omega^2 L^2} \, \lfloor \tan^{-1} \omega L/R$$

Therefore

$$i(t) = \sum_{\substack{n=1 \\ (n \text{ odd})}}^{\infty} \frac{40}{n\pi} \frac{(-1)^{(n-1)/2}}{\sqrt{R^2 + (n\omega_T L)^2}} \cos(n\omega_T t - \tan^{-1} n\omega_T L/R)$$

Using Eq. 4.48

$$\langle p \rangle = \sum_{\substack{n=1 \\ (n \text{ odd})}}^{\infty} \left(\frac{40}{n\pi}\right)^2 \frac{1}{2\sqrt{R^2 + (n\omega_T L)^2}} \cos(\tan^{-1} n\omega_T L/R)$$

$$\langle p \rangle = 800 \sum_{\substack{n=1 \\ (n \text{ odd})}}^{\infty} \frac{1}{n^2 \pi^2} \frac{1}{\sqrt{R^2 + (n\omega_T L)^2}} \frac{R}{\sqrt{R^2 + (n\omega_T L)^2}} \quad (4.52)$$

$$\langle p \rangle = 800 \sum_{\substack{n=1 \\ (n \text{ odd})}}^{\infty} \frac{1}{n^2 \pi^2} \frac{1}{1 + (n\pi/10)^2}$$

$$\langle p \rangle = 80 \text{ W}$$

On the other hand

$$\langle p \rangle = I_{\text{eff}}^2 R = I_{\text{rms}}^2 R$$

and it is easy to see from Eq. 4.52 that this will lead to the same result.

Example 9

What average power is dissipated in a 10-Ω resistor if

$$i(t) = 10 + 5\cos(3t) + 2\cos(\pi t) + \cos(1000t)$$

We have

$$I_{\text{eff}} = \sqrt{100 + \tfrac{25}{2} + \tfrac{4}{2} + \tfrac{1}{2}} = 10.72$$

$$\langle p \rangle = (I_{\text{eff}})^2 10 = 1150 \text{ W}$$

Example 10

The voltage waveform in Example 8 is applied to a 10-Ω resistor. Find the average power in two ways. First, we have by definition

$$\langle p \rangle = \frac{1}{T} \int_{-T/2}^{T/2} p(t)\, dt = \frac{1}{T} \int_{-T/2}^{T/2} \frac{v^2(t)}{R}\, dt$$

$$\langle p \rangle = \frac{1}{20} \int_{-10}^{10} \frac{100}{10}\, dt = 10 \text{ W}$$

Using Eqs. 4.50 and 4.51

$$\langle p \rangle = \frac{1}{10} V_{\text{eff}}^2 = \frac{1}{10} \sum_{\substack{n=1 \\ (n \text{ odd})}}^{\infty} a_n^2/2 = \frac{1}{20} \sum_{\substack{n=1 \\ (n \text{ odd})}}^{\infty} \frac{1600}{n^2 \pi^2}$$

$$\langle p \rangle = \frac{1600}{20\pi^2} \sum_{\substack{n=1 \\ (n \text{ odd})}}^{\infty} \frac{1}{n^2} = \frac{80}{\pi^2}\left[1 + \frac{1}{9} + \frac{1}{25} + \cdots\right]$$

$$\langle p \rangle = \frac{80}{\pi^2}\left[\frac{\pi^2}{8}\right] = 10 \text{ W}$$

If we now return to Eq. 4.31, we see that the rms error can be interpreted as the square root of the difference between the average power and the average power for the first M terms in the Fourier series expansion of $f(t)$ when $f(t)$ is taken to be the voltage or current associated with a 1-Ω resistor.

4.7 THE DISCRETE FOURIER SERIES

The complex Fourier series is given by Eqs. 4.23 and 4.24 for *continuous* signals:

$$f(t) = \sum_{m=-\infty}^{\infty} C_m e^{jm2\pi t/T} \quad (\omega_T = 2\pi/T) \tag{4.23}$$

$$C_m = \frac{1}{T}\int_{-T/2}^{T/2} f(t) e^{-jm2\pi t/T}\, dt \tag{4.24}$$

The index has been changed from n to m for reasons that will become obvious in what follows. The relationship between these equations is based on the orthogonality of exponentials with imaginary arguments. That is

$$\int_{-T/2}^{T/2} e^{jn2\pi t/T} e^{-jp2\pi t/T}\, dt \quad \begin{cases} n = 0, 1, 2, \ldots \\ p = 0, 1, 2, \ldots \end{cases}$$

$$= \int_{-T/2}^{T/2} e^{j(n-p)2\pi t/T}\, dt = \begin{cases} T & n = p \\ 0 & n \neq p \end{cases} \tag{4.53}$$

This result is more obvious than its counterpart for the sine and cosine functions, but results in the same advantages.

In the same way, when dealing with discrete samples (and we must shortly), a discrete version of orthogonality exists:

$$\sum_{m=0}^{N-1} e^{j2\pi(n-p)m/N} = \begin{cases} N & n = p \\ 0 & n \neq p \end{cases} \tag{4.54}$$

where $n = 0, 1, 2, \ldots$, and $p = 0, 1, 2, \ldots$. This result is easily established by writing it as the sum of a geometric progression:

$$\sum_{m=0}^{N-1} e^{j2\pi(n-p)m/N} = \frac{1 - e^{j2\pi(n-p)}}{1 - e^{j2\pi(n-p)/N}} = \begin{cases} N & n = p \\ 0 & n \neq p \end{cases} \tag{4.55}$$

Also, we need to consider the sum

$$\sum_{m=-(N-1)/2}^{(N-1)/2} e^{j2\pi(n-p)m/N}$$

Changing the index, $m = q - (N-1)/2$, gives

$$\sum_{q=0}^{N-1} e^{j2\pi(n-p)q/N} e^{+j2\pi(n-p)(N-1)/(2N)} = e^{+j2\pi(n-p)(N-1)/(2N)} \sum_{q=0}^{N-1} e^{j2\pi(n-p)q/N}$$

$$= \begin{cases} N & n = p \\ 0 & n \neq p \end{cases} \tag{4.56}$$

We used Eq. 4.54 to establish the last result.

There are several things to consider when we want to actually use Eqs. 4.23 and 4.24 with *numbers*. First of all, the sum in Eq. 4.23 cannot include an infinite number of terms. It must be a *finite* sum. Secondly, the integral in Eq. 4.24 must be approximated by a *finite* sum. Thirdly, we can only obtain numbers for *samples* of $f(t)$. For the sake of simplicity we take these samples at $t = 0$, $\pm T/N$, $\pm 2T/N, \ldots, \pm(N-1)T/(2N)$ for N samples. Notice that

$$(N-1)T/(2N) = (1 - 1/N)T/2 \to T/2$$

for $N \to \infty$, so we are taking these samples in the single period $-T/2 < t < T/2$. The N samples are $f(nT/N)$, $n = 0, \pm 1, \pm 2, \ldots, \pm(N-1)/2$. The present formulation requires that N be odd, but we will remove this restriction later. We can do this because the limits in Eq. 4.24 need not be $-T/2, T/2$. The only requirement is that one complete period is covered by the limits, and the limits $0, T$ (for example) will serve as well.

Consider Fig. 4.12. We want to approximate C_m in Eq. 4.24 with a finite sum. It is obvious from the definition of an integral as the limit of a sum that

$$C_m = \frac{1}{T} \int_{-T/2}^{T/2} f(t)e^{-jm2\pi t/T}\, dt \approx \frac{1}{T} \sum_{n=-(N-1)/2}^{(N-1)/2} f(nT/N)e^{-jm2\pi n/N}(T/N)$$

since $t = nT/N$ at the sampling instants. Thus

$$C_m \approx \frac{1}{N} \sum_{n=-(N-1)/2}^{(N-1)/2} f(nT/N)e^{-j2\pi mn/N} \tag{4.57}$$

This approximation becomes exact in the limit as $N \to \infty$.

Remembering Eq. 4.56 and keeping one eye on the exponential in Eq. 4.57, we go back to Eq. 4.23. We want to approximate the infinite sum with a finite sum, remembering that we can only be interested in (or use) N sample values of $f(t)$. Thus, if N is large enough (as in the preceding paragraph), we can write for Eq. 4.23 that

$$f(nT/N) \approx \sum_{m=-(N-1)/2}^{(N-1)/2} C_m e^{j2\pi nm/N} \qquad (t = nT/N) \tag{4.58}$$

It is not exactly obvious what we have done at this point. Is there a one-to-one correspondence between Eq. 4.57 and 4.58? In order to answer this question, it is necessary to remove the approximate signs ($\approx$) in Eqs. 4.57 and 4.58 and

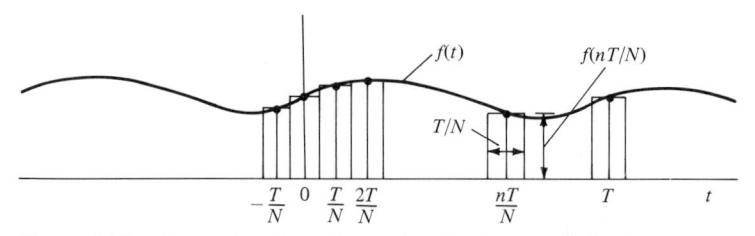

Figure 4.12. Approximation of complex Fourier coefficients.

simultaneously *define* the coefficients thus obtained to be

$$F_m \text{ (in 4.57)} \quad f_n \text{ (in 4.58)} \tag{4.59}$$

Thus, we now have

$$F_m = \frac{1}{N} \sum_{n=-(N-1)/2}^{(N-1)/2} f_n e^{-j2\pi mn/N} \tag{4.60}$$

and

$$f_n = \sum_{m=-(N-1)/2}^{(N-1)/2} F_m e^{j2\pi nm/N} \tag{4.61}$$

In order to find out about the correspondence of these equations, or, put another way, to find out if they *stand alone*, we substitute Eq. 4.60 (changing the index from n to p) into 4.61:

$$f_n = \sum_{m=-(N-1)/2}^{(N-1)/2} \frac{1}{N} \sum_{p=-(N-1)/2}^{(n-1)/2} f_p e^{-j2\pi mp/N} e^{j2\pi nm/N}$$

Combining the exponents and interchanging the order of summation

$$f_n = \frac{1}{N} \sum_{p=-(N-1)/2}^{(N-1)/2} f_p \sum_{m=-(N-1)/2}^{(N-1)/2} e^{j2\pi(n-p)m/N}$$

But according to Eq. 4.56 the inner sum is 0 unless $n = p$, in which case it is N. Therefore, we have

$$f_n = \frac{1}{N} f_n N = f_n$$

and we have established the fact that F_m (Eq. 4.60) and f_n (Eq. 4.61) do in fact correspond, and

$$f_n \leftrightarrow F_m \tag{4.62}$$

stands alone. Even though the development of this pair started with the Fourier series, we now merely have a pair of *periodic sequences* of numbers with period N. We call 4.62 the discrete Fourier series[2] (DFS) pair.

The periodicity is easily established. In Eq. 4.60

$$e^{-j2\pi(m+N)n/N} = e^{-j2\pi n} e^{-j2\pi mn/N} = e^{-j2\pi mn/N}$$

so

$$F_{m+N} = F_m$$

while in Eq. 4.61

$$e^{j2\pi(n+N)m/N} = e^{j2\pi m} e^{j2\pi nm/N} = e^{j2\pi nm/N}$$

so

$$f_{n+N} = f_n$$

Because of this periodicity, Eqs. 4.61 and 4.60 can be put in forms where N may be even *or* odd:

$$f_n = \sum_{m=0}^{N-1} F_m e^{j2\pi nm/N} \qquad (4.63)$$

$$F_m = \frac{1}{N} \sum_{n=0}^{N-1} f_n e^{-j2\pi mn/N} \qquad (4.64)$$

Notice the similarity in these two equations.

It is intuitively obvious that the value of N required for numerical computation depends on the *smoothness* of the periodic function $f(t)$ that we are working with. That is to say, N will depend on the *highest frequency* present in $f(t)$. This implies that if enough samples are taken (N large enough), then Eq. 4.64 gives the Fourier coefficients *exactly*.

Suppose that the periodic function $f(t)$ is limited (*band-limited*) to a highest frequency $\omega_h = M\omega_T$, $M = 0, 1, 2, \ldots$. Then Eq. 4.23 is exactly

$$f(t) = \sum_{m=-M}^{M} C_m e^{jm2\pi t/T} \qquad -M \leq m \leq M \qquad (4.65)$$

and Eq. 4.61 is

$$f_n = \sum_{m=-M}^{M} F_m e^{j2\pi nm/N} = \sum_{m=-M}^{M} C_m e^{j2\pi nm/N} \qquad (4.66)$$

Comparing Eqs. 4.65 and 4.66 we see that the samples $f_n = f(nT/N)$ of $f(t)$ and the coefficients C_m in Eq. 4.66 satisfy a set of N equations, and, furthermore, *if the total number of these coefficients $(2M + 1)$ is less than or equal to N*

$$N \geq 2M + 1 \qquad (4.67)$$

then the set of equations can be solved for the unknown C_m. Thus, if $f(t)$ contains no harmonics higher than the Mth ($\omega_h = M\omega_T$) then Eq. 4.60 or 4.64 will give C_m *exactly* in terms of the samples f_n of $f(t)$. The error (called the *aliasing error*) introduced by *undersampling* (failure to satisfy 4.67) will be investigated briefly here and in the next chapter in connection with the *sampling theorem* and the closely related *discrete Fourier transform* (DFT).

Recall that earlier in this chapter, in connection with the development of the Fourier series, it was pointed out that the use of the Euler formulas for calculating the coefficients of the series was highly advantageous for several reasons. In terms of simultaneous equations, and their solution by matrix inversion, we obtained inversion in a very simple manner, *even in infinite order*, because of the orthogonality of the trigonometric functions. The same thing is true here. Writing Eqs. 4.63 and 4.64 in matrix form

$$\mathbf{f} = \mathbf{E}^+ \mathbf{F} \qquad (4.68)$$

and

$$\mathbf{F} = \mathbf{E}^-\mathbf{f} \tag{4.69}$$

where **f** is the column vector:

$$\mathbf{f} = \begin{bmatrix} f_0 \\ f_1 \\ \vdots \\ f_{N-1} \end{bmatrix} \tag{4.70}$$

$\mathbf{E}^+$ is the square ($N \times N$) matrix:

$$\mathbf{E}^+ = \begin{bmatrix} 1 & 1 & 1 & \cdots & 1 \\ 1 & e^{j2\pi/N} & e^{j4\pi/N} & \cdots & e^{j2\pi(N-1)/N} \\ 1 & e^{j4\pi/N} & e^{j8\pi/N} & \cdots & e^{j4\pi(N-1)/N} \\ \vdots & & & & \\ 1 & e^{j2\pi(N-1)/N} & & \cdots & e^{j2\pi(N-1)^2/N} \end{bmatrix} \tag{4.71}$$

F is the column vector:

$$\mathbf{F} = \begin{bmatrix} F_0 \\ F_1 \\ \vdots \\ F_{N-1} \end{bmatrix} \tag{4.72}$$

and $\mathbf{E}^-$ is the square matrix:

$$\mathbf{E}^- = \frac{1}{N}\begin{bmatrix} 1 & 1 & 1 & \cdots & 1 \\ 1 & e^{-j2\pi/N} & e^{-j4\pi/N} & \cdots & e^{-j2\pi(N-1)/N} \\ 1 & e^{-j4\pi/N} & e^{-j8\pi/N} & \cdots & e^{-j4\pi(N-1)/N} \\ \vdots & & & & \\ 1 & e^{-j2\pi(N-1)/N} & & \cdots & e^{-j2\pi(N-1)^2/N} \end{bmatrix} \tag{4.73}$$

It is easy to show (because of the orthogonality again) that

$$\mathbf{E}^+\mathbf{E}^- = \mathbf{I} \quad (identity\ matrix) \tag{4.74}$$

Thus, premultiplying Eq. 4.68 by $\mathbf{E}^-$

$$\mathbf{E}^-\mathbf{f} = \mathbf{E}^-\mathbf{E}^+\mathbf{F} = \mathbf{IF} = \mathbf{F}$$

in agreement with Eq. 4.69. Premultiplying Eq. 4.69 by $\mathbf{E}^+$

$$\mathbf{E}^+\mathbf{F} = \mathbf{E}^+\mathbf{E}^-\mathbf{f} = \mathbf{If} = \mathbf{f}$$

in agreement with Eq. 4.68. This is, of course, just a different way of showing what we have already illustrated.

The use of samples of $f(t)$, that is, $f_n = f(nT/N)$, is equivalent to the "*point-matching*" scheme that was also mentioned earlier in the chapter. We used matching points in Sec. 4.2 that were slightly different than nT/N (used here), but these can be chosen any way we wish. The point is that we now have a simple way of approximating the Fourier coefficients (Eq. 4.64) in terms of

156 PHASORS AND FOURIER SERIES

the values of $f(t)$ at the matching points (that is, f_n), and, furthermore, the coefficients calculated in this manner will be exactly the same as those given by the Euler formulas[†] if N is large enough. The highest frequency in the $f(t)$ in Sec. 4.2 was generally infinity, so naturally this would require that $N \to \infty$. It seems that we have come full circle in the development of the Fourier series.

Example 11

We would like to consider a very simple example involving sampling. Suppose that $f(t) = 10\cos(2\pi t) = 5e^{j2\pi t} + 5e^{-j2\pi t}$. Thus, by inspection (or Eq. 4.24), we have

$$C_m = \begin{cases} 5 & m = \pm 1 \\ 0 & m \neq \pm 1 \end{cases} \tag{4.75}$$

This is certainly a band-limited signal since it is obvious that $\omega_h = 2\pi/T = 2\pi$. Thus, we have $T = 1$ and $M = 1$. We should obtain the exact Fourier coefficients if (Eq. 4.67) $N \geq 3$. We want to examine this using the matrix forms. We start with the case $N = 2$.

(1) $N = 2$, $f_n = f(nT/N) = f(n/2) = 10\cos(n\pi) = 10(-1)^n$

$$\begin{bmatrix} F_0 \\ F_1 \end{bmatrix} = \frac{1}{2}\begin{bmatrix} 1 & 1 \\ 1 & e^{-j\pi} \end{bmatrix}\begin{bmatrix} 10 \\ -10 \end{bmatrix} \tag{4.69}$$

Thus

$$F_0 = \tfrac{1}{2}(10 - 10) = 0$$
$$F_1 = \tfrac{1}{2}(10 + 10) = 10$$

These results are correct in the sense that they give the samples back correctly:

$$\begin{bmatrix} f_0 \\ f_1 \end{bmatrix} = \begin{bmatrix} 1 & 1 \\ 1 & e^{j\pi} \end{bmatrix}\begin{bmatrix} 0 \\ 10 \end{bmatrix} \tag{4.68}$$

$$f_0 = 10$$
$$f_1 = -10$$

but they do not give the correct Fourier coefficients because we undersampled.

(2) $N = 3$, $f_n = f(n/3) = 10\cos(2n\pi/3)$

$$\begin{bmatrix} F_0 \\ F_1 \\ F_3 \end{bmatrix} = \frac{1}{3}\begin{bmatrix} 1 & 1 & 1 \\ 1 & e^{-j2\pi/3} & e^{-j4\pi/3} \\ 1 & e^{-j4\pi/3} & e^{-j8\pi/3} \end{bmatrix}\begin{bmatrix} 10 \\ -5 \\ -5 \end{bmatrix}$$

$$F_0 = \tfrac{1}{3}(10 - 5 - 5) = 0$$
$$F_1 = \tfrac{1}{3}(10 - 5e^{-j2\pi/3} - 5e^{-j4\pi/3}) = 5$$
$$F_2 = \tfrac{1}{3}(10 - 5e^{-j4\pi/3} - 5e^{-j8\pi/3}) = 5$$

[†] As we have already seen, these coefficients minimize the rms error.

Notice that $F_2 = 5$, but $C_2 = 0$. This is the way it must be because of the periodicity

$$F_{-1} = F_{-1+3} = F_2 = 5$$

which is correct. In this case ($N = 3$) we sampled at the minimum rate.

(3) $N = 4$, $f_n = f(n/4) = 10 \cos(n\pi/2)$

$$\begin{bmatrix} F_0 \\ F_1 \\ F_2 \\ F_3 \end{bmatrix} = \frac{1}{4} \begin{bmatrix} 1 & 1 & 1 & 1 \\ 1 & e^{-j\pi/2} & e^{-j\pi} & e^{-j3\pi/2} \\ 1 & e^{-j\pi} & e^{-j2\pi} & e^{-j3\pi} \\ 1 & e^{-j3\pi/2} & e^{-j3\pi} & e^{-j9\pi/3} \end{bmatrix} \begin{bmatrix} 10 \\ 0 \\ -10 \\ 0 \end{bmatrix}$$

$F_0 = \frac{1}{4}(10 - 10) = 0$
$F_1 = \frac{1}{4}(10 + 10) = 5$
$F_2 = \frac{1}{4}(10 - 10) = 0$
$F_3 = \frac{1}{4}(10 + 10) = 5$ and $F_3 = F_{3+4} = F_1 = 5$

Notice here that

$$F_{-1} = F_{-1+4} = F_3 = 5 \quad \text{and} \quad F_{-3} = F_{-3+4} = F_1 = 5$$

as shown in Fig. 4.13.

The difference between F_m (Eq. 4.64) and the Fourier coefficients C_m (Eq. 4.24) is called the *aliasing error*:

$$\varepsilon_m = F_m - C_m = \frac{1}{N} \sum_{n=0}^{N-1} f_n e^{-j2\pi mn/N} - C_m \tag{4.76}$$

and, as we saw in Example 11, a set of *aliased coefficients* is developed. For the $N = 3$ case of Example 11 we have $F_1 = C_1 = 5$, $F_2 = F_{-1} = C_{-1} = 5$ as aliased coefficients for $0 \le n \le N - 1 = 2$. For $N = 4$ (see Fig. 4.13) we have $F_1 = C_1 = 5$, $F_3 = F_{-1} = C_{-1} = 5$ for $0 \le n \le N - 1 = 3$. Also, for $N = 2$ (undersampling)

$\varepsilon_0 = F_0 - C_0 = 0 - 0 = 0$
$\varepsilon_1 = F_1 - C_1 = 10 - 5 = 5$

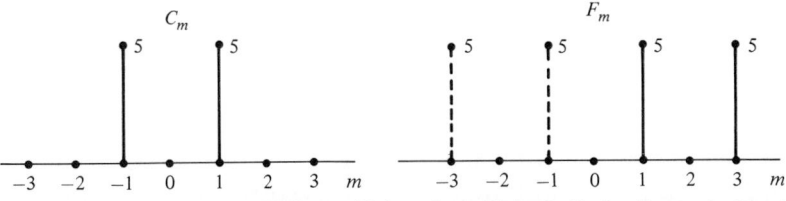

Figure 4.13. Fourier coefficients (C_m) and coefficients F_m for Example 11 with $N = 4$.

Example 12

Find f_n and $f(t)$ for the F_m given below ($N = 6$) assuming that $T = 1$, $\omega_h = 4\pi$ ($M = 2$):

$$\mathbf{F} = \begin{bmatrix} 0 \\ 0.5 \\ 0.25 \\ 0 \\ 0.25 \\ 0.5 \end{bmatrix} \quad (2M + 1 = 5, N > 2M + 1 = 5)$$

Using Eq. 4.68 (skipping the details)

$$\mathbf{f} = \begin{bmatrix} 1.5 \\ 0.25 \\ -0.75 \\ -0.5 \\ -0.75 \\ 0.25 \end{bmatrix} \tag{4.77}$$

Since $N > 2M + 1$, we have

$$\left.\begin{array}{l} C_0 = F_0 = 0 \\ C_1 = F_1 = 0.5 = C_{-1} \\ C_2 = F_2 = 0.25 = C_{-2} \end{array}\right\} \quad (M = 2)$$

Thus, we have exactly

$$f(t) = \sum_{m=-M}^{M} C_m e^{jm\omega T t} = \sum_{m=-2}^{2} C_m e^{jm2\pi t}$$

$$f(t) = 0.5e^{j2\pi t} + 0.5e^{-j2\pi t} + 0.25e^{j4\pi t} + 0.5e^{-j4\pi t}$$

$$f(t) = \cos(2\pi t) + 0.5\cos(4\pi t)$$

Comparing the samples of $f(t)$ (that is, f_n) as given by Eq. 4.77 to

$$f(nT/N) = f(n/6) = \cos(n\pi/3) + 0.5\cos(2n\pi/3)$$
$$f(0/6) = 1 + 0.5 = 1.5 = f_0$$
$$f(1/6) = \cos(\pi/3) + 0.5\cos(2\pi/3) = +0.5 - 0.25 = 0.25 = f_1$$
$$f(2/6) = \cos(2\pi/3) + 0.5\cos(4\pi/3) = -0.5 - 0.25 = -0.75 = f_2$$
$$f(3/6) = \cos(\pi) + 0.5\cos(2\pi) = -0.5 = f_3$$
$$f(4/6) = \cos(4\pi/3) + 0.5\cos(8\pi/3) = -0.5 - 0.25 = -0.75 = f_4$$
$$f(5/6) = \cos(5\pi/3) + 0.5\cos(10\pi/3) = +0.5 - 0.25 = 0.25 = f_5$$

4.7 THE DISCRETE FOURIER SERIES

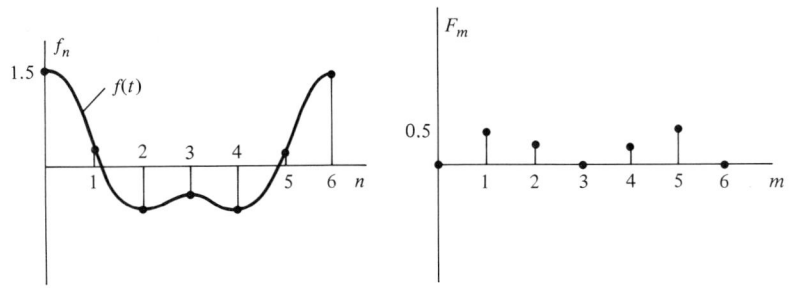

Figure 4.14. Coefficients f_n and F_m for Example 12 ($N = 6$).

Results are shown in Fig. 4.14.

Example 13

Find the DFS of order 12 for the sequence

$$F_m = \begin{cases} 1 & 0 \le m \le 4 \\ 0 & 5 \le m \le 11 \end{cases} \quad (N = 12)$$

Using Eq. 4.63

$$f_n = \sum_{m=0}^{11} F_m e^{j2\pi nm/12} = \sum_{m=0}^{4} e^{j\pi nm/6}$$

$$f_n = \frac{1 - e^{j5n\pi/6}}{1 - e^{jn\pi/6}} = e^{jn\pi/3} \frac{\sin(5n\pi/12)}{\sin(n\pi/12)}$$

$$|f_n| = \left|\frac{\sin(5n\pi/12)}{\sin(n\pi/12)}\right|$$

Results are shown in Fig. 4.15. Notice that the f_n are (in general) complex.

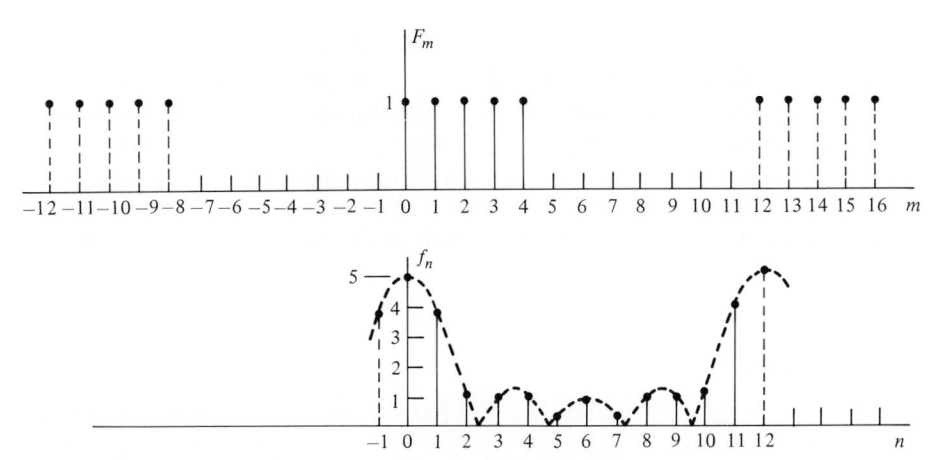

Figure 4.15. DFS for the given F_m in Example 13 ($N = 12$).

4.8 CONCLUDING REMARKS

We began this chapter with a review of the very powerful phasor concept applicable in those engineering problems where we are interested in the steady-state response only, and where the excitation is sinusoidal. In order to extend this concept, we introduced the trigonometric series for periodic functions of time, and, in particular, showed that if the coefficients are calculated according to the Euler formulas, which are very simple, then the resulting series is a Fourier trigonometric series. In addition to having very simple coefficient behavior, this series gives the least mean square error in finite order when compared to the function being expanded. The Fourier series appeared in several forms, and is valid for any independent variable which might appear in engineering problems.

Since the Fourier series applies to a function which is periodic over all time (a steady-state condition), and since it contains only sinusoidal terms, it was a simple matter to utilize the principle of superposition for linear systems and extend the phasor concept to treat those steady-state cases where the excitation is periodic, but not sinusoidal.

We next proceeded to a brief discussion of average power, whose definition should have already been familiar to most of us. This definition was applied to those linear systems where the excitation is periodic, but not sinusoidal, and we found that the average power could be found in series form.

The frequency spectrum of a periodic time function is *discrete*. That is, it generally contains a zero-frequency term, a fundamental frequency term, a second harmonic term, and so on. In the next chapter we will investigate the idea of the frequency spectrum more closely and discover that general time functions may have components throughout the frequency spectrum. In other words, they may have *continuous* spectra.

We explored a method for the numerical treatment of Fourier series that was actually independent of the continuous Fourier series itself. As it turned out, exact numerical results were possible only if the number of discrete samples we were working with was equal to (or greater than) twice the highest harmonic number in the periodic function plus one ($N \geq 2M + 1$). This can be realized in any practical system, although the calculations may be time consuming if M (and therefore N) is large. This is exactly what we should expect. If the signals in the system are very fine in detail (not smooth), requiring high resolution, we must look at them very closely. We will briefly examine a method [the *fast Fourier transform* (FFT)] to reduce the computation time in the next chapter.

PROBLEMS

1. Replace the battery and switch in Fig. 4.16 with the voltage source $v_s(t) = 100 \cos(10^3 t)$.

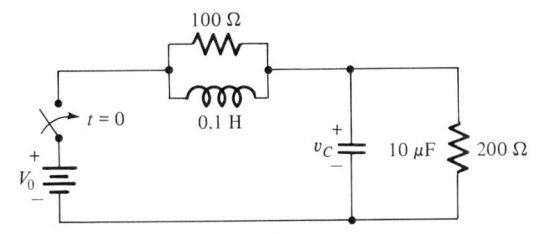

Figure 4.16. Circuit for Prob. 1.

(a) Find the transfer function, $H(10^3) = V_C(10^3)/(100)$.
(b) Find $v_C(t)$.

2. Find the transfer function for Fig. 4.2 if the "output" is the phasor voltage across the inductor, $V_L = j\omega L I_L$.

3. (a) Find the transfer function, $H(\omega) = V(\omega)/I_s(\omega)$, in Fig. 4.17.

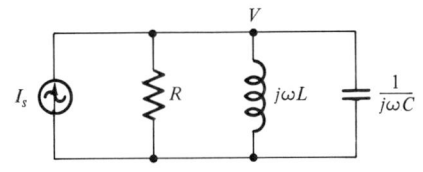

Figure 4.17. Parallel RLC network.

(b) Find $v(t)$ if $i_s(t) = 10\cos(10^3 t)$, $R = 10^3\,\Omega$, $L = 10$ mH, and $C = 100\,\mu\text{F}$.
(c) What kind of second-order system is this? See Sec. 2.7.

4. Use the point-matching scheme of Sec. 4.2 to approximate $\sin(2\pi t)$ as follows:

$$f(t) = \sin(2\pi t) \approx \sum_{n=1}^{N} B_n \sin(n 2\pi t) \qquad T = 1$$

with matching points $t_m = (m - \frac{1}{2})/(2N)$, $m = 1, 2, \ldots, N$ for (a) $N = 1$, (b) $N = 2$, and (c) $N = 3$. Why is an "exact" answer obtained in each case?

5. Repeat Prob. 4 for the $f(t)$ (an odd function) shown in Fig. 4.18.

6. Find the *Fourier* trigonometric series representation for the function shown in Fig. 4.18.

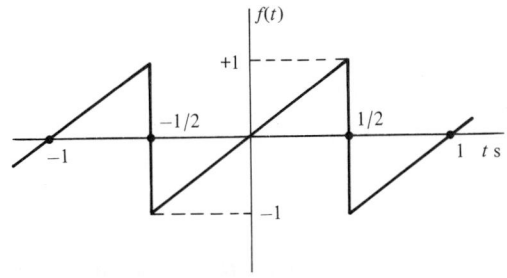

Figure 4.18 $f(t)$ for Prob. 5 ($T = 1$ s).

162 PHASORS AND FOURIER SERIES

7. Find the Fourier series for $f(t)$ in Fig. 4.19 using the form given by Eq. 4.20.

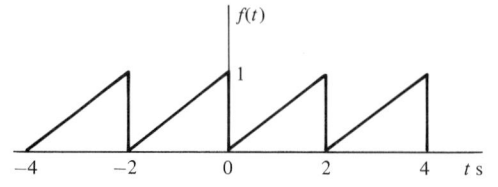

Figure 4.19. $f(t)$ for Prob. 7 ($T = 2$ s).

8. Verify Eq. 4.19: $b_n = \frac{2}{T}\int_{-T/2}^{T/2} f(t) \sin(n2\pi t/T)\, dt$

9. Repeat Example 2 with $T = 40$ s and $a_0/2 = 0$.

10. Find the Fourier series for the half-wave-rectified cosine voltage shown in Fig. 4.20 ($T = 1/60$).

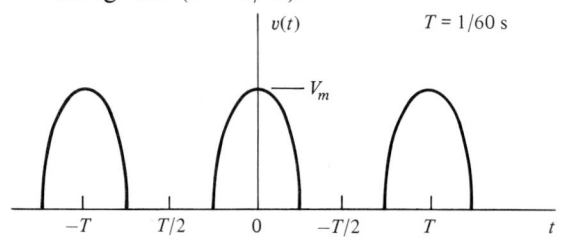

Figure 4.20. Half-wave-rectified cosine voltage.

11. (a) Expand $f(x) = x^2$, $0 < x < 2\pi$, in a Fourier series using a period of 2π.
    (b) Put $x = 0$ in the series and prove that
    $$\sum_{n=1}^{\infty} \frac{1}{n^2} = \frac{\pi^2}{6}$$

12. Find the Fourier series for the full-wave-rectified sine voltage (Fig. 4.21) without integrating ($T = \frac{1}{120}$). See Example 6.

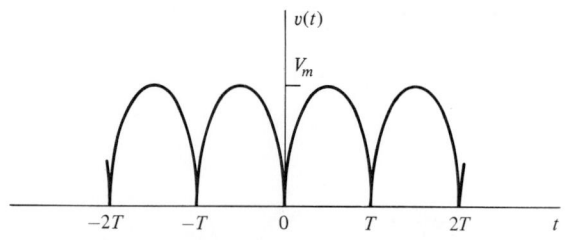

Figure 4.21. Full-wave-rectified cosine voltage.

13. Find the rms error in Prob. 10 for $M = 6$.
14. The desired radiation pattern of a certain antenna array is shown in Fig. 4.22 as a function of azimuth angle ϕ.
    (a) Express $f(\phi)$ as a Fourier series.

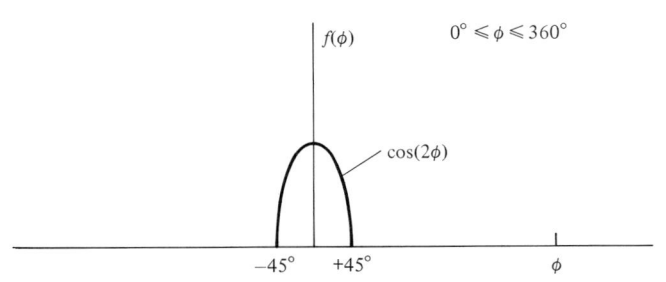

Figure 4.22. $f(\phi)$ versus ϕ.

(b) If the antenna is only able to radiate the first four terms in the series, sketch the radiation pattern.
15. Repeat Example 6 using the voltage in Prob. 10.
16. In order to improve the action of the circuit in Example 7 as a frequency doubler, suppose that the Q is increased to 200.
    (a) What is R?
    (b) Find $v_o(t)$.
17. For the simple RL network shown in Fig. 4.23 find (a) the unit-impulse response [for $i(t)$] and (b) the unit-step response [for $i(t)$].

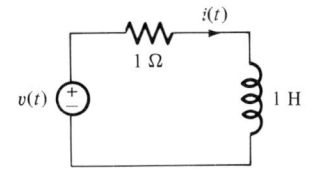

Figure 4.23. RL series network.

18. In order to show how the unit-impulse response for Fig. 4.23 can be determined (approximately) in the laboratory, imagine the *even* periodic rectangular pulse of width 0.1 s (short compared to the $L/R = 1$ s time constant) and period $T = 10$ s (long compared to L/R) for $v(t)$. If the pulse amplitude is 10 V, then the pulse area is unity. Find the Fourier series for $v(t)$; find $i(t)$; plot $i(t)$ versus t; and compare the result to that in Prob. 17(a).
19. The unit-step response for Fig. 4.23 can be determined approximately using $v(t)$ as in Fig. 4.24 ($T = 10L/R$). Find the Fourier series for $v(t)$; find $i(t)$; plot $i(t)$ versus t; and compare the result to that in Prob. 17(b).

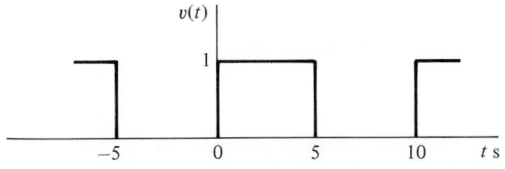

Figure 4.24. Square wave used for finding the unit-step response of the RL series network ($L/R = 1$).

20. Find the complex Fourier series for the periodic train of impulses

$$f(t) = \sum_{n=-\infty}^{\infty} \delta(t - nT)$$

21. (a) Show that the transfer function, $H(\omega) = V_0(\omega)/V(\omega)$ for the twin $-T$ or "notch" filter shown in Fig. 4.25 is

$$H(\omega) = \frac{-\omega^2 + (1/RC)^2}{-\omega^2 + j\omega(4/RC) + (1/RC)^2}$$

(b) What value of C gives $H(\omega) = 0$ when $R = 10^3 \, \Omega$ and $\omega = 2\pi \times 10^3$?
(c) Sketch the magnitude and phase of $H(\omega)$ versus ω.

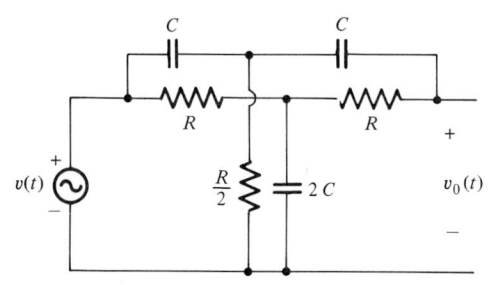

Figure 4.25. Twin-T filter.

22. What is the maximum value of L which can be used in the shunt-peaking circuit of Fig. 4.26 if $|H(\omega)| \leq R$, $H(\omega) = V_0(\omega)/I_s(\omega)$?

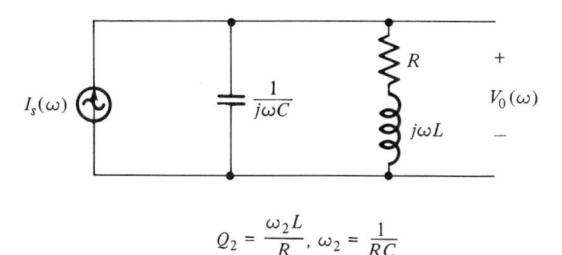

$$Q_2 = \frac{\omega_2 L}{R}, \quad \omega_2 = \frac{1}{RC}$$

Figure 4.26. Shunt-peaking coupling network.

23. The sawtooth waveform of Fig. 4.19 (with $T = 10^{-6}$ s) is applied as the input to the shunt-peaking network of Fig. 4.26 with $R = 5$ K and $C = 10$ pF. Let $i_s(t)$ be given in mA.
   (a) Sketch $v_0(t)$ versus t with $L = 0$.
   (b) Sketch $v_0(t)$ versus t with L equal to the value calculated in Prob. 22.
24. Find the average values for the waveforms of Figs. 4.18, 4.19, 4.20, 4.21, 4.22, and 4.24.
25. Find the rms (effective) values for the waveforms in Prob. 24.
26. Find the (time) average power dissipated in 1 Ω of resistance by the waveform of Fig. 4.19 if it is a current with 1-A peak value. Use two methods.

27. Find the average power dissipated in 10 Ω of resistance if the current passing through it is

$$i(t) = 2\sin(t) + \cos(1.4t) + 3 + 2\sin(2.5t) \text{ A}$$

How much heat is released in 1 hour?

28. (a) Plot $|H(\omega)|$ versus ω for the parallel *RLC* network of Fig. 4.17.
    (b) If the *bandwidth* is measured between the *two* radian frequencies where $|H(\omega)| = |H(\omega)|_{max}/\sqrt{2}$, what is the bandwidth?
    (c) What is the bandwidth in terms of $Q_0 = R/(\omega_0 L)$ $(\omega_0 = 1/\sqrt{LC})$?

29. Plot $|H(\omega)|$ versus ω for the shunt-peaking network of Fig. 4.26 for $R = 5$ K, $C = 10$ pF, and $Q_2 = 0, 0.25, \sqrt{2} - 1, 0.5, 0.75, 1.0$. Comment on the bandwidth as measured at $|H(\omega)| = R/\sqrt{2}$.

30. (a) If $\omega = 1/\sqrt{2LC}$ in Fig. 4.26, plot $|V_0/I|$ and the phase of V_0/I versus R for $0 \leq R < \infty$. Let $\sqrt{L/C} = 5$ K.
    (b) What practical use can this network have?

31. Why is the rms error more meaningful than just *average* error?

32. (a) Find the DFS ($N = 16$) for the sequence

$$f_n = \begin{cases} 1 & 0 \leq n \leq 4 \\ 0 & 4 \leq n \leq 15 \end{cases}$$

using Eq. 4.64. Obtain a closed form.

(b) Find the complex Fourier coefficients C_m for the periodic function

$$f(t) = u(t) - u(t - T/4) \qquad 0 \leq t \leq T$$

and compare to F_m in (a).

References

1. Harrington, R. F. *Field Computation by Moment Methods*. New York: Macmillan, 1968.
2. Papoulis, A. *Circuits and Systems*. New York: Holt, Rinehart and Winston, 1980.

Chapter 5
Fourier Transform
Techniques

In the preceding chapter we learned how to analyze a time-invariant linear system when excited by a series of sinusoidal functions. This technique is based on the relatively simple phasor concept and applies whether the frequencies of the sinusoids are harmonically related, as in a Fourier series, or not. The frequency spectrum of a series of sinusoids is, of course, discrete, containing terms only at certain frequencies. In this chapter we will introduce a technique whereby a function of time, if it meets certain conditions,[†] can be *transformed* into a function of frequency. The linear *transformation* which accomplishes this is called the *Fourier transform*, and it produces the spectrum of the time function. This concept is extremely useful in the analysis of communication systems.

From the standpoint of linear-systems analysis in general, the Fourier transform has other uses. We will find, for example, that some operations which occur in systems are more easily performed in the frequency domain than in the time domain. Once these operations have been performed in the frequency domain, we can *inverse Fourier transform* back to the time domain. The Fourier transform converts the differential equation of a linear system into an algebraic equation which is usually relatively easy to solve. Any difficulty encountered

---

[†] All time functions of engineering interest will meet these conditions.

5.1 EXTENSION OF THE FOURIER SERIES

usually occurs in performing the inverse Fourier transformation. Recall that the use of phasors also converted a differential equation into an algebraic equation. This is no accident, because the Fourier transform technique can be thought of as merely an extension of the Fourier techniques already studied.

5.1 EXTENSION OF THE FOURIER SERIES

Consider the periodic rectangular pulse shown in Fig. 5.1(a). The complex Fourier series for this pulse is

$$f(t) = \sum_{n=-\infty}^{\infty} C_n e^{jn\omega_T t} \tag{5.1}$$

where

$$C_n = \frac{1}{T} \int_{-T/2}^{T/2} f(t) e^{-jn\omega_T t} \, dt \tag{5.2}$$

$$C_n = \frac{A}{T} \int_{-\tau/2}^{\tau/2} e^{-jn\omega_T t} \, dt = \frac{2A}{T} \int_0^{\tau/2} \cos(n\omega_T t) \, dt$$

$$C_n = \frac{2A}{n\omega_T T} \sin(n\omega_T \tau/2)$$

or

$$C_n = \frac{A\tau}{T} \frac{\sin(n\omega_T \tau/2)}{n\omega_T \tau/2} \equiv \frac{A\tau}{T} Sa(n\omega_T \tau/2) \tag{5.3}$$

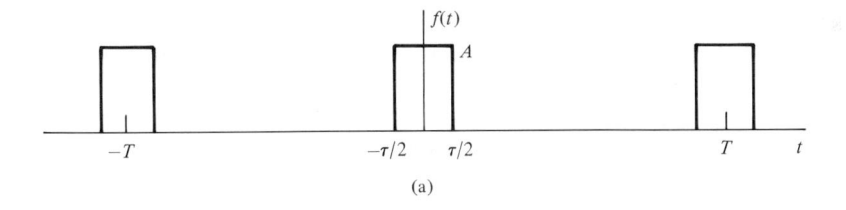

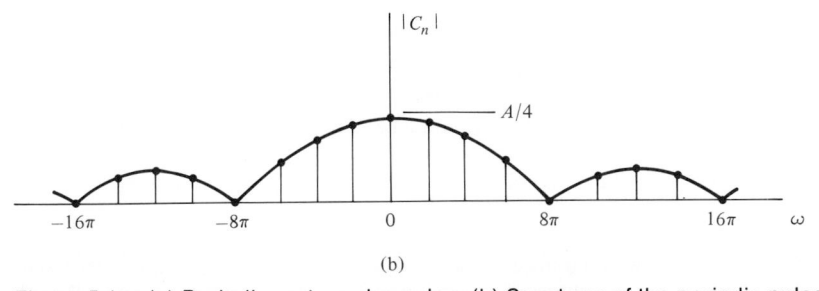

Figure 5.1. (a) Periodic rectangular pulse. (b) Spectrum of the periodic pulse $|C_n|$ versus ω, $T = 4\tau = 1$ s.

168 FOURIER TRANSFORM TECHNIQUES

where $Sa(x) = \sin(x)/x$ is called the sampling function. It is frequently encountered in communication theory. Thus

$$f(t) = \sum_{n=-\infty}^{\infty} \frac{A\tau}{T} Sa(n\omega_T \tau/2) e^{jn\omega_T t} \tag{5.4}$$

If we wish to examine the relative amplitudes of the coefficients of the series, we would quite naturally plot C_n versus ω. We must remember, however, that C_n occurs only at $n\omega_T$, so we will not obtain a continuous plot, but rather a discrete spectrum. We use vertical lines in Fig. 5.1(b) to indicate $|C_n|$ (C_n is real in our present example), and we choose $T = 4\tau = 1$ s, so that

$$C_n = \frac{A}{4} \frac{\sin(n\pi/4)}{n\pi/4}$$

or

$$|C_n| = \frac{A}{4} \left| \frac{\sin(n\pi/4)}{n\pi/4} \right|$$

Now suppose we increase T to $T = 16\tau = 4$ s, so we are quadrupling the period, or decreasing the fundamental frequency by a factor of four, and now

$$|C_n| = \frac{A}{16} \left| \frac{\sin(n\pi/16)}{n\pi/16} \right|$$

The spectrum for this case is plotted in Fig. 5.2. Notice that the zero amplitude coefficients still occur at frequencies $\omega = \pm 8\pi, \pm 16\pi$, and so on, and, also, the spectrum is four times as dense, as well as one-fourth as large in amplitude.

It is not difficult to imagine that if we allow T to approach infinity, the pulse is no longer repetitive, but becomes a "one-shot," or transient, pulse. The spectrum becomes "completely dense" or continuous,[†] and all frequencies are present even though their amplitudes become smaller and smaller. Equations 5.1 and 5.2, when combined, yield

$$f(t) = \sum_{n=-\infty}^{\infty} \left\{ \frac{1}{T} \int_{-T/2}^{T/2} f(x) e^{-jn\omega_T x} dx \right\} e^{jn\omega_T t}$$

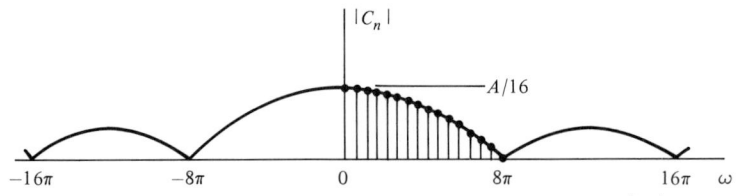

Figure 5.2. Spectrum for the repetitive pulse of Fig. 4.1(a), $|C_n|$ versus ω, $T = 16\tau = 4$ s.

[†] It is worthwhile to point out at this time that when we want to do *numerical* work with the results of this section we *must* resort to the *discrete* spectrum and not the continuous spectrum. We will in fact, need to use what is almost identical to the discrete Fourier series (DFS) of Sec. 4.7.

or

$$f(t) = \frac{1}{2\pi} \sum_{n=-\infty}^{\infty} \left\{ \int_{-T/2}^{T/2} f(x)e^{-jn\omega_T x} dx \right\} e^{jn\omega_T t} \frac{2\pi}{T} \quad (5.5)$$

Notice particularly that removing the term $1/T$ from within the braces as in Eq. 5.5, insofar as the pulse of Fig. 5.1 is concerned, means that the braced term (the inner integral) *will now not go to 0 as T goes to* ∞. Now, if we allow T to approach infinity, and simultaneously imagine $2\pi/T = \omega_T$ becoming differential (continuous) frequency $d\omega$, $n\omega_T$ will become the continuous variable ω. The summation becomes an infinite integral (the limit of a sum). Thus, in the limit we obtain

$$f(t) = \frac{1}{2\pi} \int_{-\infty}^{\infty} \left\{ \int_{-\infty}^{\infty} f(x)e^{-j\omega x} dx \right\} e^{jt\omega} d\omega \quad (5.6)$$

which is known as the *Fourier integral*. The quantity within braces is a function of ω (or $j\omega$) which we call $F(\omega)$, *the spectrum of f(t)*, the *spectral density of f(t)*, or the *Fourier transform* of $f(t)$. Thus

$$\boxed{F(\omega) = \int_{-\infty}^{\infty} f(t)e^{-j\omega t} dt} \quad (Fourier\ transform) \quad (5.7)$$

$$\boxed{f(t) = \frac{1}{2\pi} \int_{-\infty}^{\infty} F(\omega)e^{jt\omega} d\omega} \quad (Inverse\ Fourier\ transform) \quad (5.8)$$

Equations 5.7 and 5.8 form a transform pair, often symbolized by

$$f(t) \leftrightarrow F(\omega) \quad (5.9)$$

or

$$F(\omega) = \mathscr{F}\{f(t)\} \quad (5.10)$$
$$f(t) = \mathscr{F}^{-1}\{F(\omega)\} \quad (5.11)$$

Equation 5.10 means "$F(\omega)$ is the Fourier transform of $f(t)$," and Eq. 5.11 means "$f(t)$ is the inverse Fourier transform of $F(\omega)$."

Example 1

What is the Fourier transform of the aperiodic pulse of Fig. 5.3(a)? Application of Eq. 5.7 gives

$$F(\omega) = \int_{-\tau/2}^{\tau/2} Ae^{-j\omega t} dt = 2A \int_{0}^{\tau/2} \cos(\omega t) dt$$

or

$$F(\omega) = 2A \frac{\sin(\omega\tau/2)}{\omega} = A\tau \frac{\sin(\omega\tau/2)}{\omega\tau/2}$$

$$F(\omega) = A\tau Sa(\omega\tau/2)$$

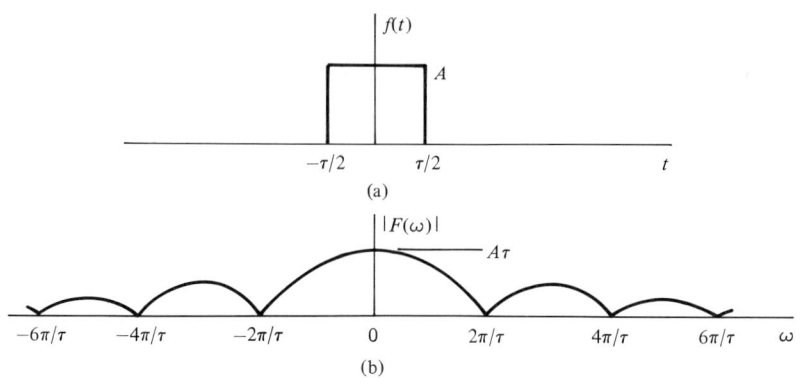

Figure 5.3. (a) Transient rectangular pulse with amplitude A and width τ. (b) Spectrum or Fourier transform, $|F(\omega)|$ versus ω.

We usually plot the spectrum in two parts: $|F(\omega)|$ versus ω, and $\underline{/F(\omega)}$ versus ω. In this example $F(\omega)$ is real, although it is negative for some values of ω, so we merely plot $|F(\omega)|$ versus ω in Fig. 5.3(b). We now have a continuous spectrum. Notice that $F(\omega) \to 0$ as $\tau \to 0$.

All functions of engineering interest possess a Fourier transform. Many functions of purely mathematical interest do not possess a Fourier transform. If $f(t)$ satisfies the Dirichlet conditions (see Sec. 4.3) in every finite interval, and if

$$\int_{-\infty}^{\infty} |f(t)|\, dt < \infty$$

then we are guaranteed that $F(\omega)$ exists. The conditions are sufficient, but not necessary, for the existence of $F(\omega)$. That is, there are functions which do not satisfy the above conditions, but do have Fourier transforms. An example (pair 5.29) is given in the next section.

It is worthwhile at this point to consider an analogy to demonstrate the usefulness of the transform idea. Addition is normally a simpler operation to perform than multiplication. Therefore, if we "transform" two numbers that are to be multiplied to logarithmic form and then add the two logarithms, we have performed the equivalent of multiplication (addition in the "logarithm domain"). Next, we inverse transform the sum of the logarithms by taking the antilog of the sum, and hence return back to the "number domain," obtaining the product. This is a familiar operation. In the same way some operations (convolution, for example) are more easily performed in the frequency domain than in the time domain. Thus, we would transform from the time domain to the frequency domain, perform the required operation, and then inverse transform from the frequency domain back to the time domain. This may at the present time seem to be a circuitous route to take, but, as we shall see, it is often the shortest and fastest.

5.2 FOURIER TRANSFORM PAIRS

We begin this section by considering the properties of Fourier transforms. This will allow us to find many transform pairs with a minimum amount of effort. Our goal is to develop a table of transform pairs.

Linearity

The transform of a linear sum of functions is the sum of the transforms of the individual functions. If $f_n(t) \leftrightarrow F_n(\omega)$, then

$$\sum_{n=1}^{N} a_n f_n(t) \leftrightarrow \sum_{n=1}^{N} a_n F_n(\omega) \tag{5.12}$$

Time Shifting[†]

The delayed function $f(t - t_0)$, which is just $f(t)$ delayed in time by t_0, has the transform indicated by

$$f(t - t_0) \leftrightarrow F(\omega) e^{-j\omega t_0} \tag{5.13}$$

where, as implied, $f(t) \leftrightarrow F(\omega)$. This result is easily verified by direct application of Eq. 5.7.

Frequency Shifting

A shift in the spectrum of a function results in

$$f(t) e^{j\omega_0 t} \leftrightarrow F(\omega - \omega_0) \tag{5.14}$$

where ω_0 is the amount of frequency shift. This result is easily verified using Eq. 5.8.

Scaling

$$f(\alpha t) \leftrightarrow \frac{1}{|\alpha|} F(\omega/\alpha) \tag{5.15}$$

where α is a real constant. This result can be verified with Eq. 5.7, but it is wise to consider the cases $\alpha < 0$ and $\alpha > 0$ separately; otherwise, the need for the absolute value sign may go unnoticed.

Symmetry or Duality

If $f(t) \leftrightarrow F(\omega)$, then

$$F(t) \leftrightarrow 2\pi f(-\omega) \tag{5.16}$$

[†] Prob. 2 at the end of the chapter asks for the proof of 5.13, 5.14, and 5.15. The reader is urged to work through these proofs.

172 FOURIER TRANSFORM TECHNIQUES

This property allows us to use a table of transform pairs backwards. For example, if we know the transform of sin(t), then we can immediately find the *inverse* transform of sin(ω) without the need for integration. Equation 5.8 gives

$$2\pi f(t) = \int_{-\infty}^{\infty} F(\omega)e^{jt\omega} d\omega = \int_{-\infty}^{\infty} F(x)e^{jtx} dx$$

or

$$2\pi f(-t) = \int_{-\infty}^{\infty} F(x)e^{-jtx} dx$$

or

$$2\pi f(-\omega) = \int_{-\infty}^{\infty} F(x)e^{-j\omega x} dx$$

or, finally

$$2\pi f(-\omega) = \int_{-\infty}^{\infty} F(t)e^{-j\omega t} dt$$

which gives pair 5.16.

Differentiation

If the transform of the derivative of $f(t)$ exists, it is given by

$$\frac{df}{dt} \leftrightarrow (j\omega)F(\omega) \tag{5.17}$$

Direct application of Eq. 5.8 shows this. In the same way, if the nth derivative of $f(t)$ has a transform which exists, it is given by

$$\frac{d^n f}{dt^n} \leftrightarrow (j\omega)^n F(\omega) \tag{5.18}$$

Before considering integration, we need to consider *convolution*. We have seen in previous chapters that convolution is extremely important in the analysis of time-invariant systems.

Convolution

If $x(t) \leftrightarrow X(\omega)$ and $h(t) \leftrightarrow H(\omega)$, then the product of $H(\omega)$ and $X(\omega)$, that we call $Y_{zs}(\omega)$, has the inverse transform $y_{zs}(t)$ given by

$$y_{zs}(t) = \mathscr{F}^{-1}\{Y_{zs}(\omega)\} = \mathscr{F}^{-1}\{H(\omega)X(\omega)\}$$

$$y_{zs}(t) = \int_{-\infty}^{\infty} x(\tau)h(t-\tau) d\tau = \int_{-\infty}^{\infty} x(t-\tau)h(\tau) d\tau \tag{5.19}$$

That is

$$\boxed{\int_{-\infty}^{\infty} x(\tau)h(t-\tau) d\tau \leftrightarrow X(\omega)H(\omega) = Y_{zs}(\omega)} \tag{5.20}$$

5.2 FOURIER TRANSFORM PAIRS

The integrals in Eqs. 5.19 and 5.20 are the convolution integrals that we have already seen. If we consider $X(\omega)$ to be the transform of the input to a time-invariant linear system, and consider $H(\omega)$ to be the transfer function of the system (or the transform of its unit-impulse response), and consider $Y_{zs}(\omega) = H(\omega)X(\omega)$ to be the transform of the zero-state response of the system, then Fig. 5.4 applies. We have seen this situation before in Chap. 2. Furthermore, Eq. 5.19 is identical with Eq. 2.59. Thus, we have an alternate way (the frequency domain) of viewing the time-invariant linear system. We will give this a closer inspection later. *Notice very carefully that we are using subscripts zs on $Y(\omega)$ to indicate that its inverse transform produces only the zero-state response, and does not include a zero-input response.* A zero-input response may, of course, be added, if necessary.

We can show that Eq. 5.19 is correct by starting with Eq. 5.7:

$$\mathcal{F}\{y_{zs}(t)\} = \mathcal{F}\left\{\int_{-\infty}^{\infty} x(\tau)h(t-\tau)\,d\tau\right\} = Y_{zs}(\omega)$$

$$= \int_{-\infty}^{\infty} \left\{\int_{-\infty}^{\infty} x(\tau)h(t-\tau)\,d\tau\right\} e^{-j\omega t}\,dt$$

Interchanging the order of integration and integrating with respect to t first

$$Y_{zs}(\omega) = \int_{-\infty}^{\infty} x(\tau)\left\{\int_{-\infty}^{\infty} h(t-\tau)e^{-j\omega t}\,dt\right\}d\tau$$

Using the time-shift (pair 5.13) property, the term within the braces is

$$\mathcal{F}\{h(t-\tau)\} = e^{-j\tau\omega}H(\omega)$$

so

$$Y_{zs}(\omega) = H(\omega)\int_{-\infty}^{\infty} x(\tau)e^{-j\tau\omega}\,d\tau$$

or

$$Y_{zs}(\omega) = H(\omega)X(\omega)$$

Thus, we have the important result that *convolution in the time domain corresponds to multiplication in the frequency domain*! See Fig. 5.4.

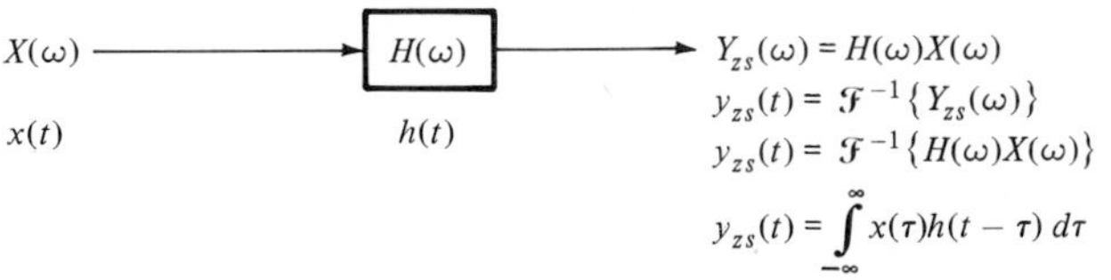

$X(\omega)$ $\longrightarrow$ $\boxed{H(\omega)}$ $\longrightarrow$ $Y_{zs}(\omega) = H(\omega)X(\omega)$
$x(t)$ $\qquad\quad$ $h(t)$ $\qquad\quad$ $y_{zs}(t) = \mathcal{F}^{-1}\{Y_{zs}(\omega)\}$
$\qquad\qquad\qquad\qquad\qquad\qquad y_{zs}(t) = \mathcal{F}^{-1}\{H(\omega)X(\omega)\}$
$\qquad\qquad\qquad\qquad\qquad\qquad y_{zs}(t) = \int_{-\infty}^{\infty} x(\tau)h(t-\tau)\,d\tau$

Figure 5.4. Time-invariant linear system with transfer function $H(\omega)$, $Y_{zs}(\omega) = H(\omega)X(\omega)$.

In the same way multiplication in the time domain (as in a *balanced modulator*†) corresponds to convolution in the frequency domain, or

$$f(t) = f_1(t)f_2(t) \leftrightarrow \frac{1}{2\pi}\int_{-\infty}^{\infty} F_1(u)F_2(\omega - u)\,du = F(\omega) \tag{5.21}$$

Correlation

If $f_1(t) \leftrightarrow F_1(\omega)$ and $f_2(t) \leftrightarrow F_2(\omega)$, then

$$f(t) = \int_{-\infty}^{\infty} f_1^*(\tau)f_2(\tau + t)\,d\tau \leftrightarrow F_1^*(\omega)F_2(\omega) = F(\omega) \tag{5.22}$$

where $F_1^*(\omega)$ is the complex conjugate of $F_1(\omega)$. The proof is similar to that used for Eq. 5.20. The integral form for $f(t)$ is known as the *correlation integral*, and if $f_1(t)$ and $f_2(t)$ are different functions, the integral is called the *cross-correlation integral*, while if $f_1(t)$ and $f_2(t)$ are identical, it is called the *auto-correlation integral*:

$$f(t) = \int_{-\infty}^{\infty} f_1^*(\tau)f_1(\tau + t)\,d\tau \leftrightarrow |F_1(\omega)|^2 = F(\omega) \tag{5.23}$$

For correlation in the frequency domain

$$f(t) = f_1^*(t)f_2(t) \leftrightarrow \frac{1}{2\pi}\int_{-\infty}^{\infty} F_1^*(u)F_2(u + \omega)\,du = F(\omega) \tag{5.24}$$

and if $f_1(t) = f_2(t)$

$$f(t) = |f_1(t)|^2 \leftrightarrow \frac{1}{2\pi}\int_{-\infty}^{\infty} F_1^*(u)F_1(u + \omega)\,du = F(\omega) \tag{5.25}$$

Pair 5.22 can be written

$$\int_{-\infty}^{\infty} f_1^*(\tau)f_2(\tau + t)\,d\tau = \frac{1}{2\pi}\int_{-\infty}^{\infty} F_1^*(\omega)F_2(\omega)e^{j\omega t}\,d\omega$$

and, in particular for $t = 0$

$$\int_{-\infty}^{\infty} f_1^*(\tau)f_2(\tau)\,d\tau = \frac{1}{2\pi}\int_{-\infty}^{\infty} F_1^*(\omega)F_2(\omega)\,d\omega \tag{5.26}$$

The last result is known as *Parseval's theorem*. If $f_1(t)$ is identical with $f_2(t)$, we have two formulas for the *energy* in a signal (voltage or current associated with a 1-Ω resistor):

$$w = \int_{-\infty}^{\infty} |f_1(\tau)|^2\,d\tau = \frac{1}{2\pi}\int_{-\infty}^{\infty} |F_1(\omega)|^2\,d\omega \tag{5.27}$$

---

† A balanced modulator can take many forms, but it basically is an analog time-domain multiplier that forms the product of two input signals as its output signal.

Unit-Impulse function[†]

The sampling property of the unit-impulse function gives

$$\mathscr{F}\{\delta(t)\} = \int_{-\infty}^{\infty} \delta(t) e^{-j\omega t}\, dt = 1$$

or

$$\delta(t) \leftrightarrow 1 \tag{5.28}$$

Since $\delta(-\omega) = \delta(\omega)$, the symmetry property (pair 5.16) of transform pairs leads to

$$1 \leftrightarrow 2\pi\delta(\omega) \tag{5.29}$$

Thus, any $f(t)$ with a *nonzero average value* ($-\infty < t < \infty$) has an impulse at $\omega = 0$ in its spectrum.

Finite Exponential

Direct application of Eq. 5.7 gives the important result

$$e^{-\alpha t} u(t) \leftrightarrow \frac{1}{\alpha + j\omega} \qquad \alpha > 0 \tag{5.30}$$

In the same way

$$e^{\alpha t} u(-t) \leftrightarrow \frac{1}{\alpha - j\omega} \qquad \alpha > 0 \tag{5.31}$$

The difference of these two results gives

$$f(t) = -e^{\alpha t} u(-t) + e^{-\alpha t} u(t) \leftrightarrow \frac{-j2\omega}{\alpha^2 + \omega^2} \tag{5.32}$$

which is plotted in Fig. 5.5. It is obvious from the graph that if $\alpha = 0$, then $f(t) = \text{signum } t = \text{sgn}(t)$, where

$$\text{sgn}(t) = \begin{cases} -1 & t < 0 \\ +1 & t > 0 \end{cases} = 2u(t) - 1 \tag{5.33}$$

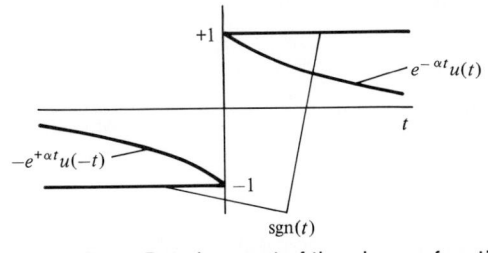

Figure 5.5. Development of the signum function.

[†] In Example 1, let $A = 1/\tau$, then let $\tau \to 0$, giving $\delta(t) \leftrightarrow 1$.

Signum(t)

Using pair 5.32 with $\alpha = 0$

$$\text{sgn}(t) \leftrightarrow 2/j\omega \tag{5.34}$$

Unit-Step Function

From Eq. 5.33 or Fig. 5.5

$$\text{sgn}(t) = 2u(t) - 1 \tag{5.35}$$

so

$$2u(t) - 1 \leftrightarrow 2/j\omega$$

or

$$u(t) - \tfrac{1}{2} \leftrightarrow 1/j\omega$$

Pair 5.29 gives

$$\tfrac{1}{2} \leftrightarrow \pi\delta(\omega)$$

Adding† the last two pairs

$$u(t) \leftrightarrow \pi\delta(\omega) + 1/j\omega \tag{5.36}$$

Using the symmetry property

$$\tfrac{1}{2}\delta(t) + j/2\pi t \leftrightarrow u(\omega) \tag{5.37}$$

and

$$j/\pi t \leftrightarrow \text{sgn}(\omega) \tag{5.38}$$

Integration

Using pair 5.35

$$u(t) - \tfrac{1}{2} \leftrightarrow 1/j\omega$$

If we call $h(t) = u(t) - \tfrac{1}{2}$, and $x(t) = f(t)$, then pair 5.20 gives

$$\int_{-\infty}^{\infty} x(\tau)h(t - \tau)\,d\tau = \mathscr{F}^{-1}\{H(\omega)X(\omega)\}$$

or

$$\int_{-\infty}^{t} f(\tau)\,d\tau - \tfrac{1}{2}\int_{-\infty}^{\infty} f(\tau)\,d\tau = \mathscr{F}^{-1}\{F(\omega)/j\omega\}$$

The second integral on the left is $F(0)$ by Eq. 5.7, so

$$\int_{-\infty}^{t} f(\tau)\,d\tau - F(0)/2 \leftrightarrow F(\omega)/j\omega$$

---

† This is possible because of pair 5.12. Also, refer to the sentence following pair 5.29 to further explain the presence of $\pi\delta(\omega)$ in pair 5.36.

But, by pair 5.29 and the sentence immediately following:

$$+F(0)/2 \leftrightarrow \pi F(0)\delta(\omega)$$

Adding the last two pairs, we have

$$\int_{-\infty}^{t} f(\tau)\,d\tau \leftrightarrow F(\omega)/j\omega + \pi F(0)\delta(\omega) \qquad (5.39)$$

Exponential, Imaginary Argument

Using pair 5.14 with $f(t) = 1$ and pair 5.29, we have

$$e^{j\omega_0 t} \leftrightarrow \pi\delta(\omega - \omega_0) \qquad (5.40)$$

The transform of $\sin(\omega_0 t)$ and $\cos(\omega_0 t)$ easily follow from pair 5.40:

$$\sin(\omega_0 t) \leftrightarrow \frac{\pi}{j}[\delta(\omega - \omega_0) - \delta(\omega + \omega_0)] \qquad (5.41)$$

and

$$\cos(\omega_0 t) \leftrightarrow \pi[\delta(\omega - \omega_0) + \delta(\omega + \omega_0)] \qquad (5.42)$$

Notice that the spectra for the last two forms are discrete as we might expect: impulses at $\omega = -\omega_0, +\omega_0$.

Periodic Function

Using pair 5.12 and 5.40

$$\sum_{n=-\infty}^{\infty} C_n e^{jn\omega_T t} \leftrightarrow 2\pi \sum_{n=-\infty}^{\infty} C_n \delta(\omega - n\omega_T) \qquad (5.43)$$

This result includes the Fourier trigonometric series when the C_n are calculated with the Euler formula (Eq. 4.24). It naturally gives a discrete spectrum.

A short table of transform pairs is included (Table 5.1). Many entries are obtained from those just obtained, and others may be added in the same way.

Comparing the time functions to their corresponding spectra reveals some interesting and useful characteristics. For example, a function which is of short duration in the time domain will possess a wide spectrum. Put in more practical terms, a *narrow* pulse, when passing through a network, will require that the network have a *large* bandwidth (wide spectrum) if the pulse is to be unaltered by the network. The most extreme case of this is, of course, the unit impulse, $\delta(t)$, which has the spectrum 1. That is, in the frequency domain it is represented by *all* frequencies in equal amounts. It is then easy to imagine the problem encountered when one is attempting to recover a desired signal by filtering in the presence of a narrow, intense pulse!

The reverse situation also exists (symmetry again). A function which is *narrow* in the frequency domain will be *wide* in the time domain. Again, the extreme case is represented by $1 \leftrightarrow 2\pi\delta(\omega)$. The everlasting $\cos(\omega_0 t)$ has a

178 FOURIER TRANSFORM TECHNIQUES

Table 5.1 FOURIER TRANSFORM PROPERTIES AND PAIRS

NO.		$f(t)$	$F(\omega)$
1	Linearity	$\sum_{n=1}^{N} a_n f_n(t)$	$\sum_{n=1}^{N} a_n F_n(\omega)$
2	Time Shift	$f(t - t_0)$	$F(\omega) e^{-j\omega t_0}$
3	Frequency Shift	$f(t) e^{j\omega_0 t}$	$F(\omega - \omega_0)$
4	Scaling	$f(\alpha t)$	$\dfrac{1}{|\alpha|} F(\omega/\alpha)$
5	Symmetry	$F(t)$	$2\pi f(-\omega)$
6	Time Differentiation	$\dfrac{df}{dt}$	$j\omega F(\omega)$
		$\dfrac{d^n f}{dt^n}$	$(j\omega)^n F(\omega)$
7	Frequency Differentiation	$(-jt) f(t)$	$\dfrac{dF}{d\omega}$
		$(-jt)^n f(t)$	$\dfrac{d^n F}{d\omega^n}$
8	Time Integration	$\int_{-\infty}^{t} f(\tau) d\tau$	$\dfrac{F(\omega)}{j\omega} + \pi F(0) \delta(\omega)$
9	Periodic Function	$\sum_{n=-\infty}^{\infty} C_n e^{jn\omega_T t}$	$2\pi \sum_{n=-\infty}^{\infty} C_n \delta(\omega - n\omega_T)$
10	Time Convolution	$\int_{-\infty}^{\infty} f_1(\tau) f_2(t - \tau) d\tau$	$F_1(\omega) F_2(\omega)$
		$\int_{-\infty}^{\infty} f_1(t - \tau) f_2(\tau) d\tau$	

5.2 FOURIER TRANSFORM PAIRS

		$f(t)$	$F(\omega)$	
11	Frequency Convolution	$f_1(t)f_2(t)$	$\dfrac{1}{2\pi}\int_{-\infty}^{\infty} F_1(u)F_2(\omega - u)\,du$ $\dfrac{1}{2\pi}\int_{-\infty}^{\infty} F_1(\omega - u)F_2(u)\,du$	
12	Time Correlation	$\int_{-\infty}^{\infty} f_1^*(\tau)f_2(t+\tau)\,d\tau$	$F_1^*(\omega)F_2(\omega)$	
		$f(t)$	$F(\omega)$	$|F(\omega)|$
13		$\delta(t)$	1	
14		$e^{-\alpha t}u(t) \quad \alpha > 0$	$\dfrac{1}{\alpha + j\omega}$	
15		$\mathrm{sgn}(t)$	$\dfrac{2}{j\omega}$	
16		$u(t)$	$\pi\delta(\omega) + 1/j\omega$	
17		$u(t+\tau/2) - u(t-\tau/2)$	$\tau \mathrm{Sa}(\omega\tau/2)$	

Table 5.1 (continued)

NO.	$f(t)$	$F(\omega)$
18	$\cos(\omega_0 t)$	$\pi[\delta(\omega - \omega_0) + \delta(\omega + \omega_0)]$
19	$\sin(\omega_0 t)$	$\dfrac{\pi}{j}[\delta(\omega - \omega_0) - \delta(\omega + \omega_0)]$
20	$u(t)\cos(\omega_0 t)$	$\dfrac{\pi}{2}[\delta(\omega - \omega_0) + \delta(\omega + \omega_0)]$ $+ j\omega/(\omega_0^2 - \omega^2)$
21	$u(t)\sin(\omega_0 t)$	$\dfrac{\pi}{2j}[\delta(\omega - \omega_0) - \delta(\omega + \omega_0)]$ $+ \omega_0/(\omega_0^2 - \omega^2)$
22	$u(t)e^{-\alpha t}\cos(\omega_0 t)$ $\alpha > 0$	$\dfrac{\alpha + j\omega}{\omega_0^2 + (\alpha + j\omega)^2}$

5.2 FOURIER TRANSFORM PAIRS

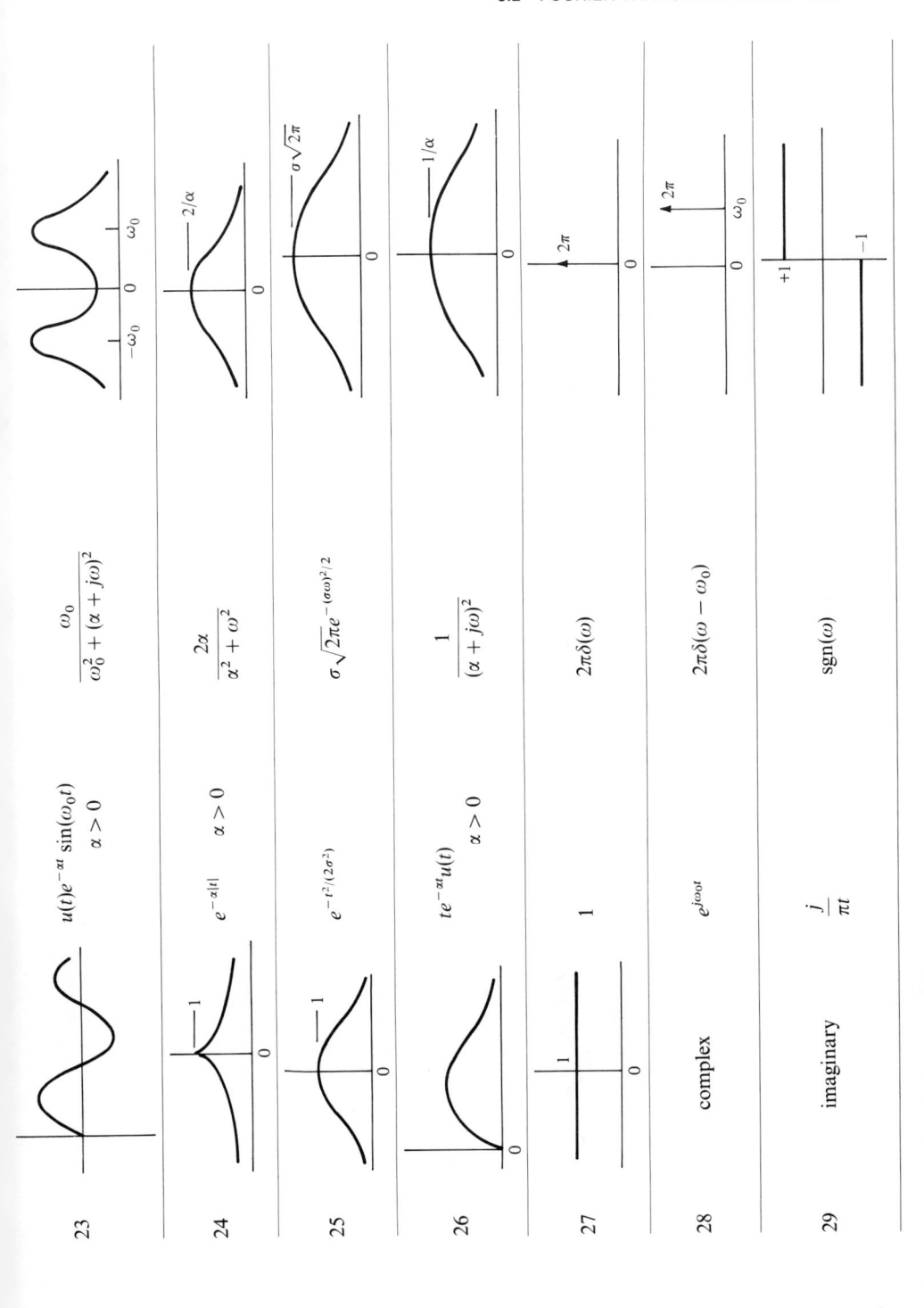

23	$u(t)e^{-\alpha t}\sin(\omega_0 t)$ $\alpha > 0$	$\dfrac{\omega_0}{\omega_0^2 + (\alpha + j\omega)^2}$	
24	$e^{-\alpha |t|}\quad \alpha > 0$	$\dfrac{2\alpha}{\alpha^2 + \omega^2}$	
25	$e^{-t^2/(2\sigma^2)}$	$\sigma\sqrt{2\pi}\,e^{-(\sigma\omega)^2/2}$	
26	$te^{-\alpha t}u(t)$ $\alpha > 0$	$\dfrac{1}{(\alpha + j\omega)^2}$	
27	1	$2\pi\delta(\omega)$	
28	$e^{j\omega_0 t}$ complex	$2\pi\delta(\omega - \omega_0)$	
29	$\dfrac{j}{\pi t}$ imaginary	$\operatorname{sgn}(\omega)$	

182 FOURIER TRANSFORM TECHNIQUES

Table 5.1 (continued)

NO.	$f(t)$	$F(\omega)$
30	complex $\dfrac{\delta(t)}{2} + \dfrac{j}{2\pi t}$	$u(\omega)$
31	$\dfrac{\omega_0}{\pi} \mathrm{Sa}(\omega_0 t)$	$u(\omega + \omega_0) - u(\omega - \omega_0)$
32	See 8, 27 $\displaystyle\int_{-\infty}^{t} f(\tau)\,d\tau - F(0)/2$	$\dfrac{F(\omega)}{j\omega}$
33	$\dfrac{1}{a}[1 - e^{-at}]u(t) - \dfrac{1}{2a} \quad a > 0$	$\dfrac{1}{j\omega(j\omega + a)}$
34	$\dfrac{1}{b-a}[e^{-at} - e^{-bt}]u(t) \quad a, b > 0$	$\dfrac{1}{(j\omega + a)(j\omega + b)}$
35	$\dfrac{1}{b-a}[-ae^{-at} + be^{-bt}]u(t) \quad a, b > 0$	$\dfrac{j\omega}{(j\omega + a)(j\omega + b)}$
36	$t[u(t) - u(t - t_0)]$	$\dfrac{e^{-j\omega t_0}(1 + j\omega t_0) - 1}{\omega^2}$
37	$\lvert t \rvert[u(t + \tau/2) - u(t - \tau/2)]$	$\dfrac{\cos(\omega\tau/2) + (\omega\tau/2)\sin(\omega\tau/2) - 1}{\omega^2/2}$

spectrum which is a pair of impulses, $\cos(\omega_0 t) \leftrightarrow \pi[\delta(\omega - \omega_0) + \delta(\omega + \omega_0)]$. Filtering in this case should be easy!

It is also worth remembering that, generally speaking, the *early* time response (or *rise* time) of a linear system is governed primarily by its response at *high* frequencies, while the *late* time response (or *fall* time) is governed primarily by its response at *low* frequencies. The initial value theorem and final value theorem in connection with the *Laplace transform* (Chapter 6) shed more light on this behavior.

As an aid to understanding the relation between $f(t)$ and its spectrum, or vice versa, it is wise to recognize that an arbitrary function can be expressed as the sum of an even part and an odd part. That is, an even function is such that $f_e(-t) = f_e(t)$, and $\cos(\omega_0 t)$ is an even function. An odd function is such that $f_o(-t) = -f_o(t)$, and $\sin(\omega_0 t)$ is an odd function. Thus, in general

$$f(t) = f_e(t) + f_o(t)$$

where

$$f_e(t) = \tfrac{1}{2}[f(t) + f(-t)]$$

and

$$f_o(t) = \tfrac{1}{2}[f(t) - f(-t)]$$

It is also true that a function may contain both real and imaginary parts:

$$f(t) = r(t) + jx(t)$$

If $f(t)$ is real, then $x(t) = 0$, and using Eq. 5.7 shows that the real part of $F(\omega)$ becomes an even function of ω, and the imaginary part of $F(\omega)$ becomes an odd function of ω. That is

$$f(t) = r(t) \leftrightarrow R(\omega) + jX(\omega) = F(\omega) = R(-\omega) - jX(-\omega)$$

so

$$F(-\omega) = R(\omega) - jX(\omega) = F^*(\omega)$$

where the asterisk means "complex conjugate." If $f(t)$ is both *real and even*, then $X(\omega) = 0$, and

$$f_e(t) = r_e(t) \leftrightarrow R_e(\omega) = R_e(-\omega) = F_e(\omega)$$

This is analogous to the situation where a real, even periodic function of time has only cosine terms in its Fourier series. If $f(t)$ is *real* and *odd*, then $R(\omega) = 0$, and

$$f_o(t) = r_o(t) \leftrightarrow jX_o(\omega) = -jX_o(-\omega) = F_o(\omega)$$

and this compares to the case where a real, odd periodic function of time has only sine terms in its Fourier series. Many times we can take advantage of special features of $f(t)$ or $F(\omega)$ to simplify the integration required to find $F(\omega)$ or $f(t)$.

Before passing on to a new topic it is worthwhile to digress and reconsider the unit-impulse function and its Fourier transform. The sampling property of the unit-impulse function gave us pair 5.28:

$$\delta(t) \leftrightarrow 1 \tag{5.28}$$

We can obtain this result from Example 1 as well. If we let $A = 1/\tau$ we have

$$F(\omega) = Sa(\omega\tau/2)$$

for the Fourier transform of the even transient pulse *whose area is unity for any* τ. In particular, if we allow τ to approach 0, we have

$$\lim_{\tau \to 0} f(t) = \delta(t) \qquad \lim_{\tau \to 0} F(\omega) = \lim_{\tau \to 0} Sa(\omega\tau/2) = 1$$

or, once again

$$\delta(t) \leftrightarrow 1$$

Now, what happens if we try to take the inverse transform of unity? From Eq. 5.8

$$\mathscr{F}^{-1}\{1\} = \frac{1}{2\pi} \int_{-\infty}^{\infty} 1 e^{jt\omega} \, d\omega$$

If we use Euler's identity, and recognize that $\sin(t\omega)$ is an odd function of ω

$$\mathscr{F}^{-1}\{1\} = \frac{1}{2\pi} \int_{-\infty}^{\infty} \cos t\omega \, d\omega = \lim_{W \to \infty} \frac{1}{2\pi} \int_{-W}^{W} \cos t\omega \, d\omega$$

$$= \lim_{W \to \infty} \frac{1}{\pi} \frac{\sin Wt}{t} = \lim_{W \to \infty} \frac{W}{\pi} Sa(Wt)$$

The last form can be recognized as having the property that at $t = 0$ the "height" is approaching ∞ while the "width" of the function is approaching 0. The area under the graph of the function is

$$\int_{-\infty}^{\infty} \left[\lim_{W \to \infty} \frac{W}{\pi} Sa(Wt) \right] dt$$

and, assuming that the order of taking the limit and integrating can be interchanged, the area is

$$\lim_{W \to \infty} \frac{W}{\pi} \int_{-\infty}^{\infty} Sa(Wt) \, dt = \lim_{W \to \infty} \frac{W}{\pi} \frac{1}{W} \int_{-\infty}^{\infty} Sa(x) \, dx$$

$$= \frac{1}{\pi} \lim_{W \to \infty} \pi = 1$$

Therefore

$$\mathscr{F}^{-1}\{1\} = \delta(t)$$

We have used the well-known result that

$$\int_0^\infty Sa(x)\,dx = Si(\infty) = \pi/2$$

where

$$\int_0^t Sa(x)\,dx = Si(t) \quad (\text{sine integral})$$

Functions of time that are discontinuous at $t = 0$ (for example) can be misinterpreted when the inverse Fourier transformation is being performed. As an example, consider

$$f(t) = e^{-at}u(t) \leftrightarrow \frac{1}{j\omega + a}$$

so that $f(0^-) = 0$, $f(0^+) = 1$, and $f(0)$ is unspecified. On the other hand,

$$f(t) = \frac{1}{2\pi}\int_{-\infty}^{\infty} \frac{e^{j t\omega}}{j\omega + a}\,d\omega$$

After rationalizing the integrand and using obvious symmetry properties,

$$f(t) = \frac{1}{\pi}\int_0^\infty \frac{a\cos(t\omega) + \omega\sin(t\omega)}{a^2 + \omega^2}\,d\omega$$

After performing the integration, the correct $f(t)$ will be obtained, but if we allow t to become zero *before* the integration is performed, we obtain

$$f(0) \stackrel{?}{=} \frac{a}{\pi}\int_0^\infty \frac{d\omega}{a^2 + \omega^2} = \frac{1}{\pi}\tan^{-1}\left(\frac{x}{a}\right)\Big|_0^\infty = \frac{1}{2}$$

which is neither $f(0^-)$ nor $f(0^+)$! The value given by the preceding result is the *midpoint* of the discontinuity. The unit step function behaves the same way. This difficulty is avoided with the *initial value theorem* in connection with the Laplace transform (Chap. 6).

5.3 THE TIME-INVARIANT LINEAR SYSTEM (AGAIN)

In the previous section we discovered another method for finding the zero-state response of a time-invariant linear system. We would now like to elaborate on this method. Consider such a system described by the Nth order ordinary linear differential equation (2.4) with constant coefficients:

$$\sum_{n=0}^{N} a_n \frac{d^n y}{dt^n} = x_f(t) = \sum_{n=0}^{M} b_n \frac{d^n x}{dt^n} \tag{5.44}$$

If more than one input $x(t)$ exists, superposition can be employed to solve the problem, so for present purposes, we consider only one input. The Fourier transform of Eq. 5.44, using pairs 5.1 and 5.18, is[†]

$$\sum_{n=0}^{N} a_n(j\omega)^n Y_{zs}(\omega) = X_f(\omega) = \sum_{n=0}^{M} b_n(j\omega)^n X(\omega) \qquad (5.45)$$

which is an Nth order *algebraic* equation [polynomial in $j\omega$], allowing simple solution for $Y_{zs}(\omega)$:

$$Y_{zs}(\omega) = \frac{\sum_{n=0}^{M} b_n(j\omega)^n}{\sum_{n=0}^{N} a_n(j\omega)^n} X(\omega) \qquad (5.46)$$

So, the transfer function is $Y_{zs}(\omega)/X(\omega)$, and

$$H(\omega) = \frac{\sum_{n=0}^{M} b_n(j\omega)^n}{\sum_{n=0}^{N} a_n(j\omega)^n} \qquad (5.47)$$

The inverse transform of $H(\omega)$ is the *causal* unit-impulse response for real systems with loss (see Prob. 37!):

$$h(t) \leftrightarrow H(\omega) \qquad (5.48)$$

and, furthermore, $H(\omega)$ can be calculated by the phasor methods of Chap. 4. That is, $H(\omega)$ is the ratio of the phasor output to the phasor input:

$$H(\omega) = \frac{Y_{zs}(\omega)}{X(\omega)} \qquad (5.49)$$

The transfer function description given by Eq. 5.47 is a ratio of polynomials, and in the usual case $M \leq N$. That is, the *degree* of the numerator polynomial is less than or equal to that of the denominator polynomial. The denominator polynomial, when set equal to 0, is the *characteristic equation*

$$a_N(j\omega)^N + a_{N-1}(j\omega)^{N-1} + \cdots + a_1(j\omega) + a_0 = 0$$

or

$$(j\omega)^N + \frac{a_{N-1}}{a_N}(j\omega)^{N-1} + \cdots + \frac{a_1}{a_N}(j\omega) + \frac{a_0}{a_N} = 0$$

or, redefining the constants

$$(j\omega)^N + C_{N-1}(j\omega)^{N-1} + \cdots + C_1(j\omega) + C_0 = 0$$

This characteristic equation (with $s = j\omega$) is identical to that given by Eq. 2.57.

[†] Notice once more that we are using subscripts zs on $Y(\omega)$ to indicate explicitly that the inverse transform we will shortly find is *the zero*-state response $y_{zs}(t)$.

5.3 THE TIME-INVARIANT LINEAR SYSTEM (AGAIN)

We can perform a partial fraction expansion (see Prob. 6) of Eq. 5.47. This is particularly easy to do when $M < N$ and when there are no repeated roots. We will consider these expansions in detail in the next chapter when the Laplace transform is introduced. Here, we merely point out that under the stated conditions Eq. 5.47 can be put in the form

$$H(\omega) = \frac{A_1}{j\omega + a_1} + \frac{A_2}{j\omega + a_2} + \cdots + \frac{A_N}{j\omega + a_N}$$

that inverse transforms to

$$h(t) = (A_1 e^{-a_1 t} + A_2 e^{-a_2 t} + \cdots + A_N e^{-a_N t}) u(t)$$

This is particularly simple.

Example 2

What is the voltage $v_1(t)$ for the circuit shown in Fig. 5.6 if $v(t) = V_0 u(t)$? From Example 5, Chap. 4, we found by phasor methods

$$H(\omega) = \frac{R_2}{R_1 + R_2 + j\omega C R_1 R_2} = \frac{1}{R_1 C} \frac{1}{\frac{R_1 + R_2}{R_1 R_2 C} + j\omega} = A \frac{1}{a + j\omega} \quad \begin{cases} A \equiv 1/R_1 C \\ a \equiv \dfrac{R_1 + R_2}{R_1 R_2 C} \end{cases}$$

and using Table 5.1 (No. 14)

$$h(t) = A e^{-at} u(t)$$

First, we obtain the solution by the time-domain method of Chap. 2, which is exactly the same as that given by Eq. 5.19:

$$y_{zs}(t) = \int_{-\infty}^{\infty} V_0 u(\tau) A e^{-a(t-\tau)} u(t-\tau) \, d\tau = v_1(t)$$

$$v_1(t) = V_0 A e^{-at} \int_0^t e^{a\tau} \, d\tau$$

$$v_1(t) = V_0 \frac{A}{a}[1 - e^{-at}] u(t)$$

The transient and steady-state parts are easy to identify. Next, we will use the frequency-domain approach:

$$Y_{zs}(\omega) = H(\omega) X(\omega) = A \frac{1}{a + j\omega} V_0 [\pi \delta(\omega) + 1/j\omega]$$

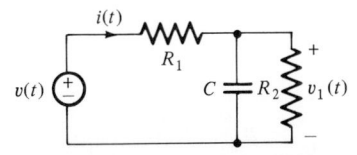

Figure 5.6. Circuit for Example 2.

since $v(t) = V_0 u(t) \leftrightarrow V_0[\pi\delta(\omega) + 1/j\omega]$. Thus

$$Y_{zs}(\omega) = \frac{\pi A V_0}{a}\delta(\omega) + AV_0\frac{1}{j\omega(j\omega + a)}$$

Then,

$$y_{zs}(t) = \mathscr{F}^{-1}\{Y_{zs}(\omega)\} = v_1(t)$$

$$v_1(t) = \frac{\pi A V_0}{a}\mathscr{F}^{-1}\{\delta(\omega)\} + AV_0\mathscr{F}^{-1}\left\{\frac{1}{j\omega(j\omega + a)}\right\}$$

Using transform pairs 27 and 33 (Table 5.1)

$$v_1(t) = \frac{\pi A V_0}{a}\frac{1}{2\pi} + AV_0\left[\frac{1}{a}(1 - e^{-at})u(t) - \frac{1}{2a}\right]$$

$$v_1(t) = V_0\frac{A}{a}[1 - e^{-at}]u(t)$$

as before. Notice that no integration was required because we were able to use the table of transform pairs.

Example 3

What is the *complete* response for Example 2 if the capacitor is already charged to V_1 V when the input voltage $v(t) = V_0 u(t)$ is applied? We now have an initial condition: $v_1(0^\pm) = V_1$. The zero-state response has already been found in Example 2. The zero-input response that must be added to form the complete response is a linear combination of the elements of the fundamental set, as we recall from Chap. 2. In this example there is only one element in the set since it is a first-order system, and it is obtained by inspection of $h(t)$:

$$y_{zi}(t) = A_1 \exp\left(-\frac{R_1 + R_2}{R_1 R_2 C}t\right) = A_1 e^{-at}$$

subject to $y(0) = V_1$. Therefore, the complete solution is

$$y(t) = y_{zi}(t) + y_{zs}(t) = v_1(t)$$

$$v_1(t) = A_1 e^{-at} + \frac{V_0 A}{a}[1 - e^{-at}] \qquad t \geq 0$$

$$V_1 = A_1 + 0, \qquad t = 0$$

Therefore

$$v_1(t) = V_1 e^{-at} + \frac{V_0 A}{a}[1 - e^{-at}] \qquad t \geq 0$$

or

$$v_1(t) = \left[V_1 - \frac{V_0 R_2}{R_1 + R_2}\right]\exp\left(-\frac{R_1 + R_2}{R_1 R_2 C}t\right) + V_0\frac{R_2}{R_1 + R_2} \qquad t \geq 0$$

If we are able to find a phasor solution for the linear system we are analyzing, then we can easily find the transfer function $H(\omega)$. Hopefully, $H(\omega)$ is a form which can be recognized in Table 5.1, or can be recognized as a combination of entries in the table, so that $h(t)$ can be obtained *without integrating*, that is, without using Eq. 5.8. Sometimes, we have no choice and must integrate. Other times, the elements in a system may be unrecognizable, or may be so complicated that a differential equation(s) cannot be easily obtained. In these cases, we may be forced to determine $h(t)$ experimentally as discussed in Chap. 2, but it is still possible to analyze the system for any input using numerical integration for the convolution integral (Eq. 5.19).

Once $H(\omega)$ is known, we can proceed to a frequency-domain solution, $Y_{zs}(\omega) = H(\omega)X(\omega)$, and if we are only interested in the spectrum of the output, we are essentially finished. If we want $y_{zs}(t)$, however, we are again faced with a choice: (1) integrate, using the inverse Fourier transform (Eq. 5.8); or (2) hopefully, use the table of transform pairs; or (3) use convolution (Eq. 5.19). If integration is necessary, and can be done analytically, then what is required should be listed in the table of transform pairs. Tables of Fourier transform pairs are notoriously skimpy, however, and we may attempt integration when it is actually not necessary. If the integration cannot be carried out analytically, then it must be done numerically. Further comments on this are found in Sec. 5.8.

5.4 PERIODIC EXCITATION

Periodic excitation of a linear system can be treated by the Fourier trigonometric series as in Chap. 4 in a straightforward manner. The output then appears as an infinite (often) series of trigonometric terms. This form is somewhat unwieldy, especially if we are interested in looking at a continuous plot of the output as a function of time. A good example of this is the simple power supply of Example 5, Chap. 4. In this section we would like to develop a closed-form solution to the same type problem.

A periodic input function $x(t)$ is by definition everlasting in time. That is, it extends from minus infinity to plus infinity in time. The output of a time-invariant linear system which has been exposed to this type input must also be periodic and everlasting in time if it is stable. For this situation, *the complete response is the zero-state response*, and Eq. 5.19 gives

$$y_{zs}(t) = \int_{-\infty}^{\infty} x(\tau)h(t-\tau)\,d\tau \tag{5.19}$$

If, in addition to being fixed and stable, the system is also causal, then $h(t-\tau)$ will *always* contain $u(t-\tau)$ as a multiplier, and in this case

$$y_{zs}(t) = \int_{-\infty}^{t} x(\tau)h(t-\tau)\,d\tau \tag{5.50}$$

Suppose that we want to find the steady-state response within the first period after $t = 0$: that is, for $0 < t < T$. Of course, any other period would serve as well because $y_{zs}(t)$ is periodic. Thus

$$y_{zs}(t) = y_{ss}(t) = \int_{-\infty}^{0} x(\tau)h(t-\tau)\,d\tau + \int_{0}^{t} x(\tau)h(t-\tau)\,d\tau \tag{5.51}$$

and

$$y_{ss}(t) = \int_{-T}^{0} x(\tau)h(t-\tau)\,d\tau + \int_{-2T}^{-T} x(\tau)h(t-\tau)\,d\tau + \int_{-3T}^{-2T} \cdots + \cdots$$
$$+ \int_{0}^{t} x(\tau)h(t-\tau)\,d\tau$$

Obvious changes of variable utilizing the periodicity, $x(\tau) = x(\tau + nT)$, $n = \pm 1, \pm 2, \ldots$, will give

$$y_{ss}(t) = \int_{-T}^{0} x(\tau)[h(t-\tau) + h(t-\tau+T) + h(t-\tau+2T) + \cdots]\,d\tau$$
$$+ \int_{0}^{t} x(\tau)h(t-\tau)\,d\tau \tag{5.52}$$

Now

$$\sum_{n=0}^{\infty} h(t-\tau+nT) = \sum_{n=0}^{\infty} \mathscr{F}^{-1}\{H(\omega)e^{-j\omega(\tau-nT)}\}$$

using transform pairs 1 and 2 in Table 5.1. So[†]

$$\sum_{n=0}^{\infty} h(t-\tau+nT) = \mathscr{F}^{-1}\left\{H(\omega)e^{-j\omega\tau}\sum_{n=0}^{\infty} e^{j\omega nT}\right\}$$

$$= \mathscr{F}^{-1}\left\{\frac{H(\omega)}{1-e^{j\omega T}}e^{-j\omega\tau}\right\} \tag{5.53}$$

So, $\sum_{n=0}^{\infty} h(t-\tau+nT)$ *always has a closed form*, although we have no guarantee that we can always find the inverse transform indicated on the right side of Eq. 5.53. Thus

$$y_{ss}(t) = \int_{-T}^{0} x(\tau) \sum_{n=0}^{\infty} h(t-\tau+nT)\,d\tau + \int_{0}^{t} x(\tau)h(t-\tau)\,d\tau \tag{5.54}$$

for $0 < t < T$.

Example 4

Find the closed-form steady-state solution for the output of the filter of Example 6, Chap. 4. We had found that

$$H(\omega) = \frac{1}{R_1 C} \frac{1}{\dfrac{R_1+R_2}{R_1 R_2 C} + j\omega} \equiv A\frac{1}{a+j\omega}$$

[†] The result is easily established by using the binomial expansion, Eq. C.25, App. C, or by simply using long division.

and
$$h(t) = Ae^{-at}u(t)$$
so
$$h(t - \tau + nT) = Ae^{-a(t-\tau+nT)}$$
Notice that the step function has been dropped, as in arriving at Eq. 5.50. Next
$$\sum_{n=0}^{\infty} h(t - \tau + nT) = Ae^{-a(t-\tau)} \sum_{n=0}^{\infty} e^{-naT} \quad \text{(binomial series)}$$
$$= A\frac{e^{-a(t-\tau)}}{1 - e^{-aT}} \quad \text{(closed form)}$$

Using Eq. 5.54
$$y_{ss}(t) = v_0(t) = \int_{-T}^{0} x(\tau)A \frac{e^{-a(t-\tau)}}{1 - e^{-aT}} d\tau + \int_{0}^{t} x(\tau)Ae^{-a(t-\tau)} d\tau$$
$$v_0(t) = \frac{Ae^{-at}}{1 - e^{-aT}} \int_{-T}^{0} x(\tau)e^{a\tau} d\tau + Ae^{-at} \int_{0}^{t} x(\tau)e^{a\tau} d\tau$$

We must now substitute
$$x(t) = v(t) = V_m|\cos(\omega_T t/2)| = V_m|\cos(\pi t/T)|$$
and perform the integration. The integration is relatively simple, but *tedious*. The result was shown in Fig. 4.9 and is given by
$$\frac{v_0(t)}{V_m/\pi} = \begin{cases} 0.796^{-2.5t/T} + 1.015\cos(\pi t/T) + 1.276\sin(\pi t/T), & 0 \le t \le T/2 \\ 9.700e^{-2.5t/T} - 1.015\cos(\pi t/T) - 1.276\sin(\pi t/T), & T/2 \le t \le T \end{cases}$$
This closed form is much better to work with (numerically speaking) than the Fourier series if, for example, we are required to plot $v_0(t)/(V_m/\pi)$ versus t, as in Fig. 4.9.

The transfer function in the preceding example was simply
$$H(\omega) = \frac{A}{j\omega + a}$$
It was pointed out in Sec. 5.3 that for transfer functions of the form given by Eq. 5.47 a partial fraction expansion gives similar terms for an Nth order system:
$$H(\omega) = \frac{A_1}{j\omega + a_1} + \frac{A_2}{j\omega + a_2} + \cdots + \frac{A_N}{j\omega + a_N}$$
or
$$h(t) = (A_1 e^{-a_1 t} + A_2 e^{-a_2 t} + \cdots + A_N e^{-a_N t})u(t)$$

192 FOURIER TRANSFORM TECHNIQUES

for $M < N$ and no repeated roots of the characteristic equation. For use in Eq. 5.54 (dropping the unit-step function as before)

$$h(t - \tau + nT) = A_1 e^{-a_1(t-\tau+nT)} + A_2 e^{-a_2(t-\tau+nT)} + \cdots + A_N e^{-a_N(t-\tau+nT)}$$

and

$$\sum_{n=0}^{\infty} h(t - \tau + nT) = A_1 \frac{e^{-a_1(t-\tau)}}{1 - e^{-a_1 T}} + A_2 \frac{e^{-a_2(t-\tau)}}{1 - e^{a_2 T}} + \cdots + A_N \frac{e^{-a_N(t-\tau)}}{1 - e^{-a_N T}}$$

Thus, Eq. 5.54 will be relatively simple to implement for this Nth order system because the infinite sum is a sum of binomial series and has simple closed forms.

It should be pointed out that even if $M = N$ and (or) some of the roots of the characteristic equation are repeated, Eq. 5.54 will still be relatively simple to work with. Consider, for example, an element of the fundamental set (from a repeated root) that has the form Ate^{-at}. The form that we are concerned with in Eq. 5.54 is

$$\sum_{n=0}^{\infty} A(t - \tau + nT) e^{-a(t-\tau+nT)} = Ae^{-a(t-\tau)} \sum_{n=0}^{\infty} (t - \tau + nT) e^{-naT}$$

$$= Ae^{-a(t-\tau)}(t - \tau) \sum_{n=0}^{\infty} e^{-naT}$$

$$+ Ae^{-a(t-\tau)} \sum_{n=0}^{\infty} nTe^{-naT}$$

$$= A(t - \tau) \frac{e^{-a(t-\tau)}}{1 - e^{-aT}} + AT \frac{e^{-a(t-\tau+T)}}{(1 - e^{-aT})^2}$$

and, once again, simple closed forms are obtained because the infinite sums are binomial series.

5.5 FREQUENCY RESPONSE

From a practical point of view, considering the fact that we wanted a low-pass filter in Example 4 that would recover the dc component while essentially rejecting the fundamental and all its harmonics, it was not even necessary to go through the calculation of $v_0(t)$. The transfer function for the system is

$$H(\omega) = A \frac{1}{a + j\omega} = 250 \frac{1}{300 + j\omega} = \frac{5}{6} \frac{1}{1 + j\omega/(300)}$$

The frequency-response characteristics can be shown in many forms, but perhaps the most useful is called the *Bode diagram*.[1] The Bode diagram consists of two parts. The first is simply $|H(\omega)|$ in decibels (that is, $20 \log_{10} |H(\omega)|$) versus normalized frequency $(\omega/300)$ on a logarithmic scale. These logarithmic scales allow a very wide range of amplitudes and frequencies to be compressed and displayed in a compact graph. Let us, for the moment, concentrate on the factor $[1 + j\omega/(300)]^{-1}$. For low frequencies $[\omega/(300) \ll 1]$ this factor is simply 1 or 0 dB. For high frequencies $[\omega/(300) \gg 1]$ it has a *magnitude* of

300/ω, so that if ω is doubled (increased by one *octave*), the magnitude is halved (*down* by 6.02 dB). If ω increases by a factor of 10 (one *decade*), the magnitude goes down by a factor of 10 (*down* by 20 dB). It is not difficult to show that on a logarithmic scale the high-frequency part of the diagram [dB versus $\log_{10}(\omega/300)$] is a straight line sloping down to the right at -6 dB per octave or -20 dB per decade. Thus, we have two *asymptotes*: one for low frequencies (a constant 0-dB horizontal line) and one for high frequencies. They intersect at $\omega/(300) = 1$ or $\omega = 300$ (called the *corner* or *break* frequency). The factor $\frac{5}{6}$ is simply a horizontal line at -1.58 dB, and this merely shifts the rest of the magnitude response downward by 1.58 dB. The asymptotic $|H(\omega)|$ versus $\omega/300$ curve is shown in Fig. 5.7 in dashed lines, while the exact curve is shown solid. Notice that the two differ by 3.01 dB at $\omega/(300) = 1$. Also notice that the asymptotic magnitude curve for $[1 + j\omega/(300)]^{+1}$ is the same as that for $[1 + j\omega/(300)]^{-1}$, *except that it breaks upward* instead of downward.

The angle of $H(\omega)$, $\underline{|H(\omega)}$, is $0°$ for $\omega/(300) \ll 1$, $-45°$ for $\omega/(300) = 1$, and $-90°$ for $\omega/(300) \gg 1$. An asymptotic curve, consisting of straight lines, can be used for the phase (angle) plot also. One common method, as shown in Fig. 5.7, is to use break frequencies at $0.2\omega/(300)$ and $5\omega/(300)$, resulting in a downward sloping line at $-90°$ for a 25 to 1 change in frequency.

The location of the fundamental frequency ω_T (or $\omega_T/300 = 240\pi/300 = 2.513$) and the harmonics is also shown in Fig. 5.7. It is obvious from this Bode diagram why this low-pass filter did not perform well: the fundamental and many of the harmonics are simply not attenuated enough by the filter. Figure 5.7 would convince us to discard this filter and design one with sharper cutoff characteristics to remove the unwanted terms without calculating $v_0(t)$.

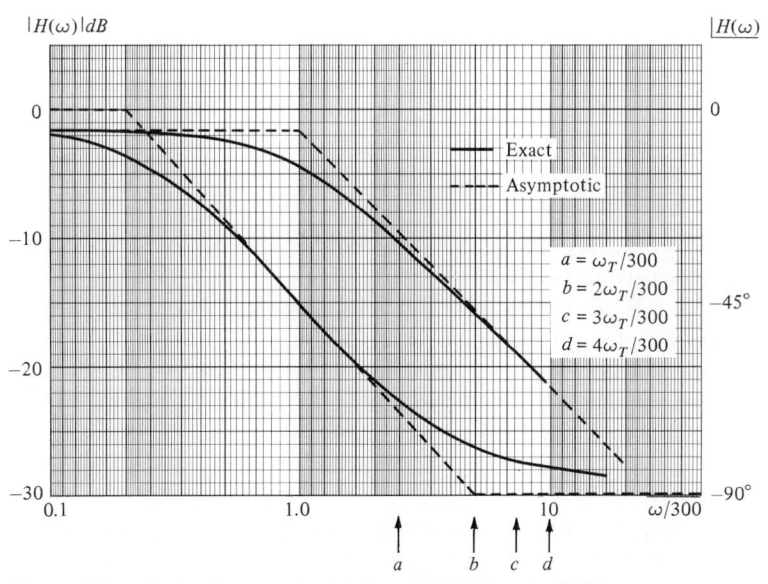

Figure 5.7. Bode diagram for $H(\omega) = (5/6)(1/1 + j\omega/300)$.

194 FOURIER TRANSFORM TECHNIQUES

Now that we know how to handle a constant term ($\frac{5}{6}$ in Example 4) and the form $(1 + j\omega/\omega_c)^{\pm 1}$ that was $[1 + j\omega/(300)]^{-1}$ in Example 4, we need to briefly examine two other forms that commonly occur in transfer functions. The first is $(j\omega)^{\pm n}, n = 0, 1, 2, \ldots$. It is easy to show that the magnitude response $(\omega^{\pm n})$ is simply a straight line with slope $\pm 20n$ dB per decade passing through 0 dB at $\omega = 1$. This is shown in Fig. 5.8(c). Also shown in the same figure is the angle of $(j\omega)^{\pm n}$ which is obviously just $\pm 90n°$. The second is the second-order form

$$[1 + j2\zeta(\omega/\omega_n) - (\omega/\omega_n)^2]^{-1}$$

This form can be obtained by factoring a constant from slightly different forms if necessary. Notice that if $j\omega$ is replaced by s, and the second-order form is set equal to 0, we obtain the characteristic equation of the system (Eq. 2.90). This is not an accident. Starting with Eq. 2.89:

$$\frac{d^2y}{dt^2} + 2\zeta\omega_n\frac{dy}{dt} + \omega_n^2 y = \omega_n^2 x \qquad (2.89)$$

and taking the Fourier transform

$$-\omega^2 Y_{zs}(\omega) + 2\zeta\omega_n j\omega Y_{zs}(\omega) + \omega_n^2 Y_{zs}(\omega) = \omega_n^2 X(\omega)$$
$$-\omega^2 + 2\zeta\omega_n j\omega + \omega_n^2 = \omega_n^2 X(\omega)/Y_{zs}(\omega)$$

or

$$H(\omega) = \frac{Y_{zs}(\omega)}{X(\omega)} = [1 + j2\zeta(\omega/\omega_n) - (\omega/\omega_n)^2]^{-1}$$

The best way to handle this form is to use ζ as a parameter (as for the unit-impulse and unit-step responses in Fig. 2.14), and let ω/ω_n be the variable. Asymptotic curves can be used, but they are of little help unless ζ is close to 1. Results are shown in Fig. 5.8(d). Notice that the magnitude response ultimately falls off with slope -40 dB per decade and the total phase shift is 180° (Why?). How would you handle the form

$$[1 + j2\zeta(\omega/\omega_n) - (\omega/\omega_n)^2]^{+1}$$

Finally notice that Bode diagrams only include $\omega > 0$. Symmetry conditions allow us to easily deduce what happens for $\omega < 0$. Frequency-response curves in the form of Bode diagrams, or otherwise, are ideally suited for numerical computation with the programmable calculator.

It is worthwhile to correlate Fig. 2.14(b), which gives the unit-step response of the second-order system ($\zeta = 0, 0.5, 1.0,$ and 2.0), with Fig. 5.8(d), which gives the frequency response of the same system. We do this in terms of the 10 to 90 percent rise time, T_r, and the 3-dB (or half-power) width of response or bandwidth, β, in hertz. Reading values from Fig. 2.14(b) and Fig. 5.8, respectively, we find that $T_r\beta \approx 0.34$ for $\zeta = 0.5$, $T_r\beta \approx 0.35$ for $\zeta = 1.0$, and $T_r\beta \approx 0.35$ for $\zeta = 2.0$. Thus, $T_r\beta$ is about 0.35 for this (and nearly every) system. This product is nearly constant. This is very revealing. It answers such questions

5.5 FREQUENCY RESPONSE   195

Figure 5.8. Bode diagrams. (a) $(1 + j\omega/\omega_c)^{-1}$. (b) $(1 + j\omega/\omega_c)^{+1}$. (c) $(j\omega)^{\pm n}$. (d) $[1 + j2\zeta(\omega/\omega_n) - (\omega/\omega_n)^2]^{-1}$.

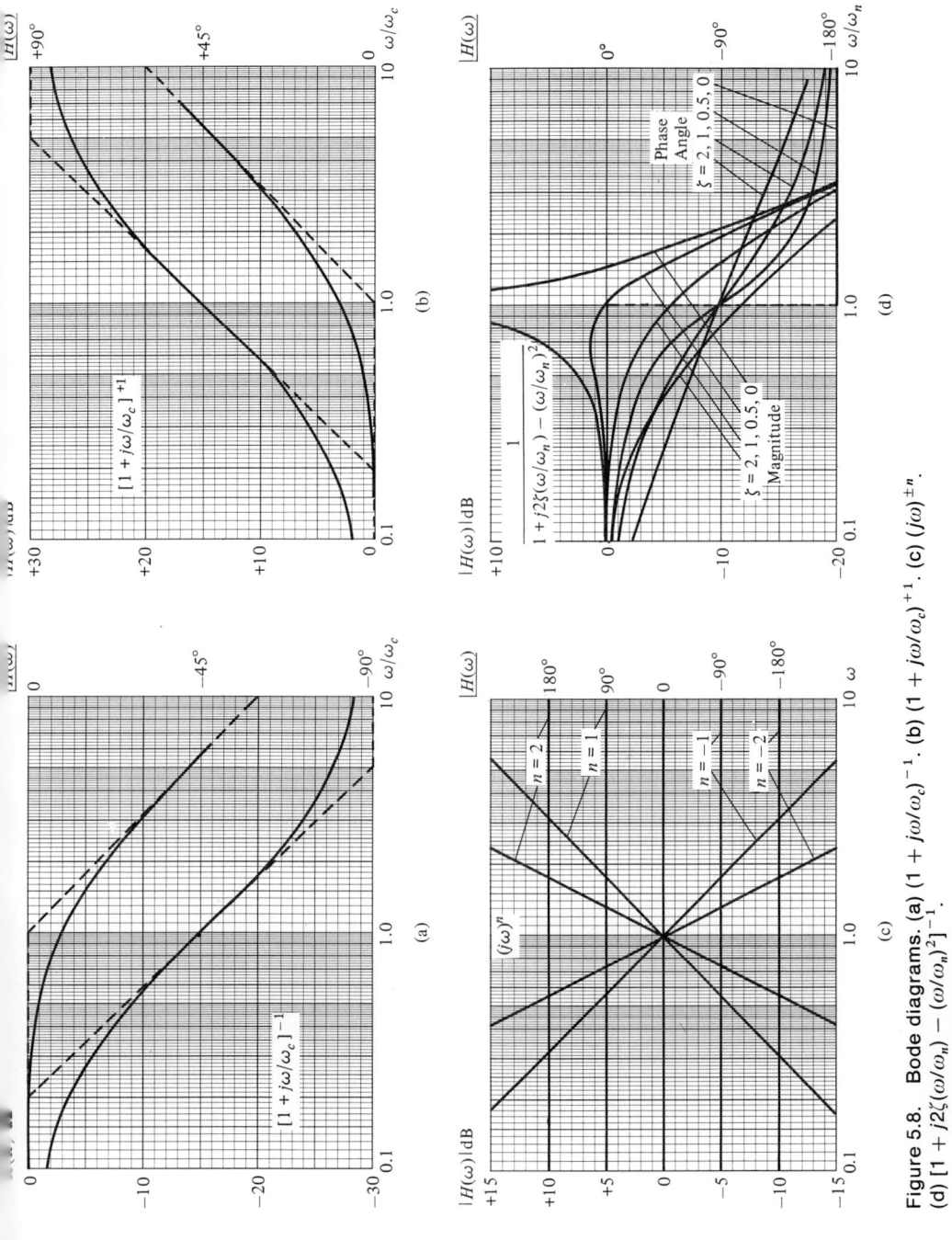

196 FOURIER TRANSFORM TECHNIQUES

as, What do I have to pay for if I try to increase the information per unit time (smaller rise time) in a communication system? The answer is simple: You pay for (or require) more bandwidth. As a rule-of-thumb idea, it is well to remember that *rise time multiplied by bandwidth is* (almost) *constant for a wide variety of systems.*

Example 5

We want to obtain the Bode diagram for

$$H(\omega) = \frac{400(j\omega)}{(2+j\omega)(-\omega^2 + j20\omega + 100)}$$

Factor 2 out of the first term in the denominator, and factor 100 out of the second term in the denominator. This gives the standard form we want:

$$H(\omega) = \frac{2(j\omega)}{(1+j\omega/2)[1 + j2(\omega/\omega_n) - (\omega/\omega_n)^2]}$$

where $\omega_n = 10$ and $\zeta = 1$. The various pieces that must be *added* for the magnitude and angle responses are shown in Fig. 5.9(a) and (b), respectively, while the overall responses (Bode diagram) are shown in Fig. 5.9(c).

In Example 7, Chap. 4, we examined a class C amplifier tuned circuit. It was a parallel RLC network. Let us now go to the frequency domain to determine more explicitly why it worked so well, even though it was relatively simple.

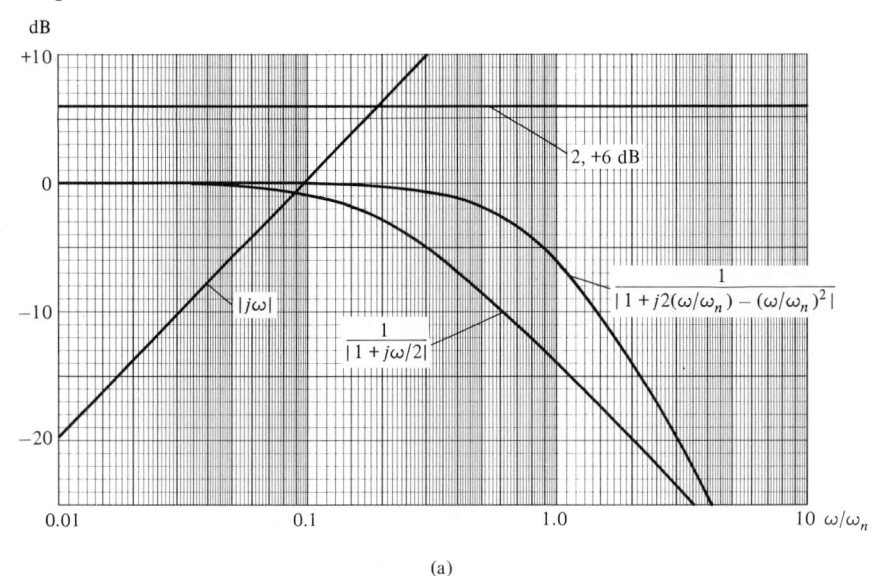

Figure 5.9. Frequency response curves. (a) Separate magnitude parts. (b) Separate angle parts. (c) Bode diagram.

5.5 FREQUENCY RESPONSE

(b)

(c)

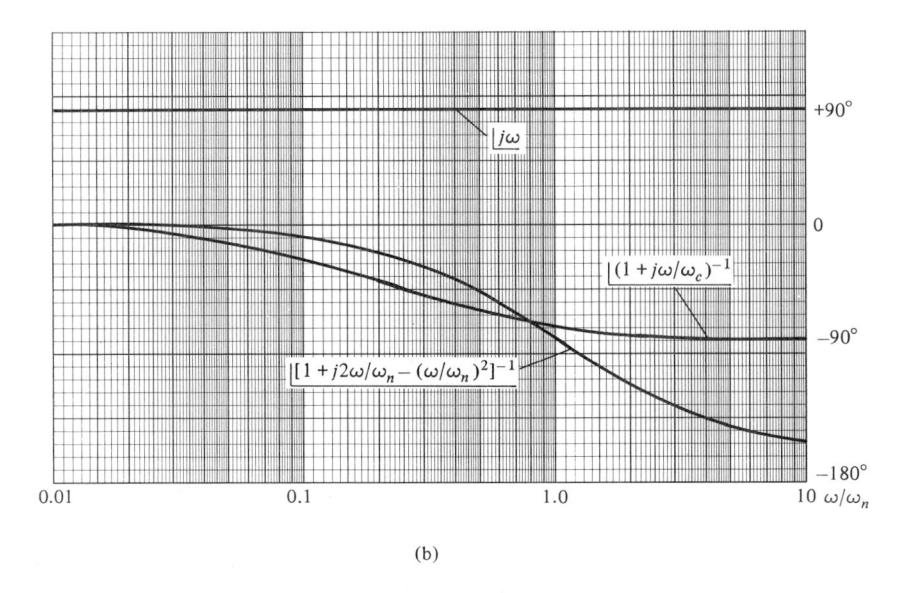

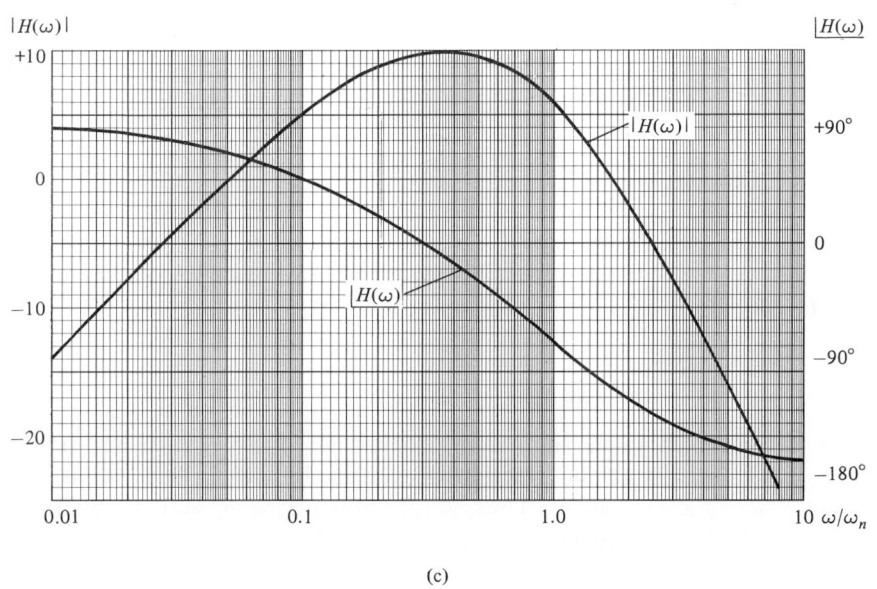

Figure 5.9 *(continued)*

Example 6

Plot $|H(\omega)|$ and $\underline{/H(\omega)}$ versus frequency for the parallel RLC network (resonant at $\omega_T = 2\pi \times 10^5$ rad/s) mentioned above. We found that

$$H(\omega) = \frac{1}{1/R + 1/j\omega L + j\omega C} = Z(\omega)$$

198 FOURIER TRANSFORM TECHNIQUES

or

$$H(\omega) = \frac{j\omega L}{1 + j\omega L/R - \omega^2 LC}$$

With $\omega_n = \omega_T = 2\pi \times 10^5$

$$H(\omega) = L\frac{j\omega}{1 + j(1/R)\sqrt{L/C}(\omega/\omega_n) - (\omega/\omega_n)^2}$$

The transfer function is now in standard form for making a Bode diagram, but before doing so, it might be wise to determine ζ. By inspection

$$\zeta = \frac{1}{2R}\sqrt{\frac{L}{C}} = \frac{1}{10^4}\sqrt{\frac{10^{-3}/(2\pi)}{10^{-6}/20\pi}} = 10^{-2}$$

which is very small and gives a very high peak in the magnitude response (Fig. 5.8).

There is another way to view this system. The *quality factor Q* is defined as 2π times the ratio of the maximum energy stored to the total energy dissipated per cycle. The half-power bandwidth, β, in this case (high Q) lies between *two* frequencies, ω_1 and ω_2. The reader should be familiar with these terms. Upon calculating the Q at the resonant frequency $(\omega_n = \omega_T)$,[2] we find

$$Q_n = \frac{R}{\omega_n L} = \omega_n RC = 2\pi \times 10^5 \times 5 \times 10^3 \times 10^{-6}/(20\pi)$$

$Q_n = 50$ (high Q)

If we now go back to the original equation for $H(\omega)$ and use some algebra, we find that

$$H(\omega) = \frac{R}{1 + jQ_n(\omega/\omega_n - \omega_n/\omega)}$$

Half power occurs when $|H(\omega)|$ is $1/\sqrt{2}$ (that is, $(1/\sqrt{2})^2 = \frac{1}{2}$, or half power) times its value (R) at the resonant frequency ($\omega = \omega_n$). This means that $|H(\omega)| = R/\sqrt{2}$ at ω_1 and ω_2. Thus, we are looking for ω_1 and ω_2, the lower and upper half-power frequencies, respectively, where

$$|H(\omega)| = \frac{R}{\sqrt{2}} = \frac{R}{\sqrt{1^2 + 1^2}} = \frac{R}{\sqrt{1 + Q_n^2(\omega/\omega_n - \omega_n/\omega)^2}}$$

Thus

$$Q_n(\omega/\omega_n - \omega_n/\omega) = \pm 1$$

or

$$Q_n(\omega_1/\omega_n - \omega_n/\omega_1) = -1 \qquad Q_n(\omega_2/\omega_n - \omega_n/\omega_2) = +1$$

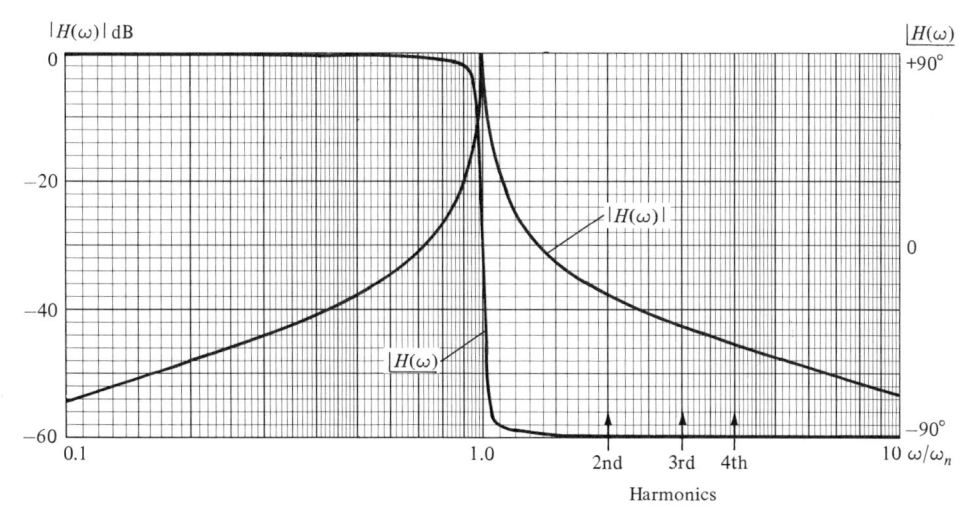

Figure 5.10. Bode diagram for Example 6.

Solving these, we have

$$\omega_1 = \omega_n\left[\sqrt{1 + (1/2Q_n)^2} - \frac{1}{2Q_n}\right]$$

$$\omega_2 = \omega_n\left[\sqrt{1 + (1/2Q_n)^2} + \frac{1}{2Q_n}\right]$$

But $\zeta = 1/2Q_n$, so

$$\omega_1 = \omega_n[\sqrt{1 + \zeta^2} - \zeta]$$

$$\omega_2 = \omega_n[\sqrt{1 + \zeta^2} + \zeta]$$

The difference is

$$\beta = \omega_2 - \omega_1 = 2\omega_n\zeta = \frac{\omega_n}{Q_n} = 0.126 \times 10^5$$

The answer to our original enquiry should now be obvious. The bandwidth is so small that all terms except the fundamental are very much attenuated. A plot of $|H(\omega)|/R$ in decibels versus ω/ω_n on a logarithmic scale (as before) and a plot of $\underline{/H(\omega)}$ in degrees versus ω/ω_n are shown in Fig. 5.10. Notice that the harmonics occur for $\omega/\omega_n = 2, 3, 4, \ldots$, and are very much attenuated.

5.6 ADDITIONAL EXAMPLES

We will now proceed with some simple examples of the use of Fourier transform methods in solving certain problems. This is intended to demonstrate that these problems can be solved by this method, although this may not be the best method in all cases. The *Laplace* transform method, to be introduced in the next chapter, is normally better suited to treat transient-type problems where switches are suddenly opened or closed and where initial conditions exist.

Example 7

Find the unit-step response of the coupling stage below. The equivalent circuit in Fig. 5.11(b) is more convenient. We have

$$V_0(\omega) = \frac{R_g V(\omega)}{R_{11} + R_g + 1/j\omega C_c} = I_0 \frac{R_{11} R_g}{R_{11} + R_g} [\pi\delta(\omega) + 1/j\omega] \frac{j\omega}{j\omega + \frac{1}{(R_{11} + R_g)C_c}}$$

or

$$V_0(\omega) = I_0 \frac{R_{11} R_g}{R_{11} + R_g} \frac{1}{j\omega + \frac{1}{(R_{11} + R_g)C_c}} = 64 \frac{1}{j\omega + 25}$$

So

$$v_0(t) = 64e^{-25t} u(t) \text{ V}$$

A Bode plot of $H(\omega)$ for this high-pass coupling network will show how well it performs at the low end of the audio spectrum. On the other hand, we can easily determine the sag (or decay) over a time equal to one period at a frequency of 60 Hz (for example):

$$v_0(\tfrac{1}{60}) = 64e^{-25/60} = 42.2 \text{ V}$$

The performance will be moderately good at low frequencies.

A 90° phase shift network is required in one method of generating a single-sideband signal for a communication system.

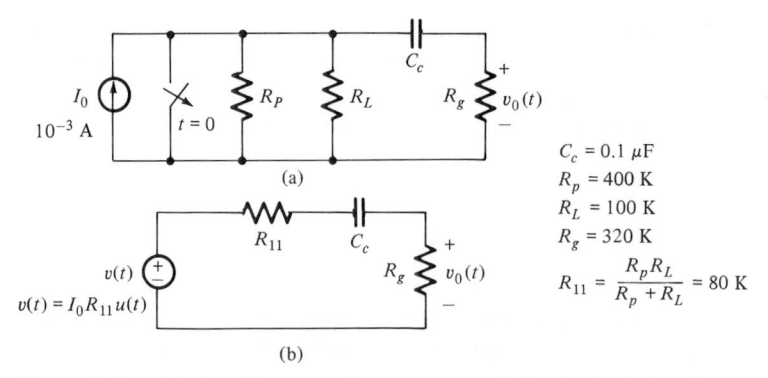

$C_c = 0.1 \ \mu\text{F}$
$R_p = 400 \text{ K}$
$R_L = 100 \text{ K}$
$R_g = 320 \text{ K}$
$R_{11} = \dfrac{R_p R_L}{R_p + R_L} = 80 \text{ K}$

$v(t) = I_0 R_{11} u(t)$

Figure 5.11. (a) Amplifier coupling network. (b) Equivalent circuit.

Example 8

Suppose we have a carrier signal $x(t) = A_0\cos(\omega_0 t)$, and we want to produce a $-90°$ phase shift in this signal. That is, we want to produce

$$y(t) = A_0\cos(\omega_0 t - \pi/2) = A_0\sin(\omega_0 t)$$

as indicated in Fig. 5.12. Actually, it is only necessary to produce this phase shift over a *band* of frequencies, but suppose we are ambitious and attempt to accomplish this feat over *all* frequencies. Inspection of Table 5.1 reveals that

$$A_0\cos(\omega_0 t) \leftrightarrow \pi[\delta(\omega - \omega_0) + \delta(\omega + \omega_0)]$$

and

$$A_0\sin(\omega_0 t) \leftrightarrow \frac{\pi}{j}[\delta(\omega - \omega_0) - \delta(\omega + \omega_0)]$$

Inspection of the preceding spectra reveals that the phase shifter must be the *frequency-domain device* with transfer function $-j\,\text{sgn}(\omega)$, because

$$-j\,\text{sgn}(\omega)\pi[\delta(\omega - \omega_0) + \delta(\omega + \omega_0)] = \frac{\pi}{j}[\delta(\omega - \omega_0) - \delta(\omega + \omega_0)]$$

as shown in Fig. 5.13. Before rejoicing, it would be wise to find the unit-impulse response of this device. Using the table we have

$$h(t) = \frac{1}{\pi t} \leftrightarrow -j\,\text{sgn}(\omega) = H(\omega)$$

which is *noncausal*, and therefore this *ideal* $-90°$ phase shifter cannot be built! A usable band-limited version can be built. This example is one that demonstrates the usefulness of the Fourier transform at its best.

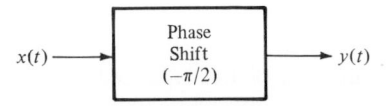

Figure 5.12. Block diagram of a $-90°$ phase shifter.

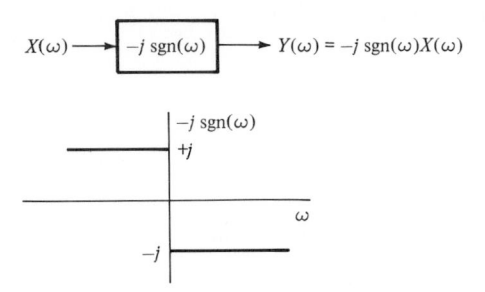

Figure 5.13. Another block diagram of the ideal $-90°$ phase shifter.

202 FOURIER TRANSFORM TECHNIQUES

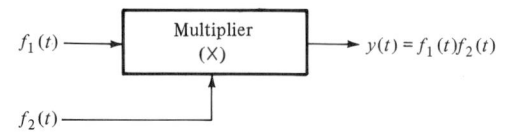

```
f₁(t) ──→ ┌─────────────┐
          │  Multiplier │ ──→ y(t) = f₁(t)f₂(t)
          │     (X)     │
f₂(t) ──→ └─────────────┘
```

Figure 5.14. Balanced modulator.

An ideal low-pass filter, $H(\omega) = u(\omega + \omega_c) - u(\omega - \omega_c)$, where ω_c is the cutoff frequency, is another example of a mathematical model which cannot be built because it is noncausal. The intuitive reason why these filters cannot be built is that the infinite slope at the cutoff frequency (for the low pass) cannot be duplicated in practice. In this regard it can be shown that for all $H(\omega)$ satisfying

$$\int_{-\infty}^{\infty} |H(\omega)|^2 \, d\omega < \infty$$

a necessary and sufficient condition on $|H(\omega)|$ in order for a filter to be physically realizable is the *Paley-Wiener* criterion:[3]

$$\int_{-\infty}^{\infty} \left| \frac{\ln|H(\omega)|}{1 + \omega^2} \right| d\omega < \infty$$

For a filter which can be built, then, $|H(\omega)|$ may not fall off to 0 faster than a function of simple exponential order, $e^{-a|\omega|}$. The attenuation may not be infinite at any finite ω.

One of the most popular and useful analog devices in systems analysis, particularly so in communication systems, is the *balanced modulator*, which is nothing more than a time-domain multiplier. A block diagram is shown in Fig. 5.14. We showed in Chap. 1 that it represents a nonlinear system.

Example 9

The balanced modulator is often used to produce a double-sideband-suppressed carrier signal. If $f_1(t) = A_0 \cos(\omega_0 t)$ is the carrier signal, and $f_2(t)$ is the modulation signal, what is the spectrum of the output, $y(t)$, of the balanced modulator? From Table 5.1 we have

$$y(t) = f_1(t)f_2(t) \leftrightarrow \frac{1}{2\pi} \int_{-\infty}^{\infty} F_1(u) F_2(\omega - u) \, du = Y(\omega)$$

where

$$F_1(\omega) = \pi A_0 [\delta(\omega - \omega_0) + \delta(\omega + \omega_0)]$$

Thus

$$Y(\omega) = \frac{1}{2\pi} \int_{-\infty}^{\infty} \pi A_0 [\delta(\omega - \omega_0) + \delta(u + \omega_0)] F_2(\omega - u) \, du$$

$$Y(\omega) = \frac{A_0}{2} [F_2(\omega - \omega_0) + F_2(\omega + \omega_0)]$$

This spectrum is the same as that of $F_2(\omega)$ itself, except that it consists of two parts: one centered at ω_0, the other centered at $-\omega_0$. Each part is altered in amplitude, *not shape*, by the scale factor $A_0/2$. $F_2(\omega)$ and $Y(\omega)$ are shown in Fig. 5.15(a) and (b), respectively, for an arbitrary $f_2(t)$. $F_2(\omega)$ is also (therefore) arbitrary, except that $F_2(0) = 0$. That is, we do not modulate down to $\omega = 0$, and the spectrum of $f_2(t)$ contains only frequencies between ω_l and ω_h. This is reflected in the spectrum of $Y(\omega)$. We see upper sidebands and lower sidebands, but no carrier term, in Fig. 5.15(b). Hence, the name double-sideband(s)-suppressed carrier signal. This example has also demonstrated what is often

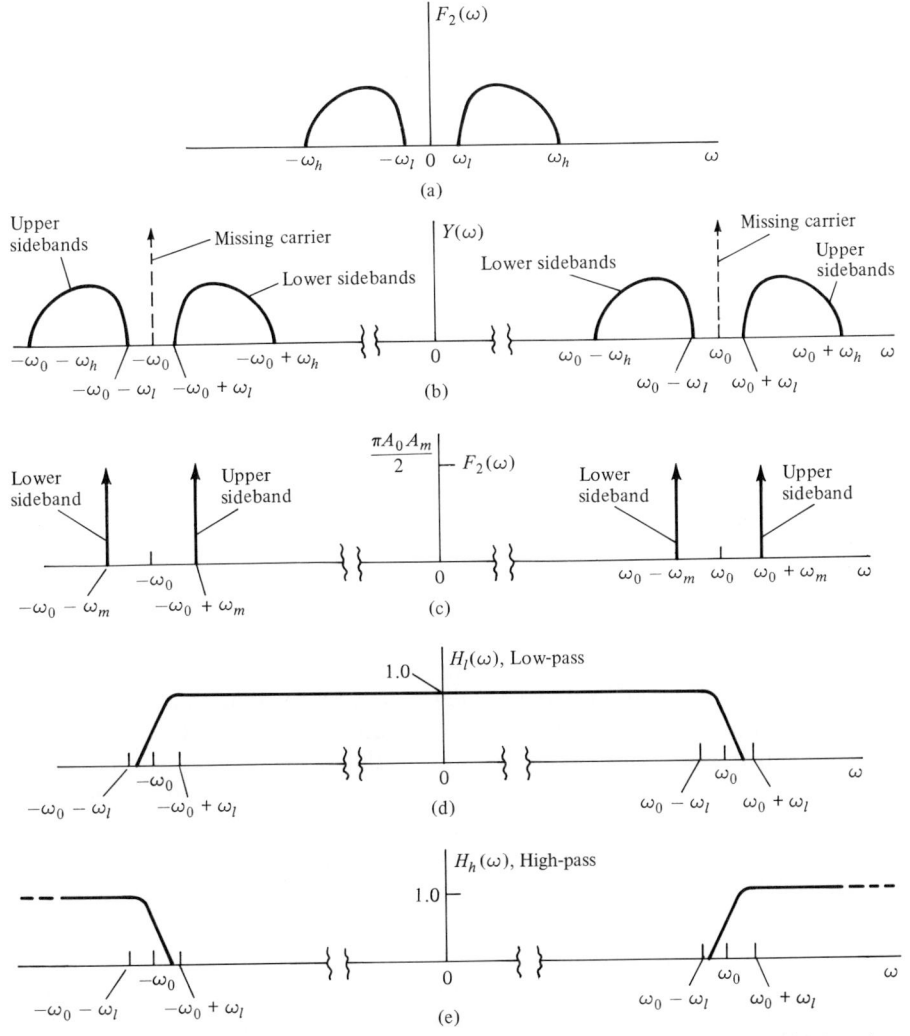

Figure 5.15. (a) $F_2(\omega)$ with $F_2(0) = 0$. (b) $Y(\omega)$, the spectrum of a double-sideband suppressed carrier signal for the modulating signal $f_2(t) \leftrightarrow F_2(\omega)$. (c) $Y(\omega)$ for $f_2(t) = A_m \cos(\omega_m t)$. (d) Low-pass filter for removing the upper sidebands. (e) High-pass filter for removing the lower sidebands.

called the *amplitude modulation operation*: $f(t)\cos(\omega_0 t) \leftrightarrow F(\omega+\omega_0)/2 + F(\omega-\omega_0)/2$.

This can perhaps be seen even more clearly if we let $f_2(t) = A_m\cos(\omega_m t)$, so that

$$F_2(\omega) = \pi A_m[\delta(\omega-\omega_m) + \delta(\omega+\omega_m)]$$

which, when used above in $Y(\omega)$ gives

$$Y(\omega) = \frac{\pi A_0 A_m}{2}[\delta(\omega-\omega_0-\omega_m) + \delta(\omega-\omega_0+\omega_m)$$
$$+ \delta(\omega+\omega_0-\omega_m) + \delta(\omega+\omega_0+\omega_m)]$$

This result is given in Fig. 5.15(c), and is a double-sideband-suppressed carrier signal.

The double-sideband-suppressed carrier signal is more efficient than the *ordinary amplitude-modulated* (AM) signal since the AM signal has a spectrum identical to that in Fig. 5.15(b), except that the carrier term (impulses at $\pm\omega_0$) is present rather than missing.[†] That is, the power in the carrier, which would be wasted in the ordinary AM signal, is not wasted in the double-sideband-suppressed carrier signal. Figure 5.15(b) clearly shows that the required bandwidth is $2\omega_h$ (that is, twice the highest frequency in the modulating signal) and is the *same* in either case. Considering a *complete* communication system, we must take note of the fact that *at a remote receiver we do not know where* (in the spectrum) *the carrier is when it is suppressed*! This is a distinct difficulty.

It is obvious from Figure 5.15(d) or (e) that either the upper or lower sidebands, respectively, can be removed with a low- or a high-pass filter,[‡] respectively. This further improves the efficiency since the required bandwidth for the resulting *single-sideband system* is only $\omega_h - \omega_l$, which is slightly less than half that ($2\omega_h$) for the ordinary AM case or the double-sideband-suppressed carrier case. Notice that the filters in either case must have *sharp cutoff characteristics*. That is, the transfer functions, $H(\omega)$, must have a transition from uniform attenuation and time delay to very large attenuation in a frequency span of $2\omega_l$. If we desire to modulate down to 50 Hz, for example, then this transition must occur over (only) 100 Hz! This requires very careful filter design and is *part* of the price to be paid for the improved efficiency.

What is the time function of a single-sideband signal? The answer to this question leads us to another method (other than filtering) of producing the single-sideband signal. Suppose that we want to retain the upper sidebands (USB) only, and thus require a high-pass filter. For the sake of simplicity in the mathematics, we use an *ideal* high-pass filter with cutoff frequency ω_0 that is sufficient but not necessary in this application. We already suspect (below

---

[†] Adding the carrier term at the output of the balanced modulator (Fig. 5.14) is one method of producing an ordinary AM signal.
[‡] Bandpass filters can also be used.

5.6 ADDITIONAL EXAMPLES

Example 8) that we cannot build an ideal high-pass filter, but we can build a high-pass filter that comes very close to performing as outlined in the preceding paragraph.

The transfer function of the ideal high-pass filter that we need is

$$H_h(\omega) = u(-\omega - \omega_0) + u(\omega - \omega_0)$$

The output of the high-pass filter is

$$Y_h(\omega) = H_h(\omega) Y(\omega)$$

$$= \frac{A_0}{2}[F_2(\omega - \omega_0) + F_2(\omega + \omega_0)][u(-\omega - \omega_0) + u(\omega - \omega_0)]$$

Upon carrying out the indicated multiplication, it is found that two of the terms are 0 leaving

$$Y_h(\omega) = \frac{A_0}{2} F_2(\omega - \omega_0)u(\omega - \omega_0) + \frac{A_0}{2} F_2(\omega + \omega_0)u(-\omega - \omega_0)$$

Since $u(-\eta) = 1 - u(\eta)$, the last equation becomes

$$Y_h(\omega) = \frac{A_0}{2}[F_2(\omega + \omega_0) - F_2(\omega + \omega_0)u(\omega + \omega_0) + F_2(\omega - \omega_0)u(\omega - \omega_0)]$$

The last two terms in the preceding equation have the same form, $F_2(\omega)u(\omega)$, and using convolution

$$\int_{-\infty}^{\infty} f_2(t - \tau)\left[\frac{1}{2}\delta(\tau) + \frac{j}{2\pi\tau}\right] d\tau \leftrightarrow F_2(\omega)u(\omega)$$

or

$$\frac{1}{2}f_2(t) + \frac{j}{2}\int_{-\infty}^{\infty} \frac{f_2(t - \tau)}{\pi\tau} d\tau \leftrightarrow F_2(\omega)u(\omega)$$

This result, together with the frequency shift property, allow us to find the time function corresponding to $Y_h(\omega)$:

$$y_h(t) = f_2(t)\frac{A_0}{2}\cos(\omega_0 t) - \frac{A_0}{2}\sin(\omega_0 t)\int_{-\infty}^{\infty} \frac{f_2(t - \tau)}{\pi\tau} d\tau \quad \text{(USB)}$$

It is left as an exercise for the reader to show that obtaining the lower sidebands (LSB) with an ideal low-pass filter gives

$$y_l(t) = f_2(t)\frac{A_0}{2}\cos(\omega_0 t) + \frac{A_0}{2}\sin(\omega_0 t)\int_{-\infty}^{\infty} \frac{f_2(t - \tau)}{\pi\tau} d\tau \quad \text{(LSB)}$$

The sum of the last two results is $f_2(t)A_0\cos(\omega_0 t)$, and this is the original double-sideband-suppressed carrier signal.

It is easy to convert $(A_0/2)\cos(\omega_0 t)$ into $(A_0/2)\sin(\omega_0 t)$ at the (single) carrier frequency ω_0 by using a $-90°$ phase shifter. Furthermore, we showed

206 FOURIER TRANSFORM TECHNIQUES

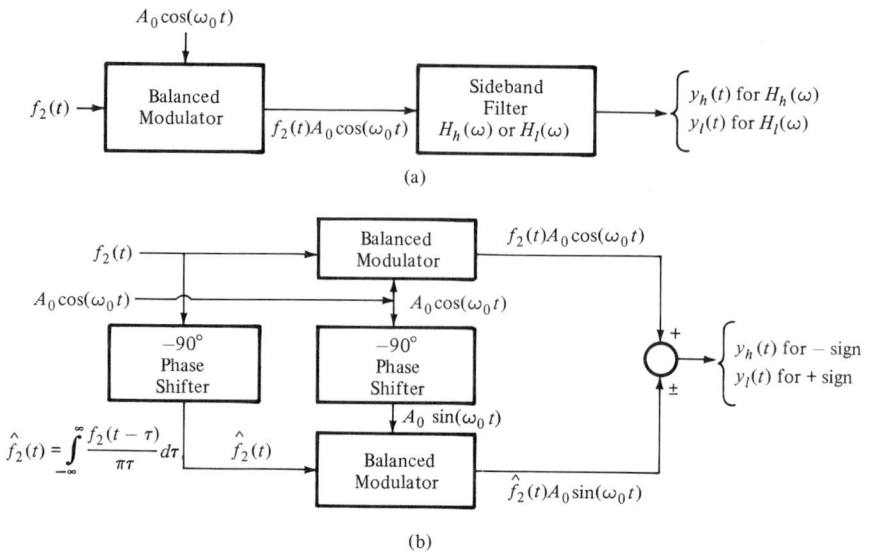

Figure 5.16. (a) Filter method for producing single-sideband AM. (b) Phase method for producing single-sideband AM.

in Example 8 that

$$\int_{-\infty}^{\infty} \frac{f_2(t-\tau)}{\pi\tau} d\tau \leftrightarrow F_2(\omega)[-j\,\text{sgn}(\omega)]$$

and this corresponded to passing $f_2(t)$ through a $-90°$ phase shifter. We can now easily construct a block diagram using $-90°$ phase shifters to produce single-sideband AM. Remember that the phase shifter that must produce $-90°$ phase shift over a *band* of frequencies is no easier to construct than the filters in the other (filter) scheme. The two single-sideband systems are shown in Fig. 5.16.

5.7 SIMULTANEOUS EQUATIONS

We next return to Example 11 of Chap. 2 where we were examining Eq. 2.110:

$$\mathbf{y} = \int_{-\infty}^{\infty} \mathbf{hx}\, d\tau \tag{2.110}$$

or

$$\mathbf{y}(t) = \int_{-\infty}^{t} \mathbf{h}(t-\tau)\mathbf{x}(\tau)\, d\tau$$

for the time-invariant causal system that was being examined. It was mentioned that Fourier (or Laplace) methods would make it much easier to find the required unit-impulse responses. The coupled equations for Fig. 2.15 were Eqs. 2.111 and 2.112, and transforming them (or simply writing the phasor forms) gives

$$I_1 = \left(\frac{1}{R} + \frac{1}{j\omega L}\right)V_1 - \left(\frac{1}{R}\right)V_2$$

$$I_2 = -\left(\frac{1}{R}\right)V_1 + \left(\frac{1}{R} + j\omega C\right)V_2$$

This set of coupled algebraic equations is as easy to solve as those obtained using the p operator (2.113, 2.114). Using the parameter values given in Example 11, Chap. 2 ($R = 40\ \Omega$, $L = 5$ H, and $C = 0.05$ F):

$$V_1 = \frac{-40\omega^2 + 20(j\omega)}{(j\omega + 4 - \sqrt{12})(j\omega + 4 + \sqrt{12})} I_1 + \frac{20(j\omega)}{(j\omega + 4 - \sqrt{12})(j\omega + 4 + \sqrt{12})} I_2$$

$$V_2 = \frac{20(j\omega)}{(j\omega + 4 - \sqrt{12})(j\omega + 4 + \sqrt{12})} I_1 + \frac{20(j\omega) + 160}{(j\omega + 4 - \sqrt{12})(j\omega + 4 + \sqrt{12})} I_2$$

or, with 2.115

$$V_1 = \frac{-40\omega^2 + 20(j\omega)}{(j\omega - s_1)(j\omega - s_2)} I_1 + \frac{20(j\omega)}{(j\omega - s_1)(j\omega - s_2)} I_2$$

$$V_2 = \frac{20(j\omega)}{(j\omega - s_1)(j\omega - s_2)} I_1 + \frac{20(j\omega) + 160}{(j\omega - s_1)(j\omega - s_2)} I_2$$

The transformed (or phasor) unit-impulse responses are

$$H_{11} = \frac{-40\omega^2 + 20(j\omega)}{(j\omega - s_1)(j\omega - s_2)} \qquad H_{12} = \frac{20(j\omega)}{(j\omega - s_1)(j\omega - s_2)}$$

$$H_{21} = \frac{20(j\omega)}{(j\omega - s_1)(j\omega - s_2)} \qquad H_{22} = \frac{20(j\omega) + 160}{(j\omega - s_1)(j\omega - s_2)}$$

Using transform pairs 6 and 34 in Table 5.1 we can obtain results that are identical to Eqs. 2.117, 2.118, 2.119, and 2.120 for h_{11}, h_{12}, h_{21}, and h_{22}, respectively.

The algebra for problems of this type is even simpler when we use the Laplace transform of the next chapter! It[†] requires that the excitation (i_1 and i_2) begin at $t = 0$, but since we are only finding unit-impulse responses for a time-invariant linear system in this example, and the impulses are applied at $t = 0$, the Laplace method works well.

5.8 THE SAMPLING THEOREM

Thus far we have primarily considered signals in analog (continuous) form in this chapter. The digital (discrete) form is also important (for numerical work, for example), and in this section we would like to determine what is necessary

---
[†] That is, the single-sided (unilateral) Laplace transforms.

208 FOURIER TRANSFORM TECHNIQUES

to convert an analog signal into a discrete one, or a discrete signal into an analog signal, without distortion if possible (see Sec. 4.7). The connection between the two types of signals is given by the *sampling theorem*. It states that a band-limited signal that has no spectral components above ω_h is uniquely determined by its values (samples) spaced at uniform intervals no greater than π/ω_h s apart. ω_h is the *highest* frequency in the signal.

Consider such a signal and its band-limited spectrum $F(\omega)$ as shown in Fig. 5.17(a) and (b). We want to sample $f(t)$ with the periodic rectangular pulse of Fig. 5.17(c) whose Fourier (series) coefficient spectrum is shown in Fig. 5.17(d). The coefficients were found earlier and given by Eq. 5.3:

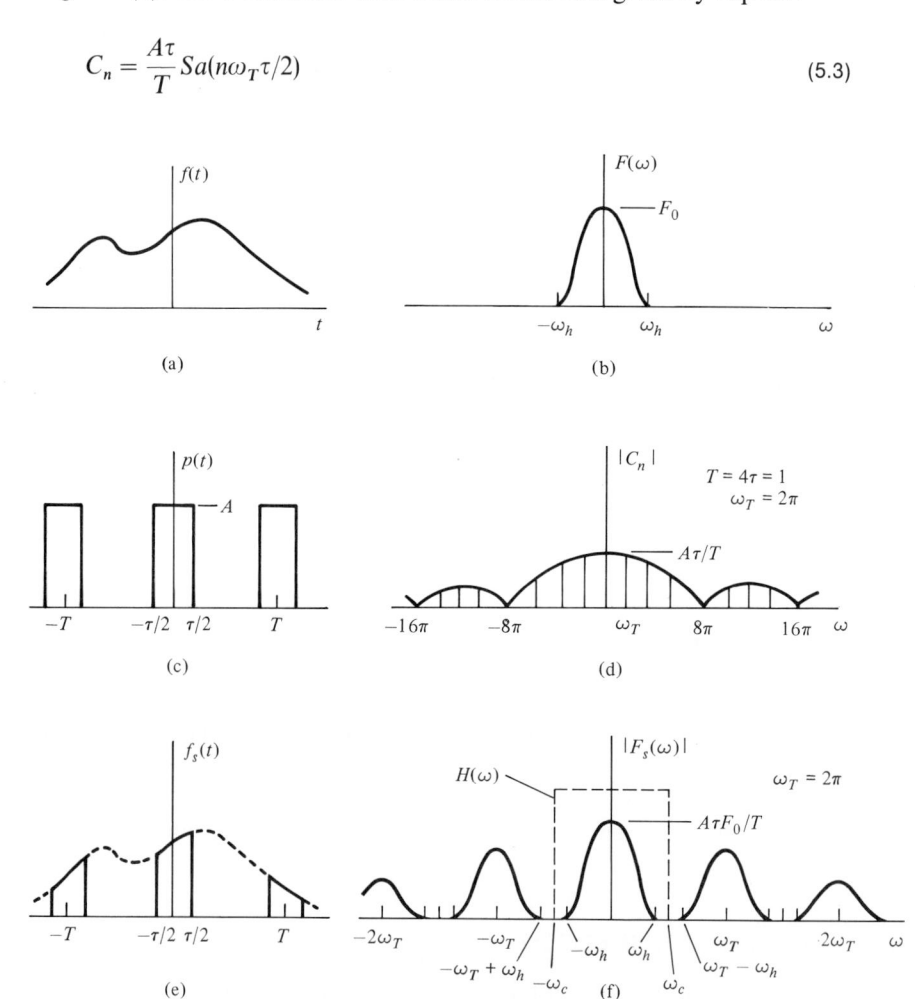

$$C_n = \frac{A\tau}{T} Sa(n\omega_T \tau/2) \tag{5.3}$$

Figure 5.17. (a) f(t). (b) Band-limited spectrum of f(t). (c) Periodic rectangular sampling pulse, p(t). (d) Fourier series coefficients of f(t), $T = 4\tau = 1$ s. (e) $f_s(t) = f(t)p(t)$. (f) $|F_s(\omega)|$. Notice the difference in the frequency scale between (d) and (f).

The sampled signal is $f_s(t)$ and is normally obtained as the product (using a balanced modulator, for example) of $f(t)$ and the periodic rectangular pulse $p(t)$:

$$f_s(t) = f(t)p(t)$$

$$= f(t) \sum_{n=-\infty}^{\infty} C_n e^{jn\omega_T t}$$

The spectrum of $f_s(t)$ is easy to obtain if we use Table 5.1 (No. 3).

$$F_s(\omega) = \sum_{n=-\infty}^{\infty} C_n F(\omega - n\omega_T)$$

Thus, we see here, as in Example 9, that the spectrum is, except for a factor C_n, a series of shifted versions of $F(\omega)$ centered at 0, $\pm\omega_T$, $\pm 2\omega_T$,.... $f_s(t)$ and $F_s(\omega)$ are shown in Fig. 5.17(e) and (f) where it can be seen that $\omega_T - \omega_h$ must be greater than ω_h to prevent overlapping and a condition called aliasing.[†] Thus, to prevent aliasing and to allow recovery of $F(\omega)$ by placing a low-pass filter around the center of $F_s(\omega)$, for example, we require

$$\omega_T - \omega_h \geq \omega_h$$

or

$$\omega_T \geq 2\omega_h$$

or, since $\omega_T = 2\pi/T$

$$\boxed{T \leq \pi/\omega_h} \tag{5.55}$$

This is the *Nyquist interval*. This result indicates that we must sample the signal at least twice within an interval of time corresponding to one period $(2\pi/\omega_h)$ for the highest frequency present. It also says that the *number of samples per second*, n_s, must be such that $n_s \geq \omega_h/\pi$. If we now return to Fig. 4.12 for periodic functions, we find that we were taking $N + 1$ samples for a period of time T (where this is the period of the periodic function). Thus, for that case $n_s = (N + 1)/T$, and, using Eq. 4.67, $N/T \geq (2M/T) + (1/T)$, or $n_s - 1/T \geq (2\pi M/\pi T) + (1/T)$, or $n_s \geq \omega_h/\pi + 2/T$. For large T this result agrees with that given by the sampling theorem.

A familiar example of undersampling is the wagon wheel in old motion pictures where the wheel *appears* to be turning in the wrong direction because the frame frequency (rate at which samples or photographs are taken) is too low in relation to the angular velocity (frequency) of the wheel. We see an alias.

[†] This term was first encountered in Sec. 4.7.

If we place an ideal low-pass filter (with a cutoff frequency ω_c) around $F_s(\omega)$, as in Fig. 5.17(f), we obtain the spectrum

$$F_s(\omega)[u(\omega + \omega_c) - u(\omega - \omega_c)]$$

$$= \sum_{n=-\infty}^{\infty} C_n F(\omega - n\omega_T)[u(\omega + \omega_c) - u(\omega - \omega_c)]$$

$$= C_0 F(\omega)$$

$$= \frac{A\tau}{T} F(\omega) \qquad n = 0, \omega_h \leq \omega_c \leq \omega_T - \omega_h \qquad (5.56)$$

or

$$\frac{A\tau}{T} f(t) \leftrightarrow \frac{A\tau}{T} F(\omega)$$

Thus, $f(t)$ (times the scale factor $A\tau/T$) is recovered.

We can obtain *impulse sampling* (mathematically speaking) by setting $A = 1/\tau$ and letting τ become 0 in Eq. 5.3. Thus, $C_n = 1/T$ and

$$p(t) = \sum_{n=-\infty}^{\infty} \delta(t - nT) = \frac{1}{T} \sum_{n=-\infty}^{\infty} e^{jn\omega_T t}$$

Then, $f_s(t)$ can be written as

$$f_s(t) = f(t) \sum_{n=-\infty}^{\infty} \delta(t - nT) = \sum_{n=-\infty}^{\infty} f(nT)\delta(t - nT)$$

But, from Eq. 5.56

$$F(\omega) = TF_s(\omega)[u(\omega + \omega_c) - u(\omega - \omega_c)] \qquad (A = 1/\tau)$$

Also

$$f_s(t) \leftrightarrow F_s(\omega) = \sum_{n=-\infty}^{\infty} f(nT)e^{-jn\omega T}$$

Thus

$$F(\omega) = T[u(\omega + \omega_c) - u(\omega - \omega_c)] \sum_{n=-\infty}^{\infty} f(nT)e^{-jn\omega T}$$

and

$$f(t) = \frac{T}{2\pi} \int_{-\omega_c}^{\omega_c} \sum_{n=-\infty}^{\infty} f(nT)e^{-jn\omega T} e^{jt\omega} d\omega$$

$$f(t) = \frac{T}{2\pi} \sum_{n=-\infty}^{\infty} f(nT) \int_{-\omega_c}^{\omega_c} e^{j(t-nT)\omega} d\omega$$

The required integration is easily carried out giving

$$f(t) = \frac{T}{2\pi} \sum_{n=-\infty}^{\infty} f(nT) 2\omega_c Sa(\omega_c t - n\omega_c T)$$

or

$$f(t) = \frac{\omega_c T}{\pi} \sum_{n=-\infty}^{\infty} f(nT) Sa(\omega_c t - n\omega_c T) \qquad (5.57)$$

If $\omega_c = \omega_h$ and $\omega_T = 2\omega_h = 2\omega_c$ (minimum sampling rate), Eq. 5.57 reduces to

$$f(t) = \sum_{n=-\infty}^{\infty} f(nT) Sa(\omega_h t - n\pi) \qquad (5.58)$$

These last two equations represent *interpolation formulas* for constructing $f(t)$ from its samples using the sampling function as an *interpolation function*. It can be seen in Eq. 5.58 that the value of $f(t)$ at any instant of time is a weighted sum of values at both earlier and later sampling times: the weighting factor being $Sa(\omega_h t - n\pi)$. Since this factor decreases slowly (as $1/n$), many terms in Eq. 5.58 may be needed to obtain acceptable results. They also obviously represent the response of an ideal low-pass filter to the sequence of samples, $f(nT)$.

In a real system the signal is never recovered without some distortion. First of all, a *real* filter may have amplitude sag and nonlinear phase characteristics across the passband. Furthermore, it has already been pointed out that a physically realizable filter cannot have infinite attenuation at any *finite* frequency. Secondly, a band-limited signal cannot be limited in time (and vice versa). This means we would be required to sample a band-limited signal *forever*. Put another way, it means that since we can realistically sample $f(t)$ only over a finite interval of time—that is, we are really sampling $f(t)[u(t) - u(t - NT)]$ (for example) whose spectrum is *not band-limited*—there will always be some overlapping and aliasing. The effects of aliasing can be reduced by filtering (band-limiting) *before* sampling, and by sampling at intervals smaller than the Nyquist interval by as much as possible.

Example 10

Assuming that an ideal low-pass filter can be used, we want to show that

$$f(t) = \cos(2\pi t) + \tfrac{1}{2}\cos(4\pi t)$$

can be recovered by sampling at the minimum rate, $\omega_T = 2\omega_h = 2\omega_c$ if $\tau = 0.1T$ and $A = 10$. Since $\omega_h = 4\pi$ for the given $f(t)$, we have $\omega_T = 8\pi$, and

$$C_n = Sa(n\pi/10)$$
$$F(\omega) = \pi[\delta(\omega + 2\pi) + \delta(\omega - 2\pi) + \tfrac{1}{2}\delta(\omega + 4\pi) + \tfrac{1}{2}\delta(\omega - 4\pi)]$$

Using Eq. 5.56, we obtain

$$F_s(\omega) = \sum_{n=-\infty}^{\infty} \{Sa(n\pi/10)[\delta(\omega + 2\pi - 8n\pi) + \delta(\omega - 2\pi - 8n\pi)$$
$$+ \tfrac{1}{2}\delta(\omega + 4\pi - 8n\pi) + \tfrac{1}{2}\delta(\omega - 4\pi - 8n\pi)]\}$$

For $\omega_c = \omega_h = 4\pi$ the ideal low-pass filter has the transfer function

$$H(\omega) = u(\omega + 4\pi) - u(\omega - 4\pi)$$

Thus, only the terms for $n = 0$ pass through the filter:

$$H(\omega)F_s(\omega) = \pi[\delta(\omega + 2\pi) + \delta(\omega - 2\pi) + \tfrac{1}{2}\delta(\omega + 4\pi) + \tfrac{1}{2}\delta(\omega - 4\pi)]$$

and

$$\mathscr{F}^{-1}\{H(\omega)F_s(\omega)\} = \cos(2\pi t) + \tfrac{1}{2}\cos(4\pi t) = f(t)$$

On the other hand, $T = 2\pi/\omega_T = \tfrac{1}{4}$

$$f(nT) = f(n/4) = \cos(n\pi/2) + \tfrac{1}{2}\cos(n\pi)$$
$$= \cos(n\pi/2) + \tfrac{1}{2}(-1)^n$$

and the interpolation formula, Eq. 5.58, gives

$$f(t) = \sum_{n=-\infty}^{\infty} [\cos(n\pi/2) + \tfrac{1}{2}(-1)^n]\frac{\sin(4\pi t - n\pi)}{4\pi t - n\pi}$$

It is not obvious by inspection that the last result is correct. Figure 5.18 shows the first nine ($n = 0, \pm 1, \pm 2, \pm 3, \pm 4$) terms in the series along with $f(t)$. As can be seen, the series correctly represents $f(t)$: that is, it is at least obvious that $f(0) = 1.5$, $f(\tfrac{1}{4}) = -0.5$, and $f(\tfrac{1}{2}) = -0.5$. These are the sample points.

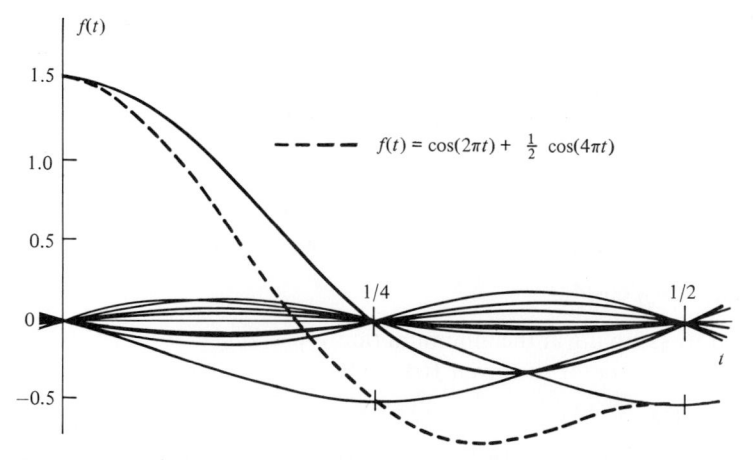

Figure 5.18. Individual terms of the interpolation formula summing to $f(t)$.

Example 11

Suppose we would like to record a piece of music by sampling the signal from a microphone pickup and then storing the sample values. The sampling frequency is 30,000 Hz, so, apparently, the highest frequency of interest in the signal is 15,000 Hz. If the recording lasts 5 minutes, then $5 \times 60 \times 30,000 = 9 \times 10^6$ samples will be required. Finally, if each sample is *quantized* [see Fig. 1.7(c)] into $128 = 2^7$ levels, then $7 \times 9 \times 10^6 = 63 \times 10^6$ binary digits (or bits) will be required.

5.9 THE DISCRETE FOURIER TRANSFORM

A discrete version of the Fourier transform is very useful in digital computing and signal processing. Its use is so important and widespread that some discussion of it and its relation to the continuous Fourier transform is necessary. It is commonly called the DFT and is closely related, as we shall see, to the DFS of Sec. 4.7. We want to find a way of treating the Fourier transform and the inverse Fourier transform that is suitable for numerical work. We begin by considering a signal that starts at some time, say $t = 0$ for convenience, and stops at some time, or is negligible after some time, say NT_s. Realistically, we always have only a finite observation time, and so this is always the case, like it or not. If we divide the interval NT_s into N subintervals each of width T_s, as in Fig. 5.19, then by definition

$$F(\omega) = \int_0^{NT_s} f(t)e^{-j\omega t}\,dt \tag{5.59}$$

It is obvious by inspection that

$$F(\omega) \approx \sum_{n=0}^{N-1} f(nT_s)e^{-j\omega nT_s}T_s \tag{5.60}$$

and this approximation becomes exactly Eq. 5.59 in the limit as $N \to \infty$, $T_s \to 0$, and $nT_s \to t$.

It is equally true that we cannot treat the continuum of ω for numerical work, but rather can only treat ω at a finite number of frequencies. For reasons that should be apparent after having gone through a similar development in Sec. 4.7, we choose this number to be N also, so that we agree to be content to

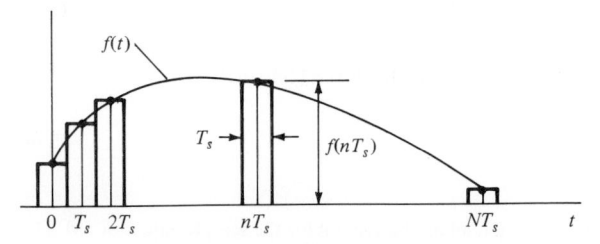

Figure 5.19. The function f(t) divided into N subintervals.

214 FOURIER TRANSFORM TECHNIQUES

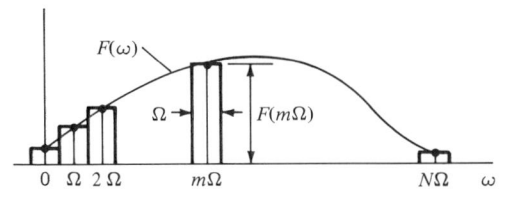

Figure 5.20. The function $F(\omega)$ divided into N subintervals.

observe with ω only at $\omega = m\Omega = m2\pi/(NT_s)$ where $\Omega = 2\pi/(NT_s)$. We will *tentatively* call the right side of 5.60 the *discrete Fourier transform* and give it the symbol[†] F_{dm}, or

$$F_{dm} = F_d(m\Omega) = T_s \sum_{n=0}^{N-1} f_d(nT_s)e^{-j2\pi mn/N} \tag{5.61}$$

Compare this result to Eq. 4.64 and ask the question, What should the inverse discrete Fourier transform (IDFT) be? According to Eq. 4.63 it should be (using *the principle of duality*)

$$f_{dn} = f_d(nT_s) = \frac{1}{NT_s} \sum_{m=0}^{N-1} F_d(m\Omega)e^{+j2\pi mn/N} \tag{5.62}$$

Consider Fig. 5.20 where $F(\omega)$ is divided into N subintervals. We have

$$f(t) \approx \frac{1}{2\pi} \int_0^{N\Omega} F(\omega)e^{jt\omega}\,d\omega \tag{5.63}$$

The approximation in 5.63 is due to the fact that if $f(t)$ exists only for a finite time as originally assumed (or imposed) for practical reasons, then $F(\omega)$ must (mathematically) exist for $-\infty < \omega < \infty$, even though its approximation cannot for practical reasons. This has already been pointed out. The *approximation for the right side of Eq. 5.63* is

$$f(t) \approx \frac{1}{2\pi} \sum_{n=0}^{N-1} F(m\Omega)e^{jtm\Omega}\Omega \tag{5.64}$$

and this becomes exactly the same as the right side of 5.63 as $N \to \infty$, $\Omega \to 0$, and $m\Omega \to \omega$. But we can only treat t at nT_s, so with $t = nT_s$ we *tentatively* call the right side of 5.64 the *inverse discrete Fourier transform* (IDFT) $f_{dn} = f_d(nT_s)$:

$$f_{dn} = f_d(nT_s) = \frac{1}{NT_s} \sum_{n=0}^{N-1} F(m\Omega)e^{j2\pi nm/N} \tag{5.65}$$

Notice that $\Omega/(2\pi) = 1/(NT_s)$. Thus, we finally have (*tentatively*)

$$F_{dm} = T_s \sum_{n=0}^{N-1} f_{dn}e^{-j2\pi mn/N} \tag{5.66}$$

[†] We also introduce the subscript d (for *discrete*) on the symbol for the sample values of $f(t)$. That is, $f(nT_s)$ becomes $f_d(nT_s)$.

and

$$f_{dn} = \frac{1}{NT_s} \sum_{m=0}^{N-1} F_{dm} e^{+j2\pi nm/N} \tag{5.67}$$

This pair of equations is analogous to the continuous Fourier transform pair. In order to make direct comparison with the continuous Fourier transform, this pair should be used. It is not necessary to prove that these stand alone and represent a discrete transform pair. We did this in Sec. 4.7. Thus

$$f_{dn} \leftrightarrow F_{dm} \tag{5.68}$$

It remains to remove the word "tentatively" that we have used to describe these results. The only difference *in form* between the pair of Eqs. 5.66 and 5.67 and the corresponding pair 4.64 and 4.63 in Sec. 4.7 is the presence of T_s in Eq. 5.66 and $1/T_s$ in Eq. 5.67. The removal of T_s from this pair of equations will in no way alter the fact that they remain a discrete transform pair (5.68). Thus, in order to obtain a DFT pair that agrees *in form* with the DFS pair, and agrees with the most widely accepted definition of the DFT pair, we take the following to be the DFT pair:

$$\boxed{F_{dm} = \sum_{n=0}^{N-1} f_{dn} e^{-j2\pi mn/N} = F_d(m\Omega)} \quad (DFT) \tag{5.69}$$

$$\boxed{f_{dn} = \frac{1}{N} \sum_{m=0}^{N-1} F_{dm} e^{j2\pi nm/N} = f_d(nT_s)} \quad (IDFT) \tag{5.70}$$

where

$$\frac{\Omega}{2\pi} = \frac{1}{NT_s} \tag{5.71}$$

Notice that (here) T_s is the spacing between samples in the time domain, whereas in Eqs. 4.63 and 4.64 T/N (where $T = 2\pi/\omega_T$ is the period of the periodic function) is the spacing between the samples in the time domain. Likewise, $\Omega = 2\pi/(NT_s)$ is the spacing between samples in the frequency domain, whereas in Eqs. 4.63 and 4.64 $\omega_T = 2\pi/T$ is the spacing between samples in the frequency domain. Both sequences (5.69 and 5.70) are periodic as we saw in Sec. 4.7. F_{dm} is periodic with period $N\Omega$, while f_{dn} has the period NT_s.

We have not yet removed the approximation sign in 5.63. What does it take to do this; or, better put, what is required in order for our original form for the IDFT (Eq. 5.67) to give us an accurate representation of the (continuous) inverse Fourier transform? Referring to Fig. 5.19 and the discussion below Eq. 5.63, we see that the intuitive answer to our question is simply that $N\Omega$ must be large enough to include all spectral components with appreciable amplitude. Since $N\Omega = 2\pi/T_s$, this simply means that T_s must be small or the

216 FOURIER TRANSFORM TECHNIQUES

sampling (radian) frequency $2\pi/T_s$ must be large. This is exactly what we would expect. We will shortly examine this by means of an example.

Since $F_d(m\Omega)$ is periodic, the approximation to the continuous transform by using the DFT in calculation will be affected by *aliasing* in a manner similar to that in the previous section. This effect can be minimized, as before, by using a high sampling rate[†] (small T_s). Also, we can add samples of zero magnitude (*augmented zeros*) to the truncated function at the end of its sequence of samples. These zeros increase NT_s and therefore decrease the harmonic spacings in the frequency domain and the effects of aliasing. The cost is increased computation time.

Example 12

Find the DFT of $f(t)$ shown below in Fig. 5.21 for the four given sample points. We have $N = 4$, $f_{d0} = 1$, $f_{d1} = 2$, $f_{d2} = 3$, and $f_{d3} = 3$. Thus, using Eq. 4.73 we have

$$NE^- = \begin{bmatrix} 1 & 1 & 1 & 1 \\ 1 & e^{-j\pi/2} & e^{-j\pi} & e^{-j3\pi/2} \\ 1 & e^{-j\pi} & e^{-j2\pi} & e^{-j3\pi} \\ 1 & e^{-j3\pi/2} & e^{-j3\pi} & e^{-j9\pi/2} \end{bmatrix} = \begin{bmatrix} 1 & 1 & 1 & 1 \\ 1 & -j & -1 & +j \\ 1 & -1 & 1 & -1 \\ 1 & +j & -1 & -j \end{bmatrix}$$

and from Eq. 5.69 we have the DFT:

$$\begin{bmatrix} F_{d0} \\ F_{d1} \\ F_{d2} \\ F_{d3} \end{bmatrix} = \begin{bmatrix} 1 & 1 & 1 & 1 \\ 1 & -j & -1 & +j \\ 1 & -1 & 1 & -1 \\ 1 & +j & -1 & -j \end{bmatrix} \begin{bmatrix} 1 \\ 2 \\ 3 \\ 3 \end{bmatrix} = \begin{bmatrix} 1 & +2 & +3 & +3 \\ 1 & -2j & -3 & +3j \\ 1 & -2 & +3 & -3 \\ 1 & +2j & -3 & -j \end{bmatrix}$$

$$\begin{bmatrix} F_{d0} \\ F_{d1} \\ F_{d2} \\ F_{d3} \end{bmatrix} = \begin{bmatrix} 9 \\ -2+j \\ -1 \\ -2-j \end{bmatrix}$$

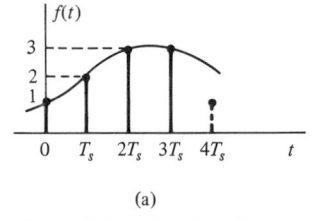

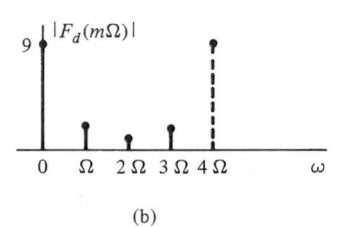

Figure 5.21. (a) Function sampled at four points. (b) Discrete Fourier transform of $f(t)$.

[†] In fact, the resolution is limited to a highest frequency corresponding to $m = N/2$ or $(N/2)\Omega = \pi/T_s$ rad/s. This agrees with the sampling theorem.

$|F_d(m\Omega)|$ is plotted in Fig. 5.20(b). Notice that

$$F_{d4} = F_{d(0+4)} = F_{d0} = 9$$

demonstrating the periodicity of $F_d(m\Omega)$.

Obviously, if we now use the IDFT, Eq. 5.70, we obtain the original samples in the time domain, but

$$f_{d4} = \tfrac{1}{4} \sum_{m=0}^{3} F_{dm} e^{-j8\pi m/4}$$

$$f_{d4} = \tfrac{1}{4}[F_{d0} + F_{d1} + F_{d2} + F_{d3}] = 1$$

which demonstrates the periodicity of f_{dn}.

In order to gain more insight into the DFT, we should look at an example where we can calculate the exact continuous Fourier transform for comparison with Eq. 5.66.

Example 13

Compute the DFT for $f(t) = u(t) - u(t-1)$ using four samples. In this case we have $NT_s = 1$, $N = 4$, and $T_s = \tfrac{1}{4}$. Therefore

$$\begin{bmatrix} F_{d0} \\ F_{d1} \\ F_{d2} \\ F_{d3} \end{bmatrix} = \begin{bmatrix} 1 & 1 & 1 & 1 \\ 1 & -j & -1 & j \\ 1 & -1 & 1 & -1 \\ 1 & j & -1 & -j \end{bmatrix} \begin{bmatrix} 1 \\ 1 \\ 1 \\ 1 \end{bmatrix} = \begin{bmatrix} 4 \\ 0 \\ 0 \\ 0 \end{bmatrix}$$

Now

$$f(t) = u(t) - u(t-1) \leftrightarrow F(\omega) = e^{-j\omega/2} Sa(\omega/2)$$

So the *continuous transform* gives (for $\omega = m\Omega = 2m\pi$)

$$F(m\Omega) = e^{-jm\pi} Sa(m\pi)$$

or

$$F(0) = 1 \qquad F(\Omega) = 0$$
$$F(2\Omega) = 0 \qquad F(3\Omega) = 0$$

whereas the DFT gives

$$F_d(0) = 4 \qquad F_d(\Omega) = 0$$
$$F_d(2\Omega) = 0 \qquad F_d(3\Omega) = 0$$

These results are consistent because we agreed to drop the T_s ($=\tfrac{1}{4}$) in going from Eq. 5.66 to 5.69. Thus, the only difference in the two sequences is a scale factor. The samples of the continuous spectrum are shown in Fig. 5.22.

If we perform the IDFT, we obtain $f_{dn} = f_d(nT_s) = 1$ for all n. If we increase N and decrease T_s, keeping $NT_s = 1$, the DFT gives us

$$F_{dm} = F_d(m\Omega) = \begin{cases} N & m = 0 \\ 0 & m = 1, 2, 3, \ldots, N-1 \end{cases}$$

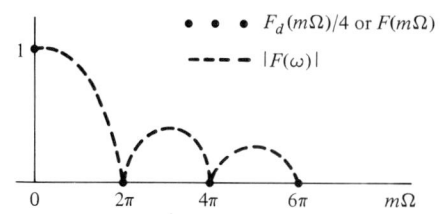

Figure 5.22. Sampled continuous spectrum of $u(t) - u(t - 1)$, $F(m\Omega)$ ($\Omega = 2\pi$).

Legend: • • • $F_d(m\Omega)/4$ or $F(m\Omega)$; - - - $|F(\omega)|$

and the IDFT continues to give us $f_d(nT_s) = 1$ for all n. This result indicates that the only frequency present is zero frequency, and this is correct since (a dc level of unity) $1 \leftrightarrow 2\pi\delta(\omega)$. It is not, however, the result that we orginally wanted.

As mentioned earlier, we can overcome the difficulty encountered in Example 13 by making $f(t)$ periodic, and this can be done by adding zeros.

Example 14

Repeat Example 13 using eight samples, the last four being added zeros. In this case we have $N = 8$, $T_s = \frac{1}{4}$, and $NT_s = 2$. Notice that the truncated $f(t)$ will become a square wave whose period is 2. The $f(nT_s)$ sequence is now 1, 1, 1, 1, 0, 0, 0, 0, and using Eq. 5.66 gives[†]

$F_{d0} = 1$ \qquad (1)

$F_{d1} = 0.653 \lfloor -67.5°$ \qquad $\left(-j\dfrac{2}{\pi}\right)$

$F_{d2} = 0$ \qquad (0)

$F_{d3} = 0.271 \lfloor -22.5°$ \qquad $\left(-j\dfrac{2}{3\pi}\right)$

$F_{d4} = 0$ \qquad (0)

$F_{d5} = 0.271 \lfloor 22.5°$ \qquad $\left(-j\dfrac{2}{5\pi}\right)$

$F_{d6} = 0$ \qquad (0)

$F_{d7} = 0.653 \lfloor 67.5°$ \qquad $\left(-j\dfrac{2}{7\pi}\right)$

Notice that the answers are not in good agreement with the samples from the *continuous* transform (in parentheses):

$F(m\Omega) = e^{-jm\pi/2} Sa(m\pi/2)$

[†] Notice that we are using Eq. 5.66 here because it includes the scale factor T_s and allows a direct comparison of the DFT with the sampled continuous spectrum.

5.9 THE DISCRETE FOURIER TRANSFORM

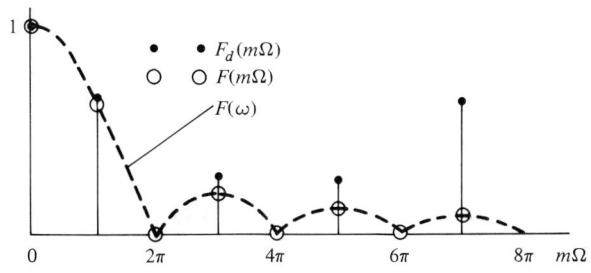

Figure 5.23. Samples of the continuous spectrum of $u(t) - u(t-1)$ and the DFT samples (multiplied by T_s).

These results are shown in Fig. 5.23. Now, suppose that we keep $NT_s = 2$, and at the same time increase the number of samples (N) maintaining the same number of ones as zeros in the $f_d(nT_s)$ sequence. In this case we have $T_s = 2/N$, and using Eq. 5.66 once more

$$F_d(m\Omega) = \frac{2}{N}[1 + e^{-j2\pi m/N} + e^{-j4\pi m/N} + \cdots + e^{-j2\pi(N/2-1)/N}]$$

$$F_d(m\Omega) = \frac{2}{N}[1 + e^{-j2\pi m/N}(1 + e^{-j2\pi m/N} + \cdots + e^{-j2\pi(N/2)m/N} + e^{-j2\pi(N/2-1)m/N} - e^{-j2\pi(N/2-1)m/N}]$$

$$F_d(m\Omega) = \frac{2}{N}\left\{1 + e^{-j2\pi m/N}\left[\frac{N}{2}F_d(m\Omega) - e^{-j2\pi(N/2-1)m/N}\right]\right\}$$

$$F_d(m\Omega) = \frac{2}{N} + F_d(m\Omega)e^{-j2\pi m/N} - \frac{2}{N}e^{-jm\pi}$$

$$F_d(m\Omega) = \frac{2}{N}\frac{1 - e^{-jm\pi}}{1 - e^{-j2m\pi/N}}$$

This reduces to

$$F_d(m\Omega) = \begin{cases} 1 & m = 0 \\ 0 & m \text{ even} \\ \frac{2}{N} - j\frac{2}{N}\operatorname{ctn}(m\pi/N) & n \text{ odd} \end{cases} \quad \text{(any } N\text{)}$$

The limit as N approaches infinity is easily shown to be

$$F_d(m\Omega) = \begin{cases} 1 & m = 0 \\ 0 & m \text{ even} \\ -j\frac{2}{m\pi} & m \text{ odd} \end{cases} = e^{-jm\pi/2}\operatorname{Sa}(m\pi/2) \quad (N \to \infty)$$

220 FOURIER TRANSFORM TECHNIQUES

Thus, for an infinite number of samples we obtain exact agreement with Eq. 5.77, and the DFT (as given by Eq. 5.66) gives the same results as the continuous Fourier transform.

What does the sampling theorem say about N in the example above?

We next discuss some of the properties of the DFT that were not mentioned earlier in connection with the DFS. First of all, the DFT is a linear transformation:

$$\mathscr{F}_d\{af_{dn1} + bf_{dn2}\} = a\mathscr{F}_d\{f_{dn1}\} + b\mathscr{F}_d\{f_{dn2}\}$$
$$= aF_{dm1} + bF_{dm2} \quad (5.72)$$

Secondly, we have seen the importance of convolution in time-invariant continuous systems, and the same is true here: that is, the IDFT of the product of the DFT's of two discrete functions of time is the convolution of the original sequences of the discrete functions. The convolution here is called *periodic, or circular, convolution* and applies to *shift-invariant* discrete systems. Let the sequences be f_{dn1} and f_{dn2}, and let their DFTs be F_{dm1} and F_{dm2}, respectively. We assume that N and T_s are the same for both sequences. If N is not the same for both sequences, we can simply add zero amplitude samples (*augmented zeros*) to the sequence with too few. T_s must be the same for both sequences in any case. Now let f_{dn3} be the IDFT of $F_{dm1} \cdot F_{dm2}$, or

$$f_{dn3} = \frac{1}{N} \sum_{m=0}^{N-1} F_{dm1} F_{dm2} e^{j2\pi nm/N} \quad (5.73)$$

Using Eq. 5.69

$$F_{dm1} = \sum_{i=0}^{N-1} f_{di1} e^{-j2\pi im/N} \quad (5.74)$$

and

$$F_{dm2} = \sum_{k=0}^{N-1} f_{dk2} e^{-j2\pi km/N} \quad (5.75)$$

so

$$f_{dn3} = \frac{1}{N} \sum_{m=0}^{N-1} \left[\sum_{i=0}^{N-1} f_{di1} e^{-j2\pi im/N} \right] \left[\sum_{k=0}^{N-1} f_{dk2} e^{-j2\pi km/N} \right] e^{j2\pi nm/N}$$

Rearranging the last equation so that the sum over k is innermost

$$f_{dn3} = \frac{1}{N} \sum_{i=0}^{N-1} \sum_{k=0}^{N-1} f_{di1} f_{dk2} \sum_{m=0}^{N-1} e^{j2\pi m(n-i-k)/N}$$

But by Eq. 4.56 the last sum in the last equation is N for $n - i - k = 0$ (or

5.9 THE DISCRETE FOURIER TRANSFORM

$i = n - k$) and 0 otherwise. Thus, we have

$$f_{dn3} = \sum_{i=0}^{N-1} f_{di1} f_{d(n-i)2} = \sum_{i=0}^{N-1} f_{d1}[iT_s] f_{d2}[(n-i)T_s] \quad (5.76)$$

or

$$f_{dn3} = \sum_{i=0}^{N-1} f_{d1}[(n-i)T_s] f_{d2}[iT_s] \quad (5.77)$$

Notice the similarities (and differences) between convolution for time-invariant continuous systems and this type convolution for *shift-invariant* discrete systems. One sequence is reversed ("folded") and indexed by an amount n relative to the other sequence, the sequences are then multiplied serially, and the products are summed ("integrated"). In order to use Eq. 5.76 or 5.77 to simulate convolution in system-analysis problems it is necessary to extend the interval with zero value samples. This avoids overlapping that would occur because of the periodicity that is present with the DFT.

We have shown that

$$\sum_{i=0}^{N-1} f_{d1}[(n-i)T_s] f_{d2}[iT_s] \leftrightarrow F_{d1}(m\Omega) F_{d2}(m\Omega) \quad (5.78)$$

or

$$\boxed{y_d(nT_s) = \sum_{i=0}^{N-1} h_d[(n-i)T_s] x_d[iT_s] \leftrightarrow H_d(m\Omega) X_d(m\Omega) = Y_d(m\Omega)} \quad (5.79)$$

so that the block diagram of Fig. 5.24 applies. The discrete linear system might be a digital filter, for example, whose *discrete transfer function* we have called $H_d(m\Omega)$. Based on prior experience with continuous systems, and having been exposed to Example 13, we should have little difficulty in finding a discrete generic input and its DFT for Fig. 5.24. If this DFT is to be an arbitrarily large sequence of unit-amplitude samples then what is its IDFT? Then

$$f_{dn} = \frac{1}{N} \sum_{n=0}^{N-1} 1 e^{j2\pi nm/N} \quad \text{(any } N\text{)}$$

or by equation 4.55

$$f_{dn} = \begin{cases} 1 & n = 0 \\ 0 & n \neq 0 \end{cases} \quad (5.80)$$

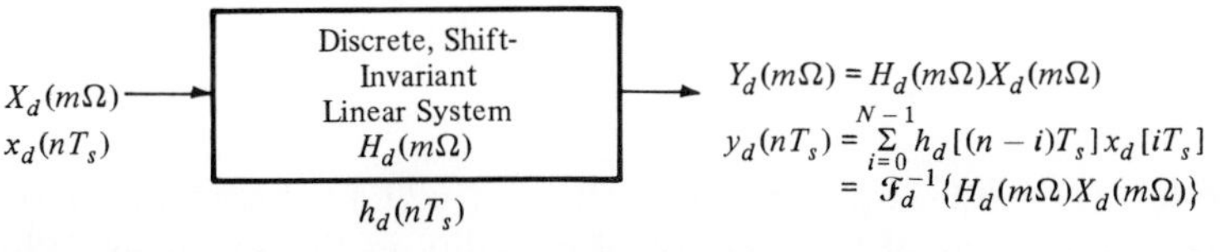

Figure 5.24. Block diagram for the analysis of a discrete, shift-invariant linear system.

222 FOURIER TRANSFORM TECHNIQUES

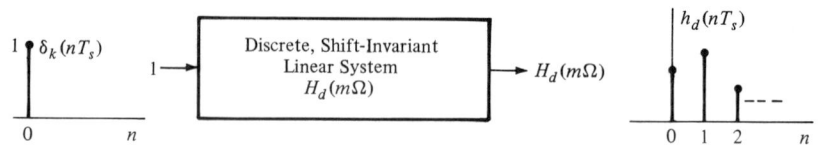

Figure 5.25. Unit-function response for the discrete, shift-invariant linear system.

It is convenient here to introduce the *Kronecker delta* or *unit function* as

$$\delta_k(i-j) = \begin{cases} 1 & i = j \\ 0 & i \neq j \end{cases} \tag{5.81}$$

Thus, we have $f_{dn} = \delta_k(nT_s)$, and finally

$$\delta_k(nT_s) \leftrightarrow 1 \tag{5.82}$$

This result is the discrete counterpart of 5.28:

$$\delta(t) \leftrightarrow 1 \tag{5.28}$$

We are thus led to call $h_d(nT_s)$ the *unit-function response* because if $X_d(m\Omega) = 1$, or $x_d(nT_s) = \delta_k(nT_s)$, then $Y_d(m\Omega) = H_d(m\Omega)$, or (Eq. 5.79)

$$y_d(nT_s) = \sum_{i=0}^{N-1} h_d[(n-i)T_s]\delta_k[iT_s] = h_d(nT_s)$$

The results are shown in Fig. 5.25.

Example 15

What is the unit-function response for the digital filter described by

$$H_d(m\Omega) = \begin{cases} 1 & 0 \leq m \leq 3 \\ 0 & 4 \leq m \leq 7 \end{cases} \quad (N = 8)$$

Using Eq. 5.70

$$h_d(nT_s) = \tfrac{1}{8} \sum_{m=0}^{7} H_{dm} e^{j2\pi nm/8}$$

$h_{d0} = 0.5$ $h_{d1} = 0.327 \underline{|67.5°}$

$h_{d2} = 0$ $h_{d3} = 0.135 \underline{|22.5°}$

$h_{d4} = 0$ $h_{d5} = 0.135 \underline{|-22.5°}$

$h_{d8} = 0$ $h_{d7} = 0.327 \underline{|-67.5°}$

Notice that $h_d(-1) = h_d(-T_s) = h_{d7}$, $h_d(-2) = h_d(-2T_s) = h_{d6}$, and so on. What are the implications of the concept of causality in this example? Why are the samples $h_d(nT_s)$ here so similar to $F_d(m\Omega)$ in Example 14?

Example 16

The discrete transformed signal

$$X_d(m\Omega) = \begin{cases} 0 & 0 \le m \le 2 \\ 4 & m = 3 \\ 1 & m = 4 \\ 0 & 5 \le m \le 7 \end{cases}$$

is applied to the filter of Example 15 ($N = 8$). What is the response $y_d(nT_s)$? We have from Eq. 5.79

$$Y_d(m\Omega) = H_d(m\Omega) X_d(m\Omega) = \begin{cases} 0 & 0 \le m \le 2 \\ 4 & m = 3 \\ 0 & 4 \le m \le 7 \end{cases}$$

and from Eq. 5.70

$$y_d(nT_s) = \tfrac{1}{8} \sum_{m=0}^{7} Y_d(m\Omega) e^{j2\pi nm/8}$$

$y_{d0} = 0.5$ $y_{d1} = 0.5 e^{j3\pi/4}$
$y_{d2} = 0.5 e^{j6\pi/4}$ $y_{d3} = 0.5 e^{j9\pi/4}$
$y_{d4} = 0.5 e^{j12\pi/4}$ $y_{d5} = 0.5 e^{j15\pi/4}$
$y_{d6} = 0.5 e^{j18\pi/4}$ $y_{d7} = 0.5 e^{j21\pi/4}$

This result was obtained by discrete "frequency-domain" analysis. Next, we will try discrete "time-domain" analysis. First, we must find $x_d(nT_s)$ (which will normally be known or given):

$$x_d(nT_s) = \tfrac{1}{8} \sum_{m=0}^{7} X_d(m\Omega) e^{j2\pi nm/8}$$

$$x_d(nT_s) = \tfrac{1}{8} [4 e^{j6n\pi/8} + e^{j8n\pi/8}]$$

$x_{d0} = \tfrac{5}{8}$ $x_{d1} = 0.595 \underline{|143.54°}$
$x_{d2} = 0.515 \underline{|75.96°}$ $x_{d3} = 0.421 \underline{|57.12°}$
$x_{d4} = -\tfrac{3}{8}$ $x_{d5} = 0.421 \underline{|-57.12°}$
$x_{d6} = 0.515 \underline{|-75.96°}$ $x_{d7} = 0.595 \underline{|-143.54°}$

Then

$$y_{dn} = \sum_{i=0}^{7} h_d(n-i) x_d(i)$$

$$y_{d0} = h_d(0) x_d(0) + h_d(-1) x_d(1) + h_d(-2) x_d(2) + h_d(-3) x_d(3) \\ + h_d(-4) x_d(4) + h_d(-5) x_d(5) + h_d(-6) x_d(6) + h_d(-7) x_d(7)$$

⋮

giving

$$y_{d0} = 0.5 \qquad y_{d1} = 0.5e^{j3\pi/4}$$
$$y_{d2} = 0.5e^{j6\pi/4} \qquad y_{d3} = 0.5e^{j9\pi/4}$$
$$y_{d4} = 0.5e^{j12\pi/4} \qquad y_{d5} = 0.5e^{j15\pi/4}$$
$$y_{d6} = 0.5e^{j18\pi/4} \qquad y_{d7} = 0.5e^{j21\pi/4}$$

as before.

There are other properties of the DFT that we could list. For example, a sequence can be considered to be even or odd, and these terms here correspond to the same terms as applied to continuous (not discrete) functions. This can be seen by periodically extending the sequences, or plotting them around a circle. We will not go into the details here since most of the results of these kinds of symmetry are either obvious or are easily deduced.

5.10 THE FAST FOURIER TRANSFORM

When numerical computations are made, N will normally be large for reasons already mentioned, and the direct use of Eq. 5.69 or 5.70 requires a total of $N(N-1)$ *complex* additions and N^2 *complex* multiplications. This can be very time consuming even for modern high-speed digital computers. Since most of this computational time is spent on the multiplications, it would be very beneficial if a way could be found to reduce the number of multiplications in computing the DFT (or IDFT). Highly efficient algorithms for doing this and implementing the DFT are generally given the name *fast Fourier transform* (FFT).

In order to see how a reduction in the number of multiplications is achieved,[4] consider Example 12 ($N=4$) and the (matrix) multiplication for finding F_{dm}. It is obvious from the second equation in this example that the number of complex multiplications[†] is $4+4+4+4 = 4^2 = 16 = N^2$, and the number of complex additions is $3+3+3+3 = 4(3) = N(N-1)$, as stated above. Now, suppose that Eq. 5.69 is rewritten [using $F_{dm} = F_d(m\Omega)$ and $f_{dn} = f_d(nT_s)$ for clarity] as

$$F_d(m\Omega) = \sum_{n=0}^{(N/2)-1} f_d(2nT_s)e^{-j2\pi m(2n)/N}$$
$$+ \sum_{n=0}^{(N/2)-1} f_d[(2n+1)T_s]e^{-j2\pi m(2n+1)/N} \qquad (5.83)$$

[†] Multiplications by unity are counted in the total number in this discussion. This is done purely in the interest of simplicity.

5.10 THE FAST FOURIER TRANSFORM

or

$$F_d(m\Omega) = \sum_{n=0}^{(N/2)-1} f_d(2nT_s)e^{-j2\pi mn/(N/2)}$$
$$+ e^{-j2\pi m/N} \sum_{n=0}^{(N/2)-1} f_d[2n+1]T_s]e^{-j2\pi mn/(N/2)} \quad (5.84)$$

for N even. Thus, we have merely expressed the DFT as the sum[†] of the DFT of the *even* elements and the DFT of the *odd* elements, and each separately has $N/2$ components with period $N/2$. Now, even though m ranges from 0 to $N-1$, *only $N/2$ values of the two DFTs in Eq. 5.84 need be computed!* The exponential in front of the second sum gives an additional $N/2$ multiplications. Thus, the total number of complex multiplications in this scheme is $2(N/2)^2 + N/2 = N^2/2 + N/2$. For large N, where $N \ll N^2$, the number of complex multiplications is cut in half and, for all practical purposes, so is the computation time.

In order to clarify this, consider the matrix form taken by Eq. 5.84 when applied to Example 12 ($N = 4$):

$$\begin{bmatrix} F_{d0} \\ F_{d1} \\ F_{d2} \\ F_{d3} \end{bmatrix} = \begin{bmatrix} +1 & +1 \\ +1 & -1 \\ +1 & +1 \\ +1 & -1 \end{bmatrix} \begin{bmatrix} f_{d0} \\ f_{d2} \end{bmatrix} + \begin{bmatrix} (+1)(+1) & (+1)(+1) \\ (-j)(+1) & (-j)(-1) \\ (-1)(+1) & (-1)(+1) \\ (+j)(+1) & (+j)(-1) \end{bmatrix} \begin{bmatrix} f_{d1} \\ f_{d3} \end{bmatrix} \quad (5.85)$$

Notice that in the first matrix on the right side of Eq. 5.85 the first and third rows are identical and the second and fourth rows are identical. Likewise, in the third matrix the first and third rows are identical, except for a minus sign, and the same is true of the second and fourth rows. If we count the number of *different* multiplications, then we have four $[(N/2)^2]$ in the first term on the right side of Eq. 5.85, while in the second term we have four $[(N/2)^2]$ plus two $(N/2)$ inside the first matrix. Thus, we have a total of ten $(= N^2/2 + N/2, N = 4)$ different multiplications, whereas we originally had sixteen ($N^2, N = 4$) in Example 12.

The Cooley-Tukey formulation[5] gives an algorithm that we can easily show in signal flow graph form for the example in the preceding paragraph ($N = 4$). It is shown in Fig. 5.26. Notice that at the second set (column) of nodes eight multiplications (counting multiplications by unity) and four summations have been performed, whereas at the third set of nodes two complex multiplications (by $-j$ and by $+j$) and four summations have been performed. Once again, ten multiplications are needed. The reader can easily verify that results of the signal flow graph and Eq. 5.85 (or Example 12) agree.

The ordering in the output is scrambled with this algorithm: F_{d1} and F_{d2} appear in reverse order. In order for the output to appear in natural order it is necessary to scramble the input order. The general way to accomplish this[6] is

---

[†] The DFT is a *linear* transformation and *superposition* applies (see Eq. 5.72).

226 FOURIER TRANSFORM TECHNIQUES

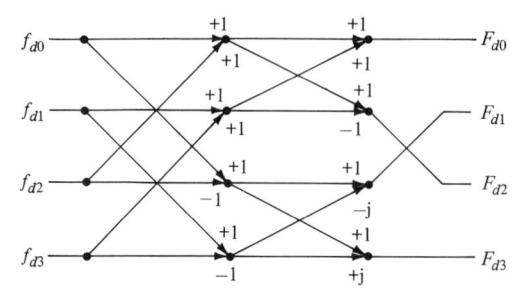

Figure 5.26. Cooley-Tukey algorithm for the FFT ($N = 4$) in signal flow graph form.

to write the *binary* form of n in *reverse* order, giving the *reversed bit order*. Thus, the natural decimal order 0, 1, 2, 3 is 00, 01, 10, 11 in binary form, and is 00, 10, 01, 11 in reversed bit order, or 0, 2, 1, 3, in decimal form. For $N = 8$, 000, 001, 010, 011, 100, 101, 110, 111 is reversed to 001, 100, 010, 110, 001, 101, 011, 111 or 0, 4, 2, 6, 1, 5, 3, 7 in decimal form. Figure 5.27 shows the reversed bit order for the input in signal flow graph form for $N = 4$.

So long as larger sequences can be further divided, the above process can be continued, and each division of the sequence gives a further reduction in computer time by a factor of (essentially) 2. *The reduction can be carried out completely if N is a power of 2* (as it was above)! In this case the required number of complex multiplications is $N \log_2 N + N/2$. For large N the number of multiplications is drastically reduced when compared to the number of multiplications (N^2) by the direct method.

We next show how the number of complex multiplications can be reduced to no more than $(N/2)\log_2 N$. We do this by merely showing the required algorithm in signal flow graph form for the case $N = 8$ and deducing the general result. It should be obvious to the attentive reader that more multiplications appear in Figs. 5.26 and 5.27 than are really necessary. For example, instead of multiplying by *both* $+j$ and $-j$, we merely need to multiply by $+j$ and then *add* and *subtract*. A block diagram (rather than a signal flow graph) shows this best, and it is given in Fig. 5.28. Notice that the reverse bit order has been used for the input to provide the natural order for the output data.

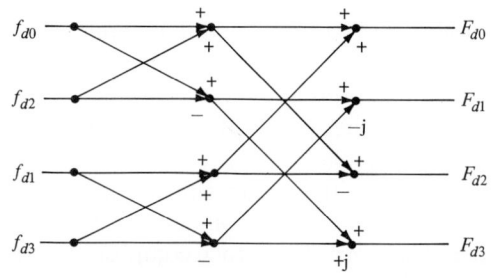

Figure 5.27. Signal flow graph of the algorithm ($N = 4$) for the FFT with reverse bit order for the input data.

5.10 THE FAST FOURIER TRANSFORM

The sub-block diagram within the dashed line is called a *butterfly*,[6] and each butterfly contains one multiplier. There are four $(N/2)$ butterflies in each column in Fig. 5.28, and, furthermore, there are three $(\log_2 N)$ columns. Thus, the number of butterflies, and therefore the number of multiplications, is twelve or $(N/2)\log_2 N$. Notice, once again, that we have counted multiplications by unity, so that there are only five multipliers actually required in Fig. 5.28. Notice also that there are really only three different multiplying factors ($e^{-j\pi/4}$, $e^{-j2\pi/4}$, $e^{-j3\pi/4}$) needed in Fig. 5.28.

Suppose that $N = 1024 = 2^{10}$. Then, using the algorithm outlined above, we find that the number of complex multiplications is no more than 5,120,

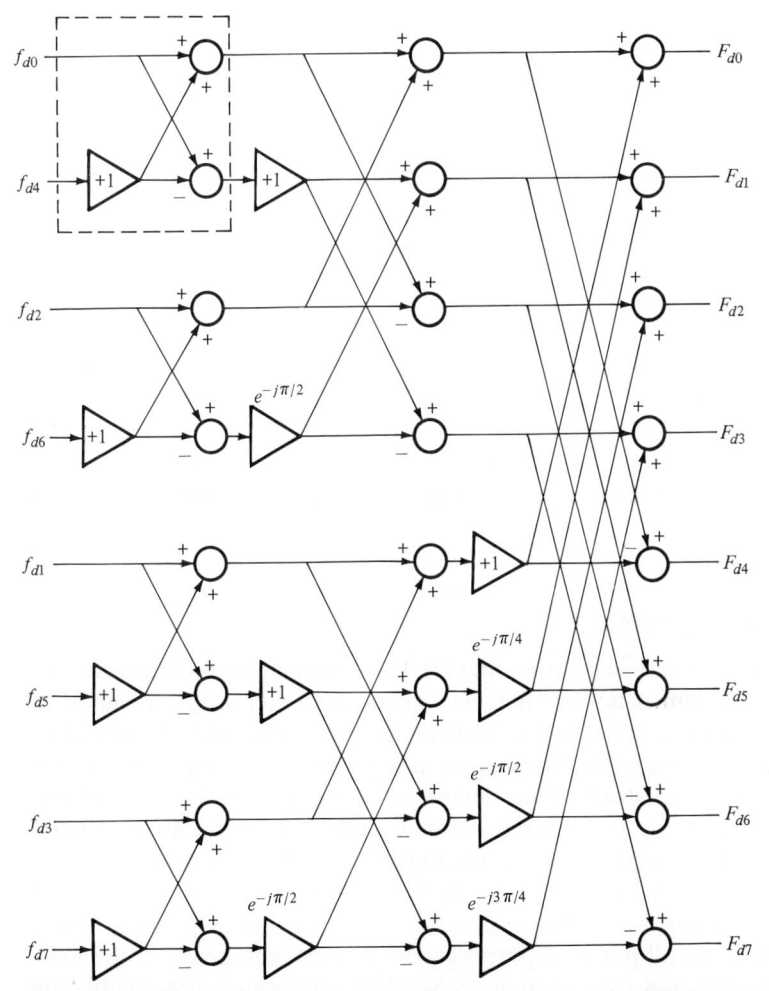

Figure 5.28. Block diagram ($N = 8$) of an algorithm for the FFT (reverse bit order for the input).

while direct implementation of the DFT requires 1,048,576 complex multiplications. In other words the computer time is reduced by a factor of about 200!

Example 17

It is desired to find the DFT for a sequence of twelve samples. Since 12 is not a power of 2, we can only divide the sequence one time. Using the results below Eq. 5.84, we find that $N^2/2 + N/2 = 144/2 + 12/2 = 78$ multiplications are required. On the other hand, if we use the same twelve samples plus four *augmenting zeros* (giving $N = 16$) with the algorithm just presented, we find that only $(N/2)\log_2 N = 8(4) = 32$ multiplications are required. Thus, for several reasons we are better off taking the latter path.

It is useful to list some guidelines[3] that are helpful in processing continuous signals with the FFT. These are based on results in this and preceding sections (including Sec. 4.7.)

1. Choose $N = 2^r$, where r is an integer. This can always be done by using augmenting zeros.
2. Both the sequences $f_d(nT_s)$ and $F_d(m\Omega)$ contain N terms.
3. The sample points corresponding to 0 and N are identical in both domains because of the periodicity.
4. Insofar as $F_d(m\Omega)$ is concerned, we consider positive frequency components to be those over 0, $N/2$, while negative frequency components are those over $N/2$, N. The same symmetry applies to the time samples for positive and negative time.
5. For *real* functions of time the positive frequency components are the complex conjugates of the negative frequency components (see Example 12).
6. The highest frequency component is increased by decreasing T_s.
7. The spacing between frequency components is decreased by adding augmenting zeros.

As has been mentioned earlier, the DFT is extremely useful in practice for such things as the simulation of filters on a digital computer or spectral analysis. In order to relate the discussion to reality, consider the problem of measuring the frequency response of a loudspeaker using a microphone pickup. This is difficult to do in the steady state, unless an anechoic chamber is available, because of reflections. If, however, we can apply a short-duration, intense, rectangular, repetitive pulse of unit area to the loudspeaker, then we are essentially applying a unit impulse. If we then let a computer perform the DFT (using an FFT algorithm) of the resultant response (microphone output), then we have found, for all practical purposes, the frequency response of the loudspeaker. This does not require an anechoic chamber because, presumably, the response had died out *before* any reflections occur.

The DFT has applications where time is not the independent variable. It may be distance or even angle. About 25 years ago, at a large southern university located near the Great Smoky Moutains, a group of researchers were analyzing circular antenna arrays. Instead of a continuous current sheet, they were constrained to consider (discrete) elements uniformly spaced around the circumference of a circular cylinder. They found that an arbitrary set of currents (samples) could be resolved into a set of *sequence* currents (transform samples). Each sequence then led to one term in a truncated Fourier series expression for the radiation pattern. Little did they realize that they were actually working with the discrete Fourier transform, and, in fact the FFT was within reach had the need for it arisen.

5.11 POWER AND ENERGY SPECTRAL DENSITIES

The energy in a signal is given by Eq. 5.27

$$w = \int_{-\infty}^{\infty} |f(\tau)|^2 \, d\tau = \frac{1}{2\pi} \int_{-\infty}^{\infty} |F(\omega)|^2 \, d\omega \tag{5.27}$$

where, as implied, $f(t) \leftrightarrow F(\omega)$, and $f(t)$ is *either* the voltage or current associated with a 1-Ω resistor. This is the customary procedure in communications work, and if the resistance is other than 1 Ω, it is a simple matter to account for it.

A *power signal* is a signal that has finite average power, but infinite energy. The everlasting periodic signal is an example of a power signal. On the other hand, an *energy signal* is a signal that has finite energy but zero average power. A "one-shot" pulse is an energy signal. The energy over a time interval T_0 (on a 1 Ω basis) for a *real* signal is

$$w_0 = \int_{-T_0/2}^{T_0/2} f^2(t) \, dt$$

while the average power over this same time interval is

$$\langle p_0 \rangle = \frac{1}{T_0} \int_{-T_0/2}^{T_0/2} f^2(t) \, dt \tag{5.86}$$

We take the *general* definition of average power to be the average over *all time*:

$$\langle p \rangle = \lim_{T_0 \to \infty} \left\{ \frac{1}{T_0} \int_{-T_0/2}^{T_0/2} f^2(t) \, dt \right\} \tag{5.87}$$

Equation 5.87 applies to random noise signals as well as to the deterministic signals implied above. Notice that for periodic signals (Sec. 4.6) Eq. 5.86 gives the average power if $T_0 = nT$, $n = 1, 2, \ldots$, and T is the period.

The *spectral power density, power density spectrum*, or, simply, *power spectrum* of a power signal is that function, $S_f(\omega)$, that gives the distribution of average power in the signal as a function of ω. In order to find an equation for $S_f(\omega)$, consider the everlasting $f(t)$ ($-\infty < t < \infty$), and let $f_0(t)$ be equal to

$f(t)$ in the interval $-T_0/2 < t < T_0/2$. Equation 5.27 gives

$$w = \int_{-\infty}^{\infty} |f_0(\tau)|^2 \, d\tau = \frac{1}{2\pi} \int_{-\infty}^{\infty} |F_0(\omega)|^2 \, d\omega \tag{5.88}$$

But

$$\int_{-\infty}^{\infty} |f_0(\tau)|^2 \, d\tau = \int_{-T_0/2}^{T_0/2} |f(\tau)|^2 \, d\tau$$

so (5.88) becomes

$$w = \int_{-T_0/2}^{T_0/2} |f(\tau)|^2 \, d\tau = \frac{1}{2\pi} \int_{-\infty}^{\infty} |F_0(\omega)|^2 \, d\omega$$

Equation 5.87 gives the average power

$$\langle p \rangle = \lim_{T_0 \to \infty} \left\{ \frac{1}{T_0} \int_{-T_0/2}^{T_0/2} |f(t)|^2 \, dt \right\} = \lim_{T_0 \to \infty} \frac{1}{2\pi} \int_{-\infty}^{\infty} \frac{|F_0(\omega)|^2}{T_0} \, d\omega$$

or

$$\langle p \rangle = \frac{1}{2\pi} \int_{-\infty}^{\infty} \lim_{T_0 \to \infty} \left\{ \frac{|F_0(\omega)|^2}{T_0} \right\} d\omega \tag{5.89}$$

The integrand of 5.89 is that quantity which when integrated (or summed) over frequency gives the average power. Thus, it is the spectral power density, or

$$S_f(\omega) = \lim_{T_0 \to \infty} \left\{ \frac{|F_0(\omega)|^2}{T_0} \right\} \qquad w - s \tag{5.90}$$

and

$$\langle p \rangle = \frac{1}{2\pi} \int_{-\infty}^{\infty} S_f(\omega) \, d\omega \tag{5.91}$$

Notice that in Eq. 5.90 all phase information is lost, and the spectral power density of $f(t)$ [that is, $S_f(\omega)$] depends only on the magnitude of $F_0(\omega)$. This, in turn, means that signals having the same Fourier transform magnitudes [for example: $\cos(\omega_0 t)$ and $\sin(\omega_0 t)$] will have the same spectral power density. Using Eq. 5.90 in a recording that lasts for T_0 seconds, for example, or even converting analog data into digital data for use with a digital computer, forms part of a study that is called *spectral analysis*.

Since it has been assumed that $f(t)$ is real, it follows that $|F_0(\omega)|^2 = F_0(\omega)F_0^*(\omega)$, and since

$$F_0(\omega) = \int_{-T_0/2}^{T_0/2} f(t) e^{-j\omega t} \, dt \tag{5.92}$$

it follows that Eq. 5.90 can be written as

$$S_f(\omega) = \lim_{T_0 \to \infty} \left\{ \frac{1}{T_0} \int_{-T_0/2}^{T_0/2} f(x) e^{-j\omega x} \, dx \int_{-T_0/2}^{T_0/2} f(t) e^{+j\omega t} \, dt \right\}$$

5.11 POWER AND ENERGY SPECTRAL DENSITIES

Letting $x = \tau + t$, or $\tau = x - t$

$$S_f(\omega) = \lim_{T_0 \to \infty} \left\{ \frac{1}{T_0} \int_{-t-T_0/2}^{-t+T_0/2} f(t+\tau) e^{-j\omega\tau} e^{-j\omega t} \, d\tau \int_{-T_0/2}^{T_0/2} f(t) e^{j\omega t} \, dt \right\}$$

or

$$S_f(\omega) = \lim_{T_0 \to \infty} \int_{-t-T_0/2}^{-t+T_0/2} \left\{ \frac{1}{T_0} \int_{-T_0/2}^{T_0/2} f(t) f(t+\tau) \, dt \right\} e^{-j\omega\tau} \, d\tau$$

or

$$S_f(\omega) = \int_{-\infty}^{\infty} \left\{ \lim_{T_0 \to \infty} \left[\frac{1}{T_0} \int_{-T_0/2}^{T_0/2} f(t) f(t+\tau) \, dt \right] \right\} e^{-j\omega\tau} \, d\tau \tag{5.93}$$

From Eq. 5.23 we recognize (interchanging t and τ)

$$\lim_{T_0 \to \infty} \int_{-T_0/2}^{T_0/2} f(t) f(t+\tau) \, dt$$

as the *autocorrelation integral*. We used this to derive Eq. 5.27. We now define the function of τ within the braces (Eq. 5.93) as the *time autocorrelation function*, $R_f(\tau)$, of $f(t)$:

$$R_f(\tau) = \lim_{T_0 \to \infty} \frac{1}{T_0} \int_{-T_0/2}^{T_0/2} f(t) f(t+\tau) \, dt \tag{5.94}$$

Equation 5.93 then becomes

$$S_f(\omega) = \int_{-\infty}^{\infty} R_f(\tau) e^{-j\omega\tau} \, d\tau \tag{5.95}$$

and therefore

$$R_f(t) \leftrightarrow S_f(\omega) \tag{5.96}$$

As pair 5.96 indicates, the spectral power density of a real power signal is the Fourier transform of the signal's autocorrelation function. Further use of these results is reserved for more advanced studies of noise and random processes.

Although it is not very difficult to show that for periodic signals

$$S_f(\omega) = 2\pi \sum_{n=-\infty}^{\infty} |C_n|^2 \delta(\omega - n\omega_T) \tag{5.97}$$

(where C_n is the coefficient in the complex Fourier series, and is given by Eq. 4.24), we shall not do so here. The proof is left as an exercise. Instead, we merely point out that when Eq. 5.97 is substituted into 5.91, we immediately obtain

$$\langle p \rangle = \sum_{n=-\infty}^{\infty} |C_n|^2 \tag{5.98}$$

a result that is consistent with those in Chap. 4.

Example 18

Find the spectral power density and average power for the periodic signal $f(t) = V_m \cos(\omega_T t)$. We have

$$f(t) = V_m \cos(\omega_T t) = \frac{V_m}{2} e^{j\omega_T t} + \frac{V_m}{2} e^{-\omega_T t}$$

Therefore, by inspection (or Eq. 4.24)

$$C_1 = C_{-1} = \frac{V_m}{2} \quad (n = \pm 1 \text{ only})$$

and, by Eq. 5.97

$$S_f(\omega) = \frac{\pi V_m^2}{2} [\delta(\omega + \omega_T) + \delta(\omega - \omega_T)] \quad w-s$$

Therefore, by either Eq. 5.91 or 5.98

$$\langle p \rangle = \frac{1}{2\pi} \int_{-\infty}^{\infty} S_f(\omega) \, d\omega = \frac{V_m^2}{2} \quad w$$

which is obviously the correct result, and is in agreement with Eq. 4.49 ($R = 1\,\Omega$). The periodic signal $f(t) = V_m \sin(\omega_T t)$ gives similar results. The autocorrelation function for $f(t) = V_m \cos(\omega_T t)$ is easily found to be

$$R_f(t) = \frac{V_m^2}{2} \cos(\omega_T t)$$

Next, suppose that the truncated function $f_0(t)$ is applied to a time-invariant linear system (a filter in frequency-domain terminology). If the transfer function of the system is $H(\omega)$, then the zero-state response is

$$Y_0(\omega) = H(\omega) X_0(\omega) \quad [X_0(\omega) \equiv F_0(\omega)]$$

The spectral power density of the response is

$$S_{fy}(\omega) = \lim_{T_0 \to \infty} \frac{1}{T_0} |Y_0(\omega)|^2 = \lim_{T_0 \to \infty} \frac{1}{T_0} |H(\omega)|^2 |X_0(\omega)|^2$$

or

$$S_{fy}(\omega) = \lim_{T_0 \to \infty} \frac{|X_0(\omega)|^2}{T_0} |H(\omega)|^2 = S_{fx}(\omega) |H(\omega)|^2$$

Thus, the power spectral density of the output signal is given by the product of the power spectral density of the input and the squared magnitude of the transfer function of the system. These concepts are useful in high-fidelity amplifier ratings, for example. The mean-square output signal is the output average power ($R = 1\,\Omega$), and is given by

$$\langle p_0 \rangle = \frac{1}{2\pi} \int_{-\infty}^{\infty} S_{fy}(\omega) \, d\omega = \frac{1}{2\pi} \int_{-\infty}^{\infty} S_{fx}(\omega) |H(\omega)|^2 \, d\omega \tag{5.99}$$

If $x_0(t) = V_m\cos(\omega_T t)$ (Example 18), for example, then

$$\langle p_0 \rangle = \frac{1}{2\pi}\int_{-\infty}^{\infty} \frac{\pi V_m^2}{2}[\delta(\omega + \omega_T) + \delta(\omega - \omega_T)]|H(\omega)|^2 d\omega$$

$$\langle p_0 \rangle = \frac{V_m^2}{4}(|H(-\omega_T)|^2 + |H(\omega_T)|^2)$$

Since $H(\omega)$ is even (Why?)

$$\langle p_0 \rangle = \frac{V_m^2}{2}|H(\omega_T)|^2$$

It is easily seen why these amplifiers are commonly rated on the basis of power-response curves that are simply $\log_{10}|H(\omega)|^2$ versus $\log_{10}\omega$.

In Eq. 5.27:

$$w = \frac{1}{2\pi}\int_{-\infty}^{\infty} |F(\omega)|^2 d\omega \quad j \quad (5.27)$$

the quantity $|F(\omega)|^2$ is logically called the *energy spectral density* since it is given in terms of joule-seconds or joules per radian per second. It is the function that describes the *relative* signal energy as a function of frequency, and whose *area* (divided by 2π) gives the signal's *total energy*.

If an energy signal is applied to a time-invariant linear system, then the zero-state response is

$$Y(\omega) = H(\omega)X(\omega)$$

and the normalized energy spectral density is

$$|Y(\omega)|^2 = |H(\omega)|^2|X(\omega)|^2$$

while the normalized energy in the output is

$$w_0 = \frac{1}{2\pi}\int_{-\infty}^{\infty} |H(\omega)|^2|X(\omega)|^2 d\omega \quad (5.100)$$

Thus, the energy density of the response is the product of the energy density of the input and the square of the magnitude of the transfer function. As in the case of the power spectral density, all phase information is lost.

Example 19

Find the input and output energies when $x(t) = u(t + 3) - u(t - 3)$ is applied to an ideal low-pass filter (gain = 1) having a cutoff frequency of 2 r/s. The energy in the input signal is ($R = 1\ \Omega$)

$$w_i = \int_{-\infty}^{\infty} |x(t)|^2 dt = \int_{-3}^{+3} dt = 6 \quad j$$

FOURIER TRANSFORM TECHNIQUES

The energy in the output signal is

$$w_0 = \frac{1}{2\pi} \int_{-\infty}^{\infty} |H(\omega)|^2 |X(\omega)|^2 \, d\omega$$

$$w_0 = \frac{1}{2\pi} \int_{-2}^{+2} |X(\omega)|^2 \, d\omega$$

Now

$$x(t) \leftrightarrow \left[\pi\delta(\omega) + \frac{1}{j\omega}\right][e^{j3\omega} - e^{-j3\omega}] = \frac{e^{j3\omega} - e^{-j3\omega}}{j\omega}$$

$$x(t) \leftrightarrow \frac{2}{\omega} \sin(3\omega) = 6Sa(3\omega)$$

Therefore

$$w_0 = \frac{1}{2\pi} \int_{-2}^{+2} 36 Sa^2(3\omega) \, d\omega = \frac{36}{\pi} \int_0^2 Sa^2(3\omega) \, d\omega$$

or with $3\omega = x$

$$w_0 = \frac{12}{\pi} \int_0^6 Sa^2(x) \, dx$$

Numerical integration with a programmable calculator gives

$$w_0 = 5.699 \quad j$$

This section is closed with the statement that *any energy signal is Fourier transformable.*

5.12 CONCLUDING REMARKS

In this chapter we have extended phasor and Fourier concepts to the continuous-frequency domain of the Fourier transform. The Fourier transform is especially useful in the analysis of time-invariant linear systems. If the system is *not time-invariant*, an algebraic equation does *not* usually result when the transform is applied to the differential equation describing the system, and no advantage is gained by its use. A time-domain solution, as described in Chap. 2, is, of course, still possible. The Fourier transform method can handle all realistic time-invariant linear systems which are described by differential equations, but many times the method becomes unwieldly and cumbersome. This is particularly true in those problems where initial conditions are present due to a switch opening or closing or an independent source suddenly changing value. These problems can best be treated by the Laplace transform method of the next chapter. The Fourier transform inherently gives the *zero-state* response, and the zero-input response must be added. Thus, the Fourier transform best handles situations where the zero-state response is the only response.

The Fourier transform is also useful whenever spectral analysis is required, such as in optics, communication systems, electromagnetics, quantum theory, noise and random variable studies, and so on. A few examples of this use have been given in this chapter, and several more are asked for in the problems at the end of this chapter.

We now have both a time-domain approach and a frequency-domain approach to linear-systems analysis. There is no distinct advantage of one method over the other, although the time-domain methods can be used for time-variable systems. It simply depends on the particular problem being investigated. Sometimes we run headlong into "conservation of difficulty," and have a tough problem no matter which way we go. Do not forget that numerical techniques and fast digital computers are available to handle difficult integrations.

A *discrete* version of the *Fourier transform* (DFT) exists, and can be used with digital methods for computation using the digital computer. The implementation of the DFT is difficult, however, if the number of samples becomes large, and it becomes inefficient. Efficient algorithms, known as the *fast Fourier transform* (FFT), have been developed. These enable the user to make very rapid calculations with the DFT.

Finally, it should be emphasized that Fourier methods have many other applications in engineering besides those discussed in this chapter. The three-(space) dimensional wave equation, for example, can be converted into a simpler two-dimensional wave equation by Fourier transforming one of the space variables. Also, as another example, the radiation field of aperture-type antennas can be obtained as the transform of the source distribution across the aperture.

PROBLEMS

1. Show that

$$\sum_{n=-\infty}^{\infty} \delta(t - nT) \leftrightarrow \omega_T \sum_{n=-\infty}^{\infty} \delta(\omega - n\omega_T) \qquad \omega_T = 2\pi/T$$

See Prob. 20, Chap. 4.

2. (a) Establish the time-shifting result (5.13):

$$f(t - t_0) \leftrightarrow F(\omega)e^{-j\omega t_0}$$

(b) Establish the frequency-shifting result (5.14):

$$f(t)e^{j\omega_0 t} \leftrightarrow F(\omega - \omega_0)$$

(c) Establish the scaling result (5.15):

$$f(\alpha t) \leftrightarrow \frac{1}{|\alpha|} F(\omega/\alpha)$$

3. Establish Eq. 5.36:
$$u(t) \leftrightarrow \pi\delta(\omega) + 1/j\omega$$
by finding
$$\mathscr{F}^{-1}\{\pi\delta(\omega) + 1/j\omega\}$$

4. Establish transform pair 20, Table 5.1:
$$u(t)\cos(\omega_0 t) \leftrightarrow \frac{\pi}{2}[\delta(\omega - \omega_0) + \delta(\omega + \omega_0)] + j\omega/(\omega_0^2 - \omega^2)$$
Use pair 11.

5. Establish transform pair 23, Table 5.1:
$$u(t)e^{-\alpha t}\sin(\omega_0 t) \leftrightarrow \frac{\omega_0}{\omega_0^2 + (\alpha + j\omega)^2}$$
Use Eq. 5.7.

6. Establish transform pair 34, Table 5.1:
$$\frac{1}{b-a}[e^{-at} - e^{-bt}]u(t) \leftrightarrow \frac{1}{(j\omega + a)(j\omega + b)} \qquad a, b > 0$$
Use a partial fraction expansion (or undetermined coefficients).

7. Establish transform pair 35, Table 5.1:
$$\frac{1}{b-a}[-ae^{-at} + be^{-bt}]u(t) \leftrightarrow \frac{j\omega}{(j\omega + a)(j\omega + b)} \qquad a, b > 0$$
Use pair 34.

8. Find the Fourier transform of
   (a) $u(t)$, by using transform pair 8 (Table 5.1),
   (b) $\delta(t + T) - \delta(t - T)$,
   (c) $te^{-\alpha|t|}$, $\alpha > 0$,
   (d) $\sin(t)\cos(t)$
   (e) $\dfrac{d^2 h}{dt^2} + 2\dfrac{dh}{dt} + h = \delta(t)$. Solve for $H(\omega)$. Find $h(t)$.

9. Find the inverse Fourier transform of
   (a) $1/j\omega$, starting with transform pair 8 (Table 5.1),
   (b) $\pi\delta(\omega) - 1/j\omega$,
   (c) $\delta(\omega + 1) + \delta(\omega - 1)$
   (d) $ASa^2(\omega\tau/2)$
   (e) $\dfrac{1}{-\omega^2 + j4\omega + 3}$. Try factoring.

10. Find the (causal) unit-impulse response for Prob. 16, Chap. 2, by using Fourier transform methods. Find $h(t)$ for
$$\frac{d^2 y}{dt^2} + 4\frac{dy}{dt} + 3y = x$$

11. Repeat Prob. 31, Chap. 2, using Fourier transform methods and Fig. 5.29.
    (a) Find $v(t)$ if $i_s(t) = 10\, u(t)$ A, $i_L(0) = v_C(0) = 0$.
    (b) Find $v(t)$ if $i_s(t) = 10 \sin(2000t)$ A.

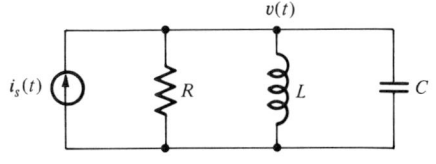

Figure 5.29. Circuit for Prob. 11.

12. Solve Prob. 33, Chap. 2, using Fourier transform methods and Fig. 5.30. Find $y_1(t)$ and $y_2(t)$ when $x_1(t) = \sin(t)$, $x_2(t) = 2\cos(3t)$, and $x_3(t) = t$.

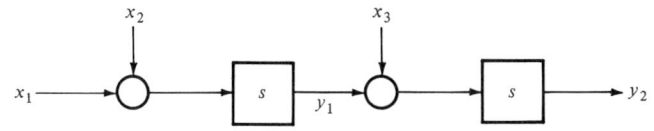

Figure 5.30. System for Prob. 12.

13. What happens when the Fourier transform is applied to the time-variable system of Prob. 34, Chap. 2?

$$t^2 \frac{d^2 y}{dt^2} + 2t \frac{dy}{dt} - 2y = 0$$

14. Solve Prob. 32, Chap. 2, using Fourier transform methods and Fig. 5.31.
    (a) Find $v(t)$, $t \geq 0$, if $v_1(t) = 60$ V.
    (b) Repeat if $v_1(t) = 60 \cos(5t)$ V.

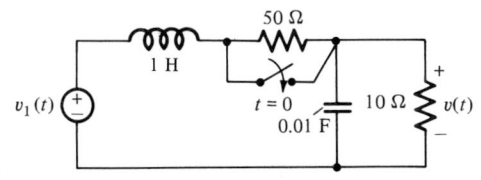

Figure 5.31. Circuit for Prob. 14.

15. Repeat Example 12, Chap. 2, using Fourier transform methods and Fig. 5.32. Find **h** if $v_1(t)$ and $v_2(t)$ are the outputs.

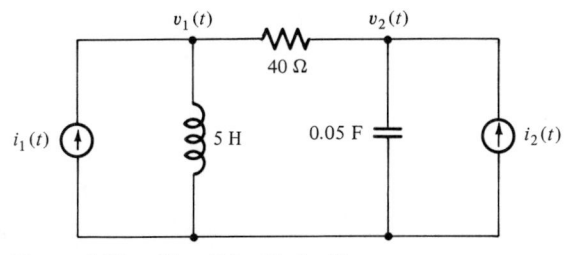

Figure 5.32. Circuit for Prob. 15.

238 FOURIER TRANSFORM TECHNIQUES

16. Repeat Example 4, Chap. 3, (second part) using Fourier transform methods and Fig. 5.33. Find $v_C(t)$ and $i_L(t)$ when $i_s(t) = 9.6e^{-4t}u(t)$ A.

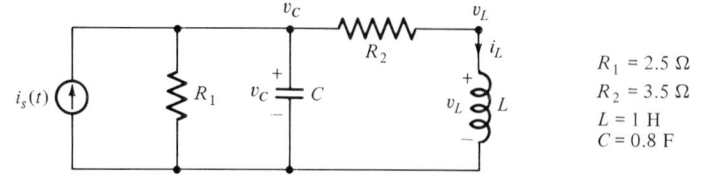

$R_1 = 2.5\ \Omega$
$R_2 = 3.5\ \Omega$
$L = 1$ H
$C = 0.8$ F

Figure 5.33.   Circuit for Prob. 16.

17. Repeat Problem 9, Chap. 3, using Fourier transform methods. Find $y_1(t)$ and $y_2(t)$ when $z_1(t) = 2\sin(2t)$ and $z_2(t) = \cos(2t)$.

$$y_1' + y_1 + y_2' = z_1$$
$$y_1' + 3y_1 - y_2' - 2y_2 = z_2$$

18. Find $v_C(t)$ in Fig. 5.34.

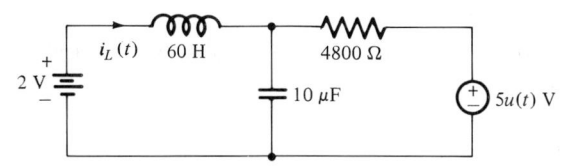

Figure 5.34.   Circuit for Prob. 18.

19. Find $i_L(t)$ in Fig. 5.34.

20. Repeat Prob. 23, Chap. 4, using the method of Sec. 5.4 in Fig. 5.35. Find $v_0(t)$ when $L = 0$ and $L = R^2C(\sqrt{2} - 1)$.

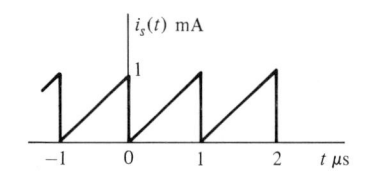

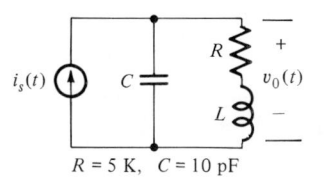

Figure 5.35.   Circuit for Prob. 20.

21. Solve the *integral equation*

$$y(x) = g(x) + \int_{-\infty}^{\infty} y(u)h(x - u)\,du$$

22. The spectrum of an *ideal* high-pass filter is

$$H(\omega) = u(-\omega - \omega_0) + u(\omega - \omega_0)$$

(a) Find $h(t)$.
(b) Explain why this filter cannot be built.

23. Plot the magnitude spectrum of the amplitude-modulated signal:
$$f(t) = A_m[1 + m_a\cos(\omega_m t)]\cos(\omega_0 t)$$
Describe the spectrum.

24. (a) Find $H(\omega)$ for the network of Figure 5.36. What kind of filter is this?
    (b) If an identical network is cascaded with the first, is the new transfer function equal to the square of the transfer function in (a)? Why?

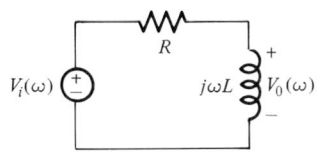

Figure 5.36. RL network.

25. Show that the three-dimensional (in space) wave equation in rectangular coordinates is converted into a simpler two-dimensional wave equation by applying the Fourier transform to any one of the three *space* variables:
$$\frac{\partial^2 u}{\partial x^2} + \frac{\partial^2 u}{\partial y^2} + \frac{\partial^2 u}{\partial z^2} - \frac{1}{a^2}\frac{\partial^2 u}{\partial t^2} = 0$$

26. Solve the boundary-value problem (heat conduction)
$$\frac{\partial \theta}{\partial t} = k\frac{\partial^2 \theta}{\partial z^2} \qquad \theta(z, 0) = f(z) \quad |\theta(z, t)| < M$$
for $-\infty < z < \infty$, $t > 0$. $\theta(z, t) =$ temperature, $k =$ diffusivity.

27. Given that
$$E_\theta = \frac{j\omega\mu}{4\pi}\frac{e^{-jkr}}{r}\sin\theta \int_{-h}^{h} I_z(z')e^{jkz'\cos\theta}\,dz'$$
for the electric field in the radiation zone from a line source of length $2h$ on the z axis, interpret the integral as a Fourier transform.

28. The voltage and current on the lossless transmission line of Fig. 5.37 (matched at the generator) are given in phasor form by
$$V(z) = \frac{V_g}{2}e^{-jkl}(e^{+jkz} + \Gamma_l e^{-jkz})$$

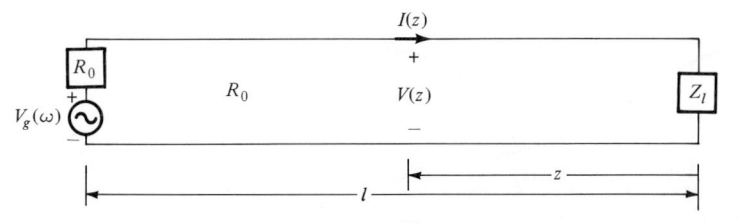

Figure 5.37. Lossless transmission line.

and

$$I(z) = \frac{V_g}{2R_0} e^{-jkl}(e^{+jkz} - \Gamma_l e^{-jkz})$$

where $k = \omega\sqrt{LC}$, $R_0 = \sqrt{L/C}$, and $\Gamma_l = (Z_l - R_0)/(Z_l + R_0)$.
(a) Find the unit-impulse response for the voltage when $Z_l = R_0$. Comment.
(b) Repeat (a) for $Z_l = R_l + j\omega L_l$
(c) Repeat (a) for $Z_l = R_l - j/(\omega C_l)$
(d) Repeat (a) for $Z_l = R_l + j\omega L_l - j/(\omega C_l)$
  Hint: Find $\Gamma_l(t)$ and use convolution.
29. Repeat Example 11 using Eq. 5.56 with $\omega_T = 4\omega_h$ and $\omega_c = 2\omega_h$. Find $f(0)$ and $f(\tfrac{1}{2})$.
30. (a) Find the (continuous) Fourier transform of $x(t) = u(t) - u(t-2)$. See Prob. 9(d).
  (b) Find the DFT of $x(t)$ using the samples

$$f(0) = 1,\ f(\tfrac{1}{2}) = 1,\ f(\tfrac{2}{2}) = 1,\ f(\tfrac{3}{2}) = 1,\ f(\tfrac{4}{2}) = 0,$$
$$f(\tfrac{5}{2}) = 0,\ f(\tfrac{6}{2}) = 0,\ \text{and}\ f(\tfrac{7}{2}) = 0, \qquad N = 8,\ T = \tfrac{1}{2}$$

31. In order to demonstrate the *discrete* version of convolution, suppose that $x(t)$ (Prob. 30) is convolved with itself.
  (a) Find the continuous function

$$y(t) = \int_{-\infty}^{\infty} x(\tau)x(t-\tau)\,d\tau$$

  (b) Find $y(n)$ using Eq. 5.73 with $N = 8$, $T = \tfrac{1}{2}$.
  (c) Repeat (b) with $N = 16$, $T = \tfrac{1}{4}$.
  (d) Plot the results of (a), (b), and (c). Comment. What happens as $T \to 0$ with $NT = 4$?
32. Repeat Prob. 31(a) and (b) using Eq. 5.72. Do these results agree with those in Prob. 31?
33. Show that a *distortionless* system is one for which $H(\omega) = Ke^{-j\omega\tau}$, where K and τ are real constants. See Prob. 28(a).
34. Derive the DFT from Eq. 5.7 by using the trapezoidal rule for integrating $f(t)$ in Fig. 5.38 if $f(nT) = f(nT + NT)$ and $\omega = 2\pi m/(NT)$.

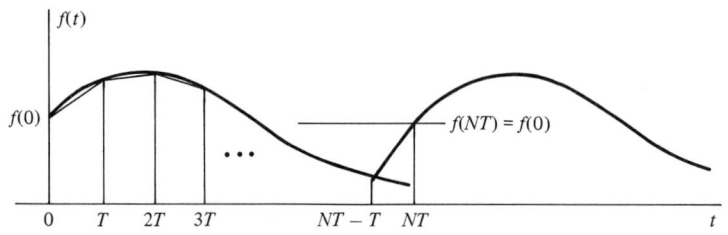

Figure 5.38. Derivation of the DFT, trapezoidal rule.

35. Approximate df/dt by using the straight-line segments in Fig. 5.38, then approximate the transform of $f(t)$ by using Eq. 5.17, written in the form

$$F(\omega) = \frac{1}{j\omega} \mathscr{F}\left\{\frac{df}{dt}\right\}$$

36. (a) If $f(nT) = f(nT + NT)$ and $\omega = 2\pi m/(NT)$, as in Prob. 34, show that $F(m\omega_T/N)$ given by the result of Prob. 35 is approximately

$$F(m\omega_T/N) \approx [Sa(m\pi/N)]^2 T \sum_{n=0}^{N-1} f(nT)e^{-j2\pi mn/N}$$

(b) Show that the approximation in (a) becomes identical to the DFT for large N.

37. Consider the second-order system modeled by Eq. 2.89:

$$\frac{d^2y}{dt^2} + 2\zeta\omega_n\frac{dy}{dt} + \omega_n^2 y = \omega_n^2 x$$

Find $H(\omega)$.

(a) Show that the (correct) unit-impulse response, obtained by setting $\alpha = \zeta\omega_n = 0$ *after* performing the inverse Fourier transformation, is

$$h(t) = \omega_n \sin(\omega_n t)u(t)$$

Compare this result to Eq. 2.101.

(b) Let $\alpha = \zeta\omega_n = 0$ *before* performing the inverse Fourier transformation, and show that the (incorrect) noncausal unit-impulse response is

$$h(t) = \frac{\omega_n}{2}\sin(\omega_n|t|)$$

Use transform pairs 19 and 21 (Table 5.1).

References

1. DiStefano, J. J., III, Stubberud, A. R., and Williams I. J. *Feedback and Control Systems*, Schaum Outline Series. New York: McGraw-Hill, 1967.
2. Hayt, W. H., and Kemmerly, J. E. *Engineering Circuit Analysis*, 3rd ed. New York: McGraw-Hill, 1978.
3. Stremler, F. G., *Introduction to Communication Systems*, 2nd ed. Reading, Mass.: Addison-Wesley, 1982.
4. McGillem, C. D., and Cooper, G. R. *Continuous and Discrete Signal and System Analysis*. New York: Holt, Rinehart and Winston, 1974.
5. Cooley, J. W., and Tukey, J. W. An Algorithm for the Machine Calculation of Complex Fourier Series. *Math Comput.* 19:297–301, April 1965.
6. Papoulis, A. *Circuits and Systems*. New York: Holt, Rinehart and Winston, 1980.

Chapter 6
The Laplace Transform

In this chapter we will extend the (linear) Fourier transform to the (linear) Laplace transform for *real* functions of time, or perhaps it would be more accurate to say that we will modify the Fourier transform to produce the Laplace transform. In doing so, we will gain an advantage in that some problems will then be easier to solve; but, all things not being perfect, we will gain some disadvantages also, and other problems will be more difficult to solve with the Laplace transform than with the Fourier transform. The Laplace transform is not a cure-all, and, as is so often the case in engineering work, we must find the right tool for the job at hand. Again, we are primarily interested in time-invariant linear systems because we will find, as in the Fourier transform, Laplace transforming a *variable coefficient* differential equation does not generally give an algebraic equation, but, instead, merely gives a new differential equation in the transformed variable. This is usually of no benefit.

Chapters 4 and 5 represent a sequence of connected developments. We went from sinusoidal excitation in the steady state (phasors) to nonsinusoidal, but periodic, excitation via the Fourier series. The response was found as the sum of the individual responses due to the individual sinusoidal excitations, and this was possible because of the linearity and the powerful principle of superposition. Next, the complex (exponential) form of the Fourier series was extended to treat nonperiodic functions by letting the period of a train of periodic rectangular pulses become larger and larger without limit. This left us with the expression for a *single* pulse in the form of a superposition (integral) of exponentials that was called the *inverse Fourier transform*:

$$f(t) = \frac{1}{2\pi} \int_{-\infty}^{\infty} F(\omega) e^{jt\omega} \, d\omega \qquad (5.8)$$

The (direct) *Fourier* transform

$$F(\omega) = \int_{-\infty}^{\infty} f(t) e^{-j\omega t} \, dt \qquad (5.7)$$

was obtained with the aid of the Fourier integral theorem and was viewed as giving the *spectrum* of $f(t)$. This pairing, $f(t) \leftrightarrow F(\omega)$, enabled us to solve a wide variety of problems for time-invariant linear systems in a manner that was, in many (but not all) cases, simpler than the direct time-domain approach.

The question now arises, What more could we possibly need? There are several things that we would like to be able to do but cannot because of limitations in the Fourier transform. First of all, even though they cannot be generated in practice, there are many functions such as the ramp, the increasing exponential, and some *random* signals that do not have Fourier transforms because these functions are not absolutely integrable. Secondly, we found that in order to treat initial energy storage, we either added a separate zero-input response to the zero-state response or provided the proper forcing function to give us the correct initial conditions. Either way, this is a rather awkward proposition. Lastly, we will obtain a simpler notation and ease of algebraic manipulation when we obtain the Laplace transformation.

6.1 FOURIER TO LAPLACE TRANSFORM

The Laplace transform can be viewed as an extension of the Fourier transform where we represent $f(t)$ as a superposition (integral) of exponentials of the form e^{st}, where $s = \sigma + j\omega$, rather than as simply a superposition of exponentials, $e^{j\omega t}$. This approach leaves us with the idea that the Fourier transform is a special case of the Laplace transform where s has simply become $j\omega$ ($\sigma = 0$). This approach is preferred because it is consistent with what has already been done in Chaps. 4 and 5, as opposed to presenting an abstract mathematical formulation in the form of a definition.

We start by finding the Fourier transform of the function $g(t) = f(t)e^{-\sigma t}$, where σ is real, and $f(t)$ is sufficiently well behaved so that $g(t)$ is absolutely integrable (that is, the Fourier transform exists). Equation 5.7 gives

$$\mathscr{F}\{g(t)\} = \int_{-\infty}^{\infty} g(t) e^{-j\omega t} \, dt = \int_{-\infty}^{\infty} f(t) e^{-(\sigma + j\omega)t} \, dt$$

Written this way, the integral gives us a function of $(\sigma + j\omega)$ rather than $j\omega$ (or simply ω, as we have been using it). Thus, we should write the last equation as

$$F_2(\sigma + j\omega) = \int_{-\infty}^{\infty} f(t) e^{-(\sigma + j\omega)t} \, dt \qquad (6.1)$$

Equation 5.8 gives $g(t)$:

$$g(t) = \frac{1}{2\pi} \int_{-\infty}^{\infty} F_2(\sigma + j\omega)e^{jt\omega} d\omega = f(t)e^{-\sigma t}$$

or multiplying by $e^{\sigma t}$

$$f(t) = \frac{1}{2\pi} \int_{-\infty}^{\infty} F_2(\sigma + j\omega)e^{(\sigma + j\omega)t} d\omega \tag{6.2}$$

If we now let s be the complex frequency: $s = \sigma + j\omega$, or $\omega = (1/j)(s - \sigma)$, then $d\omega = (1/j) ds$, and when $\omega = \pm\infty$, $s = \sigma \pm j\infty$. In this case Eqs. 6.1 and 6.2 can be written as

$$\boxed{F_2(s) = \int_{-\infty}^{\infty} f(t)e^{-st} dt} \quad \text{(bilateral Laplace transform)} \tag{6.3}$$

and

$$\boxed{f(t) = \frac{1}{2\pi j} \int_{\sigma - j\infty}^{\sigma + j\infty} F_2(s)e^{ts} ds} \quad \begin{array}{l}\text{(inverse bilateral}\\ \text{Laplace transform)}\end{array} \tag{6.4}$$

Equations 6.3 and 6.4 form the *complex Fourier transform pair* or the *bilateral (two-sided) Laplace transform pair*:

$$F_2(s) = \mathscr{L}_2\{f(t)\}$$
$$f(t) = \mathscr{L}_2^{-1}\{F_2(s)\}$$

The subscript 2 refers to *two*-sided in order to differentiate between the bilateral Laplace transform and the unilateral Laplace transform to follow shortly.

It should be obvious that the bilateral Laplace transform is obtained directly from the Fourier transform by simply replacing $j\omega$ with s *for those $f(t)$ that are absolutely integrable*. Otherwise, Eq. 6.3 must be used. Notice that we have already achieved part of what we wanted. Because of the built-in *convergence factor* $e^{-\sigma t}$ in Eq. 6.1 (or 6.3), the Laplace transform will exist for many functions that do not possess Fourier transforms. We also can imagine that it will be easier to algebraically manipulate s than $j\omega$.

The bilateral Laplace transform exists if the defining integral exists (is finite). The values of s for which this occurs can be determined by considering the negative-time and positive-time portions of Eq. 6.3 separately:

$$F_2(s) = \int_{-\infty}^{0} f(t)e^{-st} dt + \int_{0}^{\infty} f(t)e^{-st} dt$$

The first integral exists when $\text{Re}\{s\} = \sigma$ is *less* than some value that we call β ($\sigma < \beta$), while the second integral exists when $\text{Re}\{s\} = \sigma$ is *greater* than some value that we call α ($\sigma > \alpha$). When both of these conditions are satisfied, then $\alpha < \sigma < \beta$, and there is a *strip* of convergence in the s plane *inside of which* $F_2(s)$ exists. Obviously, if there is no strip, then $F_2(s)$ does not exist. Such a strip is shown in Fig. 6.1. Notice that α is governed by the positive-time portion of $f(t)$, whereas β is governed by the negative-time portion of $f(t)$.

6.1 FOURIER TO LAPLACE TRANSFORM

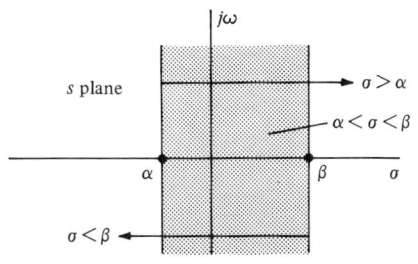

Figure 6.1. Strip of convergence ($\alpha < \sigma < \beta$) for $F_2(s)$.

There are no singularities or poles (values of s such that $|F(s)| \to \infty$) in the strip of convergence since $F_2(s)$ converges in the strip. For functions of time, $f(t)$, that are nonzero for $t < 0$ only, then β alone determines the resultant semifinite strip of convergence, and all the poles of $F_2(s)$ lie to the *right* of β. In the same way, if $f(t)$ is nonzero for $t > 0$ only, then α alone determines the semi-infinite strip of convergence, and all of the poles of $F_2(s)$ lie to the left of α. Thus, the positive-time portion of $f(t)$ contributes poles to the left of the region of convergence, and the negative-time portion of $f(t)$ contributes poles to the right of the region of convergence. This knowledge in very helpful in finding *inverse* bilateral Laplace transforms (6.4). Notice also that the region of convergence *must* include the $j\omega$ axis (strictly speaking) if the Fourier transform exists, and if this is the case (as has already been pointed out), we merely replace s with $j\omega$ (or vice versa). Figure 6.2 shows several functions of time and the region of convergence for each. Notice that the Fourier transform exists for the second, third, fourth, and seventh of these.

Take the function $e^{bt}u(-t) + e^{-at}u(-t)$, $a > 0$, $b > 0$ (the sum of the third and fourth cases of Fig. 6.2). We have

$$F_2(s) = \int_{-\infty}^{0} e^{bt}e^{-st} dt + \int_{0}^{\infty} e^{-at}e^{-st} dt$$

$$F_2(s) = \int_{-\infty}^{0} e^{-(s-b)t} dt + \int_{0}^{\infty} e^{-(s+a)t} dt$$

$$F_2(s) = -\frac{1}{s-b} + \frac{1}{s+a}$$

provided that $-a < \sigma < b$, which is the strip resulting from the overlap of those in the third and fourth cases of Fig. 6.2.

Next, take the function e^{at}, $a > 0$:

$$F_2(s) = \int_{-\infty}^{0} e^{-(s-a)t} dt + \int_{0}^{\infty} e^{-(s-a)t} dt$$

The first integral converges for $\sigma < a$, while the second converges for $\sigma > a$. There is no overlapping and the bilateral Laplace transform does not exist.

Finally, take the function $u(-t) + e^{-at}u(t)$, $a > 0$:

$$F_2(s) = \int_{-\infty}^{0} e^{-st} dt + \int_{0}^{\infty} e^{-(s+a)t} dt$$

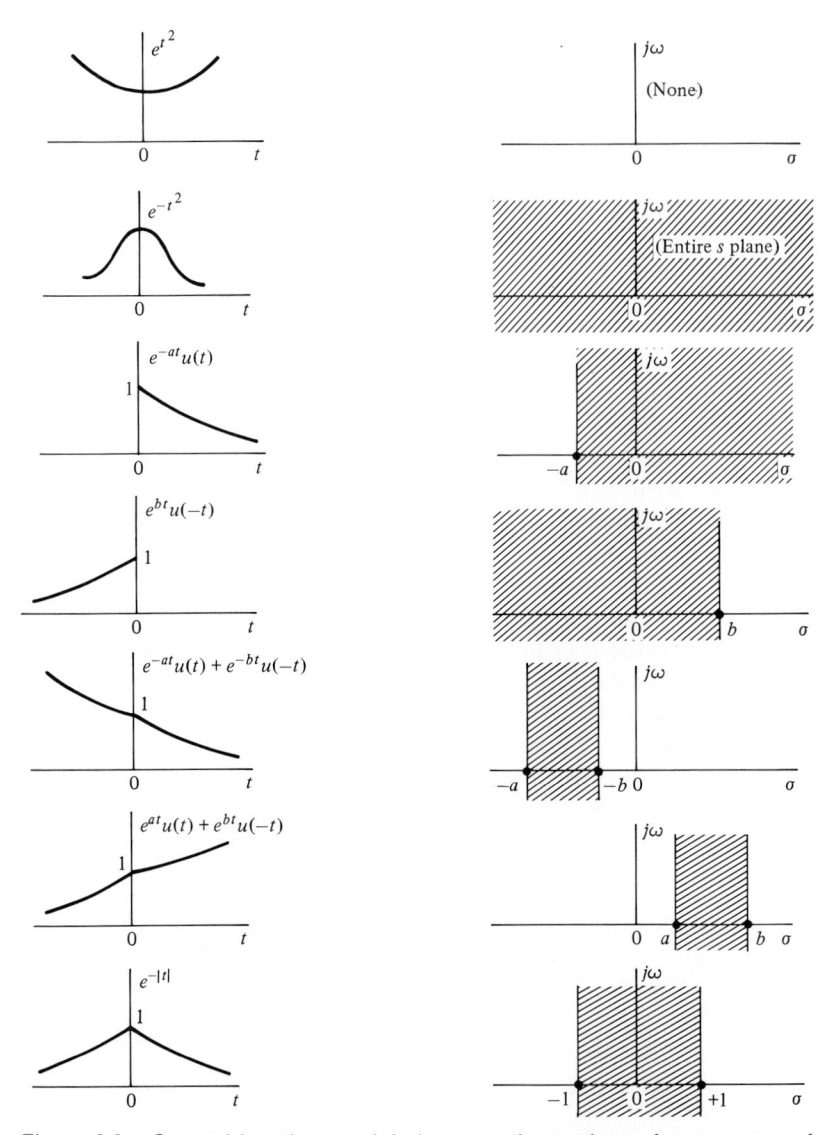

Figure 6.2. Several functions and their respective regions of convergence for $F_2(s)$.

The first integral converges for $\sigma < 0$, while the second converges for $\sigma > -a$. Thus, $-a < \sigma < 0$, and the $j\omega$ axis is excluded from the region of convergence. Strictly speaking then, the Fourier transform does not exist: at least *sufficient* conditions have not been satisfied. We did find in Chap. 5 that the Fourier transform is $F(\omega) = \pi\delta(\omega) - 1/(j\omega) + 1/(j\omega + a)$. Here, the bilateral Laplace transform is

$$F_2(s) = -1/s + 1/(s+a)$$

6.1 FOURIER TO LAPLACE TRANSFORM

Notice that

$$F_2(s)|_{s=j\omega} \neq F(\omega)$$

We have seen that the positive- and negative-time parts of the time function are best handled separately when finding the bilateral Laplace transform. The same is unfortunately necessary when finding the inverse bilateral Laplace transform. Furthermore, there is an ambiguity, or lack of uniqueness, when finding the inverse transform, *if the region of convergence is not specified*. Consider, for example,

$$e^{-\alpha t}u(t) \leftrightarrow \frac{1}{s+\alpha} \qquad \sigma > -\alpha$$

and

$$-e^{-\alpha t}u(-t) \leftrightarrow \frac{1}{s+\alpha} \qquad \sigma < -\alpha$$

The *only* distinguishing feature between the two functions of s is their different regions of convergence. Usually the difficulty with ambiguity can be resolved by the practical consideration that time functions that increase without limit for $t \to \pm\infty$ do not exist in real problems, and that time function that is bounded and practical can be selected as the correct one even though others may be possible.

The forms of $F_2(s)$ that nearly always occur in the analysis of linear time-invariant systems are

$$F_2(s) = \frac{A_1}{s+\alpha_1} + \frac{A_2}{s+\alpha_2} + \frac{A_3}{s+\alpha_3} + \cdots$$

where some of the poles, $s = -\alpha_n$, will lie in the left half of the s plane, and some will lie in the right half of the s plane. We already know that those lying in the left half correspond to exponentials that decay for positive time, while those that lie in the right half correspond to exponentials that die out for $t \to -\infty$. Therefore, realistic time functions for the inverse bilateral Laplace transform can be selected if (1) all terms that come from left-hand poles are assumed to yield time functions that exist *only* for $t \geq 0$ [i.e., multiplied by $u(t)$] and (2) all terms that come from right-hand poles are assumed to yield time functions that exist *only* for $t \leq 0$ [i.e., multiplied by $u(-t)$]. If $F_2(s) = 1/(s+\alpha)$, as in the preceding paragraph, we *choose*

$$e^{-\alpha t} \leftrightarrow \frac{1}{s+\alpha} \qquad \sigma > -\alpha$$

For $F_2(s) = -2/(s-2) + 4/(s+3)$ we choose

$$2e^{2t}u(-t) + 4e^{-3t}u(t) \leftrightarrow \frac{-2}{s-2} + \frac{4}{s+3}$$

as the only possibility *based on physical grounds*, even though other possibilities exist. We have used

$$\mathscr{L}_2\{2e^{2t}u(-t)\} = \int_{-\infty}^{\infty} 2e^{2t}u(-t)e^{-st}\,dt$$

$$= 2\int_{-\infty}^{0} e^{-(s-2)t}\,dt$$

$$= -\frac{2}{s-2} \qquad \sigma > 2$$

and

$$\mathscr{L}_2\{4e^{-3t}u(t)\} = \int_{-\infty}^{\infty} 4e^{-3t}u(t)e^{-st}\,dt$$

$$= 4\int_{0}^{\infty} e^{-(s+3)t}\,dt$$

$$= \frac{4}{s+3} \qquad \sigma < -3$$

Although the bilateral Laplace transform is useful for such things as Wiener filtering, it is not our primary concern here.[†] In the vast majority of practical applications, the time functions begin at some finite time that can always (for time-invariant systems) be chosen to be $t = 0$ for convenience. Results can be shifted to the correct starting time later. The particular Laplace transform that results in this case is called the *unilateral* (one-sided) *Laplace transform*, and from Eqs. 6.3 and 6.4 it and the inverse transform are given by

$$\boxed{F(s) = \int_0^{\infty} f(t)e^{-st}\,dt} \quad \text{(unilateral Laplace transform)} \qquad (6.5)$$

and

$$\boxed{f(t) = \frac{1}{2\pi j}\int_{\sigma-j\infty}^{\sigma+j\infty} F(s)e^{ts}\,ds} \quad \begin{array}{l}\text{(inverse unilateral}\\ \text{Laplace transform)}\end{array} \qquad (6.6)$$

respectively. Notice that we are not using a subscript [1, on $F(s)$] here. Thus, the pair

$$f(t) \leftrightarrow F(s) \qquad (6.7)$$

implies

$$F(s) = \mathscr{L}\{f(t)\} \qquad (6.8)$$

and

$$f(t) = \mathscr{L}^{-1}\{F(s)\} \qquad (6.9)$$

[†] Some of the important properties of the bilateral Laplace transform are listed in App. F for the interested reader.

6.1 FOURIER TO LAPLACE TRANSFORM

We now have a situation where poles in the left half of the s plane give exponentially decreasing functions of time, and poles in the right half of the s plane give exponentially increasing time functions. Furthermore, the inverse Laplace transform will be unique without specifying the region of convergence. This is helpful.

Notice that in using Eq. 6.5 $f(t)$ will always be $f(t)u(t)$, that is, starting at $t = 0$, whether we like it or not. This means that the unilateral Laplace transform creates a transient or switching situation in most cases. This is an advantage in some situations and a disadvantage in others. For example, in closing a switch (at $t = 0$) in a series RL network with a battery, we have a situation ideally suited for the Laplace transform because the battery and closing switch can be replaced by $V_0 u(t)$ for $t > 0$. On the other hand, if we want the steady-state response of a filter when excited by a periodic waveform, the transient response (caused by the excitation necessarily starting at $t = 0$ with the Laplace transform) is extremely difficult to subtract from the total response.[†] Using the Laplace transform on this latter problem is like "putting a round peg in a square hole." The Fourier transform or bilateral Laplace transform can better treat this problem. Section 5.4 was devoted to the solution of this kind of problem.

A further comment needs to be made about the lower limit of Eq. 6.5, and it pertains to the unit-impulse function, $\delta(t)$. We would very much like for it to be *included* in the present situation so that the Laplace transform of $\delta(t)$ is not 0. Since we have already defined the unit impulse to be an even function, $\delta(t) = \delta(-t)$, *we have no choice* but to make the lower limit in Eq. 6.5 be 0^-. That is

$$F(s) = \int_{0^-}^{\infty} f(t) e^{-st} \, dt \tag{6.10}$$

and the unit impulse is *within* the limits of integration, and we have the Laplace pair $\delta(t) \leftrightarrow 1$ because of the sampling property of the unit-impulse function.

The inverse unilateral (or bilateral) Laplace transform is given by Eq. 6.6. It requires that an integration be performed along a line to the right (a distance σ) and parallel to the imaginary axis in the complex plane! This sounds rather ominous since most students at this level have no idea how to integrate in the complex plane. There is no cause for concern, however, because we will avoid this integral entirely. We will, instead, find the required inverse transforms in a table of pairs, much as we did for Fourier transforms in the last chapter.

We have already seen that $\text{Re}\{s\} = \sigma$ must be large enough to insure that $F(s)$ in Eq. 6.5 does exist. In order to allow the interchange of certain limit processes to follow, we will require *absolute* convergence:

$$\lim_{T \to \infty} \int_0^T |f(t)| e^{-t} \, dt < \infty$$

[†] Recall from Chap. 2 that the total response is the sum of the transient response and the steady-state response.

250 THE LAPLACE TRANSFORM

Notice that there remain functions such as e^{t^2} or t^t which are not Laplace transformable regardless of low large we make σ. Unless stated to the contrary, we will henceforth be concerned only with the *unilateral* Laplace transform.

6.2 LAPLACE TRANSFORM PAIRS

In this section we will follow the same procedure as that used for Fourier transforms. We will find Laplace transform pairs and use these to generate other pairs. In this way a useful table can be built up rather quickly. Keep in mind that all functions that we consider are either explicitly or *implicitly* multiplied by $u(t)$.

Linearity

The transform of a linear sum of functions is the sum of the transforms of the individual functions. If $f_n(t) \leftrightarrow F_n(s)$, then

$$\sum_{n=1}^{N} a_n f_n(t) \leftrightarrow \sum_{n=1}^{N} a_n F_n(s) \tag{6.11}$$

Time Shifting

The delayed function $f(t - t_0)u(t - t_0)$, which is $f(t)u(t)$ delayed by t_0, has the transform indicated by[†]

$$f(t - t_0)u(t - t_0) \leftrightarrow F(s)e^{-st_0} \tag{6.12}$$

Notice that in this case the insertion of the step function is *necessary* because $f(t - t_0)u(t - t_0)$ is generally not equal to $f(t - t_0)$!

Shift in s

$$e^{-\alpha t}f(t) \leftrightarrow F(s + \alpha) \tag{6.13}$$

This pair follows almost immediately after substitution into Eq. 6.10.

Scaling

$$f(\alpha t) \leftrightarrow (1/\alpha)F(s/\alpha) \tag{6.14}$$

This also follows from Eq. 6.10.

Differentiation

If the transform of the derivative of $f(t)$ exists, it is given by

$$\frac{df}{dt} \leftrightarrow sF(s) - f(0^-) \tag{6.15}$$

[†] This pair is easily established with Eq. 6.5 and an obvious change of variable.

Starting with Eq. 6.10 and integrating by parts with $u = f(t)$ and $dv = e^{-st}\,dt$, we have

$$F(s) = -f(t)\frac{e^{-st}}{s}\bigg|_{0^-}^{\infty} + \frac{1}{s}\int_{0^-}^{\infty}\frac{df}{dt}e^{-st}\,dt$$

$$F(s) = \frac{f(0^-)}{s} + \frac{1}{s}\mathcal{L}\left\{\frac{df}{dt}\right\}$$

or

$$\mathcal{L}\left\{\frac{df}{dt}\right\} = sF(s) - f(0^-)$$

which is pair 6.15. Notice that $f(0^-)$ is the limit of $f(t)$ as t approaches 0 *from the left*. In the same way

$$\begin{aligned}\mathcal{L}\left\{\frac{d^2f}{dt^2}\right\} &= s\mathcal{L}\left\{\frac{df}{dt}\right\} - f'(0^-) \\ &= s[sF(s) - f(0^-)] - f'(0^-) \\ &= s^2 F(s) - sf(0^-) - f'(0^-)\end{aligned} \tag{6.16}$$

where $f'(0^-)$ is the limit of the derivative of $f(t)$ as t approaches 0 from the left. It is easy to extend these results to the nth derivative:

$$\frac{d^n f}{dt^n} \leftrightarrow s^n F(s) - \sum_{k=1}^{n} s^{n-k} f^{(k-1)}(0^-) \tag{6.17}$$

Integration

$$\int_0^t f(x)\,dx \leftrightarrow \frac{F(s)}{s} \tag{6.18}$$

This pair can be verified by once again starting with Eq. 6.10

$$\mathcal{L}\left\{\int_0^t f(x)\,dx\right\} = \int_{0^-}^{\infty}\left[\int_0^t f(x)\,dx\right] e^{-st}\,dt$$

Integrating by parts again with

$$u = \int_0^t f(x)\,dx \qquad du = f(t)\,dt$$

and

$$dv = e^{-st}\,dt \qquad v = -\frac{e^{-st}}{s}$$

gives

$$\mathcal{L}\left\{\int_0^t f(x)\,dx\right\} = -\frac{e^{-st}}{s}\int_0^t f(x)\,dx\bigg|_{0^-}^{\infty} + \frac{1}{s}\int_{0^-}^{\infty} f(t) e^{-st}\,dt = \frac{F(s)}{s}$$

which is pair 6.18. Also

$$\int_0^t f(x)\,dx = \int f(x)\,dx \Big|_{x=t} - \int f(x)\,dx \Big|_{x=0}$$
$$= f^{-1}(t) - f^{-1}(0)$$

where $f^{-1}(t)$ is the *indefinite integral* or *antiderivative* of $f(t)$. Therefore

$$f^{-1}(t) = \int f(x)\,dx \Big|_{x=t} = \int f(t)\,dt = \int_0^t f(x)\,dx + f^{-1}(0)$$

and

$$\mathscr{L}\left\{\int f(t)\,dt\right\} = \frac{F(s)}{s} + \mathscr{L}\{f^{-1}(0)\}$$

$$= \frac{F(s)}{s} + f^{-1}(0)\int_{0^-}^\infty e^{-st}\,dt$$

$$= \frac{F(s)}{s} + \frac{f^{-1}(0)}{s}$$

Therefore

$$\int f(t)\,dt \leftrightarrow \frac{F(s)}{s} + \frac{f^{-1}(0)}{s} \tag{6.19}$$

Pair 6.18 and 6.19 can be extended in the same way that pair 6.15 was extended.

Unit-Step Function

The last step in arriving at pair 6.19 (or simply using Eq. 6.10) shows that (for $\sigma > 0$)

$$u(t) \leftrightarrow \frac{1}{s} \quad \text{(pole at the origin in the } s \text{ plane)} \tag{6.20}$$

Let us pause here and reflect on our choice of 0^- as the lower limit in the defining equation for the Laplace transform, Eq. 6.10. We know from Eq. 2.31 that

$$\frac{d}{dt}u(t) = \delta(t)$$

and this is consistent with pair 6.15, for that result gives

$$\frac{d}{dt}u(t) \leftrightarrow s \cdot \frac{1}{s} - u(0^-) = 1 - 0 = 1$$

and

$$\frac{d}{dt}u(t) = \delta(t) \leftrightarrow 1$$

as previously verified. Now suppose (as some writers have done) that we insist that $\mathcal{L}\{\delta(t)\} = 1$ and the lower limit on Eq. 6.10 be 0^+. In this case pair 6.15 becomes

$$\frac{df}{dt} \leftrightarrow sF(s) - f(0^+)$$

so

$$\frac{d}{dt}u(t) \leftrightarrow s \cdot \frac{1}{s} - u(0^+) = 1 - 1 = 0$$

and

$$\frac{d}{dt}u(t) = \delta(t) \leftrightarrow 1$$

This is nonsense! Finally, if we insist that the lower limit on Eq. 6.10 be 0^+, and we are simultaneously smart enough to recognize that the unit impulse is then *outside* the limits on Eq. 6.10, we have

$$\frac{d}{dt}u(t) \leftrightarrow s \cdot \frac{1}{s} - u(0^+) = 1 - 1 = 0$$

and

$$\frac{d}{dt}u(t) = \delta(t) \leftrightarrow 0$$

This result is consistent and correct for the stated conditions, even though the use of the unit impulse is no longer included. Thus, the lower limit on Eq. 6.10 is whatever we *define* it to be. Since we want to use the unit impulse, we define the limit to be 0^-.

Multiplication[†] by t

If $f(t) \leftrightarrow F(s)$, then

$$tf(t) \leftrightarrow -\frac{d}{ds}F(s) \qquad (6.21)$$

Division by t

If $f(t) \leftrightarrow F(s)$, then

$$\frac{f(t)}{t} \leftrightarrow \int_s^\infty F(\eta)\, d\eta \qquad (6.22)$$

[†] The proof of pairs 6.21 and 6.22 is asked for in the first two problems at the end of the chapter.

Exponential[†]

$$e^{-\alpha t} \leftrightarrow \frac{1}{s+\alpha} \quad \text{(pole at } s = -\alpha \text{ in the } s \text{ plane)} \tag{6.23}$$

This result holds even for $\alpha < 0$ because we can always find a $\sigma > -\alpha$ large enough to insure that the integral in Eq. 6.10 converges (exists). If $\alpha = j\omega_0$

$$e^{-j\omega_0 t} \leftrightarrow \frac{1}{s+j\omega_0} \quad \text{(pole at } s = -j\omega_0 \text{ in the } s \text{ plane)}$$

while if $\alpha = -j\omega_0$

$$e^{j\omega_0 t} \leftrightarrow \frac{1}{s-j\omega_0} \quad \text{(pole at } s = +j\omega_0 \text{ in the } s \text{ plane)}$$

Trigonometric Functions

The sum of the last two pairs divided by 2 gives

$$\cos(\omega_0 t) \leftrightarrow \frac{s}{s^2+\omega_0^2} \quad \text{(poles at } s = \pm j\omega_0 \text{ in the } s \text{ plane)} \tag{6.24}$$

It is just as easy to show that

$$\sin(\omega_0 t) \leftrightarrow \frac{\omega_0}{s^2+\omega_0^2} \quad \text{(poles at } s = \pm j\omega_0 \text{ in the } s \text{ plane)} \tag{6.25}$$

Hyperbolic Functions

Using pair 6.23 we can easily show that for $\sigma > \alpha$

$$\cosh(\alpha t) \leftrightarrow \frac{s}{s^2-\alpha^2} \quad \text{(poles at } s = \pm \alpha \text{ in the } s \text{ plane)} \tag{6.26}$$

and

$$\sinh(\alpha t) \leftrightarrow \frac{\alpha}{s^2-\alpha^2} \quad \text{(poles at } s = \pm \alpha \text{ in the } s \text{ plane)} \tag{6.27}$$

Power of t

$$t^n \leftrightarrow \frac{n!}{s^{n+1}} \quad \text{(multiple pole at the origin in the } s \text{ plane)} \tag{6.28}$$

where $n = 0, 1, 2, \ldots$, and factorial $n = n! = n(n-1)(n-2)\cdots 1$. Pair 6.18 with $f(x) = u(x)$ [or $f(t) = u(t)$] gives

$$\int_0^t 1\, dx = tu(t) \leftrightarrow \frac{1/s}{s} = \frac{1}{s^2}$$

[†] The proof of pair 6.23 follows directly from Eq. 6.10 when $\sigma > -\alpha$. The *strip of convergence* is a semi-infinite plane.

Thus

$$tu(t) \leftrightarrow \frac{1}{s^2}$$

Now, using pair 6.21 with $f(t) = tu(t)$

$$t^2 u(t) \leftrightarrow -\frac{d}{ds}\left(\frac{1}{s^2}\right) = \frac{2}{s^3}$$

Using pair 6.21 again with $f(t) = t^2 u(t)$

$$t^3 u(t) \leftrightarrow -\frac{d}{ds}\left(\frac{2}{s^3}\right) = \frac{(3)(2)}{s^4}$$

Continuing this process gives the *general* result given by pair 6.28. Notice that it requires that $\sigma > 0$ when t^n is used directly in Eq. 6.10.

Convolution

If $x(t) \leftrightarrow X(s)$, and $h(t) \leftrightarrow H(s)$, then

$$y_{zs}(t) = \int_{0^-}^{t} x(\tau) h(t - \tau)\, d\tau \leftrightarrow X(s)H(s) \tag{6.29}$$

or

$$y_{zs}(t) = \int_{0^-}^{t} x(t - \tau) h(\tau)\, d\tau \leftrightarrow X(s)H(s) \tag{6.30}$$

Notice, once again returning to system concepts, that the block diagram of Fig. 6.3 applies. $H(s)$ is called the transfer function as before, and $h(t)$ is its inverse Laplace transform. In other words $h(t)$ is the unit-impulse response of the system.

Pair 6.31 is obtained from pair 6.30 with a simple change of variable. We now proceed to verify pair 6.30, although, based on results in Chap. 5, it is certainly believable. We have

$$\mathscr{L}\left\{\int_{0^-}^{t} x(\tau) h(t - \tau)\, d\tau\right\} = \int_{0^-}^{\infty} \left\{\int_{0^-}^{t} x(\tau) h(t - \tau)\, d\tau\right\} e^{-st}\, dt$$

Since $h(t)$ is actually $h(t)u(t)$ for the Laplace transform, $h(t - \tau)$ is actually $h(t - \tau)u(t - \tau)$. This is, of course, the reason why the upper limit of the inner

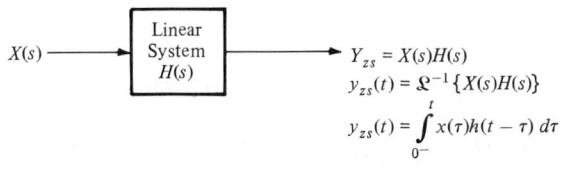

$X(s)$ → [Linear System $H(s)$] → $Y_{zs} = X(s)H(s)$
$y_{zs}(t) = \mathscr{L}^{-1}\{X(s)H(s)\}$
$y_{zs}(t) = \int_{0^-}^{t} x(\tau) h(t - \tau)\, d\tau$

Figure 6.3. Linear system in terms of s.

integral is t instead of infinity. We can now insert $h(t - \tau)u(t - \tau)$ instead of $h(t - \tau)$ and also change the upper limit to infinity without changing the last equation at all:

$$\mathscr{L}\left\{\int_{0^-}^{t} x(\tau)h(t - \tau)\, d\tau\right\} = \int_{0^-}^{\infty}\left\{\int_{0^-}^{\infty} x(\tau)h(t - \tau)u(t - \tau)\, d\tau\right\}e^{-st}\, dt$$

Since both $X(s)$ and $H(s)$ exist by the starting assumption, the order of integration with respect to τ and t can be interchanged giving

$$\mathscr{L}\left\{\int_{0^-}^{t} x(\tau)h(t - \tau)\, d\tau\right\} = \int_{0^-}^{\infty} x(\tau)\left\{\int_{0^-}^{\infty} h(t - \tau)u(t - \tau)e^{-st}\, dt\right\} d\tau$$

$$= \int_{0^-}^{\infty} x(\tau)\left\{\int_{\tau}^{\infty} h(t - \tau)e^{-st}\, dt\right\} d\tau$$

The last step should be carefully verified. Next, we make the simple change of variable in the inner integral:

$$t - \tau = \eta \qquad dt = d\eta$$

giving

$$\mathscr{L}\left\{\int_{0}^{t} x(\tau)h(t - \tau)\, d\tau\right\} = \int_{0^-}^{\infty} x(\tau)\left\{\int_{0}^{\infty} h(\eta)e^{-s(\eta + \tau)}\, d\eta\right\} d\tau$$

$$= \int_{0^-}^{\infty} x(\tau)e^{-s\tau}\, d\tau \int_{0}^{\infty} h(\eta)e^{-s\eta}\, d\eta$$

Now, as we have already stated, we want to include the unit impulse, so the lower limit on the second integral should be 0^-. This results in

$$\mathscr{L}\left\{\int_{0^-}^{t} x(\tau)h(t - \tau)\, d\tau\right\} = \int_{0^-}^{\infty} x(\tau)e^{-s\tau}\, d\tau \int_{0^-}^{\infty} h(\eta)e^{-s\eta}\, d\eta$$

$$= X(s)H(s)$$

This verifies pair 6.30.

Many times we can check a transform *before going through the inversion process*. This is useful because it can avoid wasted labor if a mistake has been made in calculating the transform. Many times we know what the initial or final value (or both) of the inverse transformed (time) function should be. This check is easily made using the *initial value* and *final value* theorems that follow.

Initial Value Theorem

If both $f(t)$ and $f'(t)$ are Laplace transformable, then the initial value of $f(t)$ is

$$\boxed{f(0^+) = \lim_{t \to 0^+} f(t) = \lim_{s \to \infty} sF(s)} \quad \text{(initial value)} \tag{6.31}$$

In order to show that Eq. 6.32 is correct, we start with pair 6.15

$$\int_{0^-}^{\infty} \frac{df}{dt} e^{-st} dt = sF(s) - f(0^-) \tag{6.32}$$

We consider two cases separately. First, if $f(t)$ is continuous at $t = 0$, then $f(0^-) = f(0^+)$. Then, if we allow s to approach infinity, we have

$$\lim_{s \to \infty} \int_{0^-}^{\infty} \frac{df}{dt} e^{-st} dt = 0 = \lim_{s \to \infty} sF(s) - f(0^-)$$

$$= \lim_{s \to \infty} sF(s) - f(0^+) \tag{6.33}$$

or

$$f(0^+) = \lim_{s \to \infty} sF(s)$$

which is the desired result. Next, if $f(t)$ is discontinuous at $t = 0$, then $f(0^-) \neq f(0^+)$. Insofar as the Laplace transform is concerned, the discontinuity can be treated by replacing $f(t)$ by $f(t)u(t)$, as in Fig. 6.4, and

$$\frac{d}{dt}[f(t)u(t)] = f(0^+)\delta(t) + \frac{df}{dt} u(t)$$

Substituting this result in Eq. 6.32, we obtain

$$\int_{0^-}^{\infty} f(0^+)\delta(t) e^{-st} dt + \int_{0^-}^{\infty} \frac{df}{dt} u(t) e^{-st} dt = sF(s) - f(0^-)$$

or, since $f(0^-) = 0$,

$$f(0^+) + \int_{0^-}^{\infty} \frac{df}{dt} e^{-st} dt = sF(s)$$

Now letting s approach infinity, we obtain

$$f(0^+) = \lim_{s \to \infty} sF(s)$$

once again! Thus, the initial value is $f(0^+)$ regardless of whether the lower limit of the integral defining the Laplace transform is 0^- or 0^+. The unit-step function is a case in point:

$$u(0^+) = \lim_{s \to \infty} s \cdot \frac{1}{s} = 1$$

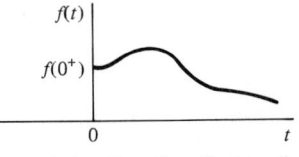

Figure 6.4. Function that is discontinuous at $t = 0$.

Final-Value Theorem

If both $f(t)$ and $f'(t)$ are Laplace transformable, then the final value of $f(t)$ is

$$\boxed{\lim_{t \to \infty} f(t) = \lim_{s \to 0} sF(s)} \quad (\textit{final value}) \tag{6.34}$$

Once again

$$\int_{0^-}^{\infty} \frac{df}{dt} e^{-st}\, dt = sF(s) - f(0^-)$$

Now let s approach 0:

$$\int_{0^-}^{\infty} \frac{df}{dt}\, dt = \lim_{s \to 0} sF(s) - f(0^-)$$

The left side of the last equation can be written

$$\lim_{t \to \infty} \int_{0^-}^{t} \frac{df}{dt}\, dt = \lim_{t \to \infty} [f(t) - f(0^-)] = \lim_{s \to 0} sF(s) - f(0^-)$$

Consequently

$$\lim_{t \to \infty} f(t) = \lim_{s \to 0} sF(s)$$

We mention in passing that in order for the final value (theorem) to be valid, $sF(s)$ must be *analytic* everywhere *in* the right half of the complex s plane $(s = \sigma + j\omega)$ and *on* the imaginary axis. Put another way, $sF(s)$ cannot have any poles (infinities) in the right half of the s plane or on the imaginary axis. Also, it should be obvious to the reader that, when applicable, the final value (theorem) is an excellent test for determining whether or not a system is stable. We know from Chap. 2 that if $h(t)$ goes to 0 as t goes to ∞, the system is stable. Thus, for a stable system, the unit-impulse response is such that

$$\boxed{\lim_{t \to \infty} h(t) = \lim_{s \to 0} sH(s) = 0} \quad (\textit{stable system}) \tag{6.35}$$

Now consider

$$F(s) = \frac{\omega_0}{s^2 + \omega_0^2} = \frac{\omega_0}{(s + j\omega_0)(s - j\omega_0)}$$

If we attempt to use the final-value theorem, we find that

$$\lim_{s \to 0} sF(s) = 0$$

but this is meaningless because the final-value theorem is not applicable in the first place since

$$sF(s) = \frac{s\omega_0}{(s+j\omega_0)(s-j\omega_0)}$$

has poles on the imaginary axis at $s = -j\omega_0$ and $s = +j\omega_0$. Of course, for this particular case

$$f(t) = \mathscr{L}^{-1}\left\{\frac{\omega_0}{s^2+\omega_0^2}\right\} = \sin(\omega_0 t)$$

and $f(t)$ has no limit as t approaches ∞.

Periodic Function

If a periodic function is described by $f_p(t)$, then

$$f_p(t) \leftrightarrow \frac{F_1(s)}{1-e^{-Ts}} \qquad (6.36)$$

where T is the period and $F_1(s)$ is the Laplace transform of the first cycle. This is not difficult to prove, but we will not do that here. Instead, we recall from Eq. 5.54 that the steady-state or zero-state response due to periodic input functions is given by

$$y_{zs}(t) = y_{ss}(t) = \int_{-T}^{0} x(\tau) \sum_{n=0}^{\infty} h(t-\tau+nT)\,d\tau + \int_{0}^{t} x(\tau)h(t-\tau)\,d\tau \qquad (5.54)$$

for $0 < t < T$. The second integral contains *both* the steady-state and transient parts $[y_{zs}(t) + y_{tr}(t)]$. Therefore

$$y_{ss}(t) = \int_{-T}^{0} x(\tau) \sum_{n=0}^{\infty} h(t-\tau+nT)\,d\tau + y_{ss}(t) + y_{tr}(t)$$

This gives the interesting result that the transient part of $y(t)$ is

$$y_{tr}(t) = -\int_{-T}^{0} x(\tau) \sum_{n=0}^{\infty} h(t-\tau+nT)\,d\tau \quad \text{(periodic function)} \qquad (6.37)$$

for $0 < t < T$.

In those cases where we are only interested in the steady state and the excitation is periodic, we will use Eq. 5.54. It is certainly possible to use Eq. 6.36 and then subtract the transient part from the complete solution to obtain the steady-state solution. This is a rather messy proposition,[1] and, as has already been pointed out, is a result of the fact that the unilateral Laplace transform has created a transient that we did not want in the first place. The Laplace transform is simply the wrong tool for analysis in this case. Remember that $f_p(t) \neq f_p(t)u(t)$, but $\mathscr{L}\{f_p(t)\} = \mathscr{L}\{f_p(t)u(t)\}$!

6.3 EXAMPLES

We will now look at some simple examples to familiarize ourselves with the pairs already developed and to develop other transform pairs. We will consider applications in the next chapter.

Example 1

Find the Laplace transform of the damped cosine:

$$f(t) = Ae^{-\alpha t}\cos(\omega_0 t + \theta)$$

Our first task is to rearrange $f(t)$ into a form such that we can utilize the pairs already developed. To that end we expand the cosine function:

$$\cos(\omega_0 t + \theta) = \cos(\theta)\cos(\omega_0 t) - \sin(\theta)\sin(\omega_0 t)$$

so that

$$f(t) = [A\cos(\theta)]e^{-\alpha t}\cos(\omega_0 t) - [A\sin(\theta)]e^{-\alpha t}\sin(\omega_0 t)$$

Each term has the general form

$$Be^{-\alpha t}f_1(t) \leftrightarrow BF_1(s + \alpha)$$

from pair 6.13. Also, using pairs 6.24 and 6.25

$$F(s) = A\cos(\theta)\frac{s + \alpha}{(s + \alpha)^2 + \omega_0^2} - A\sin(\theta)\frac{\omega_0}{(s + \alpha)^2 + \omega_0^2}$$

$$F(s) = A\frac{(s + \alpha)\cos(\theta) - \omega_0\sin(\theta)}{(s + \alpha)^2 + \omega_0^2}$$

Example 2

Find the Laplace transform of the pulse shown in Fig. 6.5.

$$f(t) = A[u(t) - u(t - T)] = Au(t) - Au(t - T)$$

so

$$F(s) = A/s - A\mathscr{L}\{u(t - T)\}$$
$$F(s) = A/s - A/se^{-sT}$$
$$F(s) = A/s[1 - e^{-sT}]$$

We have used the time-shifting pair 6.12.

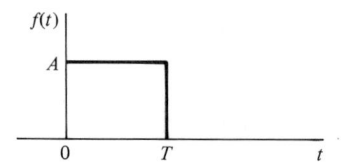

Figure 6.5. Pulse $A[u(t) - u(t - T)]$.

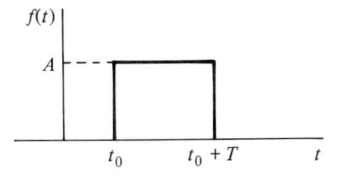

Figure 6.6. Delayed pulse.

Example 3

Find the Laplace transform of the delayed pulse shown in Fig. 6.6. Using pair 6.12 once more with the results of Example 2 we have

$$F(s) = A/s[1 - e^{-sT}]e^{-st_0}$$

Example 4

Find the Laplace transform of the triangular pulse shown in Fig. 6.7. The function is

$$f(t) = \frac{A}{T}t[u(t) - u(t-T)] + \left[-\frac{A}{T}t + 2A\right][u(t-T) - u(t-2T)]$$

The first term represents the straight line starting at 0 and stopping at $t = T$. The second term represents another straight line starting at $t = T$ and stopping at $t = 2T$. We must arrange the last equation into forms we can treat, so we first expand it:

$$f(t) = \frac{A}{T}tu(t) - \frac{A}{T}tu(t-T) - \frac{A}{T}tu(t-T) + \frac{A}{T}tu(t-2T)$$
$$+ 2Au(t-T) - 2Au(t-2T)$$

Now, the first term is in a form we can immediately treat. The second (and third) term can be put into a form we can treat if we replace t by $t - T$ and then compensate by adding a term to offset that change. The fourth term can be similarly treated. We have

$$f(t) = \frac{A}{T}tu(t) - \frac{2A}{T}(t-T) - \frac{2A}{T}Tu(t-T) + \frac{A}{T}(t-2T)u(t-2T)$$
$$+ \frac{2A}{T}Tu(t-2T) + 2Au(t-T) - 2Au(t-2T)$$

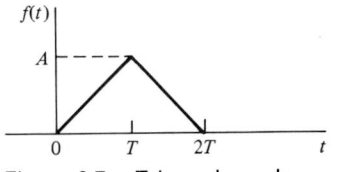

Figure 6.7. Triangular pulse.

Each term is now in a form which can easily be transformed (after canceling the third and sixth terms and fifth and seventh terms):

$$F(s) = \frac{A}{T}\frac{1}{s^2} - \frac{2A}{T}\frac{1}{s^2}e^{-Ts} + \frac{A}{T}\frac{1}{s^2}e^{-2Ts}$$

$$F(s) = \frac{A}{Ts^2}[1 - 2e^{-Ts} + e^{-2Ts}]$$

$$F(s) = \frac{A}{T}\left[\frac{1 - e^{-Ts}}{s}\right]^2$$

Example 5

Find the Laplace transform of $t^2e^{-2t}u(t)$. Using pairs 6.13 and 6.28

$$t^2 e^{-2t} \leftrightarrow \left.\frac{2}{s^3}\right|_{s+2} = \frac{2}{(s+2)^3}$$

Example 6

Find the inverse Laplace transform of $[s(s+a)]^{-1}$. We have by pair 6.18

$$\int_0^t e^{-ax}\,dx \leftrightarrow \frac{1}{s}\frac{1}{s+a} = \frac{1}{s(s+a)}$$

Therefore

$$\left.\frac{e^{-ax}}{-a}\right|_0^t \leftrightarrow \frac{1}{s(s+a)}$$

or

$$\frac{1 - e^{-at}}{a}u(t) \leftrightarrow \frac{1}{s(s+a)}$$

Example 7

Find the inverse Laplace transform of

$$F(s) = \frac{2s + 7}{s + 2}$$

Here, we notice that the degree (in s) of the numerator is the *same* as that of the denominator, forming what is called an *improper fraction*.[†] This is indicative of the presence of an impulse in $f(t)$. The simplest procedure is that of rearranging

---

[†] An improper fraction occurs when the degree of the numerator polynomial is greater than or equal to that of the denominator polynomial. Otherwise, it is a *proper fraction*.

$F(s)$ into a constant [whose inverse transform is that constant times $\delta(t)$] plus a proper fraction. That is

$$F(s) = 2\frac{s + 7/2}{s + 2} = 2\frac{s + 2 + 3/2}{s + 2} = 2 + \frac{3}{s + 2}$$

Therefore

$$f(t) = 2\delta(t) + 3e^{-2t}u(t)$$

The $F(s)$ which we must invert in linear-systems analysis is quite often in the form of the ratio of polynomials in s simply because linear differential equations for time-invariant systems lead naturally to that form. This will become more apparent in the next chapter when the transform is directly applied to the differential equations. When the ratio of the polynomials is in the form of a proper fraction, a formal procedure, called the *Heaviside expansion theorem* or the method of *undetermined coefficients*, can be applied. If the fraction is improper it can be converted to proper form as in Example 7.

Rather than becoming involved in a lengthy formal discussion of the Heaviside expansion, we will simply explain it by means of examples because in essence it is very simple.

Example 8

Find

$$\mathscr{L}^{-1}\left\{\frac{s + 3}{2s^2 + 6s + 4}\right\}$$

We first must factor the denominator:

$$F(s) = \frac{1}{2}\frac{s + 3}{(s + 2)(s + 1)}$$

and then we assume that $F(s)$ can be expressed as

$$F(s) = \frac{1}{2}\left[\frac{A_1}{s + 2} + \frac{A_2}{s + 1}\right] = \frac{1}{2}\frac{s + 3}{(s + 2)(s + 1)}$$

Our task is that of determining the coefficients A_1 and A_2. In order to obtain A_1, we multiply both sides of the preceding equation by $s + 2$:

$$A_1 + A_2\frac{s + 2}{s + 1} = \frac{(s + 3)(s + 2)}{(s + 2)(s + 1)} = \frac{s + 3}{s + 1}$$

The term containing the coefficient A_2 disappears if we set $s = -2$, so

$$A_1 = \frac{s + 3}{s + 1}\bigg|_{s=-2} = \frac{-2 + 3}{-2 + 1} = -1$$

In the same way we multiply by $s + 1$ to obtain A_2

$$A_1 \frac{s+1}{s+2} + A_2 = \frac{(s+3)(s+1)}{(s+2)(s+1)} = \frac{s+3}{s+2}$$

so

$$A_2 = \left.\frac{s+3}{s+2}\right|_{s=-1} = \frac{-1+3}{-1+2} = 2$$

Therefore

$$F(s) = \frac{1}{2}\left[\frac{-1}{s+2} + \frac{2}{s+1}\right]$$

and

$$f(t) = \tfrac{1}{2}[-e^{-2t} + 2e^{-t}]u(t)$$

The scheme is simple and works so long as the roots are distinct. When the function of s that is to be inverted is the ratio of polynomials in s, and the degree (M) of the numerator polynomial is less than that (N) of the denominator (a proper fraction)

$$F(s) = \frac{\sum_{n=0}^{M} c_n s^n}{\sum_{n=0}^{N} d_n s^n} = \frac{N(s)}{D(s)} \qquad M < N \qquad (6.38)$$

then

$$F(s) = \sum_{p=1}^{N} \frac{1}{s+s_p}\left[\frac{(s+s_p)N(s)}{D(s)}\right]_{s=-s_p} \qquad (6.39)$$

or[2]

$$F(s) = \sum_{p=1}^{N} \frac{N(-s_p)}{(s+s_p)D'(-s_p)} \qquad (6.40)$$

for distinct (not repeated) roots, $-s_p$. The $F(s)$ given by Eq. 6.39 or 6.40 is particularly easy to invert:

$$f(t) = \sum_{p=1}^{N} \left[\frac{(s+s_p)N(s)}{D(s)}\right]_{s=-s_p} e^{-s_p t} \qquad (6.41)$$

or

$$f(t) = \sum_{p=1}^{N} \frac{N(-s_p)}{D'(-s_p)} e^{-s_p t} \qquad (6.42)$$

The case of repeated roots will be considered shortly.

Example 9

Find

$$f(t) = \mathcal{L}^{-1}\left\{\frac{se^{-s}}{s^2 + 2s + 10}\right\}$$

The exponential merely indicates a time shift, so

$$f(t) = \mathcal{L}^{-1}\left\{\frac{s}{s^2 + 2s + 10}\right\}\bigg|_{t \to t-1}$$

and we can concentrate on

$$f_1(t) = \mathcal{L}^{-1}\left\{\frac{s}{s^2 + 2s + 10}\right\} = \mathcal{L}^{-1}\{F_1(s)\}$$

After factoring

$$F_1(s) = \frac{s}{(s + 1 + j3)(s + 1 - j3)} = \frac{A_1}{s + 1 + j3} + \frac{A_2}{s + 1 - j3}$$

so

$$\frac{s}{s + 1 - j3} = A_1 + A_2 \frac{s + 1 + j3}{s + 1 - j3}$$

With $s = -1 - j3$

$$\frac{-1 - j3}{-1 - j3 + 1 - j3} = A_1$$

or

$$A_1 = \tfrac{1}{2} - j\tfrac{1}{6}$$

In the same way

$$\frac{s}{s + 1 + j3} = A_1 \frac{s + 1 - j3}{s + 1 + j3} + A_2$$

and with $s = -1 + j3$

$$\frac{-1 + j3}{-1 + j3 + 1 + j3} = A_2$$

or

$$A_2 = \tfrac{1}{2} + j\tfrac{1}{6} = A_1^*$$

As a matter of fact, *the coefficients of conjugate roots will always be conjugates* (for real functions of time), so the last step was actually unnecessary! We now have

$$F_1(s) = \frac{1/2 - j(1/6)}{s + 1 + j3} + \frac{1/2 + j(1/6)}{s + 1 - j3}$$

so

$$f_1(t) = (\tfrac{1}{2} - j\tfrac{1}{6})e^{-(1+j3)t} + (\tfrac{1}{2} + j\tfrac{1}{6})e^{-(1-j3)t}$$
$$f_1(t) = e^{-t}[\tfrac{1}{2}(e^{-j3t} + e^{+j3t}) - j\tfrac{1}{6}(e^{-j3t} - e^{+j3t})]$$
$$f_1(t) = e^{-t}[\cos(3t) - \tfrac{1}{3}\sin(3t)]u(t)$$

Finally

$$f(t) = f_1(t)|_{t \to t-1}$$

so

$$f(t) = e^{-(t-1)}[\cos(3t - 3) - \tfrac{1}{3}\sin(3t - 3)]u(t - 1)$$

Do not forget to apply the time shift to $u(t)$! The presence of e^{-t} in $f_1(t)$ suggests a shortcut for the type problem we have here. Consider

$$F_1(s) = \frac{s}{s^2 + 2s + 10} = \frac{s + 1 - 1}{(s + 1)^2 + 9}$$

$$F_1(s) = \frac{(s + 1)}{(s + 1)^2 + 3^2} - \frac{1}{(s + 1)^2 + 3^2}$$

Therefore

$$f_1(t) = e^{-t}\cos(3t)u(t) - e^{-t}\sin(3t)u(t)$$

as before.

Example 10

Find

$$f(t) = \mathscr{L}^{-1}\left\{\frac{s + 3}{s(s + 2)^2(s + 1)}\right\}$$

Here, we have a root of multiplicity two at $s = -2$. That is, it is a double root, so we must express $F(s)$ as[†]

$$F(s) = \frac{A_1}{s} + \frac{A_2}{(s + 2)^2} + \frac{A_3}{s + 2} + \frac{A_4}{s + 1}$$

[†] Based on what happened in Chap. 2 for repeated roots, we should be able to predict the form of $F(s)$.

6.3 EXAMPLES

In general, if we have a root of multiplicity p at $s = -\alpha$, then we must expand that part of $F(s)$ as p terms,

$$\frac{A_p}{(s+\alpha)^p} + \frac{A_{p-1}}{(s+\alpha)^{p-1}} + \cdots + \frac{A_0}{s+\alpha}$$

For the present example

$$\frac{s+3}{s(s+2)^2(s+1)} = \frac{A_1}{s} + \frac{A_2}{(s+2)^2} + \frac{A_3}{s+2} + \frac{A_4}{s+1}$$

We can determine A_1, A_2, and A_4 as before. That is

$$A_1 = \left.\frac{s+3}{(s+2)^2(s+1)}\right|_{s=0} = \frac{3}{4(1)} = \frac{3}{4}$$

$$A_2 = \left.\frac{s+3}{s(s+1)}\right|_{s=-2} = \frac{1}{-2(-1)} = \frac{1}{2}$$

and

$$A_4 = \left.\frac{s+3}{s(s+2)^2}\right|_{s=-1} = \frac{2}{-1(1)} = -2$$

How do we determine A_3? The answer can be found in what we did to find A_2. In order to determine A_2, we first multiplied both sides of $F(s)$ by $(s+2)^2$:

$$\frac{s+3}{s(s+1)} = A_1\frac{(s+2)^2}{s} + A_2 + A_3(s+2) + A_4\frac{(s+2)^2}{s+1}$$

The term containing A_3 can now be isolated if we *first* differentiate with respect to s, and *then* set $s = -2$:

$$\left.\frac{d}{ds}\left[\frac{s+3}{s(s+1)}\right]\right|_{s=-2} = \left.A_1\frac{s(2)(s+2) - (s+2)^2}{s^2}\right|_{s=-2}$$

$$+ A_3 + \left.A_4\frac{2(s+1)(s+2) - (s+2)^2}{(s+1)^2}\right|_{s=-2}$$

or

$$A_3 = \left.\frac{d}{ds}\left[\frac{s+3}{s(s+1)}\right]\right|_{s=-2} = \left.\frac{s(s+1) - (s+3)(2s+1)}{s^2(s+1)^2}\right|_{s=-2}$$

$$A_3 = \tfrac{5}{4}$$

Thus

$$F(s) = \frac{3/4}{s} + \frac{1/2}{(s+2)^2} + \frac{5/4}{s+2} - \frac{2}{s+1}$$

and

$$f(t) = \tfrac{3}{4}u(t) + \tfrac{1}{2}te^{-2t}u(t) + \tfrac{5}{4}e^{-2t}u(t) - 2e^{-t}u(t)$$

We were not able to find A_3 in the same way that we determined the other coefficients because multiplying by $s + 2$ (in order to isolate A_3) leaves A_2 divided by $s + 2$, and this term would go to infinity if we then set $s = -2$. The scheme that was suggested certainly works, but differentiation can be avoided if we clear fractions and then *equate coefficients of like powers of s*, or if we simply *substitute numerical values of s* to give us numerical equations. For the present example

$$\frac{s+3}{s(s+2)^2(s+1)} = \frac{3/4}{s} + \frac{1/2}{(s+2)^2} + \frac{A_3}{(s+2)} + \frac{-2}{s+1}$$

A convenient value of s to choose is $s = 2$ (not $s = 0$, -1, or -2) giving

$$\frac{5}{2(16)(3)} = \frac{3}{8} + \frac{1}{32} + \frac{A_3}{4} - \frac{2}{3}$$

or $A_3 = \frac{5}{4}$ as before.

Example 11

The sine integral of t

$$\text{Si}(t) = \int_0^t \text{Sa}(x)\,dx = \int_0^t \frac{\sin(x)}{x}\,dx$$

occurs frequently in engineering work. What is its Laplace transform? We have

$$F(s) = \mathscr{L}\left\{\int_0^t \text{Sa}(x)\,dx\right\} = \frac{1}{s}\mathscr{L}\{\text{Sa}(t)\}$$

by pair 6.18. Thus

$$F(s) = \frac{1}{s}\mathscr{L}\left\{\frac{\sin(t)}{t}\right\} = \frac{1}{s}\int_s^\infty \frac{ds}{s^2+1}$$

by pair 6.22. Finally

$$F(s) = \frac{1}{s}\tan^{-1}(s)\Big|_s^\infty = \frac{1}{s}\left[\frac{\pi}{2} - \tan^{-1} s\right]$$

or, in a simpler form

$$F(s) = (1/s)\tan^{-1}(1/s)$$

6.4 A TABLE OF LAPLACE TRANSFORM PAIRS

The Laplace transform is used so frequently by engineers that tables of transform pairs are found in almost all mathematics handbooks and textbooks related to linear-systems analysis. These tables can also be used for finding definite integrals in many cases. Under certain conditions entries in a table of

Laplace transform pairs can be used as Fourier transform pairs. This is sometimes very helpful since tables of Fourier transform pairs are somewhat limited. The Campbell and Foster table[†] is probably the most complete. If $F(s)$ has no poles (infinities) on the imaginary axis or in the right half of the s plane, then the existence of an $f(t)$ which is Fourier transformable is assured.[‡] We merely replace s by $j\omega$ to obtain $F(\omega)$. The resulting $f(t)$ will, of course, be 0 for $t < 0$. For example

$$e^{-\alpha t}u(t) \leftrightarrow \frac{1}{s+\alpha} \quad (Laplace)$$

and the single pole of $F(s)$ occurs at $s = -\alpha$. Now, so long as $\alpha > 0$, this pole is on the negative real axis, and

$$e^{-\alpha t}u(t) \leftrightarrow \frac{1}{j\omega+\alpha} \quad \alpha > 0 \quad (Fourier)$$

On the other hand

$$u(t) \leftrightarrow 1/s$$

and the single pole occurs at $s = 0$ which is on the imaginary axis. Thus, we cannot simply replace s by $j\omega$ to obtain $F(\omega)$. We know from Chap. 5, in fact, that

$$u(t) \leftrightarrow \frac{1}{j\omega} + \pi\delta(\omega) \quad (Fourier)$$

It is not so simple to go from a Fourier table of pairs to a Laplace table of pairs. Fortunately, we rarely need to do this.

Table 6.1 is a short table of Laplace transform pairs. These transform pairs, plus many others we have no need of here, can be found in Cheng[1] and most of the references at the end of this chapter.

In this table

$\text{erf}(x) = $ Error function
$\text{erfc}(x) = $ Complementary error function
$J_0(x) = $ Bessel function, first kind, order zero
$J_1(x) = $ Bessel function, first kind, order one
$I_0(x) = $ Modified Bessel function, first kind, order zero
$I_1(x) = $ Modified Bessel function, first kind, order one

[†] Bell System Monograph B584.
[‡] It has already been pointed out in connection with the bilateral Laplace transform that if the strip of convergence includes the imaginary axis, then

$$F(\omega) = F_2(s)|_{s=j\omega}$$

The same thing is true with respect to the unilateral Laplace transform: if the strip of convergence includes the imaginary axis, then

$$F(\omega) = F(s)|_{s=j\omega}$$

Table 6.1 LAPLACE TRANSFORM PROPERTIES AND PAIRS

NO.		$f(t), t \geq 0$	$F(s)$
1	Linearity	$\sum_{n=1}^{N} a_n f_n(t)$	$\sum_{n=1}^{N} a_n F_n(s)$
2	Time Shift	$f(t - t_0)u(t - t_0)$	$F(s)e^{-st_0}$
3	Shift in s	$f(t)e^{-\alpha t}$	$F(s + \alpha)$
4	Scaling	$f(\alpha t)$	$(1/\alpha)F(s/\alpha)$
5	Time Differentiation	$\dfrac{df}{dt}$	$sF(s) - f(0^-)$
		$\dfrac{d^n f}{dt^n}$	$s^n F(s) - \sum_{k=1}^{n} s^{n-k} f^{k-1}(0^-)$
6	Time Integration	$\int_0^t f(\tau)\,d\tau$	$\dfrac{F(s)}{s}$
7	Indefinite Integral	$\int f(t)\,dt$	$\dfrac{F(s)}{s} + \dfrac{f^{-1}(0)}{s}$
8	Periodic Function	$f_p(t)u(t)$	$\dfrac{F_1(s)}{1 - e^{-Ts}}$ $F_1(s) = \mathcal{L}\{\text{First Cycle}\}$
9	Time Convolution	$\int_{0^-}^{t} f_1(\tau)f_2(t - \tau)\,d\tau$ $\int_{0^-}^{t} f_1(t - \tau)f_2(\tau)\,d\tau$	$F_1(s)F_2(s)$
10	Multiplication by t	$tf(t)$	$-\dfrac{d}{ds}F(s)$
11	Division by t	$f(t)/t$	$\int_s^{\infty} F(s)\,ds$
12	Unit Impulse	$\delta(t)$	1
13	Unit Step	$u(t)$	$1/s$
14	$\dfrac{t^{n-1}}{(n-1)!} \quad n = 1, 2, 3, \ldots$		$\dfrac{1}{s^n}$
15	$t^{k-1} \quad k > 0$		$\dfrac{\Gamma(k)}{s^k} \qquad \Gamma = \begin{cases}\text{Gamma} \\ \text{Function}\end{cases}$
16	$\dfrac{1}{t^{1/2}}$		$\sqrt{\pi/s}$
17	$t^{1/2}$		$\dfrac{\sqrt{\pi}}{2}s^{-3/2}$
18	e^{-at}		$\dfrac{1}{s + a}$
19	$\dfrac{1}{b - a}(e^{-at} - e^{-bt})$		$\dfrac{1}{(s + a)(s + b)}$

Table 6.1 (continued)

NO.	$f(t), t \geq 0$	$F(s)$
20	$\dfrac{1}{a-b}(ae^{-at} - be^{-bt})$	$\dfrac{s}{(s+a)(s+b)}$
21	$\dfrac{e^{-at}}{(b-a)(c-a)} + \dfrac{e^{-bt}}{(a-b)(c-b)} + \dfrac{e^{-ct}}{(a-c)(b-c)}$	$\dfrac{1}{(s+a)(s+b)(s+c)}$
22	$\dfrac{-ae^{-at}}{(b-a)(c-a)} + \dfrac{-be^{-bt}}{(a-b)(c-b)} + \dfrac{-ce^{-ct}}{(a-c)(b-c)}$	$\dfrac{s}{(s+a)(s+b)(s+c)}$
23	$\dfrac{a^2 e^{-at}}{(b-a)(c-a)} + \dfrac{b^2 e^{-bt}}{(a-b)(c-b)} + \dfrac{c^2 e^{-ct}}{(a-c)(b-c)}$	$\dfrac{s^2}{(s+a)(s+b)(s+c)}$
24	te^{-at}	$\dfrac{1}{(s+a)^2}$
25	$(1 - at)e^{-at}$	$\dfrac{s}{(s+a)^2}$
26	$\dfrac{e^{-at}}{(a-b)^2} + \dfrac{(a-b)t - 1}{(a-b)^2}e^{-bt}$	$\dfrac{1}{(s+a)(s+b)^2}$
27	$-\dfrac{ae^{-at}}{(a-b)^2} - \dfrac{b(a-b)t - a}{(a-b)^2}e^{-bt}$	$\dfrac{s}{(s+a)(s+b)^2}$
28	$\dfrac{a^2 e^{-at}}{(a-b)^2} + \dfrac{b^2(a-b)t + b^2 - 2ab}{(a-b)^2}e^{-bt}$	$\dfrac{s^2}{(s+a)(s+b)^2}$
29	$\dfrac{t^{n-1}e^{-at}}{(n-1)!}$	$\dfrac{1}{(s+a)^n}$
30	$\sin(\omega_0 t)$	$\dfrac{\omega_0}{s^2 + \omega_0^2}$
31	$\cos(\omega_0 t)$	$\dfrac{s}{s^2 + \omega_0^2}$
32	$\sinh(at)$	$\dfrac{a}{s^2 - a^2}$
33	$\cosh(at)$	$\dfrac{s}{s^2 - a^2}$
34	$\dfrac{1}{a^2 + b^2}\left[e^{-at} + \dfrac{\sqrt{a^2 + b^2}}{b}\sin(bt - \theta)\right]$ $\theta = \tan^{-1}(b/a)$	$\dfrac{1}{(s+a)(s^2 + b^2)}$
35	$\dfrac{1}{b^2 - a^2}(\cos at - \cos bt)$	$\dfrac{s}{(s^2 + a^2)(s^2 + b^2)}$
36	$\dfrac{1}{2a^3}(\sin at - at \cos at)$	$\dfrac{1}{(s^2 + a^2)^2}$
37	$\sin at \cosh at - \cos at \sinh at$	$\dfrac{4a^3}{s^4 + 4a^4}$

Table 6.1 (continued)

NO.	$f(t)$, $t \geq 0$	$F(s)$
38	$\dfrac{1}{2a^3}(\sinh at - \sin at)$	$\dfrac{1}{s^4 - a^4}$
39	$\dfrac{1}{\sqrt{\pi t}}$	$\dfrac{1}{\sqrt{s}}$
40	$\dfrac{1}{\sqrt{a}} e^{at} \operatorname{erf}(\sqrt{at})$	$\dfrac{1}{(s-a)\sqrt{s}}$
41	$\dfrac{1}{\sqrt{a}} \operatorname{erf}(\sqrt{at})$	$\dfrac{1}{s\sqrt{s+a}}$
42	$\dfrac{1}{\sqrt{\pi t}} - \sqrt{a} e^{at} \operatorname{erfc}(\sqrt{at})$	$\dfrac{1}{\sqrt{s} + \sqrt{a}}$
43	$e^{at} \operatorname{erfc}(\sqrt{at})$	$\dfrac{1}{\sqrt{s}(\sqrt{s} + \sqrt{a})}$
44	$J_0(at)$	$\dfrac{1}{\sqrt{s^2 + a^2}}$
45	$I_0(at)$	$\dfrac{1}{\sqrt{s^2 - a^2}}$
46	$e^{-at/2} I_0(at/2)$	$\dfrac{1}{\sqrt{s}\sqrt{s+a}}$
47	$\dfrac{1}{at} J_1(at)$	$\dfrac{1}{s + \sqrt{s^2 + a^2}}$
48	$\dfrac{1}{a} J_1(at)$	$\dfrac{1}{(s + \sqrt{s^2 + a^2})\sqrt{s^2 + a^2}}$
49	$\dfrac{a}{2\sqrt{\pi t^3}} e^{-a^2/4t}$	$e^{-a\sqrt{s}}$
50	$\operatorname{erfc}\left(\dfrac{a}{2\sqrt{t}}\right)$	$\dfrac{1}{s} e^{-a\sqrt{s}}$
51	$\dfrac{1}{\sqrt{\pi t}} e^{-a^2/4t}$	$\dfrac{1}{\sqrt{s}} e^{-a\sqrt{s}}$
52	$\dfrac{e^{-\zeta \omega_n t}}{2\omega_n \sqrt{\zeta^2 - 1}}(e^{+\omega_n(\zeta^2 - 1)^{1/2} t} - e^{-\omega_n(\zeta^2 - 1)^{1/2} t})$	$\dfrac{1}{s^2 + 2\zeta\omega_n s + \omega_n^2}$ $\zeta > 1$ (overdamped)
53	$t e^{-\omega_n t}$	$\dfrac{1}{s^2 + 2\zeta\omega_n s + \omega_n^2}$ $\zeta = 1$ (critically damped)

6.5 CONCLUDING REMARKS

Table 6.1 (continued)

NO.	$f(t),\ t \geq 0$	$F(s)$
54	$\dfrac{e^{-\zeta\omega_n t}}{\omega_n\sqrt{1-\zeta^2}}\sin(\omega_n\sqrt{1-\zeta^2}\,t)$	$\dfrac{1}{s^2 + 2\zeta\omega_n s + \omega_n^2}$ $\zeta = 0$ (no damping)
55	$\dfrac{1}{\omega_n}\sin(\omega_n t)$	$\dfrac{1}{s^2 + \omega_n^2}$ $\zeta = 0$ (no damping)
56	$\dfrac{1}{t}e^{-[(a+b)/2]t}I_1\left(\dfrac{a-b}{2}t\right)$	$\dfrac{\sqrt{s+a}-\sqrt{s+b}}{\sqrt{s+a}+\sqrt{s+b}}$
57	$\delta(t-t_0) - a^2 t_0\dfrac{J_1(a\sqrt{t^2-t_0^2})}{\sqrt{t^2-t_0^2}}u(t-t_0)$	$e^{-t_0(a^2+s^2)^{1/2}}$
58	$e^{-[(a+b)/2]t_0}\delta(t-t_0)$ $\quad + \left(\dfrac{a-b}{2}\right)^2 t_0 e^{-[(a+b)/2]t}$ $\quad \times \dfrac{I_1\left(\dfrac{a-b}{2}\sqrt{t^2-t_0^2}\right)}{\dfrac{a-b}{2}\sqrt{t^2-t_0^2}}u(t-t_0)$	$e^{-t_0[(s+a)(s+b)]^{1/2}}$
59	$e^{-[(a+b)/2]t_0}\delta(t-t_0)$ $\quad + \left(\dfrac{a-b}{2}\right)^2 te^{-[(a+b)/2]t}$ $\quad \times \dfrac{I_1\left(\dfrac{a-b}{2}\sqrt{t^2-t_0^2}\right)}{\dfrac{a-b}{2}\sqrt{t^2-t_0^2}}u(t-t_0)$ $\quad - \left(\dfrac{a-b}{2}\right)e^{-[(a+b)/2]t}$ $\quad \times I_0\left(\dfrac{a-b}{2}\sqrt{t^2-t_0^2}\right)u(t-t_0)$	$\dfrac{e^{-t_0[(s+a)(s+b)]^{1/2}}}{\sqrt{(s+a)/(s+b)}}$

Pairs 56 through 59 are useful for plane wave propagation in lossy media, or on lossy two-wire transmission lines, or propagation in one-conductor waveguides.

6.5 CONCLUDING REMARKS

We first introduced the bilateral Laplace transform in this chapter. The bilateral Laplace transform of $f(t)$ can be thought of as a representation of the function as a continuous superposition of exponential functions e^{st}, where

274 THE LAPLACE TRANSFORM

$s = \sigma + j\omega$ is a complex frequency. This is an extension of the idea that the Fourier transform is a representation of the function as a continuous superposition of exponential functions $e^{j\omega t}$. This view gives at least some feeling as to what is going on in the analysis process other than an abstract mathematical formulation. The inverse bilateral Laplace transform is not unique unless the region of convergence is completely specified, but, fortunately, in most problems of a practical nature this difficulty can be circumvented using a knowledge of what must occur based on physical grounds. We then observed that we were usually interested in causal systems whose inputs begin at some finite time, say $t = 0$ for convenience. This gave us the very popular unilateral Laplace transform that does not suffer from the uniqueness problem mentioned above.

The Laplace transform can treat a wider class of functions than the Fourier transform, and this offers simplification in analysis. These functions must all begin at $t = 0$ for use with the unilateral transform. This, and the fact that initial conditions are given *explicitly* in the Laplace transform formulation of integro-differential equations, make it ideal for *transient* analysis. This will be pursued in the next chapter. Finally, it should be mentioned that, in all seriousness, one of the greatest advantages of the Laplace transformation over the Fourier transformation in linear-systems analysis is that it is simpler to algebraically manipulate s than it is $j\omega$! This will be apparent in the next chapter if it has not already become so.

The main thrust of this chapter was directed toward building a table of transform pairs that can be advantageously used in the analysis of linear systems. Our experience with Fourier analysis should be of much help in Laplace analysis.

PROBLEMS

1. Establish pair 6.21 in the text:

$$tf(t) \leftrightarrow -\frac{d}{ds}F(s)$$

2. Establish pair 6.22 in the text:

$$f(t)/t \leftrightarrow \int_s^\infty F(\eta)\, d\eta$$

3. Find the Laplace transform of
(a) $5e^{3t}$
(b) $5t - 3$
(c) $(t+1)u(t-1)$
(d) $\sin(t)[u(t) - u(t-1)]$
(e) $(t+1)e^{-t}$
(f) $e^{t(t-1)}$
(g) $(t^2+1)u(t-1)$
(h) $\int_0^t x \sin(x)\, dx$

PROBLEMS 275

4. Find the inverse Laplace transform of

(a) $\dfrac{3}{s+2}$

(b) $\dfrac{1}{2s-7}$

(c) $\dfrac{1}{s^3}$

(d) $\dfrac{s^2}{s^2+a^2}$

(e) $\dfrac{3s-2}{4s^2+3s+5}$

(f) $\dfrac{s+2}{s^2+4s+4}$

(g) $\dfrac{e^{-2s}}{s^2+3s+4}$

(h) $\dfrac{1}{s^3+1}$

5. If the functions in Probs. 3 and 4 represent $h(t)$ and $H(s)$, respectively, where $H(s)$ is the transfer function of a linear system, which are *stable* systems?

6. The function shown in Fig. 6.7:

$$f(t) = \begin{cases} At/T & 0 \leq t \leq T \\ -At/T + 2A & T \leq t \leq 2T \end{cases}$$

is the result of convolving another function $g(t)$ with itself. What is $g(t)$? Sketch $g(t)$.

7. (a) Use Eq. 6.37 to determine the transient part (only) of $v_0(t)/(V_m/\pi)$ in Example 4, Chap. 5 (or Example 6, Chap. 4) if a switch that closes at $t = 0$ is placed in series with the voltage source of Fig. 6.8.
   (b) Using the result of part (a) and Example 4, Chap. 5, what is $v_0(t)$ at $t = 0$?

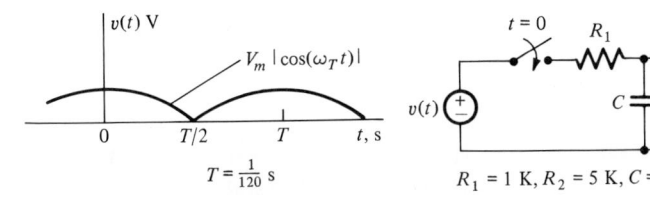

Figure 6.8. Circuit for Problem 7.

8. (a) If the gamma function is defined as

$$\Gamma(k) = \int_0^\infty x^{k-1} e^{-x}\, dx \qquad k > 0$$

show that

$$t^{k-1} \leftrightarrow \dfrac{\Gamma(k)}{s^k} \qquad k > 0$$

   (b) Show that $\Gamma(k+1) = k\Gamma(k)$.

9. (a) Find $\mathscr{L}\{\sin(t)/t\}$.
   (b) Find $\mathscr{L}\{\sin(at)/t\}$.

(c) Show that
$$\text{Si}(\infty) = \int_0^\infty \frac{\sin x}{x}\,dx = \pi/2$$

10. (a) Use pair 5, Table 6.1, to show that
$$\mathscr{L}\{\sin(at)\} = \frac{a}{s^2 + a^2}$$

(b) Repeat (a) starting with
$$\mathscr{L}\{u(t)\cos(at)\} = \frac{s}{s^2 + a^2}$$
and then differentiating $u(t)\cos(at)$.

11. (a) The Bessel function, first kind, order zero, is given in series form by
$$J_0(t) = 1 - \frac{t^2}{2^2} + \frac{t^4}{2^2 4^2} - \frac{t^6}{2^2 4^2 6^2} + \cdots$$
Find $\mathscr{L}\{J_0(t)\}$.

(b) Find $\mathscr{L}\{J_0(at)\}$.

(c) Show that in integral form
$$J_0(t) = \frac{1}{\pi}\int_0^\pi \cos(t\sin x)\,dx$$

(d) Find
$$\mathscr{L}\{J_1(t)\}$$
if $J_1(t) = -dJ_0(t)/dt$. $J_1(t)$ is a Bessel function, first kind, order one.

(e) If $J_0(jt) = I_0(t)$, find $\mathscr{L}\{I_0(t)\}$. $I_0(t)$ is a *modified* Bessel function, first kind, order zero.

12. Evaluate $f(0^+)$ for the following $f(t)$ using the initial-value theorem:
(a) $4 - \cos t$; (b) $2t + 4$; (c) $t^n e^{-3t}$; (d) $I_0(at)$.

13. Evaluate $f(\infty)$ for the following $f(t)$ using the final-value theorem:
(a) $1 + e^{-2t}\cos(t)$; (b) $t^n e^{-3t}$; (c) $\cosh(t)$; (d) $J_0(at)$.

14. (a) What is the transfer function, $H(s)$, for the linear system consisting of a simple (time) differentiator?

(b) Evaluate
$$\mathscr{L}^{-1}\left\{\frac{s^2}{s+a}\right\}$$

15. Find the Laplace transform of
$$f(t) = \sin(\pi t)[u(t) - u(t-1)]$$

16. Use convolution to evaluate $\mathscr{L}^{-1}\{F(s)\}$ for

(a) $\dfrac{s}{(s^2+a^2)^2}$ (b) $\dfrac{1}{s(s+1)}$ (c) $\dfrac{e^{-s}}{s+3}$

17. Show that
$$\sin(t) = \int_0^t J_0(x) J_0(t-x)\, dx$$

18. Find
$$\mathcal{L}^{-1}\left\{\frac{1}{\sqrt{s^2-4s+20}}\right\}$$

19. Use convolution to establish pair 6, Table 6.1:
$$\int_0^t f(\tau)\, d\tau \leftrightarrow F(s)/s$$

20. Use the Laplace transform to solve the differential equation.
$$\frac{d^2y}{dt^2} + 5\frac{dy}{dt} + 4y = x(t) \qquad t>0,\ y(0)=y'(0)=0$$

21. Solve the *integral equation* for $y(t)$ using the Laplace transform:
$$y(t) = f(t) + \int_0^t y(x) h(t-x)\, dx, \qquad t>0$$

22. Solve the differential equation below for the time-variable linear system with $y(0)=1$, $y'(0)=0$. Show that $y(t) = J_0(t)$.
$$t\frac{d^2y}{dt^2} + \frac{dy}{dt} + ty = 0 \qquad t>0$$

23. (a) Apply the trapezoidal rule to find the Laplace transform of $f(t)$, $0 \le t \le NT$, using strips of width T.
    (b) Using $F(s)$ found in (a) show that
$$\frac{T}{2}[f(0^+)g(t) + f(NT)g(t-NT)]$$
$$+ T\sum_{n=1}^{N-1} f(nT)g(t-nT) \leftrightarrow F(s)G(s)$$

24. Find the Laplace transform of the stair-step function shown in Fig. 6.9.

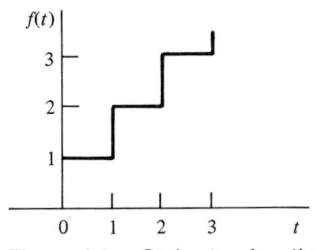

Figure 6.9. Stair-step function.

25. Show that
$$f'(t)|_{t=0^+} = \lim_{s \to \infty} [s^2 F(s) - sf(0^+)]$$

References

1. Cheng, D. K. *Analysis of Linear Systems*. Reading, Mass.: Addison-Wesley, 1959; McGillem, C. D., and Cooper, G. R. *Continuous and Discrete Signal and System Analysis*. New York: Holt, Rinehart and Winston, 1974.
2. Carslaw, H. S., and Jaeger, J. C. *Operational Methods in Applied Mathematics*. New York: Dover Publications, 1963.

Chapter 7
Applications of the Laplace Transform

We introduced the Laplace transformation in Chap. 6 as an extension of the Fourier transformation, and we primarily considered the unilateral Laplace transform that is limited to functions starting at $t = 0$. While this may not be appropriate for those situations where we want to analyze a time-invariant system that has been excited for all time, it is, as we shall shortly see, ideal for those cases where a switch is suddenly closed or opened, or where an excitation is *suddenly* applied to a linear system. The procedure for analysis is essentially the same as that for Fourier methods. We first obtain the differential (or integro-differential) equations describing the time-invariant linear system from the fundamental laws that govern it. Next, we Laplace transform the differential equation, obtaining an algebraic equation that is usually easy to solve for the response. The last step is that of inverse transforming back to the time domain using the table of pairs developed in Chap. 6.

7.1 SOLUTION OF DIFFERENTIAL EQUATIONS

Consider the time-invariant linear system with a single input $x(t)$ and output $y(t)$. If it can be described by a linear differential equation, then the coefficients will be constant, and for an Nth-order system the description is given by Eq. 2.4:

$$\sum_{n=0}^{N} a_n \frac{d^n y}{dt^n} = x_f(t) = \sum_{n=0}^{M} b_n \frac{d^n x}{dt^n} \qquad (2.4)$$

where

$$x_f(t) = \sum_{n=0}^{M} b_n \frac{d^n x}{dt^n}$$

is the forcing function. Using pairs 1 and 5 in Table 6.1, we Laplace transform the differential equation term-by-term to obtain

$$\sum_{n=0}^{N} a_n s^n Y(s) - \sum_{n=1}^{N} a_n \left\{ \sum_{k=1}^{n} y^{k-1}(0^-) s^{n-k} \right\}$$

$$= \sum_{n=0}^{M} b_n s^n X(s) - \sum_{n=1}^{M} b_n \left\{ \sum_{k=1}^{n} x^{k-1}(0^-) s^{n-k} \right\}$$

where $y^{k-1}(0^-)$ is the $(k-1)$th derivative of $y(t)$ evaluated at $t = 0^-$. We can easily solve for $Y(s)$ algebraically:

$$Y(s) = \left[\frac{\sum_{n=0}^{M} b_n s^n}{\sum_{n=0}^{N} a_n s^n} X(s) - \frac{\sum_{n=1}^{M} b_n \left\{ \sum_{k=1}^{n} x^{k-1}(0^-) s^{n-k} \right\}}{\sum_{n=0}^{N} a_n s^n} \right]$$

$$+ \left[\frac{\sum_{n=1}^{N} a_n \left\{ \sum_{k=1}^{n} y^{k-1}(0^-) s^{n-k} \right\}}{\sum_{n=0}^{N} a_n s^n} \right] \qquad (7.1)$$

or

$$Y(s) = Y_{zs}(s) + Y_{zi}(s)$$

The inverse Laplace transform of the first term on the right of Eq. 7.1 is the zero-state response, while the inverse transform of the second term on the right is the zero-input response. Notice that the Laplace transform has *explicitly* given a place to put the initial conditions! Recall that the Fourier method does not do this,[†] and we were required to *add* a zero-input response to show them explicitly. The transfer function is defined as the ratio of the transform of the output to the transform of the input when all initial conditions are 0: that is,

---

[†] Refer to Eq. 5.44 where the differential equation is in terms of $y_{zs}(t)$, the zero-state response, only.

when in the zero state. Thus, from Eq. 7.1

$$\boxed{H(s) = \frac{Y_{zs}(s)}{X(s)} = \frac{\sum_{n=0}^{M} b_n s^n}{\sum_{n=0}^{N} a_n s^n}} \qquad x^{k-1}(0^-) = y^{k-1}(0^-) = 0 \qquad (7.2)$$

We also have

$$Y_{zi}(s) = \frac{\sum_{n=1}^{N} a_n \left\{ \sum_{k=1}^{n} y^{k-1}(0^-) s^{n-k} \right\}}{\sum_{n=0}^{N} a_n s^n} \qquad (7.3)$$

and

$$Y_{zs}(s) = \frac{\sum_{n=0}^{M} b_n s^n}{\sum_{n=0}^{N} a_n s^n} X(s) - \frac{\sum_{n=1}^{M} b_n \left\{ \sum_{k=1}^{n} x^{k-1}(0^-) s^{n-k} \right\}}{\sum_{n=0}^{N} a_n s^n} \qquad (7.4)$$

Thus, we have obtained the complete response for a system starting at $t = 0$.

The transfer function $H(s)$ is just $H(\omega)$ with $j\omega$ replaced by s (see Eq. 5.47), so it can be found by *phasor methods*. In order to simplify matters further, we can even replace $j\omega$ by s *before* we attempt to perform the algebra. This can be done on a block diagram of the system, or on the circuit diagram if the system is represented by an electric circuit. This is an important simplification.

The numerator of Eq. 7.1 can be called the *total excitation function* because it consists of the *transformed forcing function* $X_f(s)$ plus the contributions due to the *initial conditions*.

It is worthwhile at this point to compare Fourier and Laplace methods for a simple second-order system:

$$\frac{d^2 y}{dt^2} + 3\frac{dy}{dt} + 2y = 2\frac{dx}{dt} + x = x_f$$

Taking the Fourier transform, we have

$$(j\omega)^2 Y_{zs}(\omega) + 3(j\omega) Y_{zs}(\omega) + 2 Y_{zs}(\omega) = 2(j\omega) X(\omega) + X(\omega)$$
$$[(j\omega)^2 + 3(j\omega) + 2] Y_{zs}(\omega) = 2[(j\omega) + \tfrac{1}{2}] X(\omega)$$

$$Y_{zs}(\omega) = 2 \frac{j\omega + 1/2}{(j\omega)^2 + 3(j\omega) + 2} X(\omega)$$

This is the transform of the zero-state response *only*, and the transform of a zero-input response (if needed) must be added to obtain the complete response.

The transfer function is

$$H(\omega) = \frac{Y_{zs}(\omega)}{X(\omega)} = 2\frac{j\omega + 1/2}{(j\omega)^2 + 3(j\omega) + 2}$$

Taking the Laplace transform, we have

$$s^2 Y(s) - sy(0^-) - y'(0^-) + 3sY(s) - 3y(0^-) + 2Y(s) = 2sX(s) - 2x(0^-) + X(s)$$

$$[s^2 + 3s + 2]Y(s) = 2[s + \tfrac{1}{2}]X(s) - 2x(0^-) + [s+3]y(0^-) + y'(0^-)$$

$$Y(s) = 2\frac{s + 1/2}{s^2 + 3s + 2}X(s) - \frac{2x(0^-)}{s^2 + 3s + 2} + \frac{(s+3)y(0^-) + y'(0^-)}{s^2 + 3s + 2}$$

Both the transform of the zero-state and zero-input responses are present:

$$Y_{zs}(s) = 2\frac{s + 1/2}{s^2 + 3s + 2}X(s) - \frac{2x(0^-)}{s^2 + 3s + 2}$$

$$Y_{zi}(s) = \frac{(s+3)y(0^-) + y'(0^-)}{s^2 + 3s + 2}$$

The s-domain form of the transfer function is obtained from $Y_{zs}(s)$ when all initial conditions involving $x(t)$ are 0 [or, from $Y(s)$ when all initial conditions involving both $x(t)$ and $y(t)$ are 0]. Thus

$$H(s) = \frac{Y_{zs}(s)}{X(s)} = 2\frac{s + 1/2}{s^2 + 3s + 2}$$

Comparing results

$$H(s) = H(\omega)\big|_{j\omega = s}$$

The last step in the process (assuming that we want a time-domain solution) is that of inverse transforming:

$$y(t) = \mathscr{L}^{-1}\{Y(s)\} \tag{7.5}$$

and in light of the right side of Eq. 7.1, this looks like a difficult procedure. Actually, it is not difficult at all, unless $X(s)$ is an unusual function of s. An example at this point best demonstrates the Laplace method.

Example 1

The switch in Fig. 7.1(a) closes at $t = 0$. Find the current $i(t)$ for all t. We first recognize that $i(t) = 0$ for $t < 0$, and since this is an inductor current, it normally cannot change instantaneously.† Thus, the required initial condition is $i(0^-) = 0$.

---

† We will later examine some idealized situations where an inductor current or capacitor voltage is *forced* to change instantaneously.

7.1 SOLUTION OF DIFFERENTIAL EQUATIONS

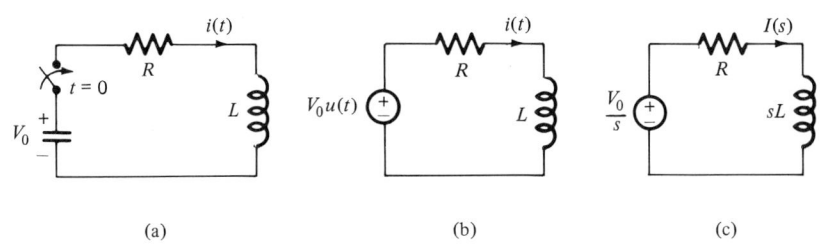

Figure 7.1. Transient RL circuit. (a) Original circuit. (b) Equivalent circuit for $i(t)$, $t > 0$. (c) s-domain version of the circuit.

The differential equation for this system is

$$v(t) = Ri + L\frac{di}{dt} \qquad t > 0$$

where the desired response is $y(t) = i(t)$, and the single input is equal to the forcing function: $x(t) = x_f(t) = v(t)$. Thus, we can write

$$V_0 u(t) = Ri(t) + L\frac{di}{dt}$$

Taking the Laplace transform of this equation results in

$$\frac{V_0}{s} = RI(s) + LsI(s) - Li(0^-) \tag{7.6}$$

Notice carefully that the s-domain version of the circuit[†] shown in Fig. 7.1(c) does not explicitly show the term $-Li(0^-)$ due to the initial condition. If we solve the circuit algebraically for $I(s)$ in the same manner as when solving for the phasor current $I(\omega)$, then $-Li(0^-)$ will be missing. If we are only interested in the zero-state response, this is fine,[‡] but if we want the complete response, we must either Laplace transform the differential equation to make the initial condition show explicitly (as we did) or we can redraw the s-domain circuit to show the initial condition. The term in question is

$$v_L(t) = L\frac{di_L}{dt} \leftrightarrow LsI_L(s) - Li_L(0^-) = V_L(s)$$

$$I_L(s) = \frac{V_L(s)}{Ls} + \frac{i_L(0^-)}{s}$$

and, as can be seen in Fig. 7.2, this equation leads to the representation of the initial condition as a voltage [Fig. 7.2(b)] or a current source [Fig. 7.2(c)]. Thus, we could redraw Fig. 7.1(c) as in Fig. 7.3(a) or (b).

[†] Obtained from the *phasor* form of the circuit by replacing $j\omega$ with s.
[‡] $I(s)$ in Fig. 7.1(c) should then be $I_{zs}(s)$, but the subscripts are usually dropped. They are also dropped in labeling block diagrams, as in Fig. 7.6 because we normally include no initial conditions when using block diagrams, and the complete response is the zero-state response.

284   APPLICATIONS OF THE LAPLACE TRANSFORM

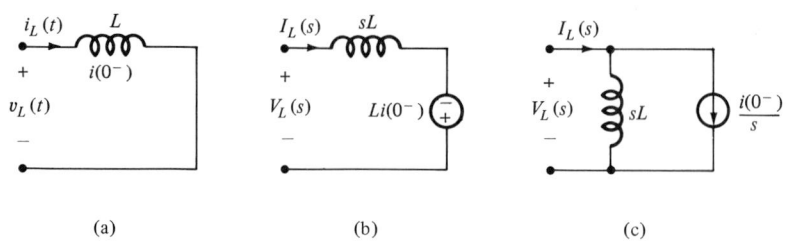

(a)   (b)   (c)

Figure 7.2.   (a) Time-domain inductor voltage. (b) s-domain inductor voltage showing the initial condition term as a voltage source. (c) s-domain inductor voltage showing the initial condition term as a current source.

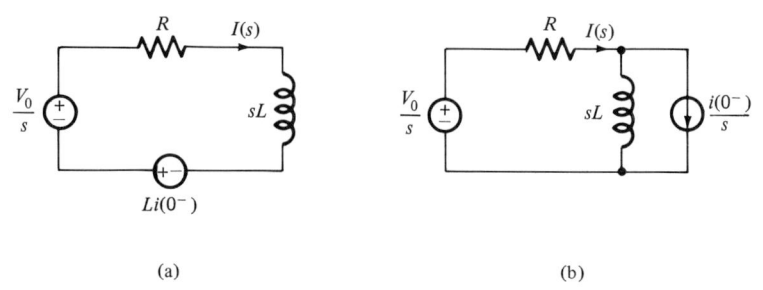

(a)   (b)

Figure 7.3.   Complete s-domain version of the RL circuit [Fig. 7.1(c)]. (a) Initial condition term as a voltage source. (b) Initial condition term as a current source.

At this point we digress briefly to point out that the same sort of thing can be done to treat the initial voltage associated with a capacitor. The pertinent equation is

$$v_c(t) = \frac{1}{C}\int i_c\, dt \leftrightarrow \frac{I_c(s)}{Cs} + \frac{v_c(0^-)}{s} = V_c(s)$$

$$I_c(s) = sCV_c(s) - Cv_c(0^-)$$

and, as can be seen in Fig. 7.4 this equation also leads to the representation of the initial condition term as a voltage source or as a current source.

It is perhaps best to simply Laplace transform the differential equation and proceed from there. In this way we do not need to keep track of the special circuits shown in Figs. 7.2 and 7.4. If we "stick to the basics" we are likely to avoid trouble.

We now return to Example 1. Since $i(0^-) = 0$ in the present example, solving for $I(s)$ in Eq. 7.6 gives

$$I(s) = \frac{V_0/s}{R + Ls} \qquad [= I_{zs}(s)]$$

which is the same form as Eq. 7.4. A better form for inversion is

$$I(s) = \frac{V_0}{L}\frac{1}{s(s + R/L)}$$

7.1 SOLUTION OF DIFFERENTIAL EQUATIONS

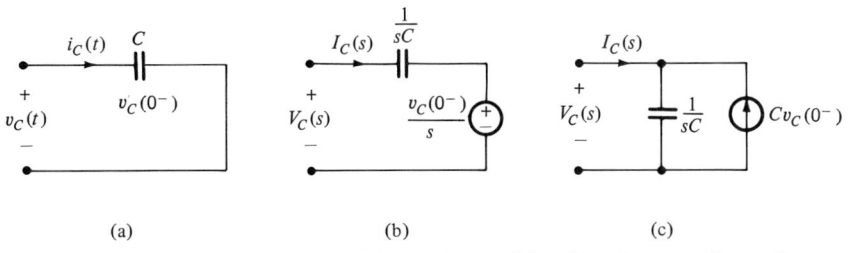

(a) (b) (c)

Figure 7.4. (a) Time-domain capacitor voltage. (b) s-domain capacitor voltage with the initial condition term as a voltage source. (c) s-domain capacitor voltage with the initial condition term as a current source.

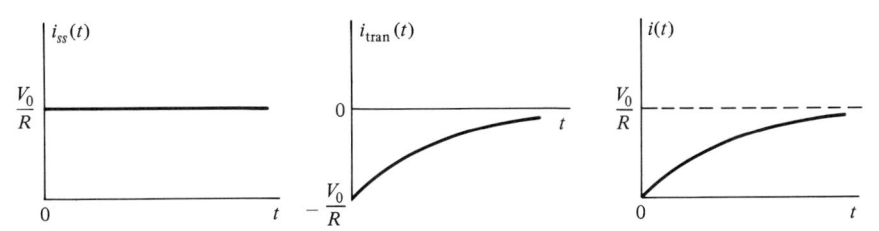

Figure 7.5. Steady-state, transient, and complete response of the circuit in Fig. 7.1.

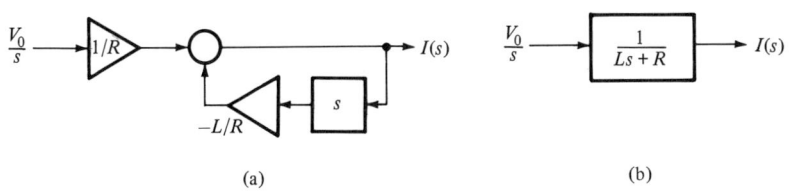

(a) (b)

Figure 7.6. (a) Block diagram of the RL series circuit: V_0/s as the input and $I(s)$ as the output. (b) Block diagram showing $H(s) = 1/(Ls + R)$.

Using Table 6.1, no. 19, with $a = 0$ and $b = R/L$ gives

$$i(t) = \frac{V_0}{R}(1 - e^{-Rt/L}) \qquad t \geq 0$$

or

$$i(t) = i_{zs}(t) = \frac{V_0}{R}(1 - e^{-Rt/L})u(t) \qquad \text{for all } t$$

a well-known answer displayed in Fig. 7.5. This is the zero-state response, and the transient part and steady-state part are easy to identify.

Before leaving this example it is well to note that the system can be represented by the block diagram[†] in Fig. 7.6(a), where the block labeled s is a

---

[†] Notice that the subscripts zs do not appear on $I(s)$ in Fig. 7.6 even though the zero-state response $I_{zs}(s)$ would be that (response) determined. This is by convention.

Example 2

Suppose the initial current in Fig. 7.1(b) is not 0, but rather is given by $i(0^-) = I_0$. Obviously this current must have come from some other source that is not shown. We need not worry here about where it came from. Equation 7.6 still applies, so

$$I(s) = \frac{V_0/s + LI_0}{R + Ls} = \frac{V_0}{L}\frac{1}{s(s + R/L)} + I_0\frac{1}{s + R/L}$$

Then

$$i(t) = \frac{V_0}{R}(1 - e^{-Rt/L})u(t) + I_0 e^{-Rt/L}u(t) \tag{7.7}$$

or

$$i(t) = i_{zs}(t) + i_{zi}(t)$$

It is informative to consider, for comparison, the other methods we have discussed in previous chapters for solving for $i(t)$. For the Fourier method we have the phasor form derived from the phasor version of the circuit:

$$V(\omega) = RI_{zs}(\omega) + j\omega L I_{zs}(\omega) \tag{7.8}$$

or

$$I_{zs}(\omega) = \frac{V(\omega)}{R + j\omega L} = \frac{V_0[\pi\delta(\omega) + 1/j\omega]}{R + j\omega L}$$

for the voltage $v(t) = V_0 u(t)$. Thus

$$I_{zs}(\omega) = \frac{V_0}{L}\frac{\pi\delta(\omega) + 1/j\omega}{j\omega + R/L} = \frac{V_0}{R}\pi\delta(\omega) + \frac{V_0}{L}\frac{1}{j\omega(j\omega + R/L)}$$

Inverse (Fourier) transforming

$$i_{zs}(t) = \frac{V_0}{2R} + \frac{V_0}{R}(1 - e^{-Rt/L})u(t) - \frac{V_0}{2R}$$

or

$$i_{zs}(t) = \frac{V_0}{R}(1 - e^{-Rt/L})u(t) \tag{7.9}$$

But this is only the zero-state response, and in order to include the initial condition, we must add the zero-input response (which is a complementary

7.1 SOLUTION OF DIFFERENTIAL EQUATIONS

function). That is, it is a solution to the homogeneous equation

$$0 = Ri + L\frac{di}{dt}$$

It is easy to obtain:

$$i_{zi}(t) = Ae^{-Rt/L}$$

Thus, the complete solution is

$$i(t) = \frac{V_0}{R}(1 - e^{-Rt/L})u(t) + Ae^{-Rt/L} \qquad t \geq 0$$

Applying the initial condition to the complete response gives $A = I_0$, so the complete solution (written for all time) is

$$i(t) = \frac{V_0}{R}(1 - e^{-Rt/L})u(t) + I_0 e^{-Rt/L}u(t)$$

in agreement with Eq. 7.7. Notice that the forcing function $v(t) = I_0 R u(-t) + V_0 u(t)$, which itself creates the given initial condition, will produce a zero-state response identical to 7.7.

A (completely) time-domain solution to this problem is also easy to obtain if we first find the unit-impulse response $h(t)$. From Fig. 7.6(b) we have

$$H(s) = \frac{I(s)}{V(s)} = \frac{I_{zs}(s)}{V(s)} = \frac{1}{R + sL} = \frac{1}{L(s + R/L)} \quad \text{or} \quad H(\omega) = \frac{I(\omega)}{V(\omega)} = \frac{1}{R + j\omega L}$$

Notice that either form is valid since the unit-impulse response is a zero-state response. Inverse transforming

$$h(t) = \frac{1}{L}e^{-Rt/L}u(t)$$

Equation 2.62 gives the zero-state response

$$i_{zs}(t) = \frac{V_0}{L}u(t)\int_0^t e^{-R(t-\tau)/L}\,d\tau$$

$$i_{zs}(t) = \frac{V_0}{L}e^{-Rt/L}u(t)\int_0^t e^{R\tau/L}\,d\tau$$

$$i_{zs}(t) = \frac{V_0}{R}e^{-Rt/L}[e^{R\tau/L}]_0^t u(t)$$

$$i_{zs}(t) = \frac{V_0}{R}(1 - e^{-Rt/L})u(t)$$

This result is the same as Eq. 7.9, and adding the zero-input response once again gives Eq. 7.7 for the complete response.

At the risk of boring the reader we will, last of all, consider the classical solution to this problem. We first need to find a particular solution or particular integral. As mentioned earlier, in Chap. 2, it is *any* solution to the inhomogeneous differential equation, and, in problems of the type presently being considered, it can usually be found by simply finding the *steady-state response*. Inspection of Fig. 7.1(b) reveals that a particular solution is $i(t) = V_0/R$, and the correctness of this can be verified by direct substitution in the differential equation ($t > 0$). Some writers prefer to call the particular solution by the name forced response, but in our present problem this is only the forced response to V_0, not $V_0 u(t)$! We choose to continue to call this response a particular solution to avoid confusion with what we have already called the forced or zero-state response. A complementary function has already been found in the Fourier method. The complete classical solution is given by the sum of the particular solution and the complementary function. It is

$$i(t) = \frac{V_0}{R} + Be^{-Rt/L} \qquad t > 0$$

We now apply the initial condition to this complete solution. We have at $t = 0$

$$I_0 = \frac{V_0}{R} + B$$

or

$$B = I_0 - V_0/R$$

Thus

$$i(t) = \frac{V_0}{R} + (I_0 - V_0/R)e^{-Rt/L} \qquad t > 0$$

$$i(t) = \frac{V_0}{R}(1 - e^{-Rt/L}) + I_0 e^{-Rt/L} \qquad t > 0$$

as before.

It is indeed convenient to have the initial conditions "built in," and this is precisely what the Laplace method does for us. We next consider another example.

Example 3

Suppose next that the switch in Fig. 7.7 has been closed for a long time and opens at $t = 0$. What is $i(t)$? Since the system is in the steady state *before* $t = 0$, $i(0^-) = V_0/R_1$, the inductor having behaved as a short circuit. The equation to be solved is

$$v(t) = L\frac{di}{dt} = -R_2 i \qquad t > 0$$

7.1 SOLUTION OF DIFFERENTIAL EQUATIONS

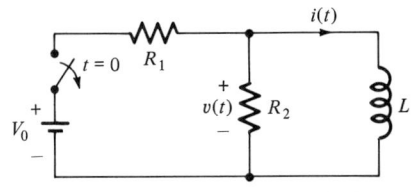

Figure 7.7. Circuit for Example 3.

or

$$0 = R_2 i + L\frac{di}{dt}$$

Thus

$$0 = R_2 I(s) + LsI(s) - Li(0^-)$$
$$Li(0^-) = (R_2 + Ls)I(s) = V_0 L/R_1$$

so

$$I(s) = \frac{V_0/R_1}{s + R_2/L}$$

Therefore

$$i(t) = \frac{V_0}{R_1} e^{-R_2 t/L} u(t)$$

This result should also be familiar. It is worth pointing out that this problem is *not* the same as in Fig. 7.8. They have the same initial conditions and the same forcing function for $t < 0$. However, for $t > 0$ the original circuit has an *open* circuit to the left of R_1, while the circuit in Fig. 7.8 has a *short* circuit to the left of R_1. Thus, in Fig. 7.7 the inductor current decays exponentially through R_2 alone, while in Fig. 7.8 it decays through the parallel combination of R_1 and R_2.

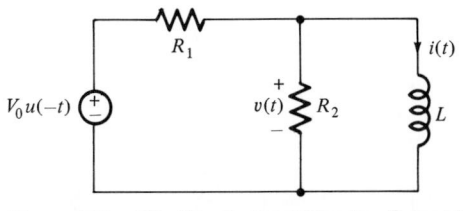

Figure 7.8. Similar (but not identical) problem to that in Fig. 7.7. For $t > 0$ the original circuit has an *open* circuit to the left of R_1, while the circuit in Fig. 7.8 has a *short* circuit to the left of R_1.

7.2 SIMULTANEOUS EQUATIONS

The advantage of the Laplace (or Fourier) transform method is enhanced further when we consider a system with several unknowns: We obtain simultaneous algebraic equations which are usually easy (but perhaps tedious) to solve. An example serves best to demonstrate how a solution is obtained.

Example 4

Find the voltage $v_L(t)$ in the circuit of Fig. 7.9. The integro-differential equations (for $t > 0$), using mesh analysis, are

$$v(t) = R_1 i_1 + L\frac{di_1}{dt} - L\frac{di_2}{dt}$$

$$0 = R_2 i_2 + \frac{1}{C}\int_0^t i_2\, dt + v_C(0) + L\frac{di_2}{dt} - L\frac{di_1}{dt}$$

Laplace transforming gives

$$V(s) = R_1 I_1(s) + LsI_1(s) - Li_1(0^-) - LsI_2(s) + Li_2(0^-)$$

$$0 = R_2 I_2(s) + \frac{1}{Cs} I_2(s) + \frac{v_C(0)}{s} + LsI_2(s) - Li_2(0^-) - LsI_1(s) + Li_1(0^-)$$

Notice that $v_C(0) \leftrightarrow v_C(0)/s$ since $v_C(0)$ is a constant. Rearranging the equations into a standard form

$$V(s) + L[i_1(0^-) - i_2(0^-)] = (R_1 + Ls)I_1(s) - LsI_2(s)$$

$$-\frac{v_C(0)}{s} - L[i_1(0^-) - i_2(0^-)] = -LsI_1(s) + \left(R_2 + Ls + \frac{1}{Cs}\right)I_2(s)$$

Notice again that the initial conditions are acting as forcing functions. In fact, we may rewrite these equations as

$$V_1(s) = (R_1 + Ls)I_1(s) - LsI_2(s)$$

$$V_2(s) = -LsI_1(s) + \left(R_2 + Ls + \frac{1}{Cs}\right)I_2(s)$$

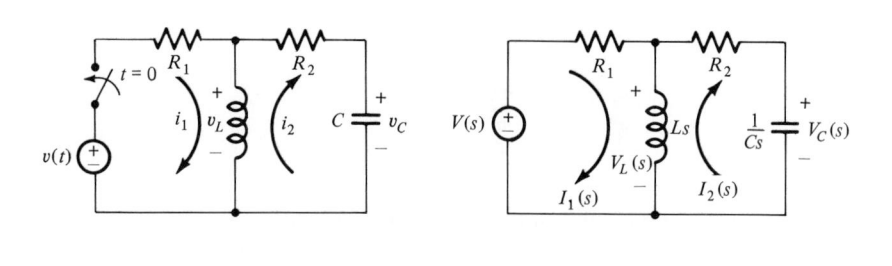

(a)

(b)

Figure 7.9. (a) Circuit to be analyzed. (b) Laplace form of the circuit.

or

$$\begin{bmatrix} V_1(s) \\ V_2(s) \end{bmatrix} = \begin{bmatrix} R_1 + Ls & -Ls \\ -Ls & R_2 + Ls + 1/(Cs) \end{bmatrix} \begin{bmatrix} I_1(s) \\ I_2(s) \end{bmatrix}$$

Thus

$$I_1(s) = \frac{V_1(s)[R_2 + Ls + (1/Cs)] + V_2(s)Ls}{\Delta(s)}$$

and

$$I_2(s) = \frac{V_1(s)Ls + V_2(s)(R_1 + Ls)}{\Delta(s)}$$

where

$$\Delta(s) = (R_1 + Ls)\left(R_2 + Ls + \frac{1}{Cs}\right) - (Ls)^2$$

Now, in the present problem, $i_1(0^-) - i_2(0^-)$ and $v_C(0^\pm) = 0$, so $V_1(s) = V(s)$ and $V_2(s) = 0$, and

$$I_1(s) = \frac{V(s)[Ls^2 + R_2 s + (1/C)]}{(R_1 + R_2)Ls^2 + [R_1 R_2 + (L/C)]s + (R_1/C)}$$

and

$$I_2(s) = \frac{V(s)Ls^2}{(R_1 + R_2)Ls^2 + [R_1 R_2 + (L/C)]s + (R_1/C)}$$

or

$$I_1(s) = \frac{1}{R_1 + R_2} \frac{V(s)[s^2 + (R_2/L)s + (1/LC)]}{s^2 + \dfrac{R_1 R_2 + L/C}{(R_1 + R_2)L} s + \dfrac{R_1}{(R_1 + R_2)LC}}$$

and

$$I_2(s) = \frac{1}{R_1 + R_2} \frac{V(s)s^2}{s^2 + \dfrac{R_1 R_2 + L/C}{(R_1 + R_2)L} s + \dfrac{R_1}{(R_1 + R_2)LC}}$$

In order to determine $v_L(t)$ we need $i_1(t) - i_2(t)$, so

$$I_1(s) - I_2(s) = \frac{R_2}{(R_1 + R_2)L} \frac{V(s)[s + 1/(R_2 C)]}{s^2 + \dfrac{R_1 R_2 + L/C}{(R_1 + R_2)L} s + \dfrac{R_1}{(R_1 + R_2)LC}}$$

Suppose that $v(t) = V_0 u(t) \leftrightarrow V_0/s$. Then

$$I_1(s) - I_2(s) = \frac{V_0 R_2}{(R_1 + R_2)Ls} \left[\frac{s + 1/(R_2 C)}{s^2 + \dfrac{R_1 R_2 + L/C}{(R_1 + R_2)L} s + \dfrac{R_1}{(R_1 + R_2)LC}} \right] \quad (7.10)$$

292 APPLICATIONS OF THE LAPLACE TRANSFORM

In order to simplify matters, suppose that $R_1 = 5\,\Omega$, $R_2 = 20\,\Omega$, $L = 0.5$ H, and $C = 2000\,\mu\text{F}$. Then

$$I_1(s) - I_2(s) = 1.6V_0 \frac{s+25}{s(s^2 + 28s + 200)}$$

$$= 1.6V_0 \frac{s+25}{s(s+14+j2)(s+14-j2)}$$

The poles of $s[I_1(s) - I_2(s)]$ are at $s = -14 - j2$ and $s = -14 + j2$, so the final-value theorem is applicable. The steady-state value of $i_1(t) - i_2(t)$ is given by the limit of $s[I_1(s) - I_2(s)]$ as s goes to 0, which, using Eq. 7.10, is

$$[i_1(t) - i_2(t)]_{ss} = V_0/R_1$$

Inspection of Fig. 7.9(b) reveals that this is correct. The final-value theorem is often used as a check on the algebra.
Now

$$V_L(s) = Ls[I_1(s) - I_2(s)] - L[i_1(0^-) - i_2(0^-)]$$
$$V_L(s) = Ls[I_1(s) - I_2(s)]$$

$$V_L(s) = 0.8V_0 \frac{s+25}{s^2 + 28s + 200} \tag{7.11}$$

We can save a lot of work by being a little tricky with the denominator of the last equation:

$$V_L(s) = 0.8V_0 \frac{s+25}{(s+14)^2 + 4}$$

or getting even smarter

$$V_L(s) = 0.8V_0 \left[\frac{(s+14)}{(s+14)^2 + 2^2} + 5.5 \frac{2}{(s+14)^2 + 2^2} \right]$$

We are finally ready to use the table of pairs:

$$v_L(t) = 0.8V_0 e^{-14t}[\cos(2t) + 5.5\sin(2t)]u(t) \tag{7.12}$$

where we have used Table 6.1, pairs 3, 31, and 30.

Some more checks on the final solution would be useful. The steady-state value of $v_L(t)$ is 0 as given by Eq. 7.12, and from Fig. 7.9(b) this is obviously correct. The initial value, as given by Eq. 7.12, is $v_L(0^+) = 0.8V_0$. The initial value, as given by the equation above 7.12 and the initial-value theorem is

$$v_L(0^+) = \lim_{s \to \infty} 0.8V_0 \frac{1 + 25/s}{1 + 28/s + 200/s^2} = 0.8V_0$$

Is this initial value consistent with Fig. 7.9(b)? The initial inductor current is 0 and the initial capacitor voltage is 0, so at $t = 0^+$ the circuit of Fig. 7.10 is

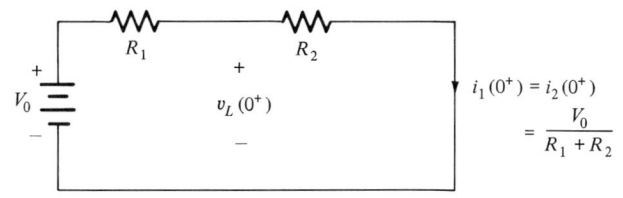

Figure 7.10. Conditions at $t = 0^+$ for Fig. 7.9(b).

applicable. Thus, we have

$$v_L(0^+) = \frac{V_0}{R_1 + R_2} R_2 = 0.8 V_0$$

7.3 APPEARANCE OF THE UNIT IMPULSE

It is easy to imagine situations where an inductor current or capacitor voltage must change value instantaneously. As we shall see, this is no cause for concern, as long as the fundamental laws are not violated.

Example 5

Find the current $i(t)$ and voltage $v_2(t)$ in the circuit of Fig. 7.11. The current through L_1 is V_0/R at $t = 0^-$, while the current through L_2 is *zero* at $t = 0^-$. We can imagine L_2 as having a very small resistance, which causes all of the battery current to flow through the switch. These are the initial conditions. The differential equation for $t > 0$ is

$$V_0 = L_1 \frac{di}{dt} + L_2 \frac{di}{dt} + Ri \qquad t > 0$$

Laplace transforming and applying the initial conditions gives

$$\frac{V_0}{s} = L_1 sI(s) - L_1 i(0^-) + L_2 sI(s) + RI(s)$$

$$\frac{V_0}{s} = (L_1 + L_2)sI(s) + RI(s) - L_1 V/R$$

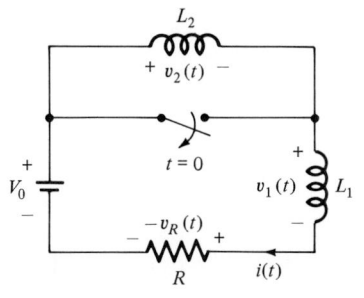

Figure 7.11. Circuit for Example 5.

or

$$I(s) = \frac{V_0/s + V_0 L_1/R}{(L_1 + L_2)s + R} = \frac{V_0}{L_1 + L_2} \frac{1}{s\left(s + \dfrac{R}{L_1 + L_2}\right)} + \frac{V_0}{R} \frac{L_1}{L_1 + L_2} \frac{1}{s + \dfrac{R}{L_1 + L_2}}$$

The inverse transform is easy to obtain:

$$i(t) = \frac{V_0}{R}(1 - e^{-Rt/(L_1 + L_2)})u(t) + \frac{V_0}{R} \frac{L_1}{L_1 + L_2} e^{-Rt/(L_1 + L_2)} u(t)$$

or

$$i(t) = \frac{V_0}{R}\left(1 - \frac{L_2}{L_1 + L_2} e^{-Rt/(L_1 + L_2)}\right) u(t) \qquad t > 0$$

For *all t*

$$i(t) = \frac{V_0}{R} - \frac{V_0}{R} \frac{L_2}{L_1 + L_2} e^{-Rt/(L_1 + L_2)} u(t)$$

while the current through L_2 for all time is

$$i_{L2} = \frac{V_0}{R}\left(1 - \frac{L_2}{L_1 + L_2} e^{-Rt/(L_1 + L_2)}\right) u(t)$$

These currents are plotted in Fig. 7.12. They are both inductor currents (identical for $t > 0$), and they are both discontinuous at $t = 0$. The question naturally arises: Have we violated anything fundamental? In order to answer this question we can take a look at Kirchhoff's first law for all t. This law, which we used (for $t > 0$) to analyze the circuit, simply states that the sum of the voltage rises (or drops) around a closed path is 0. More importantly, however, we should remember that this law is just a restatement of *conservation of energy*,[1] which is certainly fundamental. Thus, *for all t* we have

$$v_1(t) = L_1 \frac{di}{dt} = -\frac{V_0}{R} \frac{L_1 L_2}{L_1 + L_2} \delta(t) + V_0 \frac{L_1 L_2}{(L_1 + L_2)^2} e^{-Rt/(L_1 + L_2)} u(t)$$

$$v_2(t) = L_2 \frac{di_{L2}}{dt} = \frac{V_0}{R} \frac{L_1 L_2}{L_1 + L_2} \delta(t) + V_0 \frac{L_2^2}{(L_1 + L_2)^2} e^{-Rt/(L_1 + L_2)} u(t)$$

$$v_R(t) = iR = V_0 - V_0 \frac{L_2}{L_1 + L_2} e^{-Rt/(L_1 + L_2)} u(t)$$

and

$$V_0 = iR + v_1(t) + v_2(t)$$

$$V_0 = V_0 + V_0 e^{-Rt/(L_1 + L_2)} u(t) \left[\frac{L_1 L_2}{(L_1 + L_2)^2} + \frac{L_2^2}{(L_1 + L_2)^2} - \frac{L_1 L_2 + L_2^2}{(L_1 + L_2)^2}\right]$$

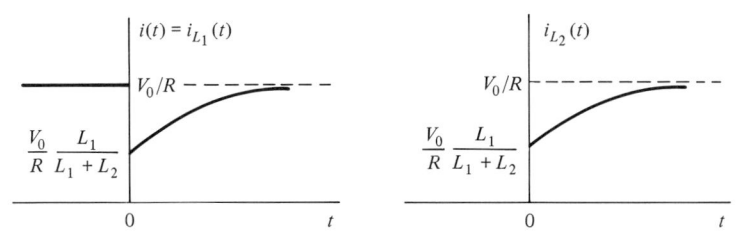

Figure 7.12. Inductor currents for Fig. 7.11.

or

$$V_0 = V_0$$

Before leaving this example we should examine the voltage $v_2(t)$, which is also the voltage across the switch. Notice that it contains $\delta(t)$. Some of us have experienced the unnerving effect of opening a switch in an inductive circuit. A pronounced arc due to breakdown of the air is often part of the (nonlinear) response!

Many of us have also (rather suddenly) discharged a charged capacitor by short circuiting its terminals. Pops and sparks are the usual result.

Example 6

Show that charge is conserved in the circuit of Fig. 7.13. C_1 is charged to V_0 V before the switch is closed.[2] That is, $v(0^-) = V_0$. Using nodal analysis we have

$$C_1 \frac{dv}{dt} + C_2 \frac{dv}{dt} + \frac{v}{R} = 0 \qquad t > 0$$

Thus

$$C_1 s V(s) - C_1 V_0 + C_2 s V(s) + V(s)/R = 0$$

Solving for $V(s)$

$$V(s) = \frac{C_1 V_0}{C_1 + C_2} \frac{1}{s + 1/R(C_1 + C_2)}$$

or

$$v(t) = \frac{C_1 V_0}{C_1 + C_2} \exp\left[-\frac{t}{R(C_1 + C_2)}\right] u(t) \tag{7.13}$$

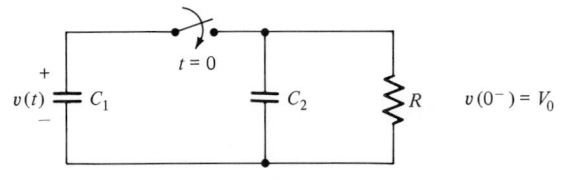

Figure 7.13. Circuit for Example 6.

Now, the voltage across C_1 for all t is

$$v_{C1}(t) = V_0 u(-t) + \frac{C_1 V_0}{C_1 + C_2} \exp\left[-\frac{t}{R(C_1+C_2)}\right] u(t)$$

and the voltage across C_2 and R for all t is simply $v(t)$ in Eq. 7.13. Kirchhoff's second law, which simply states that the sum of the currents entering (or leaving) a node must be 0, is nothing more than a restatement of *conservation of charge*. Thus, we need to show, for example, that the sum of the currents leaving the node at the top of C_2 is 0 for all t. These currents are

$$C_1 \frac{dv_{C1}}{dt} = -\frac{C_1 C_2 V_0}{C_1 + C_2} \delta(t) - \left(\frac{C_1}{C_1+C_2}\right)^2 \frac{V_0}{R} \exp\left[-\frac{t}{R(C_1+C_2)}\right] u(t)$$

$$C_2 \frac{dv}{dt} = \frac{C_1 C_2 V_0}{C_1 + C_2} \delta(t) - \frac{C_1 C_2}{(C_1+C_2)^2} \frac{V_0}{R} \exp\left[-\frac{t}{R(C_1+C_2)}\right] u(t)$$

$$\frac{V}{R} = \frac{C_1}{C_1+C_2} \frac{V_0}{R} \exp\left[-\frac{t}{R(C_1+C_2)}\right] u(t)$$

Adding these currents

$$-\frac{C_1^2}{(C_1+C_2)^2} - \frac{C_1 C_2}{(C_1+C_2)^2} + \frac{C_1^2 + C_1 C_2}{(C_1+C_2)^2} = 0 \quad 0 = 0$$

Notice that the capacitor voltages adjust instantaneously at $t = 0$ to the value $C_1 V_1/(C_1 + C_2)$ at $t = 0^+$, and this is accompanied by an instantaneous transfer of charge through the impulsive current. As a matter of fact, the charge on C_2 increases from 0 to $C_1 C_2 V_0/(C_1 + C_2) = 2V_0/3$ C (the moment or strength of the impulsive current) instantaneously. The charge on C_1 decreases by the same amount. We can now understand the pops and sparks!

Actually, there is at least one other interesting feature to this example that might escape detection by all but the most inquisitive investigators. What about the energy? At $t = 0^+$ the energy stored is

$$w(0^+) = \frac{1}{2}(C_1 + C_2)\left(\frac{C_1 V_0}{C_1+C_2}\right)^2 = \frac{1}{2}\frac{C_1^2}{C_1+C_2} V_0^2$$

while the energy ultimately dissipated in R is

$$w_R = \int_{0^+}^{\infty} p_R(t)\,dt = \int_{0^+}^{\infty} \frac{[v(t)]^2}{R}\,dt$$

$$w_R = \int_{0^+}^{\infty} \left[\frac{C_1}{C_1+C_2}\right]^2 \frac{V_0^2}{R} \exp\left[-\frac{2t}{R(C_1+C_2)}\right] dt$$

$$w_R = \frac{1}{2}\frac{C_1^2}{C_1+C_2} V_0^2$$

7.3 APPEARANCE OF THE UNIT IMPULSE

Thus, it seems that everything is as it should be until we remember that at $t = 0^-$ the energy stored in C_1 is

$$w_{C_1}(0^-) = \tfrac{1}{2}C_1 V_0^2$$

and *apparently*

$$\frac{1}{2}C_1 V_0^2 - \frac{1}{2}\frac{C_1^2}{C_1 + C_2}V_0^2 = \frac{1}{2}\frac{C_1 C_2}{C_1 + C_2}V_0^2$$

Joules of energy has disappeared! This, of course, cannot happen, but how can we account for it?

If we allow R in Fig. 7.13 to become infinite, then *for all t*

$$v(t) = V_0 u(-t) + V_0 \frac{C_1}{C_1 + C_2} u(t)$$

where we have used Eq. 7.13 with $R \to \infty$. The initial energy at $t = 0^-$ is

$$w(0^-) = \tfrac{1}{2}C_1 V_0^2$$

while the final energy at $t = 0^+$ (or $t \to \infty$, for that matter) is

$$w(0^+) = \frac{1}{2}(C_1 + C_2)\frac{C_1^2}{(C_1 + C_2)^2}V_0^2 = \frac{1}{2}\frac{C_1^2}{C_1 + C_2}V_0^2$$

and, as before, apparently

$$w = \frac{1}{2}\frac{C_1 C_2}{(C_1 + C_2)}V_0^2 \qquad (7.14)$$

Joules of energy has vanished. The reason for removing R is that we now have a simpler circuit to analyze and can avoid some very tedious algebra.

Consider the circuit of Fig. 7.14. The interconnecting wires in any circuit have some resistance, however small it might be, and we have accounted for it by placing R_1 in the circuit.

Example 7

Find the energy dissipated in R_1 of Fig. 7.14. Using mesh analysis we have

$$\frac{1}{C_1}\int_0^t i\,dt - V_0 + \frac{1}{C_2}\int_0^t i\,dt + R_1 i(t) = 0$$

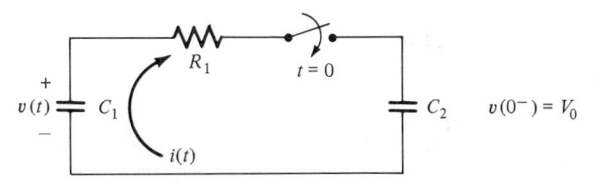

Figure 7.14. Circuit for Example 7.

or

$$\frac{1}{C_1 s}I(s) - \frac{V_0}{s} + \frac{1}{C_2 s}I(s) + R_1 I(s) = 0$$

or

$$I(s) = \frac{V_0}{R_1 s + (1/C_1) + (1/C_2)} = \frac{V_0}{R_1}\frac{1}{s + (1/R_1 C_s)}$$

so

$$i(t) = \frac{V_0}{R_1}\exp\left(-\frac{t}{R_1 C_s}\right)u(t)$$

where

$$\frac{1}{C_s} = \frac{1}{C_1} + \frac{1}{C_2}$$

Thus, the energy dissipated in R_1 is

$$w_{R1} = \int_0^\infty [i(t)]^2 R_1\, dt = \frac{V_0^2}{R_1}\int_0^\infty \exp\left(-\frac{2t}{R_1 C_s}\right) dt$$

$$w_{R1} = \frac{1}{2}C_s V_0^2 = \frac{1}{2}\frac{C_1 C_2}{C_1 + C_2}V_0^2$$

This is precisely the missing amount of energy in Eq. 7.14 and in Example 6. Notice that this dissipated energy (in R_1) is independent of R_1! Even if $R_1 = 0$, this amount of energy is dissipated, and we are relieved to find that energy is conserved.

Before moving on to a new topic it is worthwhile to examine conservation of energy in a magnetically coupled circuit.[3]

Example 8

Find $i_1(t)$ and $i_2(t)$ in the circuit of Fig. 7.15. Recalling that the self and mutual terms have opposite signs when both currents do not enter the dotted terminals,

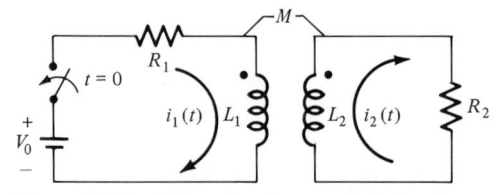

Figure 7.15. Magnetically coupled circuit.

and using mesh analysis, we have

$$V_0 u(t) - R_1 i_1 - L_1 \frac{di_1}{dt} + M \frac{di_2}{dt} = 0$$

$$R_2 i_2 + L_2 \frac{di_2}{dt} - M \frac{di_1}{dt} = 0$$

or

$$\frac{V_0}{s} = (L_1 s + R_1) I_1(s) - M s I_2(s) \tag{7.15}$$

$$0 = -M s I_1(s) + (L_2 s + R_2) I_2(s) \tag{7.16}$$

where $i_1(0^-) = i_2(0^-) = 0$.
Solving for $I_2(s)$

$$I_2(s) = V_0 \frac{M}{(L_1 L_2 - M^2) s^2 + (R_1 L_2 + R_2 L_1) s + R_1 R_2}$$

$$I_2(s) = \frac{V_0}{L_1 L_2 - M^2} \frac{M}{s^2 + \frac{R_1 L_2 + R_2 L_1}{L_1 L_2 - M^2} s + \frac{R_1 R_2}{L_1 L_2 - M^2}}$$

or

$$I_2(s) = \frac{V_0}{L_1 L_2 - M^2} \frac{M}{(s+a)(s+b)}$$

where

$$a = \frac{1}{2(L_1 L_2 - M^2)}$$
$$\times \{R_1 L_2 + R_2 L_1 + [(R_1 L_2 + R_2 L_1)^2 - 4(R_1 R_2)(L_1 L_2 - M^2)]^{1/2}\}$$

$$b = \frac{1}{2(L_1 L_2 - M^2)}$$
$$\times \{R_1 L_2 + R_2 L_1 - [(R_1 L_2 + R_2 L_1)^2 - 4(R_1 R_2)(L_1 L_2 - M^2)]^{1/2}\}$$

Therefore

$$i_2(t) = \frac{V_0 M}{L_1 L_2 - M^2} \frac{1}{(a-b)} (e^{-at} - e^{-bt}) u(t)$$

In the same way

$$i_1(t) = \frac{V_0 L_2}{L_1 L_2 - M^2} \left[\frac{(R_2/L_2 - a) e^{-at}}{a(a-b)} - \frac{(R_2/L_2 - b) e^{-bt}}{b(a-b)} + \frac{R_2/L_2}{ab} \right] u(t)$$

300 APPLICATIONS OF THE LAPLACE TRANSFORM

Notice that both $i_1(0^+)$ and $i_2(0^+)$ are 0, so both $i_1(t)$ and $i_2(t)$ are continuous at $t = 0$, and furthermore, energy is conserved at $t = 0$.

What happens for unity coupling, $M^2 = L_1 L_2$? Equations 7.15 and 7.16 still hold, but $I_2(s)$ and $I_1(s)$ reduce to

$$I_2(s) = \frac{V_0 M}{R_1 L_2 + R_2 L_1} \cdot \frac{1}{s + \dfrac{R_1 R_2}{R_1 L_2 + R_2 L_1}}$$

and

$$I_1(s) = \frac{V_0 L_2}{R_1 L_2 + R_2 L_1} \cdot \frac{s + R_2/L_2}{s\left(s + \dfrac{R_1 R_2}{R_1 L_2 + R_2 L_1}\right)}$$

so

$$i_2(t) = \frac{V_0 M \alpha_1}{R_1 R_2} e^{-\alpha_1 t} u(t)$$

and

$$i_1(t) = \frac{V_0 L_2 \alpha_1}{R_1 R_2}\left[\frac{R_2}{\alpha_1 L_2} - \left(\frac{R_2}{\alpha_1 L_2} - 1\right)e^{-\alpha_1 t}\right] u(t)$$

where

$$\alpha_1 = \frac{R_1 R_2}{R_1 L_2 + R_2 L_1}$$

For unity coupling, neither $i_1(0^+)$ nor $i_2(0^+)$ are 0, but the stored magnetic energy at $t = 0^+$,

$$w(0^+) = \tfrac{1}{2} L_1 [i_1(0^+)]^2 + \tfrac{1}{2} L_2 [i_2(0^+)]^2 - M i_1(0^+) i_2(0^+)$$

must be 0. We have

$$i_2(0^+) = \frac{V_0 M \alpha_1}{R_1 R_2}$$

and

$$i_1(0^+) = \frac{V_0 L_2 \alpha_1}{R_1 R_2}$$

so

$$w(0^+) = \frac{V_0^2 \alpha_1^2}{R_1^2 R_2^2}\left[\frac{1}{2} L_1 L_2^2 + \frac{1}{2} L_1 L_2^2 - L_1 L_2^2\right]$$

$$w(0^+) = 0$$

Finally, notice that for unity coupling only *one* exponential transient term appears in $i_2(t)$.

7.4 OTHER EXAMPLES

The examples that we have considered so far in this chapter have been concerned with electric circuits having lumped parameters, voltage sources, current sources, and switches. These inevitably lead to polynomial forms in s and, as we have seen, are relatively easy to treat. The linear differential equation with constant coefficients is, of course, the starting point that leads to these forms, and although the engineer is very likely to encounter these frequently, there are other forms that he or she may encounter also. It would be wise to consider some other forms as well.

The lossless transmission line is a two-dimensional problem, but it is not difficult to analyze in phasor form, and this will most likely be done at the junior or senior level in most electrical engineering curricula. In the interest of brevity we will not present a derivation here but will simply present the results.[4] The lossless line driven by the voltage source V_g (with internal impedance Z_g) and terminated in the load impedance Z_l is shown in Fig. 7.16. The line is described as having a characteristic impedance $R_0 = \sqrt{L/C}$, where L is the *distributed* series inductance per unit length and C is the *distributed* shunt capacitance per unit length. R_0 is the input impedance for an *infinite* length of line (such that there is no reflection from the load end). The coefficients of reflection at the generator end and load end are

$$\Gamma_g = \frac{Z_g - R_0}{Z_g + R_0} \qquad \Gamma_l = \frac{Z_l - R_0}{Z_l + R_0}$$

and represent the ratio of the *incident* phasor voltage wave to the *reflected* phasor voltage wave at each end. A traveling wave that leaves the generator travels at a velocity given by $u = 1/\sqrt{LC}$ until it reaches the load end, whereupon it is partially reflected (if $\Gamma_l \neq 0$) and partially absorbed. The reflected wave travels back toward the generator where it too is partially reflected (if $\Gamma_g \neq 0$) and partially absorbed. This process repeats until the resistance in Z_g or Z_l (or both), which represents the only loss mechanism, finally reduces the amplitude of the waves to a negligible value.

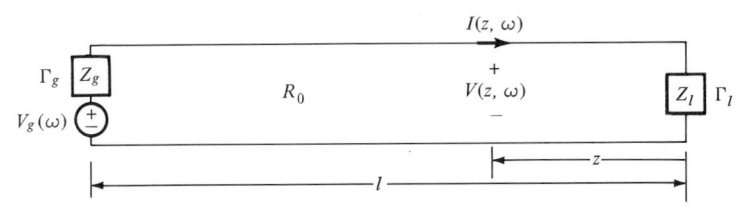

Figure 7.16. Lossless transmission line.

The preceding narrative is easy to verify given the phasor voltage at a point on the line (Fig. 7.16). The infinite series of reflections described above is a geometric series, and therefore has a closed form:

$$V(z, \omega) = \frac{V_g R_0}{R_0 + Z_g} \frac{e^{-jkl}}{1 - \Gamma_l \Gamma_g e^{-j2kl}} (e^{jkz} + \Gamma_l e^{-jkz}) \tag{7.17}$$

where $k = \omega/u = \omega\sqrt{LC}$. If the generator is "matched to the line," then $Z_g = R_0$, $\Gamma_g = 0$, and

$$V(z, \omega) = \frac{V_g(\omega)}{2} e^{-j\omega l\sqrt{LC}} (e^{j\omega z\sqrt{LC}} + \Gamma_l(\omega) e^{-j\omega z\sqrt{LC}}) \tag{7.18}$$

In this case if there is a reflection at the load end, this reflected wave is completely absorbed ($\Gamma_g = 0$) when it arrives back at the generator. We will investigate this simpler case.

Example 9

Find the unit-impulse response for the lossless line of Fig. 7.16 when $Z_g = R_0$ or $\Gamma_g = 0$. Fourier methods will work as well as Laplace methods for this example, but since we are concerned with the Laplace transform in this chapter, we replace $j\omega$ by s in Eq. 7.18:

$$V(z, s) = \frac{V_g(s)}{2} e^{-sl\sqrt{LC}} (e^{sz\sqrt{LC}} + \Gamma_l(s) e^{-sz\sqrt{LC}}) \tag{7.19}$$

Now, $V_g(s) = 1 \leftrightarrow \delta(t)$ for the unit-impulse response, so

$$V(z, s) = \frac{1}{2} e^{-s(l-z)\sqrt{LC}} + \frac{\Gamma_l(s)}{2} e^{-s(l+z)\sqrt{LC}}$$

Using pair 2 (time shift) in Table 6.1 we have

$$v(z, t) = \tfrac{1}{2}\delta[t - (l - z)\sqrt{LC}] u[t - (l - z)\sqrt{LC}]$$
$$+ \tfrac{1}{2}\Gamma_l[t - (l + z)\sqrt{LC}] u[t - (l + z)\sqrt{LC}] \tag{7.20}$$

The first term in Eq. 7.20 is the impulsive traveling wave with amplitude 0.5 because the *unit* impulse at the generator divides equally between $Z_g = R_0$ and the input impedance R_0 to the line (whose length *appears* to be infinite, and the incident impulsive wave does not know otherwise until it reaches the load). It is easy to see that this term is a traveling wave. Take $z = 0$, for example. In this case the impulse occurs at $t = l\sqrt{LC}$ which is exactly the time required for the impulse to travel from the generator to the load.

The second term in Eq. 7.20 has a character that depends on $\Gamma_l(s) \leftrightarrow \Gamma_l(t)$. Suppose, for example, that $Z_l = 0$ (short circuit) and $\Gamma_l(s) = -1$. In this case $\Gamma_l(t) = -\delta(t)$.

$$v(z, t) = \tfrac{1}{2}\delta[t - (l - z)\sqrt{LC}] u[t - (l - z)\sqrt{LC}]$$
$$- \tfrac{1}{2}\delta[t - (l + z)\sqrt{LC}] u[t - (l + z)\sqrt{LC}] \tag{7.21}$$

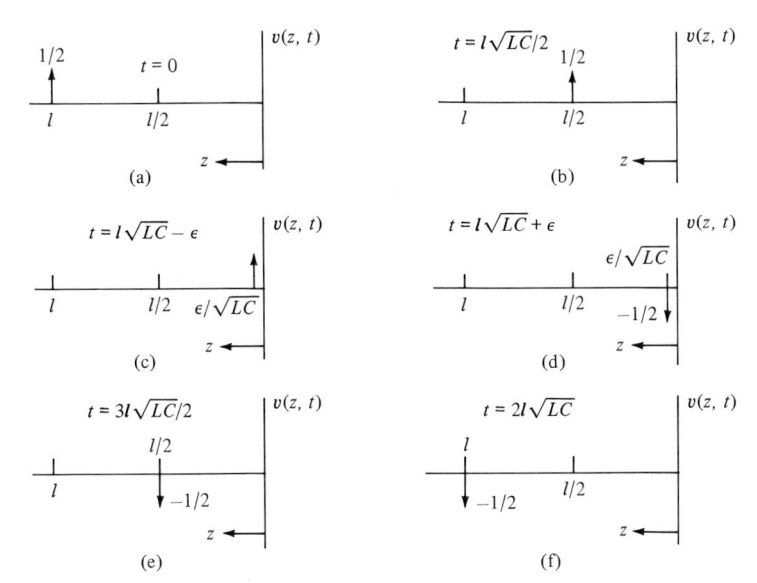

Figure 7.17. Sequence of plots, v(z, t) versus z, for (a) $t = 0$, (b) $t = l\sqrt{LC}/2$, (c) $t = l\sqrt{LC} - \epsilon$, $\epsilon \ll l\sqrt{LC}$, (d) $t = l\sqrt{LC} + \epsilon$, (e) $t = 3l\sqrt{LC}/2$, (f) $t = 2l\sqrt{LC}$.

and the second term is a *completely* reflected impulsive wave (with sign change) traveling back toward the generator. A series of plots of $v(z, t)$ versus z for various times t (like "snapshots") is shown in Fig. 7.17. Notice that $z \geq 0$, and remember that the impulse exists where its argument is 0.

Example 10

Repeat Example 9 for $\Gamma_g = 0$ and $\Gamma_l = -1$ when $v_g(t)$ is the pulse shown in Fig. 7.18. Equation 7.19 gives

$$V(z, s) = \tfrac{1}{2}V_g(s)e^{-s(l-z)\sqrt{LC}} - \tfrac{1}{2}V_g(s)e^{-s(l+z)\sqrt{LC}}$$

Inverse transforming, we obtain

$$v(z, t) = \tfrac{1}{2}v_g[t - (l - z)\sqrt{LC}]u[t - (l - z)\sqrt{LC}] \\ - \tfrac{1}{2}v_g[t - (l + z)\sqrt{LC}]u[t - (l + z)\sqrt{LC}]$$

Just as in Fig. 7.17, we can imagine the pulse traveling from the generator to the short circuit at the end, being reflected with sign change, and ultimately

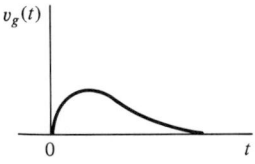

Figure 7.18. Generator pulse, $v_g(t)$, for Example 10.

being absorbed in $Z_g = R_0$ after it travels back to the generator. Notice that the pulse shape is not altered (no distortion).

If $Z_g = R_0$, and, in addition, $Z_l = R_0$, so that $\Gamma_l = 0$ and the load is "matched to the line," there is no reflection at the load, and Eq. 7.19 gives

$$V(z, s) = \tfrac{1}{2}V_g(s)e^{-s(l-z)\sqrt{LC}}$$

or

$$v(z, t) = \tfrac{1}{2}v_g[t - (l - z)\sqrt{LC}]u[t - (l - z)\sqrt{LC}]$$

Thus, in this case, the pulse travels (half amplitude) *undistorted*, but with time shift or *delay*, from the generator to the load where it is absorbed.

Example 11

Repeat Example 9 for $\Gamma_g = 0$, $Z_l = R_l + j\omega L_l$. We have

$$\Gamma_l(s) = \frac{Z_l(s) - R_0}{Z_l(s) + R_0} = \frac{R_l + sL_l - R_0}{R_l + sL_l + R_0} = 1 - \frac{2R_0}{R_l} \frac{1}{s + (R_l + R_0)/L_l}$$

so

$$\Gamma_l(t) = \delta(t) - \frac{2R_0}{R_l} e^{-(R_l + R_0)t/L_l} u(t)$$

Equation 7.20 gives

$$v(z, t) = \frac{1}{2}\delta[t - (l - z)\sqrt{LC}]u[t - (l - z)\sqrt{LC}]$$

$$+ \frac{1}{2}\delta[t - (l + z)\sqrt{LC}]u[t - (l + z)\sqrt{LC}]$$

$$- \frac{R_0}{R_l} \exp\left\{-\frac{R_l + R_0}{L_l}[t - (l + z)\sqrt{LC}]\right\} u[t - (l + z)\sqrt{LC}]$$

The third term in this result shows that the reflected wave is now "dispersed." Had we been using a pulse (instead of an impulse) from the generator, the shape of the reflected wave would be altered.

A wave traveling in the positive z direction on a transmission line *with loss*, that is, with distributed series resistance per unit length, R, and distributed shunt conductance per unit length, G, has the form

$$V(z, \omega) = V_0(\omega)\exp[-\sqrt{(R + j\omega L)(G + j\omega C)}z]$$

or

$$V(z, s) = V_0(s)\exp[-\sqrt{(R + sL)(G + sC)}z]$$
$$V(z, s) = V_0(s)\exp[-\sqrt{(s + R/L)(s + G/C)}\sqrt{LC}z] \tag{7.22}$$

The unit-impulse response $[V_0(s) = 1]$ is

$$v(z, t) = \mathscr{L}^{-1}\{\exp[-\sqrt{(s + R/L)(s + G/C)}\sqrt{LC}z]\}$$

and this is given by pair 58, Table 6.1. Rather than writing down the rather lengthy time-domain form, we observe from the table that this impulse response contains a *damped* impulse *plus* a term that represents dispersion. This dispersion will cause *distortion* of a pulse traveling along the line. Furthermore, if it is given that $\lim I_1(x)/x = \frac{1}{2}$ for $x \to 0$, then the dispersion is removed if $a = b$ or $R/L = G/C$. A transmission line with $R/L = G/C$ (usually obtained by periodically *adding* series inductors) is called a *distortionless line*. In this case

$$v(z, t) = e^{-(R/R_0)z}\delta(t - z\sqrt{LC})$$

which is only a *damped* impulse, so that a *pulse* would be decreasing in amplitude (attenuated) as it traveled along the line but would maintain its shape. Hence it is a distortionless line.

As another example of the use of the Laplace transform in solving a two-dimensional boundary value problem, consider the *heat conduction equation*

$$\frac{\partial \theta}{\partial t} = k \frac{\partial^2 \theta}{\partial z^2} \quad z > 0 \quad t > 0 \tag{7.23}$$

where θ is the *temperature* and k is *diffusivity*, a parameter that is intrinsic to the material in which the temperature is to be found. We have assumed that the solid is semi-infinite in order to keep the discussion as simple as possible.

Example 12

Find the temperature in the semi-infinite solid shown in Fig. 7.19. We are essentially assuming that $\partial/\partial x = \partial/\partial y \equiv 0$. The initial temperature is 0; that is, $\theta(z, 0^-) = 0$, and at $t = 0$ a step change in temperature θ_0 is applied at the surface $z = 0$; that is, $\theta(0, t) = \theta_0 u(t)$. The temperature must be bounded for all z and t; that is, $|\theta(z, t)| < M$. Laplace transforming Eq. 7.23 gives

$$s\Theta(z, s) - \theta(z, 0^-) = k \frac{d^2\Theta}{dz^2}$$

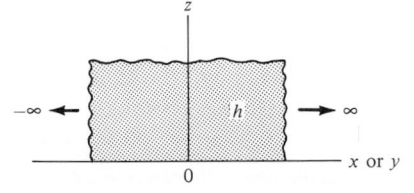

Figure 7.19. Semi-infinite heat-conducting solid.

or

$$s\Theta(z, s) = k\frac{d^2\Theta}{dz^2}$$

or

$$\frac{d^2\Theta}{dz^2} - \frac{s}{k}\Theta = 0 \tag{7.24}$$

The solution to this *second-order ordinary* linear differential equation is easy to obtain.[†] The elements of the fundamental set are $\cosh(\sqrt{s/k}\,z)$, $\sinh(\sqrt{s/k}\,z)$, $e^{\sqrt{s/k}\,z}$, and $e^{-\sqrt{s/k}\,z}$. Of these, the last is the only one that can keep $\Theta(z, s)$ bounded. Thus, the solution to Eq. 7.24 must be

$$\Theta(z, s) = C_1 e^{-\sqrt{s/k}\,z} \tag{7.25}$$

Since $\theta(0, t) = \theta_0 u(t)$, we have

$$\Theta(0, s) = \theta_0/s$$

Equation 7.25 gives

$$\Theta(0, s) = C_1$$

Thus

$$C_1 = \theta_0/s$$

and Eq. 7.25 becomes

$$\Theta(z, s) = \theta_0 \frac{e^{-\sqrt{s/k}\,z}}{s}$$

Inverse transforming using pair 50 (Table 6.1) gives

$$\theta(z, t) = \theta_0 \operatorname{erfc}\left(\frac{z}{2\sqrt{kt}}\right)$$

Replacing the complementary error function with its defining relation

$$\theta(z, t) = \theta_0 \left\{1 - \frac{2}{\sqrt{\pi}} \int_0^{z/(2\sqrt{kt})} e^{-v^2}\, dv\right\} \tag{7.26}$$

It is a tabulated function.[5]

The examples in this section were presented to demonstrate other kinds of problems that can be treated with the Laplace transform. We did not intend to elaborate on transmission lines or heat conduction. Having the right transform pairs in a convenient table is of tremendous value in linear-systems analysis.

[†] Actually, we could perform a second Laplace transformation, this time with respect to z, in order to solve Eq. 7.24. This is not necessary, however.

7.5 STABILITY AND FEEDBACK

In this section we would like to determine limitations on the form of the transfer function $H(s)$ that corresponds to a time-invariant linear system that can actually be built or can occur naturally. It has already been pointed out in Sec. 2.10 that, as far as the time domain is concerned, a physically realizable system must be *causal* (whether it is a time-invariant system or not). This requirement is stated in Eq. 2.47:

$$h_f(t, \tau) = y_{zi}(t, \tau)u(t - \tau) = \begin{cases} 0 & t < \tau \\ y_{zi}(t, \tau) & t > \tau \end{cases} \quad (2.47)$$

for a unit-impulse forcing function applied at $t = \tau$, $\delta(t - \tau)$. Notice (Eq. 2.46b) that if $h_f(t, \tau)$ is causal, then $h(t, \tau)$ is clearly causal. $y_{zi}(t, \tau)$ is a zero-input response (complementary function) that satisfies Eq. 2.40 and the set of initial conditions 2.44. In simple terms a causal system is one that cannot respond to an applied unit impulse *before* the impulse is applied. If the system is time-invariant, then Eq. 2.128 applies, and ($\tau = 0$)

$$h_f(t) = y_{zi}(t)u(t) = \begin{cases} 0 & t < 0 \\ y_{zi}(t) & t > 0 \end{cases} \quad (2.128)$$

and (Eq. 2.46d) if $h_f(t)$ is causal, then $h(t)$ is causal. If $h(t) \leftrightarrow H(s)$, where $H(s)$ is the *unilateral Laplace transform* of $h(t)$, then the system is *automatically* causal since $h(t) = 0$ for $t < 0$.

It was also pointed out in Sec. 2.10 that for a stable system the unit-impulse response must approach 0 as t approaches ∞:

$$h_f(t, \tau) \to 0 \quad h(t, \tau) \to 0 \quad \text{for} \quad t \to \infty$$
$$y_{zi}(t, \tau) \to 0 \quad \text{for} \quad t \to \infty \quad (2.127)$$

or for a time-invariant system

$$h_f(t) \to 0 \quad h(t) \to 0 \quad \text{for} \quad t \to \infty$$
$$y_{zi}(t) \to 0 \quad \text{for} \quad t \to \infty \quad (2.129)$$

For a time-invariant linear system described by an Nth-order linear differential equation with constant coefficients, we have[†]

$$y_{zi}(t) = D_1 e^{s_1 t} + D_2 e^{s_2 t} + \cdots + D_N e^{s_N t} \quad (7.27)$$

We first obtained Eq. 7.27 in Chap. 2 by simply *guessing* a solution of the form e^{st} to the homogeneous (zero-input) differential equation. This gave (and always will give) us the characteristic equation

$$s^N + C_{N-1} s^{N-1} + \cdots + C_1 s + C_0 = 0 \quad (7.28)$$

Comparing Eqs. 2.129, 7.27, and 7.28 leads us to the conclusion that if the system we have described is to be stable, then the real parts of all the roots of

---
[†] Refer to the footnote below Eq. 2.129 concerning repeated roots.

308 APPLICATIONS OF THE LAPLACE TRANSFORM

the characteristic equation must be negative, because in that case, every term in Eq. 7.27 dies out exponentially. This conclusion was reached in Chap. 2.

Notice carefully that in Chap. 2 s was in *no way associated with the Laplace transform*! It was just a number (complex, perhaps). We can certainly associate it with the Laplace transform now, however, because the characteristic equation (obtained from the transfer function, Eq. 7.2) is

$$\sum_{n=0}^{N} a_n s^n = 0 \quad (s = \sigma + j\omega) \tag{7.29}$$

and is identical ($a_n = C_n$, $a_N = 1$) with Eq. 7.28. We can now concur with the conclusion we reached in Sec. 2.10. The roots of the *characteristic equation must lie in the left half of the complex s plane* ($s = \sigma + j\omega$) for a stable system.

Another statement regarding the stability of a system is that every bounded input must produce a bounded output, and a sufficient condition for this to occur is given by Eq. 2.130:

$$\int_0^\infty |h(t)|\, dt \leq K_2 < \infty \tag{2.130}$$

In terms of $H(s)$ we merely need to examine the $h(t)$ that correspond to $H(s)$.

A special case, leading to what is called a *marginally* stable system, occurs when the roots of the characteristic equation lie *on* the imaginary ($j\omega$) axis in the s plane. Take, for example, the case $h(t) = u(t) \leftrightarrow 1/s = H(s)$ that is represented by a pole at the origin ($s = 0$) in the s plane. $h(t)$ does not increase, nor does it decrease, but if a unit-step input (a bounded input) is applied to such a system, the response is $tu(t) \leftrightarrow 1/s^2$, which is *not bounded*. In the same way, consider

$$H(s) = \frac{1}{s^2 + \omega_0^2} = \frac{1}{(s + j\omega_0)(s - j\omega_0)}$$

giving conjugate poles on the imaginary axis at $s = \pm j\omega_0$. If an input sinusoid with frequency $\omega = \omega_0$ is applied to this lossless second-order system, then the response is

$$\frac{1}{s^2 + \omega_0^2} \cdot \frac{\omega_0}{s^2 + \omega_0^2} = \frac{\omega_0}{(s^2 + \omega_0^2)^2}$$

and by pair 36, Table 6.1, this inverse transforms to

$$\frac{1}{2\omega_0^2}[\sin(\omega_0 t) - \omega_0 t \cos(\omega_0 t)]u(t)$$

which is not bounded. Thus, if the system transfer function has *repeated* poles on the imaginary axis [$H(s) = 1/s^2$ or $H(s) = 1/(s^2 + \omega_0^2)^2$] it will always be unstable, regardless of the input.

In summary then, we require that if a system is to be at least marginally stable, and if it is time-invariant and lumped-parameter, then in terms of $H(s)$,

it must have the following characteristics:[6]
1. no poles in the right half of the s plane,
2. no repeated poles on the imaginary axis, and
3. the degree of the numerator polynomial of $H(s)$ cannot exceed the degree of the denominator polynomial of $H(s)$.

The last requirement can be seen by taking as an example $H(s) = s$ and the bounded input $\cos(\omega_0 t) \leftrightarrow s/(s^2 + \omega_0^2)$ whose response is $s^2/(s^2 + \omega_0^2)$, or

$$\frac{s^2 + \omega_0^2 - \omega_0^2}{s^2 + \omega_0^2} = 1 - \frac{\omega_0^2}{s^2 + \omega_0^2}$$

or, inverse transforming, the response is

$$\delta(t) - \omega_0 \sin(\omega_0 t) u(t)$$

which is certainly not bounded. The three conditions listed are necessary and sufficient for determining whether $H(s)$ corresponds to a stable system.

The loci of the roots of the characteristic equation [poles of $H(s)$] for the second-order system are shown in Fig. 2.21 as an example of what we have said. The corresponding unit-impulse responses are shown in Fig. 2.14(a).

If we are not interested in *exactly where* the poles of $H(s)$ are located, but only if any of them are in the right half of the s plane (requirement 1), then the Routh criterion (App. D) is particularly simple to use.

The *canonical* form of a feedback system is extremely important in the analysis and design of *control systems*. The s-domain block diagram is shown in Fig. 7.20. The conventional symbols used in the canonical form of a feedback system are shown in Fig. 7.20(c). Notice carefully that for this generally accepted form $H(s)$ is not the overall transfer function! For Fig. 7.20(c):

$G(s)$ = forward transfer function, open-loop transfer function, or the *process*;

$H(s)$ = feedback transfer function;

$E(s)$ = transform of the *error* signal;

$R(s)$ = transform of the *reference* signal;

$C(s)$ = transform of *controlled* (output) signal;

$G(s)H(s)$ = *loop* transfer function; and

$T(s) = C(s)/R(s)$ = overall closed-loop transfer function.

If we assume that there is no loading effect between blocks; that is, if we assume that the transfer function for the individual (isolated) blocks remain unchanged when the blocks are interconnected, then the overall transfer function is easy to obtain. The output of the summer, $E(s)$, in Fig. 7.20(c) is $R(s) \pm H(s)C(s)$ and this term, multiplied by $G(s)$, must be $C(s)$. Thus

$$C(s) = G(s)[R(s) \pm H(s)C(s)]$$

310 APPLICATIONS OF THE LAPLACE TRANSFORM

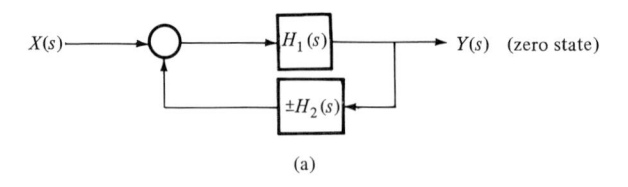

Figure 7.20. (a) Canonical form of a feedback system. (b) Overall transfer function block diagram for (a). (c) Conventional symbols for the canonical feedback system. (d) Transfer function for (c).

Solving algebraically for $C(s)$, we obtain the *zero-state* response

$$C(s) = \frac{G(s)}{1 \mp G(s)H(s)} R(s) \quad [= C_{zs}(s)] \tag{7.30}$$

or

$$\boxed{T(s) = \frac{C(s)}{R(s)} = \frac{G(s)}{1 \mp G(s)H(s)}} \quad \text{(closed-loop transfer function)} \tag{7.31}$$

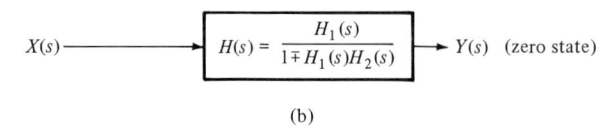

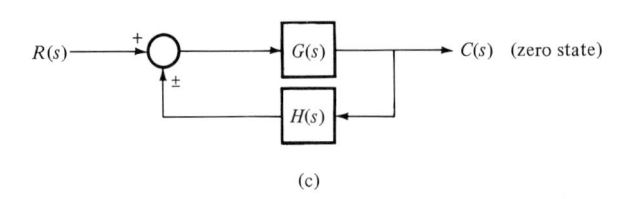

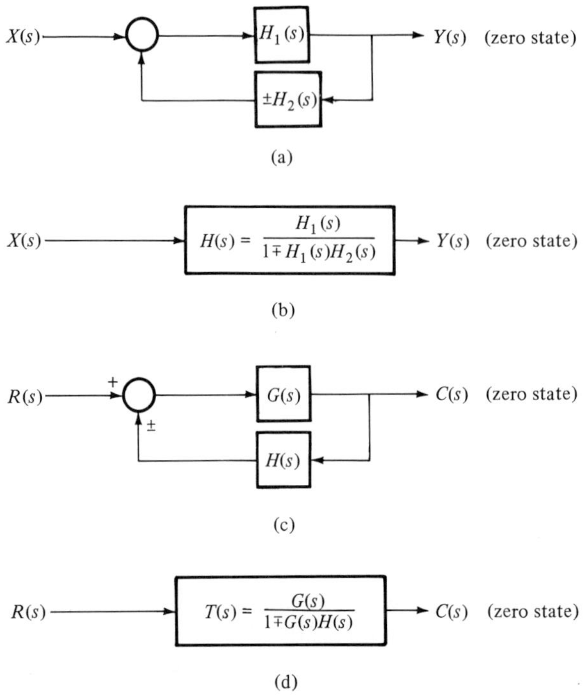

It is important to recognize that normally we would have little difficulty in reducing a feedback system to canonical form [Fig. 7.20(c)]. Three common methods are:[†]

1. block diagram reduction,
2. signal-flow graph reduction, and
3. Mason's gain formula.

[†] See Dorf in the list of references at the end of the chapter. These methods are all relatively simple to use, but are beyond the scope of this introductory treatment. A signal flow graph was used in Figs. 5.26 and 5.27.

7.5 STABILITY AND FEEDBACK

The signal flow graph is a streamlined (simplified) version of the block diagram, and Mason's gain formula enables one to write down the overall transfer function (gain) upon inspection of the signal flow graph for the system. We leave a detailed discussion of these methods for a first course in control systems.

The characteristic equation for the system is obtained from Eq. 7.31:

$$1 \mp G(s)H(s) = 0 \tag{7.32}$$

and in the study of the stability of feedback systems (for which previous remarks apply) one is interested in the *location* of the roots of this equation.

Example 13

Suppose that $G(s) = 1/[(s-2)(s+3)]$ and $\pm H(s) = -K$ (a constant) in Fig. 7.20(c). We can think of the block $H(s)$ as representing an amplifier or attenuator with 180° of phase shift. Thus, we immediately see that if $K = 0$, the system is unstable because

$$T(s) = G(s) = \frac{1}{(s-2)(s+3)} \quad (K=0)$$

and $T(s)$ has a pole ($s = 2$) in the right half of the s plane. For any K the characteristic equation is

$$1 + \frac{K}{(s-2)(s+3)} = 0$$

or

$$(s-2)(s+3) + K = 0$$

or

$$s^2 + s + (K-6) = 0 \tag{7.33}$$

Thus, we have a second-order system, and we would like to know the *locus* of the roots of Eq. 7.33 for any K. This is called the *root locus*. Factoring gives

$$s_1 = -\tfrac{1}{2} + \tfrac{1}{2}\sqrt{25 - 4K}$$
$$s_2 = -\tfrac{1}{2} - \tfrac{1}{2}\sqrt{25 - 4K}$$

We can immediately conclude that the system is unstable for all $K < 0$ since $s_1 > 0$. The critical point is reached when $s_1 = 0$ for which $\sqrt{25 - 4K} = 1$, or $K = 6$. If $K > 6$ then both s_1 and s_2 lie in the left half of the s plane. When $\sqrt{25 - 4K} = 0$, or $K = \tfrac{25}{4}$, we have $s_1 = s_2 = -\tfrac{1}{2}$ (double pole), and for $K > \tfrac{25}{4}$, we have conjugate poles with real parts equal to $-\tfrac{1}{2}$. This *root locus* is shown in Fig. 7.21. We conclude that the system is stable for $K > 6$.

The reader can also easily check this result using the Routh criterion (App. D).

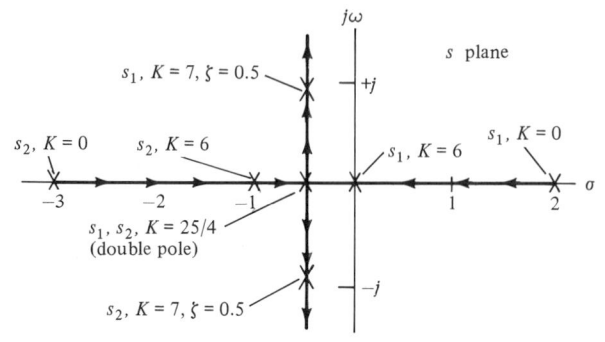

Figure 7.21. Root locus for Example 13 for $0 < K < \infty$.

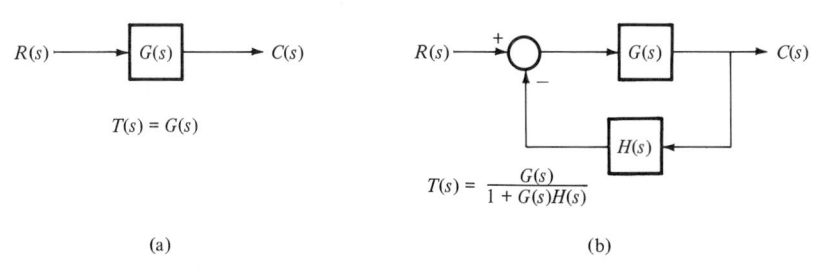

Figure 7.22. (a) Open-loop system. (b) Closed-loop system with negative feedback.

We might also ask the question: What value of K gives $\zeta = 0.5$ for this second-order system? Comparing Eqs. 2.90 and 7.33 shows that $2\zeta\omega_n = \omega_n = 1$ and $\omega_n^2 = K - 6 = 1$. Thus, $K = 7$, and for this value of K

$$s_1 = -\frac{1}{2} + j\frac{\sqrt{3}}{2}$$

$$s_2 = -\frac{1}{2} - j\frac{\sqrt{3}}{2}$$

It should be mentioned that very well defined techniques exist for rapidly sketching the *root locus*. They are beyond the scope of our treatment but are very valuable in the analysis and design of control systems.

Before leaving this section it is worthwhile to compare an open-loop system with a closed-loop system to illustrate some of the characteristics and advantages of introducing feedback.[7] An open-loop and closed-loop (negative feedback) system are shown in Fig. 7.22.

Example 14

Generally speaking, an open-loop system with transfer function, or process, $G(s)$ is subject to changes of all kinds that affect a control process in (more or less) a direct manner. A closed-loop system, however, can detect these changes in the process and attempt to correct the output. That is, if $G(s)H(s) \gg 1$ for all s of interest, then according to Eq. 7.31 (+ sign)

$$C(s) \approx \frac{1}{H(s)} R(s) \qquad (7.34)$$

and $C(s)$ depends on $H(s)$ rather than the process $G(s)$. Notice carefully that the requirement $G(s)H(s) \gg 1$ may make the system *unstable*. Increasing the loop transfer function $G(s)H(s)$, nevertheless, *reduces the effect of variations of parameters in the process*. This is extremely important for control systems, and in order to demonstrate it explicitly consider the effect of a small change (ΔG) in G. For the open-loop system

$$C + \Delta C = (G + \Delta G)R = GR + (\Delta G)R$$

and subtracting $C = GR$ we have

$$\Delta C = (\Delta G)R \qquad \text{(open loop)} \qquad (7.35)$$

For the closed-loop system

$$C + \Delta C = \frac{G + \Delta G}{1 + (G + \Delta G)H} R$$

and subtracting C as given in Eq. 7.31 (+ sign) we have after some simple algebra

$$\Delta C = \frac{\Delta G}{[1 + GH + (\Delta G)H][1 + GH]} R$$

or since $G \gg \Delta G$,

$$\Delta C = \frac{\Delta G}{(1 + GH)^2} R \qquad \text{(closed-loop)} \qquad (7.36)$$

Comparing Eqs. 7.35 and 7.36, we see that ΔC is reduced by the factor $(1 + GH)^2$ for the closed-loop system relative to the open-loop system! As pointed out above, ordinarily, $1 + GH \gg 1$ over the range of s of interest.

The *sensitivity* of the control system is defined as the ratio of the *percentage* change in the transfer function to the *percentage* change in some other function (of s) within the system. For example, the sensitivity to changes in $G(s)$ is

$$S_G = \frac{\Delta T/T}{\Delta G/G}$$

or, for small changes,

$$S_G = \frac{\partial T/T}{\partial G/G} = \frac{\partial T}{\partial G} \frac{G}{T} \qquad (7.37)$$

Thus, the sensitivity of the open-loop system to changes in G is *always* unity since $T = G$ for an open-loop system.

Example 14

Suppose that for the canonical feedback system of Fig. 7.20(c) the process $G(s)$ is an amplifier with gain A (assumed constant over the frequency range of interest), while the feedback transfer function $H(s)$ is a constant K (negative feedback). Thus,

$$T(s) = \frac{A}{1 + AK}$$

and

$$S_A = \frac{\partial T}{\partial A} \frac{A}{T}$$

or, skipping the details

$$S_A = \frac{1}{1 + AK} \tag{7.38}$$

Thus, for large A and large AK, the sensitivity of the amplifier with feedback is considerably reduced over the open-loop sensitivity (unity).

On the other hand the sensitivity to changes in K is

$$S_K = \frac{\partial T}{\partial K} \frac{K}{T}$$

or

$$S_K = -\frac{AK}{1 + AK} \tag{7.39}$$

and for large AK, $S_K \approx -1$. The last result shows very clearly that the *elements making up $H(s)$ ($= K$ here) must remain constant.* Otherwise, changes in K affect the output in a direct way. Put another way, in order to spend money for components for this amplifier in a wise manner, we should not skimp on the elements in $H(s)$. See Prob. 14 at the end of the chapter.

A control system is designed to provide a desired response, and the transient response is inevitably a part. If the transient response is not satisfactory, it must be altered. The *only* way this can be accomplished for an open-loop system is to alter $G(s)$, but for a closed-loop system we can alter $H(s)$ and leave $G(s)$ unchanged.

Example 15

An armature-controlled dc motor is shown in Fig. 7.28 (Prob. 4). It is easy to show that (see Prob. 37)

$$\Omega(s)/V_i(s) = \frac{K_t}{(Is + D)(L_a s + R_a) + K_t K_f}$$

or, if we ignore the armature inductance

$$\Omega(s)/V_i(s) = \frac{K_t}{(Is + D)R_a + K_t K_f}$$

The last result can be written

$$\Omega(s)/V_i(s) = \frac{K_1}{\tau_1 s + 1} \tag{7.40}$$

where

$$K_1 = \frac{K_t}{R_a D + K_t K_f}$$

and

$$\tau_1 = \frac{R_a I}{R_a D + K_t K_f} \quad \text{(time constant)} \tag{7.41}$$

In many practical applications the load inertia I is large and dominates the time constant τ_1. Now, suppose a unit-step voltage is applied so that

$$\Omega(s) = \frac{K_1}{s(\tau_1 s + 1)} = \frac{K_1/\tau_1}{s(s + 1/\tau_1)}$$

or

$$\omega(t) = K_1(1 - e^{-t/\tau_1}) \quad \text{(speed)} \tag{7.42}$$

If this response is too slow, and we want to decrease τ_1, then, according to Eq. 7.41, we can only increase K_t, and this simply requires a *larger* dc motor.

On the other hand we can accomplish the same thing in a more efficient manner by using feedback. The classical way to accomplish this is to use a tachometer, whose transfer function is ideally

$$H(s) = \frac{V(s)}{\Omega(s)} = K \tag{7.43}$$

and provides an *output voltage* for an *input angular velocity*. Thus, the tachometer generates a voltage proportional to shaft speed that can then be subtracted from the input and amplified. A relatively *inexpensive* (for low-power) solid-state amplifier can be used since it is in the forward loop (Fig. 7.23). The new

316 APPLICATIONS OF THE LAPLACE TRANSFORM

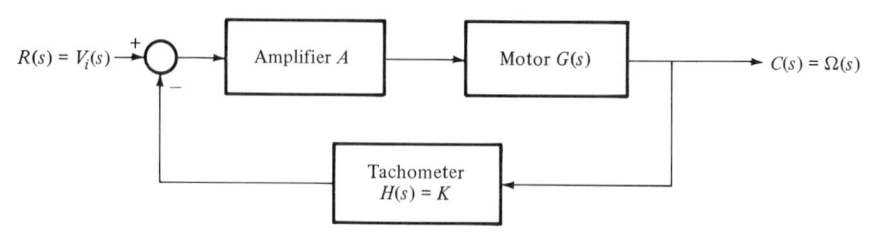

```
R(s) = V_i(s) →(+−)→ [Amplifier A] → [Motor G(s)] → C(s) = Ω(s)
                    ↑                                    |
                    └──[Tachometer H(s) = K]←────────────┘
```

Figure 7.23. Closed-loop speed control system.

(closed-loop) transfer function is

$$T(s) = \frac{C(s)}{R(s)} = \frac{AG(s)}{1 + AG(s)H(s)} = \frac{AG(s)}{1 + AKG(s)} \tag{7.44}$$

or, using Eq. 7.40 for $G(s)$

$$\frac{\Omega(s)}{R(s)} = \frac{AK_1}{\tau_1 s + 1 + AKK_1} = \frac{K_4}{\tau_2 s + 1} \tag{7.45}$$

where

$$K_4 = \frac{AK_1}{1 + AKK_1}$$

and

$$\tau_2 = \frac{\tau_1}{1 + AKK_1} \tag{7.46}$$

For a unit-step voltage

$$\Omega(s) = \frac{K_4}{s(\tau_2 s + 1)} = \frac{K_4/\tau_2}{s(s + 1/\tau_2)}$$

or

$$\omega(t) = K_4(1 - e^{-t/\tau_2}) \tag{7.47}$$

The new time constant τ_2 is normally much less (7.46) than τ_1 and can be decreased by increasing the amplifier gain A or the tachometer constant K. Thus, the use of feedback offers an efficient way to increase the speed of response of this speed control system.

Have we paid for this increased speed of response with increased sensitivity? For the open-loop system the transfer function can be written

$$T(s) = \frac{K_t}{B(s) + K_t K_f} \quad \text{(open-loop)}$$

from the equation above (7.40). Thus, for changes in the motor constant

$$S_{K_t} = \frac{\partial T}{\partial K_t} \frac{K_t}{T} = \frac{B(s)}{B(s) + K_t K_f} \quad \text{(open-loop)} \tag{7.48}$$

7.5 STABILITY AND FEEDBACK

The transfer function for the closed-loop system is obtained from Eq. 7.45 and those preceding, and it can be put in the form

$$T(s) = \frac{AK_t}{B(s) + (1 + A)K_t K_f} \quad \text{(closed-loop)}$$

Thus

$$S_{K_t} = \frac{B(s)}{B(s) + (1 + A)K_t K_f} \quad \text{(closed-loop)} \tag{7.49}$$

and comparing Eqs. 7.48 and 7.49 we see that insofar as the motor constant K_t is concerned, the sensitivity has decreased. This is an additional benefit.

Another benefit of feedback is the reduction in the effect of extraneous signals such as noise (unless, of course, the unwanted signal enters with the input). Suppose that the mechanical load on the armature-controlled motor of Example 15 is an antenna that is subject to wind loading such that the load torque is now given by

$$T_l(s) = T_m(s) - T_w(s)$$

where $T_m(s)$ is the motor torque given by

$$T_m(s) = K_t I_a(s)$$

The load torque is related to the angular velocity by

$$\Omega(s) = \frac{1}{Is + D} T_l(s)$$

while, if $L_a = 0$ (as before),

$$V_i(s) = R_a I_a(s) + V_b(s)$$

The block diagram of this system without tachometer feedback is asked for in Prob. 37 ($L_a \neq 0$). Here, it is sufficient to find $\Omega(s)/T_w(s)$ when $V_i(s) = 0$ in order to determine the effect of the wind loading. Solving the preceding equation gives

$$\frac{\Omega(s)}{T_w(s)} = \frac{-1/(Is + D)}{1 + K_t K_f/(Is + D)R_a} = \frac{-1}{Is + D + K_t K_f/R_a}$$

and for a sudden wind gust in the form of a unit step of torque:

$$\Omega(s) = \frac{-1}{s(Is + D + K_t K_f/R_a)}$$

The steady-state value of $\omega(t)$ caused by this gust is found using the final-value theorem:

$$\omega(\infty) = \lim_{s \to 0} \frac{-1}{Is + D + K_t K_f/R_a} = \frac{-1}{D + K_t K_f/R_a} \tag{7.50}$$

318 APPLICATIONS OF THE LAPLACE TRANSFORM

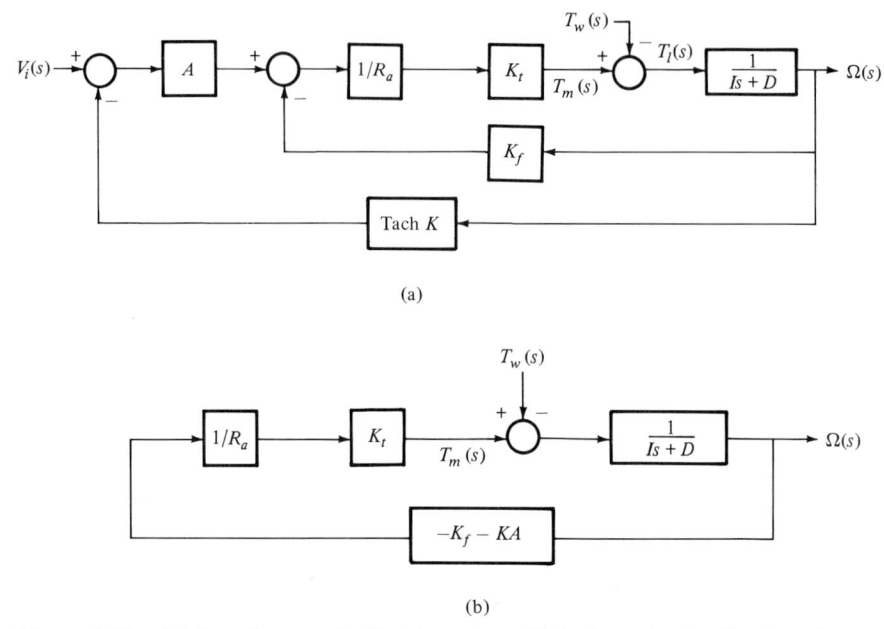

Figure 7.24. (a) Armature-controlled dc motor with tachometer feedback and wind loading. (b) Reduced form of (a) for $V_i(s) = 0$.

Example 16

Compare the steady-state speed found in Eq. 7.50 with that for the same system with tachometer feedback shown in Fig. 7.24. If we were looking for $\Omega(s)/V_i(s)$, we would have serious problems. We could solve (simultaneously) the given equations algebraically, or reduce the block diagram, or convert the block diagram to a signal flow graph and use Mason's gain formula.[†] Fortunately we are seeking $\Omega(s)/T_w(s)$ with $V_i(s) = 0$, which means that the first summer in Fig. 7.24 is not needed. We must, however, keep the minus sign from the tachometer, and this can be done by simply changing the tachometer constant to $-K$. Having done this, A and $-K$ are in cascade so that one block with transfer function $-KA$ suffices. But this then leaves us with parallel feedback paths whose transfer functions are additive. Thus, we can *reduce* the block diagram of Fig. 7.24(a) to that of Fig. 7.24(b). It is then very simple to further reduce Fig. 7.24(b) to canonical form since by inspection

$$G(s) = \frac{1}{Is + D}$$

and

$$H(s) = -K_t(K_f + KA)/R_a$$

[†] Except that we decided not to study these techniques.

so that

$$\frac{\Omega(s)}{-T_w(s)} = \frac{G(s)}{1 - G(s)H(s)} = \frac{1/(Is+D)}{1 + \dfrac{K_t(K_f + KA)}{(Is+D)R_a}}$$

or

$$\frac{\Omega(s)}{T_w(s)} = \frac{-1}{Is + D + (K_f + KA)K_t/R_a}$$

If $T_w(s) = 1/s$, as in the case for no-tachometer feedback, then

$$\Omega(s) = \frac{-1}{s[Is + D + (K_f + KA)K_t/R_a]}$$

and

$$\omega(\infty) = \frac{-1}{D + (K_f + KA)K_t/R_a} \qquad (7.51)$$

by the final-value theorem. A comparison of Eqs. 7.50 and 7.51 shows that $(KA \gg 1)$ the effect of wind gust is much less for the closed-loop case.

Briefly summarizing, we have seen that feedback can reduce the effect of spurious changes in the parameters that determine the process $G(s)$. It can afford better control of the transient response and can reduce the effect (on the output) of unwanted signals such as noise. The steady-state error in a system can be significantly reduced by using feedback. This is demonstrated in the solution to Prob. 38. These are not the only benefits gained by using feedback but are certainly major ones. Do not forget that stability must also be considered, and many times we are forced to compromise some characteristics to gain or ensure stability.

7.6 STATE VARIABLES (AGAIN)

It was pointed out in Chap. 3 that the use of state variables provides us with a very organized way of treating a linear system with several inputs and (or) outputs. Although the primary use of state variables is in time-invariant linear systems where we desire a time-domain numerical solution, it is still worthwhile to look at state variables in the complex frequency (s) domain.

The normal-form equations can be written as the matrix equation

$$\mathbf{q}'(t) = \mathbf{A}\mathbf{q}(t) + \mathbf{B}\mathbf{x}(t) \qquad (7.52)$$

as in Chap. 3. Taking the Laplace transform of Eq. 3.34 leaves us with the transformed matrix equation

$$s\mathbf{Q}(s) - \mathbf{q}(0^-) = \mathbf{A}\mathbf{Y}(s) + \mathbf{B}\mathbf{X}(s) \qquad (7.53)$$

Transposing, and using the identity matrix for proper matrix manipulation

$$s\mathbf{Q}(s) - \mathbf{AQ}(s) = \mathbf{BX}(s) + \mathbf{q}(0^-)$$

or

$$(s\mathbf{I} - \mathbf{A})\mathbf{Q}(s) = \mathbf{BX}(s) + \mathbf{q}(0^-) \tag{7.54}$$

$$\mathbf{Q}(s) = (s\mathbf{I} - \mathbf{A})^{-1}\mathbf{BX}(s) + (s\mathbf{I} - \mathbf{A})^{-1}\mathbf{q}(0^-) \tag{7.55}$$

Notice the need for matrix inversion, and also notice that Eq. 7.55 is an algebraic equation. An example best demonstrates the use of Eq. 7.55.

Example 17

Repeat Example 5 (second part), Chap. 3, where $i_s = 9.6e^{-4t}u(t)$, $x_1 = i_s$, $x_2 = 0$, and

$$\mathbf{A} = \begin{bmatrix} -0.5 & -1.25 \\ 1 & -3.5 \end{bmatrix} \qquad \mathbf{B} = \begin{bmatrix} 1.25 \\ 0 \end{bmatrix}$$

We have

$$(s\mathbf{I} - \mathbf{A}) = \begin{bmatrix} s & 0 \\ 0 & s \end{bmatrix} - \begin{bmatrix} -0.5 & -1.25 \\ 1 & -3.5 \end{bmatrix} = \begin{bmatrix} s+0.5 & 1.25 \\ -1 & s+3.5 \end{bmatrix}$$

and

$$(s\mathbf{I} - \mathbf{A})^{-1} = \frac{1}{s^2 + 4s + 3}\begin{bmatrix} s+3.5 & -1.25 \\ 1 & s+0.5 \end{bmatrix}$$

Notice that the denominator of the preceding equation, when set equal to 0, is the characteristic equation. Using Eq. 7.55 we have

$$\mathbf{BX}(s) = \begin{bmatrix} 1.25 \\ 0 \end{bmatrix}[X_1(s)] = \begin{bmatrix} 1.25X_1(s) \\ 0 \end{bmatrix} = \begin{bmatrix} 1.25I_s(s) \\ 0 \end{bmatrix}$$

$$\mathbf{Q}(s) = \frac{1}{s^2 + 4s + 3}\begin{bmatrix} s+3.5 & -1.25 \\ 1 & s+0.5 \end{bmatrix}\begin{bmatrix} 1.25I_s(s) \\ 0 \end{bmatrix}$$

$$+ \frac{1}{s^2 + 4s + 3}\begin{bmatrix} s+3.5 & -1.25 \\ 1 & s+0.5 \end{bmatrix}\begin{bmatrix} 0 \\ 0 \end{bmatrix}$$

Carrying out the matrix multiplication, with $I_s(s) = 9.6/(s+4)$:

$$Q_1(s) = 12\frac{s+3.5}{(s+1)(s+3)(s+4)}$$

$$Q_2(s) = 12\frac{1}{(s+1)(s+3)(s+4)}$$

Using the partial fraction expansion or the table of pairs

$$q_1(t) = v_C(t) = (5e^{-t} - 3e^{-3t} - 2e^{-4t})u(t)$$
$$q_2(t) = i_L(t) = (2e^{-t} - 6e^{-3t} + 4e^{-4t})u(t)$$

which agrees with the results in Example 5, Chap. 3.

This method may look simple and concise, but do not forget that the matrix $(s\mathbf{I} - \mathbf{A})$ must be inverted. This inversion, or its equivalent, must be performed, however, regardless of what scheme we use to solve the simultaneous equations of the system, so we are led back to the main advantage of state-variable analysis: *organization*.

7.7 CONCLUDING REMARKS

The Laplace transform is a very popular and widely used technique for analyzing time-invariant linear systems. Its main advantage over other schemes lies in the fact that it explicitly leaves a place for initial conditions to be inserted, and this is ideal for transient-type problems. It is also very convenient to describe the behavior of a linear system in terms of s. The control system engineer, for example, is interested in the stability of a system, and much of his analysis and design is carried out using s-domain techniques. Nyquist and root-locus analysis are two examples, and these can be studied successfully if the material in the preceding chapters has been mastered. A good knowledge of the Laplace transform, including a respect for its advantages and its disadvantages, opens the way for further study of linear systems.

The examples of this chapter were primarily those that involve electric circuits with time (t) as the lone independent variable. There are many other situations that arise where the Laplace transform is useful, and some of these are examined in the problems at the end of this chapter. The reader is encouraged to investigate these to obtain a deeper understanding of the Laplace transform technique.

PROBLEMS

1. (a) Find the transfer function of the lead compensator shown in Fig. 7.25.
   (b) Find its unit-impulse response.

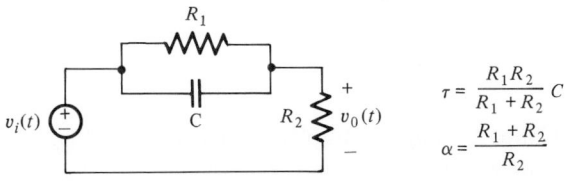

$$\tau = \frac{R_1 R_2}{R_1 + R_2} C$$

$$\alpha = \frac{R_1 + R_2}{R_2}$$

Figure 7.25. Lead compensator.

322 APPLICATIONS OF THE LAPLACE TRANSFORM

2. Find the transfer function of the lag compensator shown in Fig. 7.26.

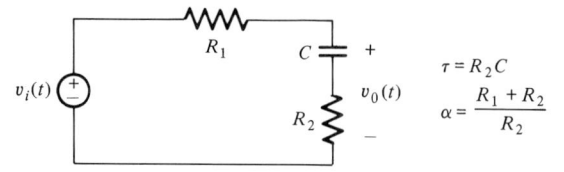

$$\tau = R_2 C$$
$$\alpha = \frac{R_1 + R_2}{R_2}$$

Figure 7.26. Lag compensator.

3. Find the transfer function of the lead-lag compensator of Fig. 7.27.

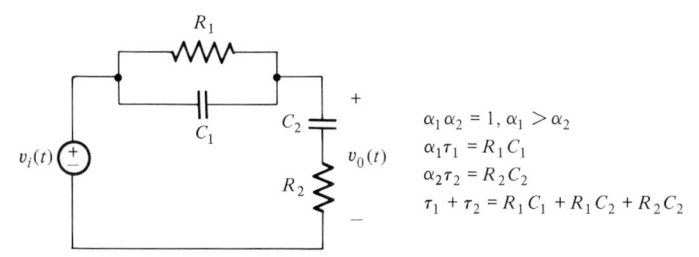

$$\alpha_1 \alpha_2 = 1, \; \alpha_1 > \alpha_2$$
$$\alpha_1 \tau_1 = R_1 C_1$$
$$\alpha_2 \tau_2 = R_2 C_2$$
$$\tau_1 + \tau_2 = R_1 C_1 + R_1 C_2 + R_2 C_2$$

Figure 7.27. Lead-lag compensator.

4. Find the unit-step response for the shaft position (θ) of the *armature-controlled dc motor* in Fig. 7.28 if the back emf is given by

$$v_b(t) = K_f (d\theta/dt)$$

where K_f is a constant, and the current-torque relation is (constant field current)

$$\tau = K_t i_a(t) = I \frac{d^2\theta}{dt^2} + D \frac{d\theta}{dt}$$

K_t is a constant, I is the combined armature-load inertia, and D is the viscous friction (see App. E).

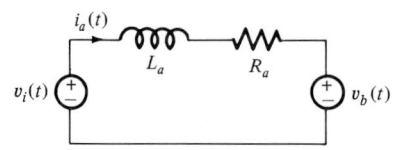

Figure 7.28. dc motor armature circuit.

5. (a) Repeat Example 8, Chap. 2, using the Laplace transform and Fig. 7.29. Find $i(t)$ for $V_0 u(t) = Ri + L\,di/dt$.
   (b) Repeat Example 9, Chap. 2. Find $v_C(t)$.

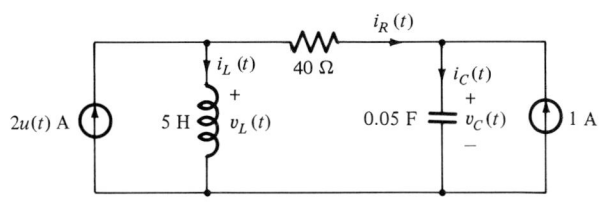

Figure 7.29. Circuit for Prob. 5(b).

6. Obtain Eq. 2.108 for Example 10, Chap. 2, using the Laplace transform, How does this method compare in ease of use to that in Chap. 2? The differential equation is

$$2\frac{d^2z}{dt^2} + 4.8\frac{dz}{dt} + 8z = 10u(t)$$

and Eq. 2.108 is

$$z_{zs}(t) = 1.25\,\{1 - 0.625e^{-1.2t}[1.2\sin(1.6t) + 1.6\cos(1.6t)]\}\,u(t)$$

7. Show that the Laplace transform of the unit-step response of a fixed linear system is

$$Y_u(s) = H(s)/s$$

8. Use the Laplace transform to solve Prob. 21, Chap. 2. Find the fundamental set for

$$\frac{d^3y}{dt^3} + 4\frac{d^2y}{dt^2} + 6\frac{dy}{dt} + 4y = x$$

9. Repeat Prob. 6, Chap. 3, using the Laplace transform. Find y_1 and y_2 when $z_1 = 4u(t)$ and $z_2 = -3u(t)$.
10. Repeat Prob. 16, Chap. 3, using the Laplace transform and Fig. 7.30. Find $v_C(t)$.

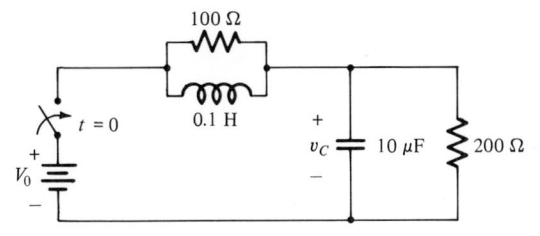

Figure 7.30. Circuit for Prob. 10.

11. (a) Plot the unit-step response for the shunt-peaking coupling network of Fig. 7.31 for $R = 5K$, $C = 10$ pF, and $Q_2 = 0, 0.25, \sqrt{2} - 1, 0.5, 0.75, 1.0$.
    (b) Find the 10 to 90 percent (of final value) *rise time* for each value of Q_2.

324 APPLICATIONS OF THE LAPLACE TRANSFORM

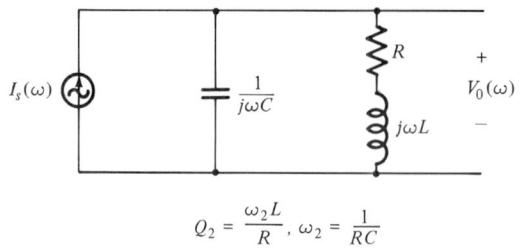

$$Q_2 = \frac{\omega_2 L}{R}, \quad \omega_2 = \frac{1}{RC}$$

Figure 7.31. Shunt-peaking network.

(c) Find the rise time-bandwidth product (see Prob. 29, Chap. 4) for each value of Q_2. Comment.

12. The sawtooth waveform of Fig. 7.32 is applied (with $T = 10^{-6}$ s) at $t = 0$ as the input to the shunt-peaking network of Fig. 7.31 with $R = 5K$, $C = 10\,\text{pF}$, and $Q_2 = \sqrt{2} - 1$. Find $v_0(t)$. Is the transient part of the response easily discernible? Compare results to those of Prob. 20, Chap. 5.

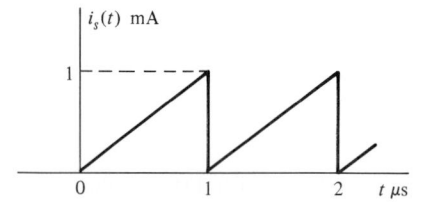

Figure 7.32. Sawtooth input.

13. (a) Find the transfer function $H(s) = Y_0(s)/Y_1(s)$ for the accelerometer of Fig. 7.33 if $d^2 y_1/dt^2$, the case acceleration, is the forcing function, and $y_0 = y_2 - y_1$ is the output (see App. E.):

$$-M\frac{d^2 y_1}{dt^2} = M\frac{d^2 y_0}{dt^2} + D\frac{dy_0}{dt} + Ky_0$$

(b) Show that $H(\omega) \approx \dfrac{\omega^2}{K/M}$ for $\omega \ll \omega_n$.

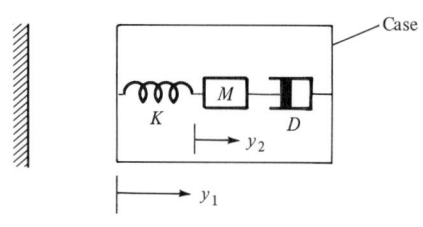

Figure 7.33. Accelerometer.

14. The symbol that is used to represent an operational amplifier (op-amp) is shown in Fig. 7.34(a). It is shown connected as an inverting amplifier in Fig. 7.34(b). If the gain (A) of the op-amp and its input impedance are both

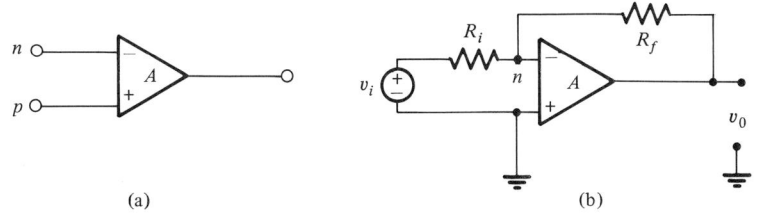

Figure 7.34. (a) Operational amplifier symbol. (b) Inverting op-amp.

very high, show that $V_o/V_i = -R_f/R_1$. Sketch the block diagram in canonical form.

15. If the inductor insulation can withstand only 4000 V, will it survive (see Fig. 7.35)?

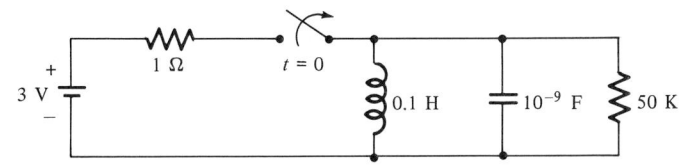

Figure 7.35. Circuit for Prob. 15.

16. A (linearized) block diagram model of a *phase-lock loop* that is widely used in communication systems is shown in Fig. 7.36. Its function is to closely maintain zero *difference* in phase between θ_o and θ_i.
    (a) Find the maximum permissible value of gain $(K_1 K_2)$ for a stable system.
    (b) Find the gain for an error (phase difference) of $1°$ ($\pi/180$ rad) when $\theta(t) = 50tu(t)$.

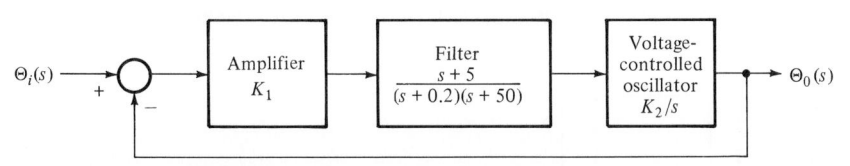

Figure 7.36. Linearized model of a phase-lock loop.

17. Repeat Example 12, transforming with t, and then *Laplace transforming* with z [i.e., $g(z) \leftrightarrow G(p)$].
18. Find $i_2(t)$, $t > 0$, in Fig. 7.37.

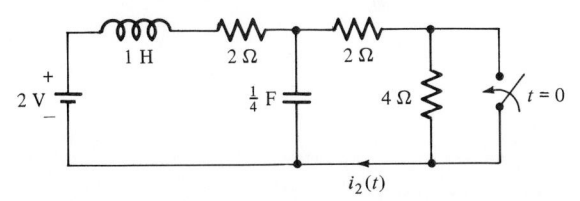

Figure 7.37. Circuit for Prob. 18.

19. (a) If $v_i(t) = V_m \cos(\omega t + \theta)$ in Fig. 7.38. Find θ such that there is no transient in $v_C(t)$, $t > 0$.
    (b) Repeat if $v_C(0) = V_0$.

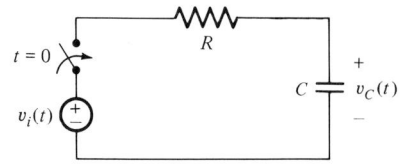

Figure 7.38. Circuit for Prob. 19.

20. What is the maximum power absorbed by the 10-Ω resistor in Figure 7.39 if $v(t) = 100u(t)$?

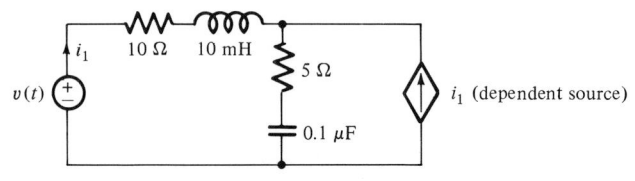

Figure 7.39. Network for Prob. 20.

21. (a) What is the transfer function in Example 8 if $v_2(t) = i_2(t)R_2$ is the output and unity coupling is assumed?
    (b) Find $v_2(t)$ if the battery is replaced by the voltage source $v_s(t) = 10\cos(\omega t)$. Find $V_2(s)$ and invert.
    (c) Repeat (b) using convolution.

22. The pulse train of Fig. 7.40(a) is applied to the RC circuit of Fig. 7.40(b). Find $v_0(t)$ for the first few applied pulses. Is the system acting as an integrator?

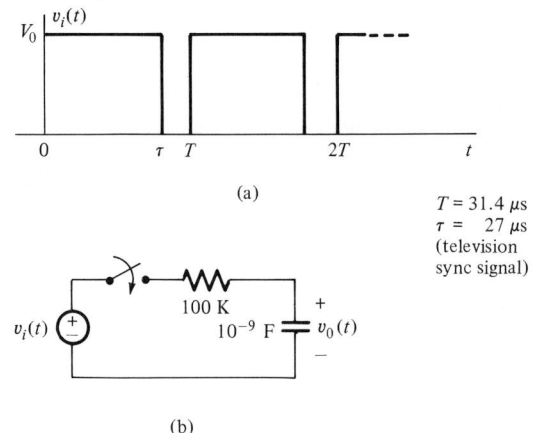

$T = 31.4\ \mu s$
$\tau = 27\ \mu s$
(television sync signal)

Figure 7.40. (a) Pulse train. (b) Integrator circuit.

23. The pulse train of Fig. 7.41(a) is applied to the CR circuit of Fig. 7.41(b).

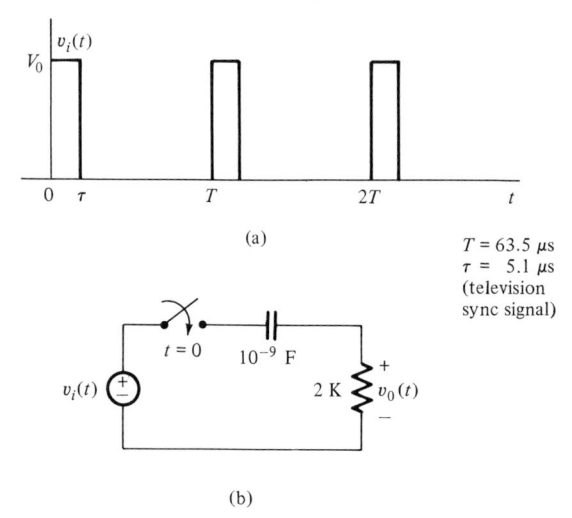

Figure 7.41. (a) Pulse train. (b) Differentiator circuit.

$T = 63.5\ \mu s$
$\tau = 5.1\ \mu s$
(television sync signal)

Find $v_0(t)$ for the first few applied pulses. Is this system acting as a differentiator?

24. The output of a time-invariant linear system with no initial energy storage is $y(t) = (3te^{-2t} + 4e^{-3t})u(t)$ when the input is $tu(t)$. What is its unit-impulse response?

25. The switch in the circuit of Fig. 7.42 closes at $t = 0$ and opens at $t = 1$ s. Find the voltage across the switch for all t. Sketch $v_s(t)$ versus t. What happens to $v_s(t)$ if R is increased indefinitely?

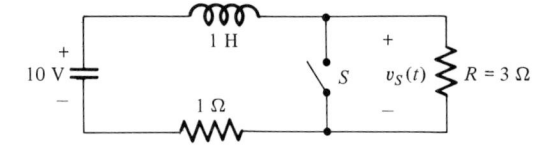

Figure 7.42. Circuit for Prob. 25.

26. Find $I_L(s) = I_1(s) - I_2(s)$ in Example 4 using the state-variable method of Sec. 7.6. Use the numbers that are given.
27. (a) Find $V_2(s)$ in Fig. 7.43 using the state-variable method of Sec. 7.6, (v_1, v_2, i_L).
    (b) Find $H(s)$.

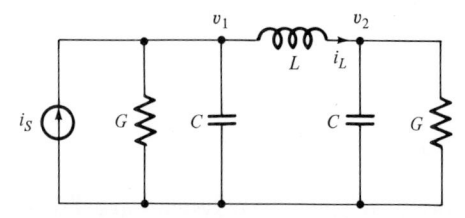

Figure 7.43. Filter circuit for Prob. 27.

328 APPLICATIONS OF THE LAPLACE TRANSFORM

28. Find $v_0(t)$ for $t > 0$ in Fig. 7.44 if $v_s(t) = 10\cos(10^3 t)$.

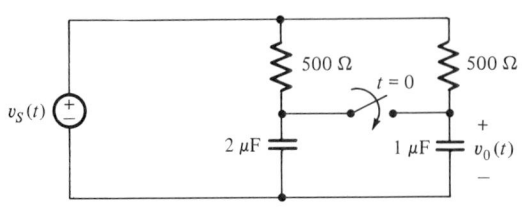

Figure 7.44. Circuit for Prob. 28.

29. (a) The sine pulse of Fig. 7.45(a) is applied to the RL circuit of Fig. 7.45(b). Find $i(t)$.
  (b) Repeat using the results of Prob. 23, Chap. 6, with $N = 4$.
  (c) Plot the results of (a) and (b).

(a)

(b)

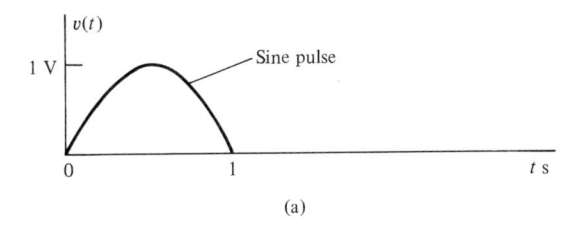

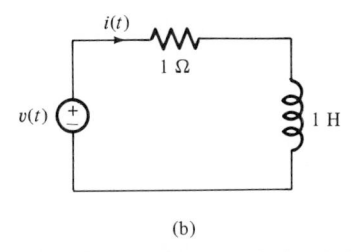

Figure 7.45. Input voltage and circuit for Prob. 29.

30. The transfer function for some linear systems are listed below. What are the roots of the characteristic equation? Plot these in the s plane. Determine which systems are stable without using the final-value theorem.

(a) $\dfrac{1}{s+a}$, $a > 0$ (b) $\dfrac{1}{(s+a)(s+b)}$, $a, b > 0$

(c) $\dfrac{s}{(s-a)(s+b)}$, $a, b > 0$ (d) $\dfrac{1}{(s+\alpha+j\beta)(s+\alpha-j\beta)}$, $\alpha, \beta > 0$

31. The roots of the characteristic equations for some linear systems are shown in the s plane in Fig. 7.46. Which are stable systems?

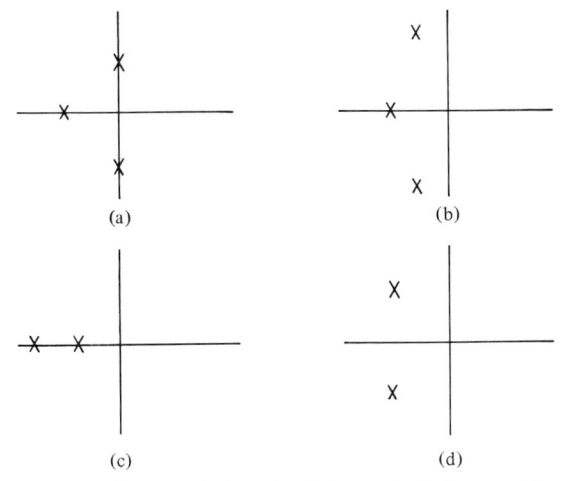

Figure 7.46. Loci of roots of characteristic equations in the s plane.

32. The transfer functions or unit-impulse reponses for some linear systems are given below. Which are stable systems?

(a) $h(t) = J_0(at)$ (b) $h(t) = I_0(at)$ (c) $H(s) = \dfrac{1}{(s^2 + a^2)^2}$

(d) $H(s) = \dfrac{1}{s\sqrt{s+a}}$ (e) $H(s) = \dfrac{e^{-a\sqrt{s}}}{s}$ (f) $h(t) = J_1(at)$

33. Solve the integral equation

$$y(t) = t^2 + \int_0^t y(\tau)\sin(t-\tau)\,d\tau$$

34. Solve the integral equation

$$t = y(t) - 2\int_0^t y(\tau)\cos(t-\tau)\,d\tau$$

35. Use the Laplace transform to find the differential equation for u for the loudspeaker of Fig. 7.47 and Eqs. E.26 and E.27 (App. E). See Prob. 41, Chap. 2.

$$L\dfrac{di}{dt} + Ri + (2\pi rBN)u = v_s(t)$$

$$M\dfrac{du}{dt} + Du + \dfrac{1}{K}\int u\,dt - (2\pi rBN)i = 0$$

Figure 7.47. Loudspeaker.

APPLICATIONS OF THE LAPLACE TRANSFORM

36. In the *field-controlled* dc motor the armature current is constant and the current-torque relation is

$$\tau = K_m i_f(t) = I\frac{d^2\theta}{dt^2} + D\frac{d\theta}{dt}$$

where K_m is a constant. See Fig. 7.48.

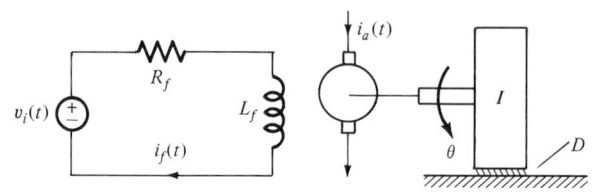

Figure 7.48. Field-controlled dc motor.

(a) Find the transfer function $H(s) = \Theta(s)/V_i(s)$.
(b) Show that the steady-state *power gain* is K_m^2/DR_f.

37. (a) Draw a block diagram (s domain) of the armature-controlled dc motor of Prob. 4.
(b) Repeat if the mechanical load is an antenna that is subject to external wind load such that the motor torque is

$$T_m(s) = T_l(s) + T_w(s)$$

where $T_l(s)$ is the load torque and $T_w(s)$ is the wind torque.
(c) Repeat (b) for the field-controlled dc motor of Prob. 36.

38. The process transfer function for a certain open-loop system is $G(s) = K/(\tau s + 1)$.
(a) Find the steady-state error for a unit-step input and $K = 1$.
(b) Repeat for the closed-loop system with the same $G(s)$ and *unity* negative feedback, $H(s) = 1$, when $K = 100$.
(c) Find the relative *changes* in the steady-state errors in (a) and (b) when K changes by 5 percent in both cases and comment on the results.

39. (a) Plot the pole-zero location for the lead compensator of Prob. 1 if $\alpha = 10$ and $\tau = 0.1$.
(b) Sketch the asymptotic Bode plots.

40. Repeat Prob. 39 for the lag compensator of Prob. 2.

41. Repeated Prob. 39 for the lead-lag compensator of Prob. 3 if $\alpha_1 = 5$, $\alpha_2 = 0.2$, $\tau_1 = 1$, and $\tau_2 = 0.1$.

42. A unity (negative) feedback system in canonical form, Fig. 7.20(c), has $G(s) = K/s^2$.
(a) Show that the closed-loop system is marginally stable (and therefore unstable) for any K.
(b) Suppose that originally $K = 10$, and a lead compensator $G_l(s)$ with $\alpha = 5$ and $\tau = 0.1$ is added in cascade with $G(s)$. K(gain) is increased to 50 to compensate for the compensator attenuation ($1/\alpha = 0.2$). Show that the compensated system is stable.

43. The canonical feedback system of Fig. 7.22(b) has $G(s) = 1/(s + \alpha)$ and $H(s) = 1/(s + \beta)$.
    (a) Prove that the closed-loop system is always stable provided α and β are both positive.
    (b) Find the differential equation that describes this linear system.
44. (a) Convert the generic second-order system of Fig. 7.49 and Eq. 2.89 into canonical feedback form with the transformed input signal first passing through a block that is a multiplier ω_n^2.
    (b) Repeat (a), obtaining a unity feedback system.

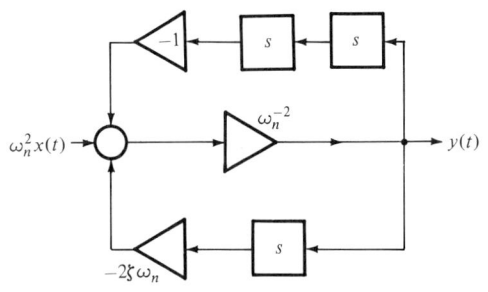

Figure 7.49. Second-order system.

45. Convert the canonical feedback system of Fig. 7.22(b) into a unity feedback system.
46. A unity (negative) feedback system has $G(s) = K/s(s + \alpha)$. Find K and α for this system such that the step response is as rapid as possible for a maximum overshoot of 4 percent and a maximum settling time of 1 s.

References

1. Hayt, W. H., Jr., and Kemmerley, J. E. *Engineering Circuit Analysis*, 3rd ed. New York: McGraw-Hill, 1978.
2. Cheng, D. K. *Analysis of Linear Systems*. Reading, Mass.: Addison-Wesley, 1959.
3. Goldman, S. *Transformation Calculus and Electrical Transients*. New York: Prentice-Hall, 1949.
4. Neff, H. P., Jr. *Basic Electromagnetic Fields*. New York: Harper & Row, 1981.
5. Spiegel, M. R. *Mathematical Handbook*, Schaum Outline Series. New York: McGraw-Hill, 1968.
6. McGillem, C. D., and Cooper, G. R. *Continuous and Discrete Signal and System Analysis*. New York: Holt, Rinehart and Winston, 1974.
7. Dorf, R. C. *Modern Control Systems*, 3rd ed. Reading, Mass.: Addison-Wesley, 1980.

Chapter 8
Discrete-Time Systems

The discrete-time linear system was introduced in Chap. 1. The independent variable t appears only at discrete instants of time: $t = nT$; $t = 0, T, 2T, \ldots$. The dependent variable $x(nT)$ has meaning only at these instants of time, hence it is a *sequence of numbers*. $x(nT)$ might be the discrete-time input signal to a discrete-time linear system, for example. There are many linear systems in which the data appear naturally in the form of discrete samples from pulses of short duration (sampled data systems). A tracking radar system and a pulse amplitude-modulated communication system are examples. Both continuous (analog) and discrete (digital) signals occur frequently in a natural form or as generated by man. Consider the temperature around us. This is certainly a continuous signal, but we are usually only aware of its discrete value at certain instants of time that are not necessarily uniformly spaced. The meteorologist may be interested in temperature on a continuous basis, but he is usually satisfied with a knowledge of its value at discrete times. Indeed, he is more than satisfied if he can sample at a high enough rate.[†] One might classify the observation of Halley's comet as a (natural) discrete event, since it appears only at uniform intervals.

[†] According to the *sampling theorem* (Chap. 5).

DISCRETE-TIME SYSTEMS

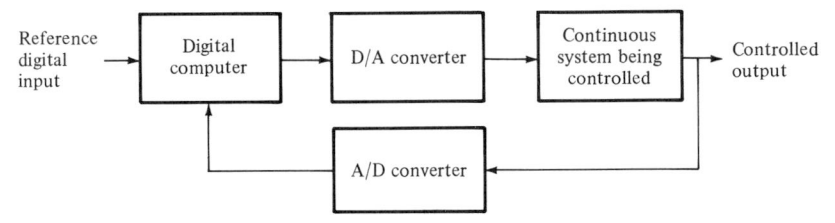

Figure 8.1. Digitally controlled system.

Continuous and discrete systems can be connected by analog-to-digital (A/D) converters or digital-to-analog (D/A) converters. These (or their equivalents) appear naturally and otherwise. The human eye, for example, has the ability to perform the D/A function. Because of *persistence of vision* the eye (and brain) interprets a motion picture as a continuous system when, in fact, it is a digital system. The display on a television screen is essentially discrete (an array of dots), but the eye interprets a continuous picture because of what is called *visual acuity*.

Our primary interest here, of course, lies in those systems that can be observed and (or) altered. A good example of this is a modern control system where the device actually performing the control is the digital computer (definitely a discrete system in itself). The system being controlled is usually a continuous-time system. Such a control system is shown in Fig. 8.1. This particular system, which is typical, requires both a D/A converter and an A/D converter as interfaces to and from, respectively, the continuous system being controlled. The A/D conversion can be accomplished by using a uniform sampler that can be thought of as simply a switch that closes for an instant every Ts (see Fig. 8.2(c)). A simple D/A converter consists of a *zero-order hold*[†] (ZOH) followed by a low-pass filter. The ZOH simply takes a (digital) sample value and holds that level until another sample is applied (see Fig. 8.2(d)). The ZOH produces a stair-step signal from a sequence of samples. The low-pass filter is required to remove the "ripple" at the output of the ZOH.

The procedure that will be followed in this chapter is identical to that followed for the continuous-time system (Chap. 2). We will find a zero-state (forced) response and a zero-input (free) response for the discrete system by first finding the *unit-function response* of the system. This corresponds to the unit-impulse response for continuous systems. The zero-state response will be given by a convolution *summation* that corresponds to convolution (integration) for continuous systems. We can then find the response for *any* input (source) sequence.

Discrete systems are often described by *linear difference equations* in much the same way that continuous systems are described by linear differential equations. We will solve these difference equations with the technique that was briefly described in the preceding paragraph. We will find a very evident

---

[†] See Prob. 38, Chap. 2.

correspondence (or analogy) between discrete and continuous systems throughout this chapter. The reader will find that the discrete-time system is actually easier to describe and analyze than the continuous-time system, even though he or she may not be as familiar with it.

8.1 THE RELATION BETWEEN CONTINUOUS-TIME AND DISCRETE-TIME SYSTEMS

The uniform sampler and zero-order hold can be combined to convert a continuous signal into a stair-step approximation of that signal. The stair-step signal then involves only *differences* in the signal. This is demonstrated in Fig. 8.2. We would now like to relate continuous systems to discrete systems by applying $x_1(t)$ to a continuous system with a transfer function $H(s)$. In order to keep the discussion as simple as possible, we will consider a first-order system:

$$\frac{dy}{dt} + ay = x(t)$$

whose unit-impulse response is

$$h(t) = e^{-at}u(t)$$

Higher-order systems will have unit-impulse responses containing exponentials (assuming no repeated roots), and can be treated (superposition) in a manner similar to what follows. The complete response is

$$y(t) = y_{zi}(t) + y_{zs}(t)$$

$$y(t) = Ae^{-at} + \int_0^t x(\tau)e^{-a(t-\tau)}\, d\tau$$

$$y(t) = Ae^{-at} + e^{-at}\int_0^t x(\tau)e^{a\tau}\, d\tau$$

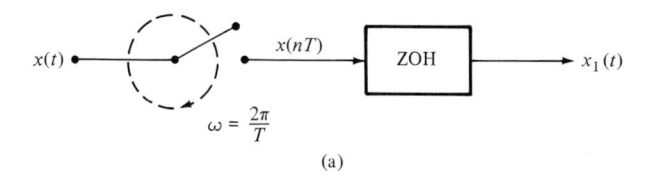

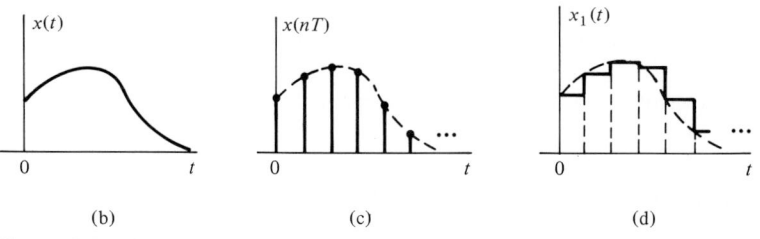

Figure 8.2. (a) Approximating $x(t)$ with a stair step. (b) Continuous signal, $x(t)$. (c) Sampled signal, $x(nT)$. (d) Stair-step approximation to $x(t)$, $x_1(t)$.

8.1 THE RELATION BETWEEN CONTINUOUS-TIME AND DISCRETE-TIME SYSTEMS

We must know $y(0) = y(0T)$, so

$$y(t) = e^{-at}y(0T) + e^{-at}\int_0^t x(\tau)e^{a\tau}\,d\tau$$

At the instant $t = T$, using Fig. 8.2(d),

$$y(1T) = e^{-aT}y(0T) + e^{-aT}\int_0^T x(0T)e^{a\tau}\,d\tau$$

which gives

$$y(1T) = e^{-aT}y(0T) + \frac{1 - e^{-aT}}{a}x(0T)$$

At the instant $t = 2T$ $(n = 2)$

$$y(2T) = e^{-2aT}y(0T) + e^{-2aT}\int_0^{2T} x(\tau)e^{a\tau}\,d\tau$$

$$= e^{-2aT}y(0T) + e^{-2aT}\int_0^T x(0T)e^{a\tau}\,d\tau + e^{-2aT}\int_T^{2T} x(1T)e^{a\tau}\,d\tau$$

which can be put in the form

$$y(2T) = e^{-aT}y(1T) + \frac{1 - e^{-aT}}{a}x(1T)$$

Continuing this process and comparing results for the response *at the various sampling times* shows that, in general

$$y(nT + T) = e^{-aT}y(nT) + \frac{1 - e^{-aT}}{a}x(nT)$$

This is called a *first-order linear difference equation* since only differences of the dependent variable appear. It also shows a *recursive* relation. Generally speaking, a *recursive system* is one for which the output $y(nT)$ depends not only on the input $x(nT)$, but also on the *past* values of the output. On the other hand a *nonrecursive* system is one in which the output depends only on the input $x(nT)$ and its past values (see Eq. 8.44).

If T is small enough

$$e^{-aT} \approx 1 - aT$$

which gives

$$y(nT + T) \approx (1 - aT)y(nT) + Tx(nT)$$

On the other hand the original differential equation can be approximated in the usual way:

$$\frac{y(nT + T) - y(nT)}{T} + ay(nT) \approx x(nT)$$

which obviously gives the *same* difference equation. Thus, the sampler and ZOH have converted the differential equation description of the system into a difference equation description. We will see this again.

8.2 THE UNIT FUNCTION AND UNIT SEQUENCE

The generic input function for discrete-time linear system analysis is the Kronecker delta or *unit function*, defined as (see Eq. 5.81)

$$\delta_k(nT) = \begin{cases} 1 & n = 0 \\ 0 & n \neq 0 \end{cases} \tag{8.1}$$

For our purposes here we will change the notation slightly. First, we let the clock period, or time between samples, T, be unity so that it is not even necessary to write it down. Thus, when we see the argument n (or perhaps some other lowercase letter such as m, but not t or τ) alone, we will know that this represents the discrete instant nT. Second, we drop the subscript k. It will be clear from the context that we mean the unit function, not the unit-impulse function. Thus, we have

$$\delta(n) = \begin{cases} 1 & n = 0 \\ 0 & n \neq 0 \end{cases} \tag{8.2}$$

or, more generally

$$\delta(n - m) = \begin{cases} 1 & n = m \\ 0 & n \neq m \end{cases} \tag{8.3}$$

as shown in Fig. 8.3. The (discrete) unit-step function or *unit* (step) *sequence* is defined as

$$u(n) = \begin{cases} 1 & n \geq 0 \\ 0 & n < 0 \end{cases} \tag{8.4}$$

or, more generally

$$u(n - m) = \begin{cases} 1 & n \geq m \\ 0 & n < m \end{cases} \tag{8.5}$$

These *uniform* sequences are shown in Fig. 8.4.

In continuous-time systems the unit-step function is obtained from the unit-impulse function by integrating (Eq. 2.32), and the unit-impulse function is the time derivative of the unit-step function (Eq. 2.31). Here the unit sequence

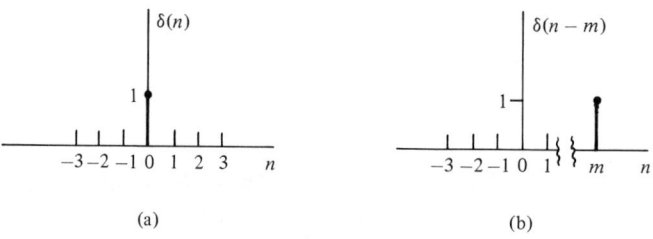

Figure 8.3 Unit function. (a) $\delta(n)$. (b) $\delta(n - m)$.

8.3 THE UNIT-FUNCTION RESPONSE

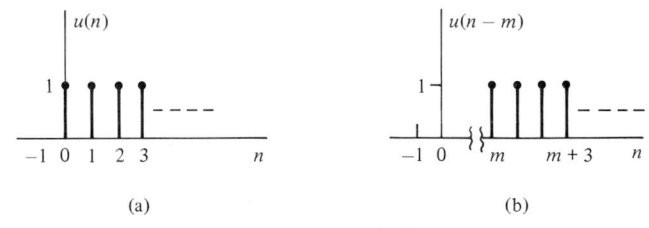

Figure 8.4. Unit sequence. (a) $u(n)$. (b) $u(n - m)$.

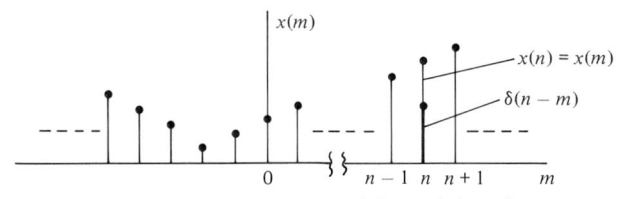

Figure 8.5. Sampling property of the unit function.

is obtained from the unit function by *summation*:

$$u(n) = \sum_{m=0}^{\infty} \delta(n - m) \tag{8.6}$$

while the unit function is the *difference* of two unit sequences:

$$\delta(n) = u(n) - u(n - 1) \tag{8.7}$$

Equations 8.6 and 8.7 are easily visualized and established by the reader.

Corresponding to the sampling property of the unit-impulse function for continuous-time systems, which we also recognized as the resolution of an arbitrary continuous function into a continuum of impulses, we need (here) a resolution of an arbitrary, but uniformly distributed, discrete-time function (or sequence) into unit functions. This is very easy. It is simply

$$\boxed{x(n) = \sum_{m=-\infty}^{\infty} x(m)\delta(n - m)} \tag{8.8}$$

as the reader can easily verify (see Fig. 8.5). The only nonzero product in the sum in Eq. 8.8 occurs for $m = n$.

8.3 THE UNIT-FUNCTION RESPONSE

Now suppose that we apply a unit function to a discrete-time system (at rest) at the instant m as shown in Fig. 8.6. For the present we simply say that the discrete system operates on the unit function to produce the *unit-function response*. We will say more about the system later. Thus, we have

$$H\{\delta(n - m)\} = h(n, m) \tag{8.9}$$

This is the response at the instant n to a unit function applied at the instant m.

The reader who has been through Chap. 2 with us may guess that the next step is that of applying a *general input sequence* $x(n)$ to the same system,

338 DISCRETE-TIME SYSTEMS

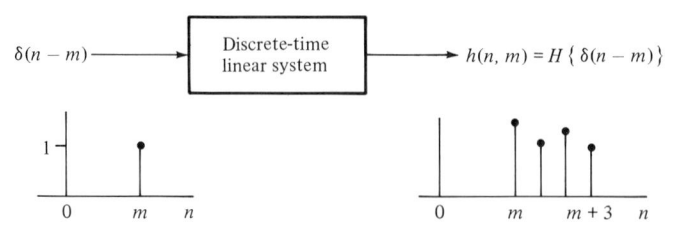

Figure 8.6. Unit function applied to a discrete-time system producing the unit-function response, $h(n, m)$.

producing the *output* or *response sequence* $y_{zs}(n)$. Notice that this is a *zero-state* (forced) response. Thus, for this situation we have (with Eq. 8.8)

$$H\{x(n)\} = H\left\{\sum_{m=-\infty}^{\infty} x(m)\delta(n-m)\right\} = y_{zs}(n) \tag{8.10}$$

But the discrete system only responds at the instant n (more will be said later about this). That is, H operates on n, not m, so

$$H\{x(n)\} = \sum_{m=-\infty}^{\infty} x(m)H\{\delta(n-m)\} = y_{zs}(n) \tag{8.11}$$

Using Eq. 8.9 the last result can be written

$$H\{x(n)\} = \sum_{m=-\infty}^{\infty} x(m)h(n, m) = y_{zs}(n)$$

Thus, we have

$$\boxed{y_{zs}(n) = \sum_{m=-\infty}^{\infty} x(m)h(n, m)} \quad \text{(general case)} \tag{8.12}$$

This situation is pictured in Fig. 8.7. We can now safely state that if we can find the unit-function response, we can find the zero-state response sequence for *any* input sequence using Eq. 8.12.

A *causal* system cannot respond before the unit function is applied, so $h(n, m) = 0$ for $m > n$, and thus

$$y_{zs}(n) = \sum_{m=-\infty}^{n} x(m)h(n, m) \quad \text{(causal system)} \tag{8.13}$$

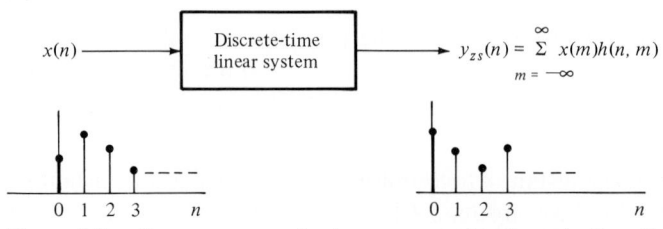

Figure 8.7. Response or output sequence of a discrete-time linear system.

8.3 THE UNIT-FUNCTION RESPONSE

A *fixed* continuous-time system is said to be *time-invariant*. Digital time invariance, or time invariance for discrete systems, is called *shift invariance*. For *shift-invariant* discrete-time systems, the unit function may be applied at *any* instant, the interval $n - m$ being all that matters, giving $h(n, m) = h(n - m)$, and thus

$$y_{zs}(n) = \sum_{m=-\infty}^{\infty} x(m)h(n-m) \quad \text{(shift-invariant system)} \tag{8.14}$$

If, in Eq. 8.14, we define a new index $l = n - m$ (m is a "dummy" index just like τ is a "dummy" variable in Eq. 2.39a), we have the equivalent form

$$y_{zs}(n) = \sum_{l=-\infty}^{\infty} x(n-l)h(l) \quad \text{(shift-invariant system)} \tag{8.15}$$

For systems that start at the instant $n = 0$ we may replace $x(n)$ with $x(n)u(n)$, or $x(m)$ with $x(m)u(m)$, so Eq. 8.13 becomes

$$y_{zs}(n) = \sum_{m=0}^{\infty} x(m)h(n, m) \quad \text{(starting at } n = 0\text{)} \tag{8.16}$$

Finally, as is often the case (certainly in this material), we have the shift-invariant and causal discrete-time linear system that starts at $n = 0$, for which

$$y_{zs}(n) = \sum_{m=0}^{n} x(m)h(n-m) \quad \begin{array}{l}\text{(shift invariant, causal,} \\ \text{and starting at } n = 0\text{)}\end{array} \tag{8.17}$$

The summation given by Eq. 8.17 is called *convolution-summation*. (See Eq. 5.77 in connection with the DFT.) Certainly the implications are clear. It involves essentially the same operations as convolution integration (Eq. 2.62). We must *fold, shift, multiply,* and *sum*.

Example 1

Let $x(n) = nu(n)$ and $h(n) = u(n) + u(n - 1)$. Find $y_{zs}(n)$ using Eq. 8.17. We have

$$y_{zs}(n) = \sum_{m=0}^{n} m[u(n-m) + u(n-m-1)]$$

Perhaps the simplest method of evaluating the sum is direct substitution:

$$y_{zs}(0) = 0(1 + 0) = 0$$
$$y_{zs}(1) = 0(1 + 1) + 1(1 + 0) = 1$$
$$y_{zs}(2) = 0(1 + 1) + 1(1 + 1) + 2(1 + 0) = 4$$
$$y_{zs}(3) = 0(1 + 1) + 1(1 + 1) + 2(1 + 1) + 3(1 + 0) = 9$$
$$\vdots$$
$$y_{zs}(n) = n^2$$

On the other hand it is worthwhile to examine convolution-summation graphically.

Example 2

Repeat Example 1 using graphical convolution-summation. We have

$$h(m) = u(m) + u(m-1)$$
$$h(-m) = u(-m) + u(-m-1) \quad \text{(folding)}$$
$$h(n-m) = u(n-m) + u(n-m-1) \quad \text{(shifting)}$$

The first two operations are shown in Fig. 8.8(b) and (c). We can now shift, multiply, and sum to find $y_{zs}(n)$. The $x(m)$ sequence is shown in Fig. 8.9. We begin by shifting by 0 (no shift), obtaining $h(-m)$, multiplying by $x(m)$, and summing. We next shift by 1, obtaining $h(1-m)$, multiply by $x(m)$, and sum to obtain $y_{zs}(1)$. The process is repeated. As can be seen in Fig. 8.10, we obtain $y_{zs}(n) = n^2$ as before.

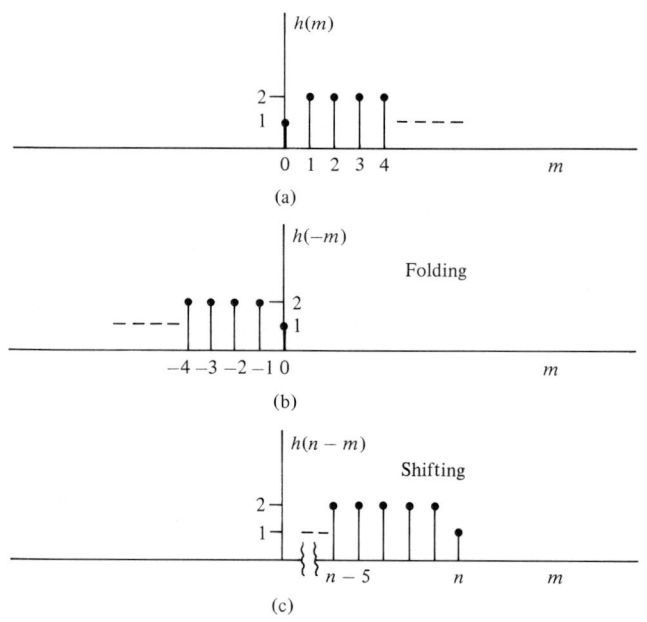

Figure 8.8. (a) Sequence $h(m)$. (b) $h(m)$ folded: $h(-m)$. (c) $h(-m)$ shifted: $h(n-m)$.

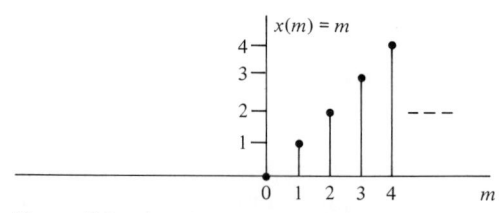

Figure 8.9. Input sequence $x(m) = m$.

8.3 THE UNIT-FUNCTION RESPONSE 341

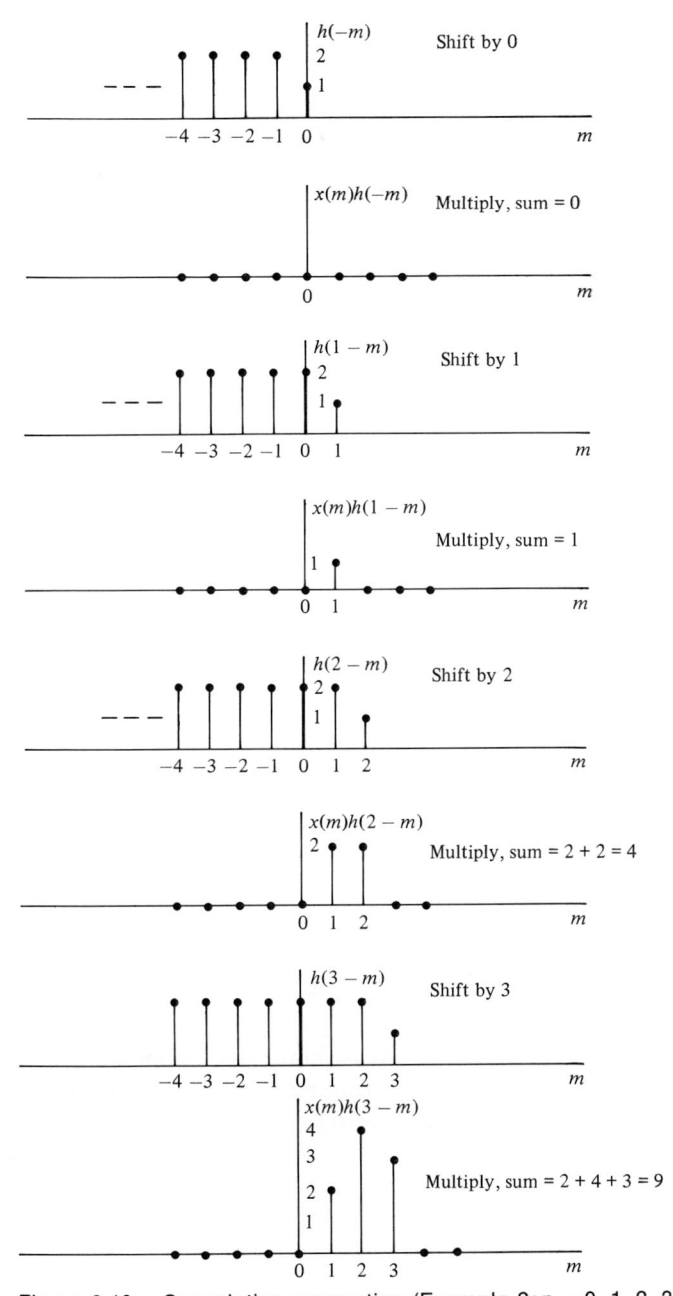

Figure 8.10. Convolution-summation (Example 2: $n = 0, 1, 2, 3$).

8.4 LINEAR DIFFERENCE EQUATIONS

Difference equations arise often in electrical, mechanical, and other systems when a structure repeats itself. An attenuator consisting of several identical resistive sections is an example, and it can best be treated by difference equations because it is not necessary to solve N differential equations for an attenuator with N sections, but, rather, it is only necessary to solve *one* difference equation for a typical section. Notice that in solving problems of the type just mentioned, t represents the discrete *position index* of the component structure being examined, not time. On the other hand we are primarily interested in discrete-*time* linear-systems analysis and we will proceed in that direction here.[†]

Difference equations do not contain derivatives of the dependent variable, but contain only *differences* of the dependent variable values at *discrete* values of the independent variable. In this material we are only interested in the shift-invariant linear system, in which case the coefficients are constant. For such a system that contains one input sequence $[x(n)]$ and one output sequence $[y(n)]$ the difference equation can be written[‡]

$$\boxed{\sum_{m=0}^{N} a_m y(n+m) = x_f(n) = \sum_{m=0}^{M} b_m x(n+m)} \qquad (8.18)$$

where $x_f(n)$ is the *forcing function sequence*. Notice the similarity in form between this equation and that for the continuous-time system (Eq. 2.4) with constant coefficients:

$$\sum_{n=0}^{N} a_n \frac{d^n y}{dt^n} = x_f(t) = \sum_{n=0}^{M} b_n \frac{d^n x}{dt^n} \qquad (2.4)$$

Equation 8.18 is Nth order because it contains an Nth-order ordinate [value of y when the independent variable is N periods away: that is, $y(n+N)$]. It is linear because it only contains first powers of the dependent variable y and contains no products of y values.

Corresponding to the continuous-time case, there are at least three ways that we can go about obtaining a solution to Eq. 8.18. The classical method is very similar to that for the continuous-time case in that the complete solution is obtained as the sum of the *complementary function* and the *particular solution*. We will not elaborate on this method, but we will present an example a little later on to compare it with the method that we do intend to exploit in this chapter; the unit-function response method. The z-transform method will be examined in Chap. 9. It basically does for discrete systems what the Laplace (or Fourier) transform does for continuous systems.

It was pointed out in the preceding section that if the unit-function response $h(n-m)$ (for shift-invariant discrete systems) could be found, then convolution-

---

[†] It was shown in Sec. 8.1 how a differential equation can be approximated by a difference equation.
[‡] Superposition can treat multiple-input-multiple-output systems. Notice carefully that M in Eq. 8.18 *is not* the number of separate inputs.

summation (Eq. 8.14) would give us the zero-state output sequence for *any* input sequence. It is simpler to find the unit-function response for the forcing function sequence $x_f(n)$. We choose to call this particular unit-function response $h_f(n - m)$, so that the equivalent of Eq. 8.14 is

$$y_{zs}(n) = \sum_{m=-\infty}^{\infty} x_f(m)h_f(n-m) \quad \text{(shift-invariant system)} \tag{8.19}$$

while the equivalent of Eq. 8.15 is ($l = m$)

$$y_{zs}(n) = \sum_{m=-\infty}^{\infty} x_f(n-m)h_f(m) \quad \text{(shift-invariant system)} \tag{8.20}$$

and the equivalent of Eq. 8.17 is

$$y_{zs}(n) = \sum_{m=0}^{n} x_f(m)h_f(n-m) \quad \begin{array}{l}\text{(shift invariant, causal,}\\ \text{and starting at } t = 0)\end{array} \tag{8.21}$$

In order to find the unit-function response for the *input* at any instant we simply apply a unit function at that instant. As an example, suppose we have a second-order system; that is, one in which the difference equation is

$$a_0 y(n) + a_1 y(n+1) + a_2 y(n+2) = x_f(n) \tag{8.22}$$

Also suppose that we want to find the unit-function response for $x(n + 2)$ and we have already found the unit-function response for $x_f(n)$ satisfying

$$a_0 h_f(n) + a_1 h_f(n+1) + a_2 h_f(n+2) = \delta(n)$$

We now want a solution for

$$a_0 y(n) + a_1 y(n+1) + a_2 y(n+2) = x(n+2) \tag{8.23}$$

so we must find the unit-function response satisfying

$$a_0 h(n) + a_1 y(n+1) + a_2 y(n+2) = \delta(n+2)$$

According to Eq. 8.20 (Eq. 8.19 works as well)

$$h(n) = \sum_{m=-\infty}^{\infty} \delta(n+2-m)h_f(m) = h_f(n+2)$$

Since this is a shift-invariant system, the last equation gives a result that is rather obvious. In general, then, for a forcing function as given in Eq. 8.18, the unit-function response for $x(n)$ is[†]

$$h(n) = \sum_{m=0}^{M} b_m h_f(n+m) \tag{8.24}$$

[†] An explicit example will be given in the next section.

Notice that because we are dealing with a shift-invariant system, Eq. 8.23 can be written ($n \to n - 2$)

$$a_0 y(n - 2) + a_1 y(n - 1) + a_2 y(n) = x(n)$$

or

$$y(n) = \frac{1}{a_2} x(n) - \frac{a_0}{a_2} y(n - 2) - \frac{a_1}{a_2} y(n - 1)$$

This result indicates a *recursive relation*: The discrete function y at the instant n (or nT) can be found in terms of the input at the *same instant* and y at the instants $n - 1$ and $n - 2$ (previous outputs). On the other hand, if $x_f(n) = x(n)$, Eq. 8.22 can be written ($n \to n - 2$)

$$a_0 y(n - 2) + a_1 y(n - 1) + a_2 y(n) = x(n - 2)$$

or

$$y(n) = \frac{1}{a_2} x(n - 2) - \frac{a_0}{a_2} y(n - 2) - \frac{a_1}{a_2} y(n - 1)$$

indicating that y at the instant n can be found in terms of the input at the instant $n - 2$ (two periods previously) and y at the instants $n - 1$ and $n - 2$. Thus, we have some degree of flexibility in how we choose to describe the system. The recursive properties will be examined in the next section.

The causal unit-function response for $x_f(n)$ is given by

$$\boxed{h_f(n) = y_{zi}(n - 1) u(n - 1)} \quad \text{(causal unit-function response)} \quad (8.25)$$

subject to

$$\left.\begin{array}{l} y_{zi}(0) = 0 \\ y_{zi}(1) = 0 \\ y_{zi}(2) = 0 \\ \quad \vdots \\ y_{zi}(N - 1) = 1/a_N \end{array}\right\} \quad (8.26)$$

where $y_{zi}(n)$ is the zero-input response (complementary function) and is a solution to the homogeneous form of Eq. 8.18:[†]

$$\sum_{m=0}^{N} a_m y_{zi}(n + m) = 0 \quad (8.27)$$

[†] There is a noncausal unit-function response also. It is given by

$$h_f(n) = -y_{zi}(n - 1) u(-n) \quad \text{(noncausal unit-function response)}$$

subject to the set 8.26. It is useful in cases where some independent variable other than time (distance, for example) is present and causality has no meaning. It can be linearly combined with the causal unit-function response (Eq. 8.25), as in Eq. 2.41 for continuous systems, if necessary.

8.4 LINEAR DIFFERENCE EQUATIONS

It is not difficult to prove that our assertions are correct. We have (Eq. 8.25)
$$h_f(n+m) = y_{zi}(n+m-1)u(n+m-1)$$
and we must show that (Eq. 8.18)
$$\sum_{m=0}^{N} a_m h_f(n+m) = \delta(n)$$
or
$$\sum_{m=0}^{N} a_m y_{zi}(n+m-1)u(n+m-1) = \delta(n) \tag{8.28}$$

For $n = 0$
$$\sum_{m=0}^{N} a_m y_{zi}(m-1)u(m-1) = a_0 y_{zi}(-1)u(-1) + \sum_{m=1}^{N-1} a_m y_{zi}(m-1)u(m-1)$$
$$+ a_N y_{zi}(N-1)u(N-1)$$
$$= 0 + 0 + 1$$

The first term is 0 since $u(-1) = 0$, the second term is 0 because of the set 8.26, and the last term is unity because of the last equation in the set 8.26. For $n = 1$
$$\sum_{m=0}^{N} a_m y_{zi}(m)u(m) \equiv 0$$
because of Eq. 8.27 (with $n = 0$). For $n = 2$
$$\sum_{m=0}^{N} a_m y_{zi}(m+1)u(m+1) \equiv 0$$
because of Eq. 8.27 (with $n = 1$). It is clear that the left side of Eq. 8.28 is unity for $n = 0$ and is 0 for $n = 1, 2, 3, \ldots$. Thus, Eq. 8.28 is established for $n \geq 0$. The left side of Eq. 8.28 is also 0 for $n < 0$. It is left as an exercise to prove that this is the case.

We must find the zero-input response, not only for Eq. 8.25, but because it is added to the zero-state response to find the complete solution. Recall that for the time-invariant (continuous-time) case (Eq. 2.4) we could *always* find the zero-input response by simply assuming the form e^{st}. The same thing occurs in the discrete-time case if we simply assume the form
$$y_{zi}(n) = z^n \tag{8.29}$$
Substituting this guess into the homogeneous difference equation (8.27) gives
$$\sum_{m=0}^{N} a_m z^{n+m} = 0$$

or

$$z^n \sum_{m=0}^{N} a_m z^m = 0$$

or, since $z^n = y_{zi}(n)$ is not zero unless we want a trivial solution

$$\boxed{\sum_{m=0}^{N} a_m z^m = 0} \quad \text{(characteristic equation)} \tag{8.30}$$

and, *again*, we have a polynomial form for the characteristic equation. The procedure that we have used in this section shows that *our problem has been reduced to that of finding the zero-input response, and it requires nothing more than finding the roots of an Nth order polynomial*. The quantity z may be complex. This is precisely the situation we wound up with for the time-invariant (continuous-time) case. Equation 8.30 has N roots so the general solution is (assuming no roots are repeated)

$$\boxed{y_{zi}(n) = \sum_{m=1}^{N} C_m (z_m)^n} \quad \text{(zero-input response)} \tag{8.31}$$

The $(z_m)^n$ make up the elements of the fundamental set. The unit-function response for the forcing function sequence $x_f(n)$ (Eq. 8.25) is (again, assuming no repeated roots)

$$\boxed{h_f(n) = u(n-1) \sum_{m=1}^{N} D_m (z_m)^{n-1}} \quad \text{(unit-function response)} \tag{8.32}$$

subject to the set of conditions 8.26. The complete solution is given by

$$y(n) = y_{zi}(n) + y_{zs}(n)$$

$$\boxed{y(n) = \sum_{m=1}^{N} C_m (z_m)^n + \sum_{m=-\infty}^{\infty} x_f(m) h_f(n-m)} \quad \text{(complete response)} \tag{8.33}$$

or

$$\boxed{y(n) = \sum_{m=1}^{N} C_m (z_m)^n + \sum_{m=-\infty}^{\infty} x(m) h(n-m)} \quad \text{(complete response)} \tag{8.34}$$

We need very much to summarize the method before looking at examples.

1. Find the *zero-input response* sequence. That is, find the roots of the characteristic equation (8.30). (The case of repeated roots will be treated later.)
2. Find the unit-function response for the forcing function sequence using Eqs. 8.18, 8.32, and 8.26. The unit-function response for an input at any instant can be found by Eq. 8.24.

3. Find the *zero-state response* sequence using the superposition-summation, Eq. 8.19 or 8.14.
4. Form the complete solution by adding the zero-input response and the zero-state response (Eq. 8.33 or 8.34).
5. Apply the given initial conditions to the complete solution to evaluate the N constants C_n.

The unit-function response method that we have just summarized may seem overly complicated and lengthy. It is a general method, however, and once a particular discrete system has been analyzed, we have a formal solution for *any* input or *forcing function* sequence. The classical method requires that we find a *new* particular solution each time the forcing function sequence is changed. This may be difficult (or impossible) to do if this sequence is not in analytic form, for example.

We mention in passing that a discrete-time system with multiple inputs and (or) outputs has a zero-state response that can be characterized by a matrix of unit-function responses. If it is shift-invariant, then

$$\mathbf{y}_{zs}(n) = \sum_{m=-\infty}^{\infty} \mathbf{h}_f(n-m)\mathbf{x}_f(m) \tag{8.35}$$

or

$$\mathbf{y}_{zs}(n) = \sum_{m=-\infty}^{\infty} \mathbf{h}(n-m)\mathbf{x}(m) \tag{8.36}$$

8.5 EXAMPLES

Now that the theory of shift-invariant discrete linear-system analysis has been presented, we need to look at some examples. Consider the block diagram of such a system as shown in Fig. 8.11. As mentioned in Chap. 1, the blocks labeled z^{-1} are shift registers that produce unit time delay. The output of the summer is $y(n+2)$:

$$y(n+2) = -3y(n+1) - 2y(n) + x_f(n)/3$$

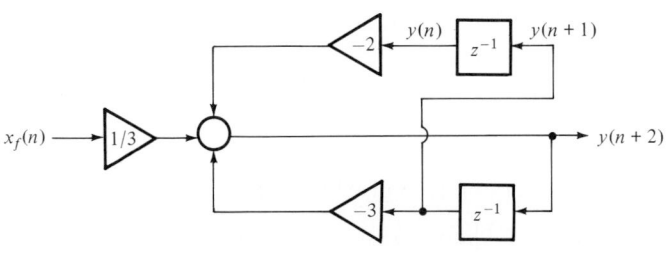

Figure 8.11. Block diagram of a discrete-time linear system (shift invariant).

or

$$3y(n + 2) + 9y(n + 1) + 6y(n) = x_f(n) \qquad (8.37)$$

Thus, it is a second-order system.

Example 3

Find the zero-input response for the system shown in Fig. 8.11. The homogeneous equation is

$$3y(n + 2) + 9y(n + 1) + 6y(n) = 0$$

The characteristic equation (8.30) is

$$3z^2 + 9z + 6 = 0$$

or

$$z^2 + 3z + 2 = 0$$

or

$$(z + 1)(z + 2) = 0$$

or

$$z_1 = -1 \qquad z_2 = -2$$

Thus, the zero-input response (Eq. 8.31) is

$$y_{zi}(n) = C_1(-1)^n + C_2(-2)^n \qquad (8.38)$$

Notice that neither term in Eq. 8.38 *decreases* as n increases, so $y_{zi}(n)$ has no transient part. It is entirely steady state and *grows larger* as n increases.

Example 4

Find the unit-function response for the forcing function sequence $x_f(n)$ of Example 3. Using the zero-input response of Eq. 8.38 (with $C_n \to D_n$) with the special set of initial conditions given by the set 8.26, we have

$$y_{zi}(n) = D_1(-1)^n + D_2(-2)^n$$
$$y_{zi}(0) = 0 \qquad y_{zi}(1) = 1/a_2 = \tfrac{1}{3}$$

Thus

$$\left.\begin{array}{l} 0 = D_1 + D_2 \\ \tfrac{1}{3} = -D_1 - 2D_2 \end{array}\right\} \begin{array}{l} D_1 = \tfrac{1}{3} \\ D_2 = -\tfrac{1}{3} \end{array}$$

Equation 8.32 gives the unit-function response for $x_f(n)$:

$$h_f(n) = \tfrac{1}{3}[(-1)^{n-1} - (-2)^{n-1}]u(n - 1) \qquad (8.39)$$

It is worthwhile to check this result. It must satisfy

$$3h_f(n+2) + 9h_f(n+1) + 6h_f(n) = \delta(n)$$

Substituting $h_f(n)$, the left side of this equation is

$$[(-1)^{n+1} - (-2)^{n+1}]u(n+1)$$
$$+ 3[(-1)^n - (-2)^n]u(n) + 2[(-1)^{n-1} - (-2)^{n-1}]u(n-1)$$
$$= (-1)^n[-u(n+1) + 3u(n) - 2u(n-1)]$$
$$+ (-2)^n[2u(n+1) - 3u(n) + u(n-1)]$$

Using Eqs. 8.3 and 8.7 this becomes

$$(-1)^n[-\delta(n+1) + 2\delta(n)] + (-2)^n[2\delta(n+1) - \delta(n)] = \delta(n) \qquad \delta(n) = \delta(n)$$

Suppose next that the forcing function sequence is $x_f(n) = x(n+2)$. The difference equation is now

$$3y(n+2) + 9y(n+1) + 6y(n) = x(n+2)$$

or

$$3y(n) + 9y(n-1) + 6y(n-2) = x(n) \qquad (8.40)$$

or

$$y(n) = \tfrac{1}{3}x(n) - 3y(n-1) - 2y(n-2)$$

as described in Fig. 8.12. Notice that this block diagram is *identical* to that in Fig. 8.11. The input was changed from the general form $x_f(n)$ to $x(n+2)$. The unit-function response for this input is easily obtained from the general formula, Eq. 8.24.

$$h(n) = b_m h_f(n+m) = h_f(n+2)$$

or

$$h(n) = \tfrac{1}{3}[(-1)^{n+1} - (-2)^{n+1}]u(n+1) \qquad (8.41)$$

where we have also used Eq. 8.39 for h_f.

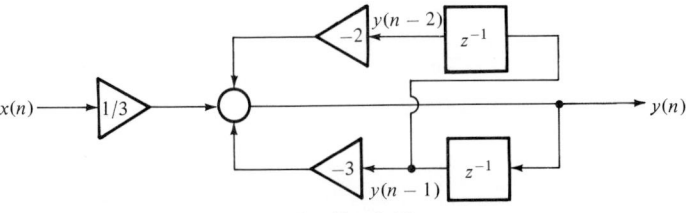

Figure 8.12. Block diagram for Eq. 8.40.

Example 5

Find the complete response for Fig. 8.12 when the input is $x(n) = 18 \cdot 2^n u(n)$. The zero-state response is obtained from Eqs. 8.14 and 8.41:

$$y_{zs}(n) = \sum_{m=-\infty}^{\infty} x(m)h(n-m)$$

$$= \sum_{m=-\infty}^{\infty} 18 \cdot 2^m u(m) \cdot \tfrac{1}{3}[(-1)^{n-m+1} - (-2)^{n-m+1}]u(n-m+1)$$

$$y_{zs}(n) = 6 \sum_{m=0}^{n+1} 2^m [(-1)^{n-m+1} - (-2)^{n-m+1}]$$

Notice the gating effect of the product $u(m)u(n-m+1)$. Notice also that $y_{zs}(-1) \equiv 0$ and $y_{zs}(-2) \equiv 0$ for this zero-state response.[†] This is as it should be. The next terms in the sequence are

$$y_{zs}(0) = 6 \sum_{m=0}^{1} 2^m [(-1)^{-m+1} - (-2)^{-m+1}]$$

$$= 6[(-1)^1 - (-2)^1] + 6 \cdot 2^1 [(-1)^0 - (-2)^0]$$

$$= 6$$

$$y_{zs}(1) = 6 \sum_{m=0}^{2} 2^m [(-1)^{-m+2} - (-2)^{-m+2}]$$

$$= 6[(-1)^2 - (-2)^2] + 6 \cdot 2^1 [(-1)^1 - (-2)^1]$$
$$+ 6 \cdot 2^2 [(-1)^0 - (-2)^0]$$

$$= -6$$

$$y_{zs}(2) = 6 \sum_{m=0}^{3} 2^m [(-1)^{-m+3} - (-2)^{-m+3}]$$

$$= 6[(-1)^3 - (-2)^3] + 6 \cdot 2^1 [(-1)^2 - (-2)^2]$$
$$+ 6 \cdot 2^2 [(-1)^1 - (-2)^1] + 6 \cdot 2^3 [(-1)^0 - (-2)^0]$$

$$y_{zs}(2) = 30$$

and so forth. The zero-input response is given by Eq. 8.38, so the complete response is

$$y(n) = C_1(-1)^n + C_2(-2)^n + 6 \sum_{m=0}^{n+1} 2^m [(-1)^{n-m+1} - (-2)^{n-m+1}] \quad (8.42)$$

Suppose that $y(-1) = 1$ and $y(-2) = 0$. With these *initial conditions* we obtain

$$y(n) = (-1)^n - 4(-2)^n + 6 \sum_{m=0}^{n+1} 2^m [(-1)^{n-m+1} - (-2)^{n-m+1}]$$

[†] We can certainly multiply $y_{zs}(n)$ by $u(n+1)$ if we want.

8.5 EXAMPLES

The complete response in Example 5 can also be found recursively. Using the equation following 8.40 we obtain the output sequence

$$3, 1, 15, 1, 63, 1, 255, \ldots \quad \text{for } n = 0, 1, 2, \ldots$$

This sequence has the *closed form*

$$y(n) = \begin{cases} 1 & n \text{ odd} \\ 2^{n+2} - 1 & n \text{ even} \end{cases}$$

The input sequence $x(n) = 18 \cdot 2^n u(n)$ was chosen for a particular reason. This is one of the forms for which it is easy to obtain a *particular solution* in the classical sense. We can find a particular solution by any means at our disposal, including an educated guess. Since the form 2^n repeats itself when n is incremented in unit steps,[†] we assume a particular solution $y_p(n) = C_3 2^n$. This gives

$$3 y_p(n) + 9 y_p(n-1) + 6 y_p(n-2) = 18 \cdot 2^n \quad n \geq 0$$

or

$$3 C_3 2^n + 9 C_3 2^{n-1} + 6 C_3 2^{n-2} = 18 \cdot 2^n$$

or multiplying by 2^{-n}

$$(3 + 9 \cdot 2^{-1} + 6 \cdot 2^{-2}) C_3 = 18$$

Therefore $C_3 = 2$, and

$$y_p(n) = 2^{n+1}$$

Notice that this *particular solution is not the same as the zero-state response*. The complete response is the sum of the particular solution and a complementary function:

$$y(n) = C_4(-1)^n + C_5(-2)^n + 2^{n+1}$$

Using the same initial conditions as before (Example 5) we have

$$y(-1) = 1 = -C_4 - \tfrac{1}{2} C_5 + 1$$
$$y(-2) = 0 = C_4 + \tfrac{1}{4} C_5 + \tfrac{1}{2}$$

or

$$\left. \begin{array}{l} 0 = -C_4 - \tfrac{1}{2} C_5 \\ -\tfrac{1}{2} = C_4 + \tfrac{1}{4} C_5 \end{array} \right\} \quad \begin{array}{l} C_4 = -1 \\ C_5 = 2 \end{array}$$

Thus

$$y(n) = (-1)^{n+1} + 2^{n+1}[1 + (-1)^n]$$

or

$$y(n) = \begin{cases} 1 & n \text{ odd} \\ 2^{n+2} - 1 & n \text{ even} \end{cases}$$

[†] Remember e^{st} for continuous-time systems?

as before. Do not be misled by the fact that this result was obtained with less effort than with the other method. As mentioned earlier, this occurred because of the special form of the input sequence.† In order to emphasize this point we repeat Example 5 with a new input sequence.

Example 6

Repeat Example 5 for $x(n) = 4nu(n)$. We have already found the unit-function response in Eq. 8.41, so

$$y_{zs}(n) = \sum_{m=-\infty}^{\infty} 4mu(m) \cdot \frac{1}{3}[(-1)^{n-m+1} - (-2)^{n-m+1}]u(n - m + 1)$$

$$y_{zs}(n) = \frac{4}{3} \sum_{m=0}^{n+1} m[(-1)^{n-m+1} - (-2)^{n-m+1}]$$

This output sequence is

$$0, \tfrac{4}{3}, -\tfrac{4}{3}, \tfrac{16}{3}, -8, \ldots \qquad \text{for } n = 0, 1, 2, \ldots$$

Is the system that we have been using in our examples a stable system? Can the unit-function response shed any light on this question?

Discrete systems or digital systems can also be *nonrecursive* (defined in Sec. 8.1) or *transversal*. Consider the general form, Eq. 8.18 with $N = 3$, $M = 2$, and $a_0 = a_1 = a_2 = 0$, for example. We then have

$$a_3 y(n + 3) = b_0 x(n) + b_1 x(n + 1) + b_2 x(n + 2) + b_3 x(n + 3) \qquad (8.43)$$

Shifting back by three units

$$a_3 y(n) = b_0 x(n - 3) + b_1 x(n - 2) + b_2 x(n - 1) + b_3 x(n) \qquad (8.44)$$

Equation 8.44 is a nonrecursive difference equation because the output merely depends on the *present* value and past three values of the input.

We now need to consider the case of repeated roots. The procedure is similar to that for the continuous-time case. The elements of the fundamental set for an Nth-order system with the mth root repeated i times are (without proof)

$$\boxed{z_1, z_2, \ldots, z_m, nz_m, n^2 z_m, \ldots, n^{i-1} z_m, z_{m+1}, \ldots, z_N} \qquad (8.45)$$

Example 7

Consider the system shown in Fig. 8.13. As can be seen

$$y(n) = y(n - 1) - \tfrac{1}{4}y(n - 2) + x(n)$$

---

† This method did, however, give us a *closed-form* answer, and this is certainly convenient.

8.5 EXAMPLES

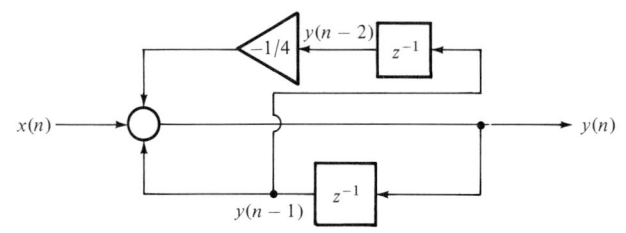

Figure 8.13. Block diagram of a second-order discrete system.

or

$$y(n) - y(n-1) + \tfrac{1}{4} y(n-2) = x(n)$$

The "standard" form that we have been using for this (shifting up two units) is

$$y(n+2) - y(n+1) + \tfrac{1}{4} y(n) = x_f(n) = x(n+2)$$

The characteristic equation is easily obtained:

$$z^2 - z + \tfrac{1}{4} = 0$$

or

$$(z - \tfrac{1}{2})(z - \tfrac{1}{2}) = 0$$

We have a *pair* of roots at $z = \tfrac{1}{2}$, and the elements of the fundamental set are given by the set 8.45. The zero-input response is

$$y_{zi}(n) = D_1 (\tfrac{1}{2})^n + n D_2 (\tfrac{1}{2})^n$$

We can find the unit-function response for $x_f(n)$ as before:

$$\left. \begin{array}{l} y_{zi}(0) = 0 = D_1 + 0 \\ y_{zi}(1) = 1 = \tfrac{1}{2} D_1 + \tfrac{1}{2} D_2 \end{array} \right\} \begin{array}{l} D_1 = 0 \\ D_2 = 2 \end{array}$$

Thus

$$h_f(n) = 2(n-1)(\tfrac{1}{2})^{n-1} u(n-1)$$

or, using Eq. 8.24

$$h(n) = h_f(n+2) = 2(n+1)(\tfrac{1}{2})^{n+1} u(n+1)$$

What can be said about the stability of this discrete system? The zero-state response is given by Eq. 8.15 ($l = m$):

$$y_{zs}(n) = 2 \sum_{m=-\infty}^{\infty} x(n-m)(m+1)\left(\frac{1}{2}\right)^{m+1} u(m+1)$$

$$y_{zs}(n) = 2 \sum_{m=-1}^{\infty} x(n-m)(m+1)\left(\frac{1}{2}\right)^{m+1}$$

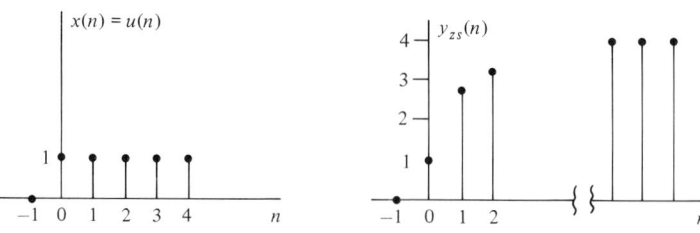

Figure 8.14. Input-output (unit-sequence response) for the system in Example 7.

If $x(n) = u(n)$ we have the *unit-sequence response*

$$y_{zs}(n) = 2 \sum_{m=-1}^{\infty} u(n-m)(m+1)\left(\frac{1}{2}\right)^{m+1}$$

$$y_{zs}(n) = 2 \sum_{m=-1}^{n} (m+1)\left(\frac{1}{2}\right)^{m+1} = 2 \sum_{l=0}^{n+1} l\left(\frac{1}{2}\right)^{l}$$

$$y_{zs}(n) = 2 \sum_{l=1}^{n+1} l\left(\frac{1}{2}\right)^{l}$$

This result is shown in Fig. 8.14. The summation is an arithmetic-geometric series,† and for large n it is given by

$$\frac{1/2}{(1-1/2)^2} = 2$$

Thus, $y_{zs}(n) \to 4$ for $n \to \infty$. This output sequence is bounded for the bounded input $u(n)$.

The unit-function response can, of course, also be determined recursively, although it may be difficult to recognize or deduce a *closed form* using this procedure. Consider the difference equation

$$y(n) + 3y(n-1) + 2y(n-2) = x(n)$$

If we want to find $h(n)$, then we are looking for a solution to

$$h(n) + 3h(n-1) + 2h(n-2) = \delta(n)$$

or

$$h(n) = \delta(n) - 3h(n-1) - 2h(n-2)$$

Thus

$$h(0) = 1 - 3(0) - 2(0) = 1 \qquad n = 0$$
$$h(1) = 0 - 3(1) - 2(0) = -3 \qquad n = 1$$
$$h(2) = 0 - 3(-3) - 2(1) = 7 \qquad n = 2$$
$$h(3) = 0 - 3(7) - 2(-3) = -15 \qquad n = 3$$
$$h(4) = 0 - 3(-15) - 2(7) = 7 \qquad n = 4$$

† See Eq. C.24, App. C.

In this (simple) case it is not too difficult to recognize that
$$h(n) = [(-1)^{n+1} - (-2)^{n+1}]u(n+1)$$
Compare this result to that for Example 4.

It is straightforward to obtain the unit-function response for *nonrecursive* (defined in Sec. 8.1) filters.

Example 8

Consider the nonrecursive filter shown in Fig. 8.15. The difference equation mathematical model is
$$y(n) = x(n) - 3x(n-1) + 2x(n-2) + x(n-3)$$
Therefore
$$h(n) = \delta(n) - 3\delta(n-1) + 2\delta(n-2) + \delta(n-3)$$
$$h(0) = 1 \quad n = 0$$
$$h(1) = -3 \quad n = 1$$
$$h(2) = 2 \quad n = 2$$
$$h(3) = 1 \quad n = 3$$
$$h(n) = 0 \quad 3 < n < 0$$

These filters are often called *Finite-duration Impulse Response* (FIR) filters for reasons that are obvious, although the unit-function response is a finite sum of weighted unit functions, *not unit impulses*.

We need to consider a discrete system where the input sequence *plus* one (or more) delayed input sequence(s) is applied. Consider the system shown in Fig. 8.16. The difference equation is
$$y(n) = -x(n) + 3x(n-2) - 3y(n-1) - 2y(n-2)$$
or
$$y(n) + 3y(n-1) + 2y(n-2) = -x(n) + 3x(n-2) \tag{8.46}$$

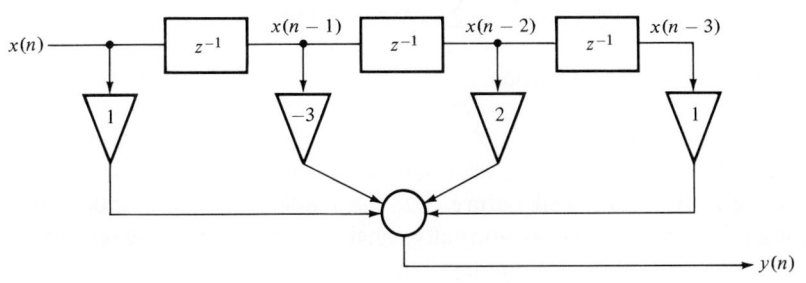

Figure 8.15. Block diagram of a nonrecursive filter.

356 DISCRETE-TIME SYSTEMS

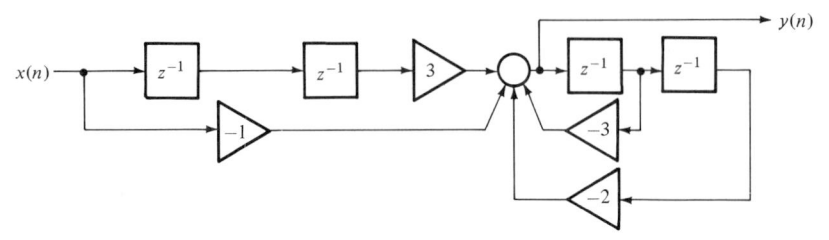

Figure 8.16. Block diagram of a discrete system.

or

$$y(n + 2) + 3y(n + 1) + 2y(n) = x_f(n) = -x(n + 2) + 3x(n) \qquad (8.47)$$

The last equation represents the *standard form* that we have been using. There are several ways that we can find $y(n)$. We can find the unit-function response for $-x(n)$ and $3x(n-2)$ *separately* and then apply superposition, or we can find the unit-function response for $-x(n) + 3x(n-2)$ as a single entity by first finding the unit-function response for $x_f(n)$. We choose the latter course in the following example.

Example 9

Find the unit-function response for $x_f(n)$ in Eq. 8.47 and then find the unit-function response for Eq. 8.46. We previously found the unit-function response for $x_f(n)$ in Eq. 8.47 in Example 4 (except for the factor $\frac{1}{3}$). It is

$$h_f(n) = [(-1)^{n-1} - (-2)^{n-1}]u(n-1)$$

Thus, we find the unit-function response for $-x(n + 2) + 3x(n)$ by either using this in Eq. 8.20 or by simply using the general result indicated by Eq. 8.24. This gives

$$h(n) = -h_f(n + 2) + 3h_f(n)$$

or

$$h(n) = -[(-1)^{n+1} - (-2)^{n+1}]u(n + 1) + 3[(-1)^{n-1} - (-2)^{n-1}]u(n-1)$$

Given $x(n)$, the zero-state response is found by superposition-summation:

$$y_{zs}(n) = \sum_{m=-\infty}^{\infty} x(m)h(n-m) = \sum_{m=-\infty}^{\infty} x(n-m)h(m)$$

and the solution proceeds as usual.

Now that the general method of solution of linear difference equations has been explored with some rather abstract examples, we will consider some examples[1] of a more practical nature that the reader can easily understand even though they are not what we normally consider to be engineering problems. The first involves a savings account (system).

Example 10

The amount of money in a savings account at the end of the nth period is given by the linear difference equation

$$y(n) = (1 + i/k)y(n-1) + x(n)$$

where

k = number of compounding periods per year,
i = yearly interest rate, and
$x(n)$ = total deposits during the nth period.

We first want to find the amount of money in the account at the end of 5 years if $k = 2$ (semi-annual compounding), $i = 0.10$ (10 percent interest); and $x(n) = 1000\delta(n)$ (an initial deposit only). Thus

$$y(n) = 1.05y(n-1) + 1000\delta(n) \qquad y(0) = 1000$$
$$y(n) = 1.05y(n-1) \qquad n > 0$$

Solving iteratively

$$y(0) = \$1000$$
$$y(1) = \$1050$$
$$y(2) = \$1102.50$$
$$\vdots$$
$$y(10) = \$1628.89$$

Next, we would like to find the amount of money in the account if, in addition to the initial $1000 investment, $100 is added to the account every 6 months. In this case

$$x(n) = \$1000\delta(n) + \$100u(n-1)$$

and

$$y(n) = 1.05y(n-1) + 100 \qquad n > 0, \quad y(0) = 1000$$

Thus

$$y(1) = \$1150$$
$$y(2) = \$1307.50$$
$$\vdots$$
$$y(10) = \$2886.68$$

It is worthwhile noting that the unit-function response for the preceding system obtained by the method of Sec. 8.4 is

$$h(n) = (1 + i/k)^n u(n)$$

and is the amount in the account after n periods *per-dollar* initial investment when nothing (but the initial amount) is added to the account. For Example 10 (first part)

$$y(n) = 1000h(n) = 1000(1.05)^n u(n)$$

so we easily obtain $y(10) = \$1628.89$ as before!

Example 11

As an example of a *time-variable*, linear discrete system, consider the evaluation of the number of combinations of k items taken n at a time. This frequently occurs in probability and statistics, and the number of combinations is given by the zero-input response:

$$y_{zi}(n) = \binom{k}{n} = \frac{k!}{n!(k-n)!} \qquad n = 0, 1, \ldots, k, \quad n \geq 0, \quad k \geq 0$$

It is then also true that

$$y_{zi}(n-1) = \frac{k!}{(n-1)!(k-n+1)!} \qquad n = 1, 2, \ldots, k+1$$

Combining the last two equations gives

$$y_{zi}(n) - \frac{k-n+1}{n} y_{zi}(n-1) = 0$$

which is a first-order, linear, time-variable, homogeneous difference equation. Transposing

$$y_{zi}(n) = \frac{k-n+1}{n} y_{zi}(n-1)$$

and this is a form that is particularly well suited for iteration and ease of calculation, compared to the original form *when n is large*.

How many combinations exist for 6 items taken 4 at a time? Notice that $y_{zi}(0) = 1$. We have

$$y_{zi}(0) = 1$$

$$y_{zi}(1) = \frac{6-1+1}{2}(1) = 6 = k$$

$$y_{zi}(2) = \frac{6-2+1}{2}(6) = 15$$

$$y_{zi}(3) = \frac{6-3+1}{3}(15) = 20$$

$$y_{zi}(4) = \frac{6-4+1}{4}(20) = 15$$

This result required a minimum of $n - 1 = 3$ multiplications and the same number of divisions. The direct calculation of $y_{zi}(4)$ requires $2n^2 - 9n + 10 = 6$ multiplications and $n - 2 = 2$ divisions! For large n the iterative scheme offers a tremendous savings in computer time.

Example 12

In a loan payment scheme (that is used by many of us in buying a home or automobile) called *amortization*, the interest rate is i, the remaining principal after n payments is $y(n)$ (the "output"), and the amount of the nth payment is $x(n)$ (the "input"). It can then be shown that (see Prob. 9)

$$y(n) = (1 + i)y(n - 1) - x(n)$$

or

$$y(n + 1) - (1 + i)y(n) = -x(n + 1) = x_f(n)$$

The unit-function response for $x_f(n)$ is easily found to be

$$h_f(n) = (1 + i)^{n-1} u(n - 1)$$

and it is interesting that this response does not go to 0 as n increases, indicating an *unstable* system. This is, after all, rather obvious. The output (remaining principal) must be *forced* to 0.

Suppose that periodic payments are made in equal amounts (a typical scheme):

$$x(n) = pu(n - 1) \qquad n = 1, 2, \ldots, N$$

where p is the (uniform) payment. The problem now is that of determining p such that the loan is repaid completely and exactly in N periods. Thus, we now have

$$y(n) = (1 + i)y(n - 1) - pu(n - 1)$$

or

$$y(n + 1) - (1 + i)y(n) = -pu(n) = x_f(n)$$

The *complete* solution for $y(n)$ is

$$y(n) = C_1(1 + i)^n + \sum_{m=-\infty}^{\infty} x_f(m) h_f(n - m)$$

$$y(n) = C_1(1 + i)^n + \sum_{m=-\infty}^{\infty} -pu(m)(1 + i)^{n-m-1} u(n - m - 1)$$

$$y(n) = C_1(1 + i)^n - pu(n - 1) \sum_{m=0}^{n-1} (1 + i)^{n-m-1}$$

$$y(n) = C_1(1 + i)^n - pu(n - 1) \sum_{l=0}^{n-1} (1 + i)^l$$

The *initial* debt is $y(0) \equiv d$, so

$$y(0) = C_1 = d$$

or

$$y(n) = (1+i)^n d - pu(n-1) \sum_{l=0}^{n-1} (1+i)^l$$

We can obtain a *closed form* by using Eq. C.23, App. C, for the *geometric series* that appears above. Thus

$$y(n) = (1+i)^n d - p \frac{(1+i)^{n-1}}{i} u(n-1)$$

If the debt is repaid in N periods, then $y(N) \equiv 0$, or

$$0 = (1+i)^N d - p \frac{(1+i)^N - 1}{i}$$

or solving for p

$$p = \frac{i(1+i)^N}{(1+i)^N - 1}$$

At 12 percent per year (or 1 percent per month) a \$12,000 automobile (\$16,000 less \$4,000 down payment) requires equal payments over 48 months:

$$p = \frac{0.01(1.01)^{48}}{(1.01)^{48} - 1} (12{,}000) = \$316.01$$

As a last example in this section we consider a situation where discrete time is not the independent variable, but position index is. It is not necessary to make any changes in notation, however.

Example 13

Find the node voltage $v(n)$ for the ladder network made up of N sections with terminating resistance R as shown in Fig. 8.17. At the $n+1$ node Kirchhoff's current law gives

$$\frac{v(n) - v(n+1)}{R} + \frac{v(n+2) - v(n+1)}{R} - \frac{v(n+1)}{aR} = 0$$

or

$$v(n+2) - (2 + 1/a)v(n+1) + v(n) = 0$$

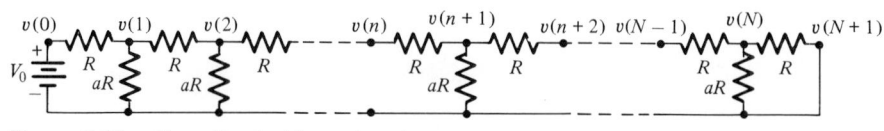

Figure 8.17. *N*-section ladder network.

Thus, we are only required to find a complementary function to obtain the complete solution. We choose $a = 1$ to simplify matters, and so

$$v(n + 2) - 3v(n + 1) + v(n) = 0$$

The characteristic equation is

$$z^2 - 3z + 1 = 0$$

So

$$z_1 = \frac{3 - \sqrt{5}}{2} \qquad z_2 = \frac{3 + \sqrt{5}}{2} \tag{8.48}$$

and

$$v(n) = C_1 z_1^n + C_2 z_2^n$$

We have two known *boundary conditions*: $v(0) = V_0$ and $v(N + 1) = 0$, so

$$\left. \begin{array}{l} V_0 = C_1 + C_2 \\ \\ 0 = C_1 z_1^{N+1} + C_2 z_2^{N+1} \end{array} \right\} \quad \begin{array}{l} C_1 = \dfrac{V_0 z_2^{N+1}}{z_2^{N+1} - z_1^{N+1}} \\ \\ C_2 = \dfrac{-V_0 z_1^{N+1}}{z_2^{N+1} - z_1^{N+1}} \end{array}$$

Thus

$$v(n) = \frac{V_0}{z_2^{N+1} - z_1^{N+1}} [z_2^{N+1} z_1^n - z_1^{N+1} z_2^n] \tag{8.49}$$

Eqs. 8.48 and 8.49 give the complete solution.

The input resistance is given by

$$R_{in} = \frac{RV_0}{V_0 - v(1)}$$

or simplifying

$$R_{in} = R \frac{z_2^{N+1} - z_1^{N+1}}{z_2^{N+1}(1 - z_1) - z_1^{N+1}(1 - z_2)}$$

$R_{in} = 1.5R$ for $N = 1$
$R_{in} = 1.6R$ for $N = 2$

It is not exactly obvious what R_{in} is for an infinite number of sections, but using a little trickery we can find it. Consider Fig. 8.18. As indicated, adding one more

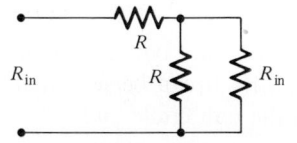

Figure 8.18. Simplified network for R_{in} ($N \to \infty$).

section to an infinite number of sections will not change the input resistance, so

$$R_{in} = R + \frac{RR_{in}}{R + R_{in}}$$

or

$$R_{in}^2 - RR_{in} - R^2 = 0$$

or

$$R_{in} = R\frac{+1 \pm \sqrt{5}}{2}$$

The minus sign cannot apply, so

$$R_{in} = \frac{\sqrt{5}+1}{2}R = 1.618R \qquad (N \to \infty)$$

8.6 THE SECOND-ORDER SYSTEM

We have used the second-order difference equation in several of the examples describing second-order systems. Because it does occur frequently in practice, we would like to formalize the description for the case where the homogeneous difference equation is

$$y(n+2) - 2ay(n+1) + y(n) = 0 \tag{8.50}$$

The characteristic equation is

$$z^2 - 2az + 1 = 0$$

and the roots are

$$z_1 = a - \sqrt{a^2 - 1} \qquad z_2 = a + \sqrt{a^2 - 1} \tag{8.51}$$

so that the zero-input response is

$$y_{zi}(n) = C_1(a - \sqrt{a^2 - 1})^n + C_2(a + \sqrt{a^2 - 1})^n \tag{8.52}$$

Consider a to be a parameter. We treat several special cases.

1. $a = 0$.

$$y_{zi}(n) = C_1(-j)^n + C_2(+j)^n \tag{8.53}$$

2. $|a| < 1$.

$$y_{zi}(n) = C_1(a - j\sqrt{1-a^2})^n + C_2(a + j\sqrt{1-a^2})^n \tag{8.54}$$

Noticing that $|z_1| = |z_2| = 1$, we can put this result in a better looking form. The case for $0 < a < 1$ is shown in Fig. 8.19. It can be seen that the roots of the characteristic equation lie on the unit circle and

$$\cos(\phi) = a \tag{8.55}$$

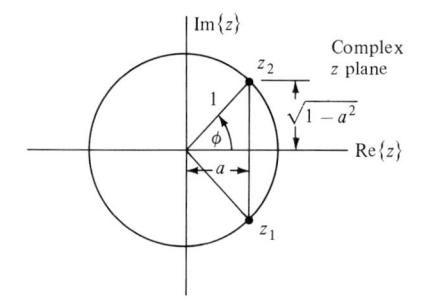

Figure 8.19. Roots of the characteristic equation in the complex z plane, $0 < a < 1$.

or
$$\sin(\phi) = \sqrt{1-a^2} \tag{8.56}$$

Rewriting Eq. 8.54
$$y_{zi}(n) = C_1(\cos\phi - j\sin\phi)^n + C_2(\cos\phi + j\sin\phi)^n$$
$$y_{zi}(n) = C_1(e^{-j\phi})^n + C_2(e^{j\phi})^n$$
$$y_{zi}(n) = C_1 e^{-jn\phi} + C_2 e^{+jn\phi} \tag{8.57}$$

The last result can also be written (see Sec. 2.6) as
$$y_{zi}(n) = C_3\cos(n\phi) + C_4\sin(n\phi) \tag{8.58}$$

3. $a = 1$.
$$z_1 = z_2 = 1$$
Here we have repeated roots, so
$$y_{zi}(n) = C_1 + nC_2 \tag{8.59}$$

4. $a = -1$.
$$z_1 = z_2 = -1$$
$$y_{zi}(n) = C_1(-1)^n + nC_2(-1)^n$$
$$y_{zi}(n) = (C_1 + nC_2)(-1)^n \tag{8.60}$$

5. $|a| > 1$. If we make the definitions
$$\cosh(\phi) = a \tag{8.61}$$
$$\sinh(\phi) = \sqrt{a^2 - 1} \tag{8.62}$$
then $\cosh^2(\phi) - \sinh^2(\phi) = 1$ as required. Thus
$$y_{zi}(n) = C_1(\cosh\phi - \sinh\phi)^n + C_2(\cosh\phi + \sinh\phi)^n$$
$$y_{zi}(n) = C_1(e^{-\phi})^n + C_2(e^{\phi})^n$$
$$y_{zi}(n) = C_1 e^{-n\phi} + C_2 e^{+n\phi} \tag{8.63}$$

or

$$y_{zi}(n) = C_3\cosh(n\phi) + C_4\sinh(n\phi) \tag{8.64}$$

Notice that for $a \leq 1$ the roots always lie on the unit circle in the complex z plane. Do the discrete systems represented by $a \leq 1$ represent stable systems? What about $|a| > 1$?

Example 14

Find the zero-state response for the discrete system of Fig. 8.20. The difference equation is

$$y(n) = x(n) - y(n-2)$$

or

$$y(n) + y(n-2) = x(n)$$

or

$$y(n+2) + y(n) = x(n+2) = x_f(n)$$

This fits case 1 (above) so

$$y_{zi}(n) = D_1(-j)^n + D_2(+j)^n$$

$$\begin{aligned} 0 &= D_1 + D_2 \\ 1 &= -jD_1 + jD_2 \end{aligned} \Biggr\} \begin{aligned} D_1 &= +j/2 \\ D_2 &= -j/2 \end{aligned}$$

$$h_f(n) = \tfrac{1}{2}j[(-j)^{n-1} - (+j)^{n-1}]u(n-1)$$
$$h_f(n) = \tfrac{1}{2}[(-j)^{n-2} - (+j)^{n-2}]u(n-1)$$
$$h(n) = h_f(n+2) = \tfrac{1}{2}[(-j)^n - (+j)^{n+2}]u(n+1)$$

$$y_{zi}(n) = \sum_{m=-\infty}^{\infty} x(m)h(n-m)$$

$$= \sum_{m=-\infty}^{\infty} 3^{-m}\cos\frac{m\pi}{2}u(m) \cdot \frac{1}{2}[(-j)^{n-m} - (+j)^{n-m+2}]u(n-m+1)$$

$$= \frac{1}{2}\cos\frac{n\pi}{2}\sum_{m=0}^{n+1} 3^{-m}[1 + (-1)^m]$$

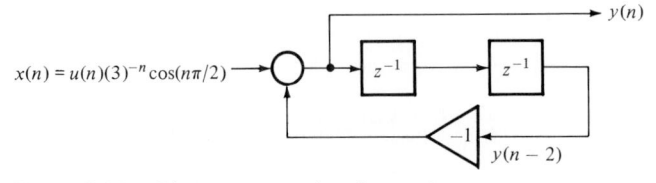

$x(n) = u(n)(3)^{-n}\cos(n\pi/2)$ → [z^{-1}] → [z^{-1}] → $y(n)$

$y(n-2)$

Figure 8.20. Discrete second-order system.

Example 15

Find the unit-function response for $a = \frac{1}{2}$ (case 2). From Eq. 8.55

$$\phi = \cos^{-1}(1/2) = \pi/3$$

while from Eq. 8.58

$$y_{zi}(n) = D_1\cos(n\pi/3) + D_2\sin(n\pi/3)$$

Thus

$$0 = D_1$$
$$1 = D_1\cos(\pi/3) + D_2\sin(\pi/3) \qquad D_2 = 2/\sqrt{3}$$

Therefore, the unit-function response for $x_f(n)$ in

$$y(n+2) - y(n+1) + y(n) = x_f(n)$$

is

$$h_f(n) = (2/\sqrt{3})\sin[(n-1)\pi/3]u(n-1)$$

8.7 FREQUENCY RESPONSE

As was seen earlier in Chap. 5, we are often interested in the frequency response of linear systems. In the continuous-time case the techniques of Bode analysis lead to the Bode diagram, and it consists of a plot of both the magnitude and phase (angle) of $H(\omega)$ versus ω. We discussed several methods for determining the transfer function $H(\omega)$, but in simple terms, if we apply $e^{j\omega t}$ to a continuous-time (time-invariant) linear system, then the response will be $H(\omega)e^{j\omega t}$ as shown in Fig. 8.21(a). Thus, $H(\omega)$ is simply the ratio of the output to the input when the input is $e^{j\omega t}$. The frequency response, or Bode diagram, gives the amplitude and phase response of the system to input sinusoids at all frequencies. These characteristics are often used to specify system performance.

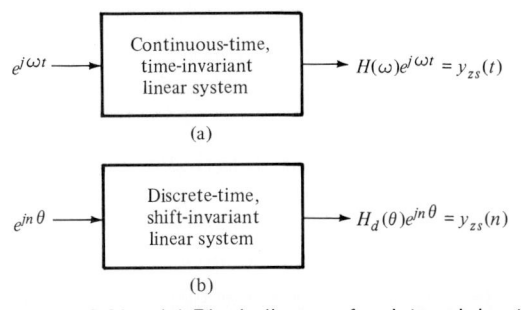

Figure 8.21. (a) Block diagram for determining the frequency response of a continuous-time, time-invariant linear system. (b) Block diagram for determining the frequency response of a discrete-time, shift-invariant linear system.

In much the same way, the engineer is often interested in the frequency response of discrete systems. Here, however, we are only interested in discrete times, $t = nT$, so we apply $e^{j\omega nT} = e^{jn\theta}$ with $\theta = \omega T$, as in Fig. 8.21(b). As we shall see, this produces the response[†] $H_d(e^{j\theta})e^{jn\theta} = H_d(\theta)e^{jn\theta}$ in all cases. Notice that this is a zero-state response, and (as in the continuous-time case) it is also a steady-state response. Since the input has been applied for all time $(-\infty < n < \infty)$, $H_d(\theta)$ is called the frequency response of the discrete system.

The zero-state response for *any* input is given by Eq. 8.15 ($l = m$):

$$y_{zs}(n) = \sum_{m=-\infty}^{\infty} x(n-m)h(m) \tag{8.65}$$

In particular, if $x(n) = e^{jn\theta}$, then

$$y_{zs}(n) = \sum_{m=-\infty}^{\infty} e^{j(n-m)\theta} h(m)$$

$$y_{zs}(n) = e^{jn\theta} \sum_{m=-\infty}^{\infty} h(m) e^{-jm\theta} \tag{8.66}$$

If we *define* the summation to be $H_d(\theta)$, then, as predicted

$$y_{zs}(n) = e^{jn\theta} H_d(\theta) \tag{8.67}$$

where

$$\boxed{H_d(\theta) \equiv \sum_{m=-\infty}^{\infty} h(m) e^{-jm\theta}} \tag{8.68}$$

This is an important result for several reasons. It gives the discrete frequency response $H_d(\theta)$ in terms of the unit-function response $h(n)$. It shows that $H_d(\theta)$ is periodic in θ with period 2π: that is, $H_d(\theta) = H_d(\theta \pm 2p\pi)$, $p = 0, 1, 2, \ldots$. Equation 8.68 has the form of a complex (exponential) Fourier series, so, given $H_d(\theta)$, we can find $h(n)$ using the Euler formula (see Chap. 4) for the Fourier coefficients:

$$h(n) = \frac{1}{2\pi} \int_{-\pi}^{\pi} H_d(\theta) e^{jn\theta} d\theta \tag{8.69}$$

These results should not be surprising. We found that a periodic continuous-time function gives a *discrete frequency* spectrum. Thus, a periodic continuous frequency function (spectrum) should result in (or from) a *discrete-time* function or sequence. The frequency response is a *continuous* function of the frequency of the *sampled* continuous waveforms that give the input sequence.

[†] We use a subscript d (for digital) to distinguish $H_d(\theta)$ and $H(\omega)$.

Equation 8.18 gives a general difference equation for shift-invariant discrete systems:

$$\sum_{m=0}^{N} a_m y(n+m) = \sum_{m=0}^{M} b_m x(n+m) \tag{8.18}$$

Since we are only interested here in the steady-state response ($-\infty < n < \infty$), $y(n) = y_{zs}(n)$, and the input is $x(n) = e^{jn\theta}$. Therefore

$$\sum_{m=0}^{N} a_m y_{zs}(n+m) = \sum_{m=0}^{M} b_m e^{j(n+m)\theta} \tag{8.70}$$

Using Eq. 8.67 in 8.70

$$\sum_{m=0}^{N} a_m e^{j(n+m)\theta} H_d(\theta) = \sum_{m=0}^{M} b_m e^{j(n+m)\theta}$$

Multiplying both sides by $e^{-jn\theta}$, and solving for $H_d(\theta)$

$$\boxed{H_d(\theta) = \frac{\sum_{m=0}^{M} b_m e^{jm\theta}}{\sum_{m=0}^{N} a_m e^{jm\theta}}} \tag{8.71}$$

This is a result that gives the frequency response in terms of the coefficients in the difference equation. Recall earlier results for the continuous-time case with $e^{j\omega t}$ and e^{st} applied to the differential equation (2.4):

$$H(\omega) = \frac{\sum_{m=0}^{M} b_m (j\omega)^m}{\sum_{m=0}^{N} a_m (j\omega)^m} \tag{5.47}$$

and

$$H(s) = \frac{\sum_{m=0}^{M} b_m s^m}{\sum_{m=0}^{N} a_m s^m} \tag{7.2}$$

respectively. An even more obvious resemblance occurs when we consider the z transform in the next chapter.

Substituting Eq. 8.71 into Eq. 8.69 gives us another way to find the unit-function response. This may not be an efficient way (numerically speaking) to find $h(n)$ because of the sum in the denominator in Eq. 8.71. In order to perform the integration we would need to expand the right side of Eq. 8.71 into a series that, when placed in Eq. 8.69, would permit term-by-term integration. This will be demonstrated for a simple case in the examples to follow.

A digital filter is a discrete-time linear system that operates on an input sequence, modifies it, and produces the output sequence. As mentioned earlier, these filters can be of the recursive or nonrecursive type. Recursive filters possess an infinite memory because they can utilize previous values of the output. For a given cutoff characteristic they usually require fewer elements than the nonrecursive type. On the other hand the phase characteristics of recursive filters are generally poorer than those for the nonrecursive type. As implied, the nonrecursive filter has a finite memory because only a finite number of delays can be used in a practical situation.

We do not intend to become deeply involved in digital filter design, but we do need to at least consider it. As a first step, suppose that we want to *digitalize* a continuous-time filter. This can be done in various ways.† We will look at one method at the present time. Let the unit-function response have sample values that equal those that are obtained when the unit-impulse response of the continuous-time filter is sampled (see Sec. 8.1). This is demonstrated here by using a simple RC low-pass filter as the continuous-time (analog) filter that we want to digitalize. This is a first-order system that we can describe with a first-order differential equation. Consequently, its digital counterpart is described by a first-order difference equation. The analog filter is shown in Fig. 8.22(a), and the transfer function is easily found to be

$$H(\omega) = \frac{\omega_2}{j\omega + \omega_2} \qquad (8.72)$$

where $\omega_2 = 1/(RC)$ is the bandwidth, or half-power frequency, of the RC filter. Its unit-impulse response is also easily found:

$$h(t) = \omega_2 e^{-\omega_2 t} u(t) \qquad (8.73)$$

Setting $t = nT$ in Eq. 8.73 (as agreed upon), we obtain

$$h(nT) = \omega_2 e^{-\omega_2 nT} u(nT) \qquad (8.74)$$

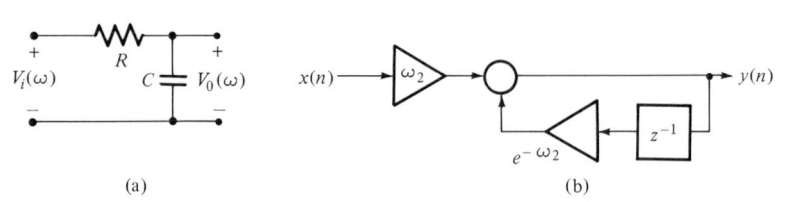

Figure 8.22. (a) Analog RC low-pass filter. (b) Digital version of the low-pass filter.

† This is called an impulse invariant (IIR) filter.

It is easy to show that

$$h(nT) - e^{-\omega_2 T} h(nT - T) = \omega_2 \delta(nT) \tag{8.75}$$

or, with $T = 1$ for simplicity

$$h(n) - e^{-\omega_2} h(n-1) = \omega_2 \delta(n) \tag{8.76}$$

Thus, the difference equation must be

$$y(n) - e^{-\omega_2} y(n-1) = \omega_2 x(n) \tag{8.77}$$

The block diagram of this digital filter is shown in Fig. 8.22(b). Much insight is gained if we compare the frequency responses of the systems shown in Fig. 8.22.

Example 16

Compare $H(\omega)$ and $H_d(\theta)$ for the filters of Fig. 8.22. The term $H(\omega)$ is given by Eq. 8.72, and $H_d(\theta)$ is obtained from Eq. 8.71:

$$H_d(\theta) = \frac{\omega_2 e^{j\theta}}{-e^{-\omega_2} + e^{j\theta}} = \frac{\omega_2}{1 - e^{-\omega_2} e^{-j\theta}}$$

Thus, with $\theta = \omega T = \omega$

$$|H_d(\theta)| = \omega_2 \left(\frac{1}{1 + e^{-2\omega_2} - 2e^{-\omega_2} \cos(\omega)} \right)^{1/2}$$

$$\underline{/H_d(\theta)} = -\tan^{-1}\left[\frac{e^{-\omega_2} \sin \omega}{1 - e^{-\omega_2} \cos(\omega)} \right]$$

$$|H(\omega)| = \left[\frac{1}{1 + (\omega/\omega_2)^2} \right]^{1/2}$$

$$\underline{/H(\omega)} = -\tan^{-1}(\omega/\omega_2)$$

Here, we choose to graph these quantities versus ω/ω_s (where ω_s is the sampling frequency, $\omega_s = 2\pi/T = 2\pi$) on a *linear* scale. That is, we are *not* using Bode diagrams because we want the periodic nature of $H_d(\theta)$ to be obvious. Three cases ($\omega_2 = 1$, 0.5, and 0.1) are shown in Fig. 8.23. Notice that the maximum response has been normalized to 0 dB in each case. The approximation involved in the digitalization process is clearly evident. The folding or *aliasing* effect is a result of undersampling. Remember that $h(t)$ in this example is *not* band-limited, and, therefore, aliasing will *always* occur even though the severity of its effect is reduced as ω_s/ω_2 is increased. Notice that the phase characteristics are, as predicted earlier, rather poor. It should also be pointed out to the reader that in Fig. 8.23 the plots of $H(\omega)$ are related to each other by a simple *frequency scaling*. This is, of course, *not* the case for the plots of $H_d(\theta)$.

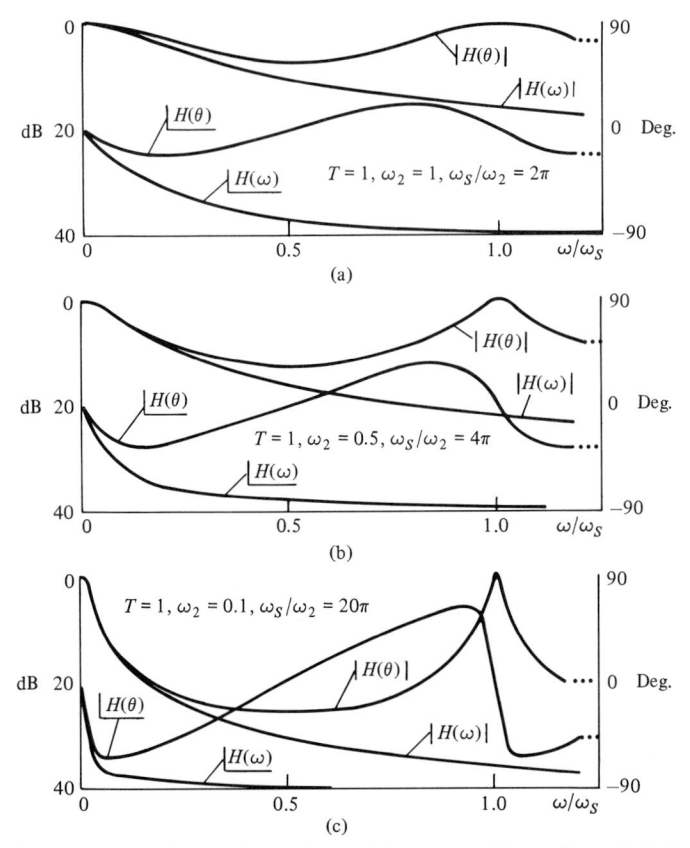

Figure 8.23. Comparison of an RC low-pass filter with a digitalized version in terms of frequency response. (a) $\omega_2 = 1$, $\omega_s/\omega_2 = 2\pi$. (b) $\omega_2 = 0.5$; $\omega_s/\omega_2 = 4\pi$. (c) $\omega_2 = 0.1$; $\omega_s/\omega_2 = 20\pi$.

Example 17

Given that

$$H_d(\theta) = \frac{\omega_2}{1 - e^{-\omega_2}e^{-j\theta}}$$

as in Example 16, find $h(n)$. As mentioned earlier, we must expand $H_d(\theta)$. Using long division (or a binomial series)[†], we obtain

$$H_d(\theta) = \omega_2(1 + e^{-\omega_2}e^{-j\theta} + e^{-2\omega_2}e^{-j2\theta} + \cdots)$$

Substituting this result into Eq. 8.69 gives

$$h(n) = 0 \quad n < 0$$
$$h(0) = \omega_2 \quad h(1) = \omega_2 e^{-\omega_2}$$
$$h(2) = \omega_2 e^{-2\omega_2}, \ldots$$

[†] See Eq. C.27, App. C.

or

$$h(n) = \omega_2 e^{-n\omega_2} u(n)$$

in agreement with Eq. 8.74.

Next, suppose that we want to *design* a nonrecursive finite length (FIR) filter.

Example 18

Design a low-pass filter with a cutoff frequency ω_0 and a sampling frequency $\omega_s = 2\pi/T = 4\omega_0$. Taking $T = 1$ again, we have $\omega_s = 2\pi$ and $\omega_0 = (1/4)\omega_s = \pi/2$. Thus, $\theta = \omega T = \omega$ and its cutoff value is $\theta_0 = \omega_0 = \pi/2$. We start by being ambitious and requesting an *ideal* low-pass filter, even though we know we are not going to achieve this. Thus, we would like to have

$$H_d(\theta) = \begin{cases} 1 & |\theta| < \pi/2 \\ 0 & |\theta| > \pi/2 \end{cases} = u(\theta + \pi/2) - u(\theta - \pi/2) \quad -\pi < \theta < \pi \quad (8.78)$$

Remember that $H_d(\theta)$ is continuous [except at $\theta = (2p+1)\pi/2$, $p = 0, \pm 1, \pm 2, \ldots$]. The unit-function response sequence for this filter is given by Eq. 8.69:

$$h(n) = \frac{1}{2\pi} \int_{-\pi}^{\pi} [u(\theta + \pi/2) - u(\theta - \pi/2)] e^{jn\theta} d\theta$$

$$= \frac{1}{2\pi} \int_{-\pi/2}^{\pi/2} e^{jn\theta} d\theta = \frac{1}{\pi} \int_0^{\pi/2} \cos(n\theta) d\theta$$

$$h(n) = \tfrac{1}{2} Sa(n\pi/2) \quad -\infty < n < \infty \quad (8.79)$$

The results for Eqs. 8.78 and 8.79 are shown in Fig. 8.24. Inspection of $h(n)$ shows that we cannot attain our goal for at least two reasons. First, we can only implement a finite number of delays, not the infinite number called for by Eq. 8.79. Secondly, $h(n)$ is not causal since it represents a response $(n < 0)$ before the unit function is applied at $n = 0$. In order to overcome the first difficulty, let us agree to be content with 11 samples (or 11 delays in the block diagram).[2] This compromise raises another question. Do we want to choose

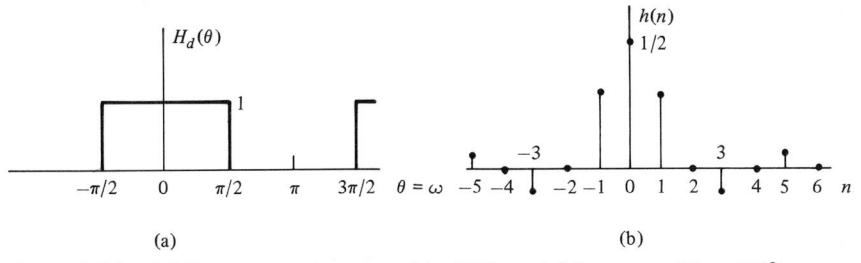

Figure 8.24. (a) Frequency response, ideal (discrete) low-pass filter, $H_d(\theta)$. (b) Unit-function response of the filter.

our 11 sample values according to Eq. 8.79 or do we want to adjust them now that we are designing an FIR filter? The answer to this question is somewhat nebulous. Certainly, if we want the $H_d(\theta)$ that we can obtain with a finite sequence to *fit* the ideal frequency response given by Eq. 8.78 with *least mean square error*, then, as we showed in Chap. 4, we *must* pick the coefficients to be the Fourier coefficients as calculated (Eq. 8.79). This may not give us exactly what we want, however. In order to find out about this, we will simply try it: The truncated sequence with Fourier coefficients is

$$h(n) = \tfrac{1}{2}\,\text{Sa}(n\pi/2) \qquad -5 \le n \le 5 \tag{8.80}$$

We can introduce a time delay to overcome the causality problem. We require that $h(n)$ start no earlier than $t = 0$ ($n = 0$), so we replace $h(n)$ in Eq. 8.80 with $h(n-5)$. Thus, we now have a new $h(n)$:

$$h(n) = \tfrac{1}{2}\,\text{Sa}[(n-5)\pi/2] \qquad 0 \le n \le 10 \tag{8.81}$$

This new truncated sequence is shown in Fig. 8.25. Notice that it is just the shifted and truncated version of the original sequence given by Eq. 8.79 and pictured in Fig. 8.24(b).

The new frequency response is given by Eqs. 8.68 and 8.81:

$$H_d(\theta) = \sum_{m=0}^{10} \tfrac{1}{2}\,\text{Sa}[(m-5)\pi/2]e^{-jm\theta} \tag{8.82}$$

Replacing $m - 5$ with n

$$H_d(\theta) = \tfrac{1}{2}e^{-j5\theta}\sum_{n=-5}^{5}\text{Sa}(n\pi/2)e^{-jn\theta} \tag{8.83}$$

This frequency response is shown in Fig. 8.26. Notice that Eq. 8.83 reduces to

$$H_d(\theta) = \tfrac{2}{\pi}e^{-j5\theta}\left[\tfrac{\pi}{4} + \cos(\theta) - \tfrac{1}{3}\cos(3\theta) + \tfrac{1}{5}\cos(5\theta)\right] \tag{8.84}$$

so

$$\underline{/H_d(\theta)} = -5\theta \tag{8.85}$$

The block diagram for Eq. 8.81 or Fig. 8.25 would be easy to obtain if approached directly, although this might not give the most economical version.

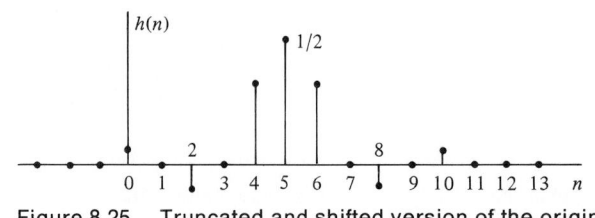

Figure 8.25. Truncated and shifted version of the original $h(n)$ [Fig. 8.22(b)].

8.7 FREQUENCY RESPONSE

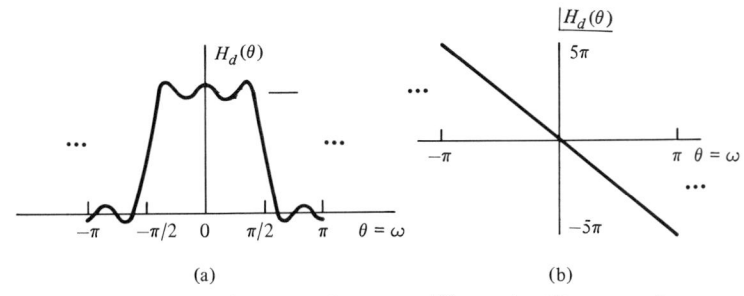

Figure 8.26. (a) $H_d(\theta)$ versus θ (less $e^{-j5\theta}$). (b) $|H_d(\theta)|$ versus θ.

The frequency response obtained from Eq. 8.84 and shown in Fig. 8.26(a) exhibits some ripple effect even though (or *because*) it represents a least mean square fit to the desired frequency response. This situation can be improved by adjusting the coefficients in the sequence given by Eq. 8.81. This is obviously equivalent to multiplying the sequence given by Eq. 8.81 by another sequence, called a *window* sequence, $w(n)$. That is, the new sequence is given by

$$h(n) = \tfrac{1}{2}\text{Sa}[(n-5)\pi/2]w(n) \qquad 0 \le n \le 10 \tag{8.86}$$

For example, a *rectangular window*

$$w(n) = \sum_{n=0}^{10} u(n)$$

which is all *ones* for $0 \le n \le 10$ produces Eq. 8.81 again and thus accomplishes nothing. The elements in the window sequence must be tapered.

Example 19

Find the frequency response of the low-pass filter of Example 18 when a *triangular* window

$$w(n) = 1 - |n-5|/6 \qquad 0 \le n \le 10 \tag{8.87}$$

is used. Using Eqs. 8.68, 8.86, and 8.87

$$H_d(\theta) = \sum_{m=0}^{10} \tfrac{1}{2}\text{Sa}[(m-5)\pi/2][1-|m-5|/6]e^{-jm\theta}$$

or, with $m - 5 = n$

$$H_d(\theta) = \frac{1}{12}e^{-j5\theta}\sum_{n=-5}^{5}\text{Sa}(n\pi/2)[6-|n|]e^{-jn\theta}$$

This reduces to

$$H_d(\theta) = \frac{1}{3\pi}e^{-j5\theta}\left[\frac{3\pi}{2} + 5\cos(\theta) - \cos(3\theta) + \frac{1}{5}\cos(5\theta)\right] \tag{8.88}$$

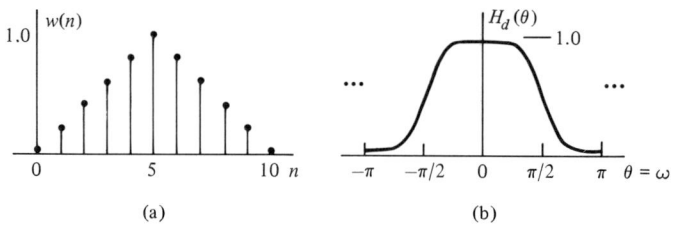

Figure 8.27

Figure 8.27. (a) Triangular window sequence. (b) Frequency response, low-pass filter (Fourier coefficients) with triangular window.

The triangular window and the frequency response obtained by using it with the Fourier series coefficients [that is, Eq. 8.88 (less the phase term $e^{-j5\theta}$)] are shown in Fig. 8.27. The ripple has virtually disappeared, but the slope of the cutoff characteristic is not as great as in Fig. 8.26(a). Also, this filter never completely "cuts off." This is the kind of trade-off that the engineer is often faced with: No single filter has *all* the desirable characteristics. We can make the filter almost cut off and essentially have unity response in the passband by using a slightly different window.

Example 20

Find the frequency response of the low-pass filter of Example 18 using the Hamming[2] window:

$$w(n) = \frac{1 + \cos[(n-5)\pi/6]}{2} \qquad 0 \le n \le 10 \qquad (8.89)$$

It is left as an exercise to show that

$$H_d(\theta) = \frac{1}{\pi} e^{-j5\theta} \left[\frac{\pi}{2} + (1 + \cos \pi/6)\cos(\theta) - \frac{1}{3}\cos(3\theta) \right.$$

$$\left. + \frac{1}{5}(1 + \cos 5\pi/6)\cos(5\theta) \right] \qquad (8.90)$$

As can be seen in Fig. 8.28, this digital filter has what one would normally consider to be excellent characteristics. It is also slightly more difficult to analyze.

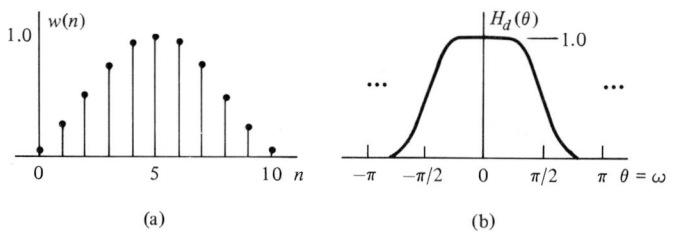

Figure 8.28. (a) Hamming window sequence. (b) Frequency response, low-pass filter (Fourier coefficients) with Hamming window.

It should be mentioned that the results for the last two examples can also be obtained in the frequency domain. That is, the $H_d(\theta)$ given by Eq. 8.82 (obtained from Eq. 8.81) can be convolved with the frequency response of the window, $W_d(\theta)$, to obtain the new frequency response:

$$H_d(\theta) = \frac{1}{2\pi} \int_{-\pi}^{\pi} \left\{ \frac{1}{2} \sum_{m=0}^{10} \text{Sa}[(m-5)\pi/2]e^{-jm\alpha} \right\} W_d(\theta - \alpha) d\alpha \qquad (8.91)$$

Notice that this is *circular* convolution, as in Sec. 5.9 in connection with the DFT. Equation 8.91 is not exactly inviting since it involves integration.

We can easily specify the elements of the digital filters from the frequency responses after we have introduced the z transform. See Example 17, Chap. 9, and the material that follows.

8.8 STATE-SPACE ANALYSIS

A discrete system in which the internal states of the system are to be used for analysis or optimization requires a model that defines the internal states. This is the state-space model. The state variables for the discrete system will be $q_1(n), q_2(n), q_3(n), \ldots$. For a system described by the general difference form (Eq. 8.18) the state of the system will normally be the *contents* of the unit delays that appear in the block diagram representation of the difference equation.† Thus, the state is the *memory* of the system. These unit delays hold the entering value of the variable for one clock cycle T. The contents of the unit delays give us the entire past behavior of the system (memory). We will find, as in the continuous-time case, that one of the greatest advantages of state-space analysis is organization, and this is due primarily to the fact that it is ideal for matrix operations.

Analogous to the situation in Chap. 3 for continuous-time systems, the *normal-form* difference equations are ($T = 1$ again)

$$\boxed{\mathbf{q}(n+1) = \mathbf{A}\mathbf{q}(n) + \mathbf{B}\mathbf{x}(n)} \qquad (8.92)$$

and the output vector is

$$\boxed{\mathbf{y}(n) = \mathbf{C}\mathbf{q}(n) + \mathbf{D}\mathbf{x}(n)} \qquad (8.93)$$

Notice that Eq. 8.92 represents N *first-order* coupled difference equations, not one Nth order difference equation. The input (sequence) vector is $\mathbf{x}(n)$, and $\mathbf{B}\mathbf{x}(n) = \mathbf{f}(n)$ may be called the forcing function (sequence) vector. If **A**, **B**, **C**, and **D** are *constant* matrices, then we have a shift-invariant system (the only case considered here). We now have at least three characterizations of a discrete-time system: difference equations, unit-function response, and, now, the matrices **A**, **B**, **C**, and **D**.

---

† Other sets of state variables are possible, as we will see in Chap. 9. Those here are called the *natural* state variables.

How are Eqs. 8.92 and 8.93 obtained? In order to answer this question we reconsider the second-order system of Example 5 (Fig. 8.12) that was characterized by the difference equation (Eq. 8.40):

$$3y(n) + 9y(n-1) + 6y(n-2) = x(n) \tag{8.40}$$

and whose unit-function response was

$$h(n) = \tfrac{1}{3}[(-1)^{n+1} - (-2)^{n+1}]u(n+1) \tag{8.94}$$

The block diagram is shown in Fig. 8.29. We define the *two* state variables for this *second-order* system to be the contents of the *two* delay elements

$$q_1(n) \equiv y(n-1) \tag{8.95}$$
$$q_2(n) \equiv y(n-2) \tag{8.96}$$

Notice that we could just as well define $q_1(n) \equiv y(n-2)$ and $q_2(n) \equiv y(n-1)$, and generally this would result in *different* **A**, **B**, **C**, and **D** matrices. Using Fig. 8.29, the recursion formulas for the state variables are

$$q_1(n+1) = y(n) = \tfrac{1}{3}x(n) - 3y(n-1) - 2y(n-2)$$
$$q_2(n+1) = y(n-1) = q_1(n)$$

or, in normal form

$$q_1(n+1) = -3q_1(n) - 2q_2(n) + \tfrac{1}{3}x(n)$$
$$q_2(n+1) = q_1(n)$$

In terms of the state variables, the output is

$$y(n) = \tfrac{1}{3}x(n) - 3q_1(n) - 2q_2(n)$$

or

$$y(n) = -3q_1(n) - 2q_2(n) + \tfrac{1}{3}x(n)$$

Thus, in matrix form, we have

$$\begin{bmatrix} q_1(n+1) \\ q_2(n+1) \end{bmatrix} = \begin{bmatrix} -3 & -2 \\ 1 & 0 \end{bmatrix} \begin{bmatrix} q_1(n) \\ q_2(n) \end{bmatrix} + \begin{bmatrix} \tfrac{1}{3} \\ 0 \end{bmatrix} [x_1(n)] \tag{8.97}$$

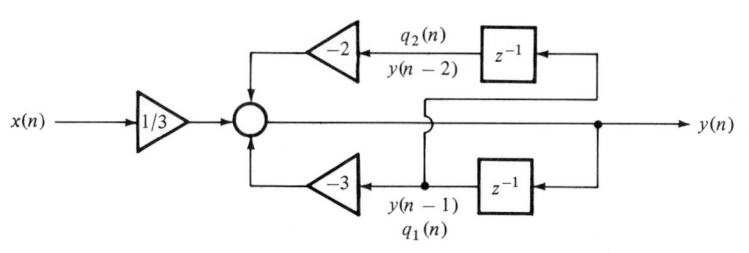

Figure 8.29. Block diagram for a second-order system (Eq. 8.40).

and

$$[y(n)] = [-3 \quad -2]\begin{bmatrix} q_1(n) \\ q_2(n) \end{bmatrix} + [\tfrac{1}{3}][x_1(n)] \tag{8.98}$$

where the *system matrix* is

$$\mathbf{A} = \begin{bmatrix} -3 & -2 \\ 1 & 0 \end{bmatrix} \tag{8.99}$$

and is $N \times N$ for N state variables. The input coefficient matrix $\mathbf{B}$ is $N \times M$ for M inputs and here is (2×1)

$$\mathbf{B} = \begin{bmatrix} \tfrac{1}{3} \\ 0 \end{bmatrix} \quad \mathbf{B}\mathbf{x}(n) = \tfrac{1}{3}x_1(n) = \mathbf{f}(n) \tag{8.100}$$

If there are P outputs then the matrix $\mathbf{C}$ is $P \times N$. Here it is (1×2)

$$\mathbf{C} = [-3 \quad -2] \tag{8.101}$$

The matrix $\mathbf{D}$ is $P \times M$, and here it is (1×1)

$$\mathbf{D} = [\tfrac{1}{3}] \tag{8.102}$$

One rather obvious and easily understood method[3] for solving Eq. 8.92 is by iteration. That is, we recursively solve for successive values of the state variables in terms of the preceding values, given the initial state $\mathbf{q}(0)$. The reader may want to consider the relations between this recursive method for the discrete system and that for the *approximate* numerical solution of the continuous system using state variables and the *difference* approximation of the derivative (Eq. 3.110). Thus

$$\mathbf{q}(1) = \mathbf{A}\mathbf{q}(0) + \mathbf{B}\mathbf{x}(0)$$
$$\mathbf{q}(2) = \mathbf{A}\mathbf{q}(1) + \mathbf{B}\mathbf{x}(1)$$
$$\quad = \mathbf{A}[\mathbf{A}\mathbf{q}(0) + \mathbf{B}\mathbf{x}(0)] + \mathbf{B}\mathbf{x}(1)$$
$$\quad = \mathbf{A}^2\mathbf{q}(0) + \mathbf{A}\mathbf{B}\mathbf{x}(0) + \mathbf{B}\mathbf{x}(1)$$
$$\mathbf{q}(3) = \mathbf{A}\mathbf{q}(2) + \mathbf{B}\mathbf{x}(2)$$
$$\quad = \mathbf{A}^3\mathbf{q}(0) + \mathbf{A}^2\mathbf{B}\mathbf{x}(0) + \mathbf{A}\mathbf{B}\mathbf{x}(1) + \mathbf{B}\mathbf{x}(2)$$

Continuing this process gives

$$\boxed{\mathbf{q}(n) = \mathbf{A}^n\mathbf{q}(0) + \sum_{m=0}^{n-1} \mathbf{A}^{n-m-1}\mathbf{B}\mathbf{x}(m)} \tag{8.103}$$

This is called the *discrete state-transition equation* and can be compared to Eq. 3.81 for the continuous-time case. $\mathbf{A}^n = \mathbf{A}\mathbf{A}\cdots\mathbf{A}$ is called the *discrete transition matrix* of the system. If the initial state occurs for $n = n_0$ (not $n = 0$), then Eq. 8.103 becomes

$$\mathbf{q}(n) = \mathbf{A}^{n-n_0}\mathbf{q}(n_0) + \sum_{m=0}^{n-n_0-1} \mathbf{A}^{n-n_0-m-1}\mathbf{B}\mathbf{x}(m+n_0) \tag{8.104}$$

378 DISCRETE-TIME SYSTEMS

Using Eqs. 8.93 and 8.103 ($n_0 = 0$), the output is

$$y(n) = \mathbf{CA}^n \mathbf{q}(0) + \sum_{m=0}^{n-1} \mathbf{CA}^{n-m-1}\mathbf{Bx}(m) + \mathbf{Dx}(n) \qquad (8.105)$$

This output or response sequence consists of three terms. The first term is due entirely to the initial state $q(0)$, and is thus the *zero-input response*. The third term is part of the output due *directly* to the input by way of direct connection, modified only by **D**. This term, together with the second term, form the *zero-state* response. As a matter of fact, the output in the zero state, when the input is the unit function $\delta(n)$, is simply the unit-function response $h(n)$. Thus, using Eq. 8.105

$$h(n) = \sum_{m=0}^{n-1} \mathbf{CA}^{n-m-1}\mathbf{B}\delta(m) + \mathbf{D}\delta(n) \qquad (8.106)$$

The second term is **D** for $n = 0$ only, while the first term is $\mathbf{CA}^{n-1}\mathbf{B}$ for $n > 0$. Notice that $\mathbf{A}^0 \equiv \mathbf{I}$ (the identity matrix). In concise form

$$h(n) = \mathbf{D}\delta(n) + \mathbf{CA}^{n-1}\mathbf{B}u(n-1) \qquad (8.107)$$

Example 21

Find $h(n)$ for the system shown in Fig. 8.29 (whose matrices were given by Eqs. 8.99 through 8.102) for $n = 0, 1, 2, 3$. Using Eq. 8.107 we have

$$h(0) = \tfrac{1}{3}$$

$$h(1) = [-3 \quad -2]\begin{bmatrix} 1 & 0 \\ 0 & 1 \end{bmatrix}\begin{bmatrix} \tfrac{1}{3} \\ 0 \end{bmatrix}$$

$$= [-3 \quad -2]\begin{bmatrix} \tfrac{1}{3} \\ 0 \end{bmatrix}$$

$$h(1) = 1$$

$$h(2) = [-3 \quad -2]\begin{bmatrix} -3 & -2 \\ 1 & 0 \end{bmatrix}\begin{bmatrix} \tfrac{1}{3} \\ 0 \end{bmatrix}$$

$$h(2) = [-3 \quad -2]\begin{bmatrix} -1 \\ \tfrac{1}{3} \end{bmatrix}$$

$$h(2) = \tfrac{7}{3}$$

$$h(3) = [-3 \quad -2]\begin{bmatrix} -3 & -2 \\ 1 & 0 \end{bmatrix}\begin{bmatrix} -3 & -2 \\ 1 & 0 \end{bmatrix}\begin{bmatrix} \tfrac{1}{3} \\ 0 \end{bmatrix}$$

$$h(3) = [-3 \quad -2]\begin{bmatrix} \tfrac{7}{3} \\ -1 \end{bmatrix}$$

$$h(3) = -5$$

8.8 STATE-SPACE ANALYSIS

These results are identical to those given by the closed form, Eq. 8.94. Given this unit-function response we could proceed to solve for the output as we did in Example 5. On the other hand we could solve for the output directly using Eq. 8.105. If we find the zero-state response alone:

$$\mathbf{y}_{zs}(n) = \sum_{m=0}^{n-1} \mathbf{CA}^{n-m-1}\mathbf{Bx}(m) + \mathbf{Dx}(n)$$

or

$$y_{zs}(0) = \tfrac{1}{3}x(0)$$

$$y_{zs}(1) = \tfrac{1}{3}x(1) + \begin{bmatrix} -3 & -2 \end{bmatrix} \begin{bmatrix} 1 & 0 \\ 0 & 1 \end{bmatrix} \begin{bmatrix} \tfrac{1}{3}x(0) \\ 0 \end{bmatrix}$$

$$y_{zs}(1) = -x(0) + \tfrac{1}{3}x(1)$$
$$\vdots$$

These results agree with those in Example 5:

$$y_{zs}(n) = \frac{1}{3}\sum_{m=0}^{n+1} x(m)[(-1)^{n-m+1} - (-2)^{n-m+1}]$$

$$y_{zs}(0) = \tfrac{1}{3}x(0)$$
$$y_{zs}(1) = -x(0) + \tfrac{1}{3}x(1)$$
$$\vdots$$

The method is not enticing, however, because of the n-fold matrix multiplication in $\mathbf{A}^n$ (Eq. 8.105).

The preceding example has demonstrated the fact that it would be highly advantageous to have a closed form for $\mathbf{A}^n$. This is possible.[2] As we saw in Chap. 3 (Eq. 3.38), the characteristic equation is obtained from

$$(\mathbf{A} - \lambda \mathbf{I})\mathbf{q} = 0 \qquad (8.108)$$

where the λ are the *eigenvalues*, and the $\mathbf{q}$ are the *eigenvectors*. It was shown that the characteristic equation is given by

$$|\mathbf{A} - \lambda \mathbf{I}| = 0 \qquad (8.109)$$

As we have already seen in Eqs. 3.43 and 8.30 ($\lambda = z$), the characteristic equation has the general polynomial form ($C_N \equiv 1$)

$$\sum_{m=0}^{N} C_m \lambda^m = 0 \quad \text{(characteristic equation)} \qquad (8.110)$$

Again, our *fundamental* problem is that of finding the N roots, $\lambda_1, \lambda_2, \ldots, \lambda_N$, of an Nth-order polynomial, and the zero-input response has the same form for *any* dependent variable, including the state variables.

Example 22

Find the eigenvalues in Example 21. We have

$$|\mathbf{A} - \lambda\mathbf{I}| = \begin{vmatrix} -3-\lambda & -2 \\ 1 & 0-\lambda \end{vmatrix} = 0$$

or

$$\lambda^2 + 3\lambda + 2 = 0$$

or

$$\lambda_1 = -1 \qquad \lambda_2 = -2$$

Equation 8.94 shows that these are correct.

We showed in Chap. 3, Eq. 3.84, that any function of a matrix can be put in the form

$$\mathbf{f}(\mathbf{A}) = \sum_{m=0}^{N-1} \beta_m \mathbf{A}^m \qquad (3.84)$$

which has only N terms! We further showed that

$$f(\lambda) = \sum_{m=0}^{N-1} \beta_m \lambda^m \qquad (8.111)$$

where the β_m are identical in the last two equations. Here, we want $\mathbf{f}(\mathbf{A}) = \mathbf{A}^n$:

$$\boxed{\mathbf{A}^n = \sum_{m=0}^{N-1} \beta_m \mathbf{A}^m} \qquad (8.112)$$

in order to avoid the n-fold multiplication in Eq. 8.105. In order to find β_m we write Eq. 8.111 as

$$\boxed{\lambda^n = \sum_{m=0}^{N-1} \beta_m \lambda^m} \qquad (8.113)$$

which is N equations (one for each eigenvalue, assuming that none are repeated) in the N unknown β_m. Thus, we have an easy way to find β_m and $\mathbf{A}^n$.

Example 23

Find $\mathbf{A}^n$ for the $\mathbf{A}$ in Example 22 ($\lambda_1 = -1, \lambda_2 = -2$). Using Eq. 8.113 we have

$$\lambda^n = \sum_{m=0}^{1} \beta_m \lambda^m = \beta_0 + \lambda\beta_1$$

and using this for each eigenvalue

$$\begin{aligned}(-1)^n &= \beta_0 - \beta_1 \\ (-2)^n &= \beta_0 - 2\beta_1\end{aligned} \bigg\} \begin{aligned}\beta_0 &= 2(-1)^n - (-2)^n \\ \beta_1 &= (-1)^n - (-2)^n\end{aligned}$$

Using Eq. 8.112

$$\mathbf{A}^n = \sum_{m=0}^{1} \beta_m \mathbf{A}^m = \beta_0 \mathbf{I} + \beta_1 \mathbf{A}$$

$$\mathbf{A}^n = \beta_0 \begin{bmatrix} 1 & 0 \\ 0 & 1 \end{bmatrix} + \beta_1 \begin{bmatrix} -3 & -2 \\ 1 & 0 \end{bmatrix} = \begin{bmatrix} \beta_0 - 3\beta_1 & -2\beta_1 \\ \beta_1 & \beta_0 \end{bmatrix}$$

Substituting β_0 and β_1 and simplifying gives

$$\mathbf{A}^n = \begin{bmatrix} -(-1)^n + 2(-2)^n & -2(-1)^n + 2(-2)^n \\ (-1)^n - (-2)^n & 2(-1)^n - (-2)^n \end{bmatrix} \tag{8.114}$$

We can quickly check and see that $\mathbf{A}^0 = \mathbf{I}$ and $\mathbf{A}^1 = \mathbf{A}$ for this result. Equation 8.114 can now be used in Eq. 8.105 to find $y(n)$ in a more efficient manner.

The case of repeated roots needs some attention. When the mth root is repeated i times, the eigenvalues are given by Eq. 8.45 with $z = \lambda$:

$$\boxed{\lambda_1, \lambda_2, \ldots, \lambda_m, n\lambda_m, n^2\lambda_m, \ldots, n^{i-1}\lambda_m, \lambda_{m+1}, \ldots, \lambda_N} \tag{8.115}$$

Instead of treating the general case, we will demonstrate how repeated roots are handled by using the second-order system. For this case Eq. 8.113 gives

$$\left. \begin{array}{l} \lambda_1^n = \beta_0 + \lambda_1 \beta_1 \\ \lambda_2^n = \beta_0 + \lambda_2 \beta_1 \end{array} \right\} \quad \begin{array}{l} \beta_0 = \dfrac{\lambda_1^n \lambda_2 - \lambda_2^n \lambda_1}{\lambda_2 - \lambda_1} \\ \beta_1 = \dfrac{\lambda_2^n - \lambda_1^n}{\lambda_2 - \lambda_1} \end{array}$$

and if $\lambda_2 = \lambda_1$, then both β_0 and β_1 are indeterminate. L'Hospital's rule gives

$$\beta_1 = \lim_{\lambda_2 \to \lambda_1} \frac{n\lambda_2^{n-1}}{1} = n\lambda_1^{n-1} \tag{8.116}$$

when numerator and denominator are differentiated with respect to λ_2. Also

$$\beta_0 = \lambda_1^n - \lambda_1 \beta_1 = \lambda_1^n - n\lambda_1^n = (1-n)\lambda_1^n \tag{8.117}$$

This process is exactly the same as using the first of the two equations above, but differentiating the second with respect to λ_2, then setting $\lambda_2 = \lambda_1$:

$$\lambda_1^n = \beta_0 + \lambda_1 \beta_1$$
$$n\lambda_1^{n-1} = \beta_1$$

which gives Eqs. 8.116 and 8.117 again. We can then find $\mathbf{A}^n$, substitute in Eq. 8.105, and find $y(n)$.

It has already been pointed out in Chap. 3 that it may be desirable in some cases to alter the internal structure of a system without changing the external description. One form might require fewer multipliers than another, and since these are relatively expensive, this would be helpful to know about.

382 DISCRETE-TIME SYSTEMS

The state-variable formulation is ideal for the application of transformations that do this sort of thing. In particular, we saw a transformation that converted **A** into a diagonal matrix and uncoupled the state-variable equations. We will now spend some time paralleling the third matrix method of Chap. 3. The development of the eigenvectors and the *modal matrix T* is *exactly* the same as that for Eqs. 3.81 through 3.88, so these equations are not repeated here. Our goal is that of uncoupling the first-order state-variable equations for the discrete case. Corresponding to Eq. 3.89 we have

$$\boxed{\mathbf{r}(n) = \mathbf{T}^{-1}\mathbf{q}(n)} \quad (8.118)$$

so

$$\mathbf{r}(n+1) = \mathbf{T}^{-1}\mathbf{q}(n+1) \quad (8.119)$$

and

$$\boxed{\mathbf{q}(n) = \mathbf{T}\mathbf{r}(n)} \quad (8.120)$$

$$\mathbf{q}(n+1) = \mathbf{T}\mathbf{r}(n+1) \quad (8.121)$$

Equation 8.92 is the normal form:

$$\mathbf{q}(n+1) = \mathbf{A}\mathbf{q}(n) + \mathbf{f}(n) \quad (8.92)$$

where

$$\mathbf{f}(n) = \mathbf{B}\mathbf{x}(n) \quad (8.122)$$

Combining Eqs. 8.120, 8.121, and 8.92

$$\mathbf{T}\mathbf{r}(n+1) = \mathbf{A}\mathbf{T}\mathbf{r}(n) + \mathbf{f}(n)$$

so

$$\mathbf{T}^{-1}\mathbf{T}\mathbf{r}(n+1) = \mathbf{T}^{-1}\mathbf{A}\mathbf{T}\mathbf{r}(n) + \mathbf{T}^{-1}\mathbf{f}(n)$$

or using Eq. 3.88

$$\mathbf{r}(n+1) = \boldsymbol{\lambda}\mathbf{r}(n) + \mathbf{f}_r(n) \quad (8.123)$$

where

$$\mathbf{f}_r(n) \equiv \mathbf{T}^{-1}\mathbf{f}(n) \quad (8.124)$$

Since $\boldsymbol{\lambda}$ is a diagonal matrix, Eq. 8.123 represents N *first-order uncoupled* difference equations! Thus, the ith equation is

$$r_i(n+1) = \lambda_i r_i(n) + f_{ri}(n)$$

whose unit-function response is easily obtained as

$$h_r(n) = \lambda_i^{n-1} u(n-1)$$

Therefore

$$r_i(n) = \sum_{m=-\infty}^{\infty} f_{ri}(m)\lambda_i^{n-m-1}u(n-m-1)$$

or

$$r_i(n) = \sum_{m=-\infty}^{n-1} f_{ri}(m)\lambda_i^{n-m-1}$$

or

$$r_i(n) = \lambda_i^{n-1} \sum_{m=-\infty}^{n-1} f_{ri}(m)\lambda_i^{-m} \tag{8.125}$$

In matrix form† the general result is

$$\boxed{\mathbf{r}(n) = \sum_{m=-\infty}^{n-1} \boldsymbol{\lambda}^{n-m-1}\mathbf{f}_r(m)} \tag{8.126}$$

When $\mathbf{r}(n)$ has been found, $\mathbf{q}(n)$ is given by Eq. 8.120:

$$\mathbf{q}(n) = \mathbf{T}\mathbf{r}(n) \tag{8.120}$$

Example 24

Find $\mathbf{T}$, $\mathbf{T}^{-1}$, $\mathbf{q}(n)$, and $\mathbf{y}(n)$ when $x(n) = 18 \cdot 2^n u(n)$ for the discrete system of Fig. 8.29 (Example 5). We found $\mathbf{A}$ earlier (Eq. 8.99), while the eigenvalues were found in Example 22. The modal matrix $\mathbf{T}$ for this second-order system is given by Eq. 3.85:

$$\mathbf{T} = \begin{bmatrix} q_1^1 & q_1^2 \\ q_2^1 & q_2^2 \end{bmatrix} \tag{3.85}$$

and the eigenvectors are found by using Eq. 3.38 for $\lambda_1, \lambda_2, \ldots, \lambda_N$ successively (assuming no repeated roots). For $N = 2$ (the present case) the eigenvectors are given by Eqs. 3.83 and 3.84:

$$\mathbf{q}^1 = \begin{bmatrix} q_1 \\ -q_1 \dfrac{a_{11}-\lambda_1}{a_{12}} \end{bmatrix} = \begin{bmatrix} q_1 \\ -q_1 \dfrac{-3-(-1)}{-2} \end{bmatrix} = \begin{bmatrix} q_1 \\ -q_1 \end{bmatrix}$$

$$\mathbf{q}^2 = \begin{bmatrix} q_1 \\ -q_1 \dfrac{a_{11}-\lambda_2}{a_{12}} \end{bmatrix} = \begin{bmatrix} q_1 \\ -q_1 \dfrac{-3-(-2)}{-2} \end{bmatrix} = \begin{bmatrix} q_1 \\ -\dfrac{1}{2}q_1 \end{bmatrix}$$

† $\boldsymbol{\lambda}$ is the diagonal matrix with elements given by $\lambda_1, \lambda_2, \ldots, \lambda_N$ (Eq. 3.87).

Normalizing the eigenvectors gives

$$\mathbf{q}^1 = \begin{bmatrix} \dfrac{1}{\sqrt{2}} \\ -\dfrac{1}{\sqrt{2}} \end{bmatrix} \qquad \mathbf{q}^2 = \begin{bmatrix} \dfrac{1}{\sqrt{1.25}} \\ -\dfrac{0.5}{\sqrt{1.25}} \end{bmatrix}$$

Thus

$$\mathbf{T} = \begin{bmatrix} \dfrac{1}{\sqrt{2}} & \dfrac{1}{\sqrt{1.25}} \\ -\dfrac{1}{\sqrt{2}} & -\dfrac{0.5}{\sqrt{1.25}} \end{bmatrix}$$

and

$$\mathbf{T}^{-1} = \begin{bmatrix} -\sqrt{2} & -2\sqrt{2} \\ 2\sqrt{1.25} & 2\sqrt{1.25} \end{bmatrix}$$

Now

$$\mathbf{f}_r(n) = \begin{bmatrix} -\sqrt{2} & -2\sqrt{2} \\ 2\sqrt{1.25} & 2\sqrt{1.25} \end{bmatrix} \begin{bmatrix} f_1(n) \\ 0 \end{bmatrix}$$

or

$$\mathbf{f}_r(n) = \begin{bmatrix} -\sqrt{2} f_1(n) \\ 2\sqrt{1.25} f_1(n) \end{bmatrix} = \begin{bmatrix} -\dfrac{\sqrt{2}}{3} x(n) \\ \dfrac{2}{3}\sqrt{1.25} x(n) \end{bmatrix}$$

since $f_1(n) = \tfrac{1}{3} x_1(n) = \tfrac{1}{3} x(n)$ in the present case. Then, using Eq. 8.126

$$r_1(n) = (-1)^{n-1} \sum_{m=-\infty}^{n-1} -\dfrac{\sqrt{2}}{3} x(m)(-1)^{-m}$$

and

$$r_2(n) = (-2)^{n-1} \sum_{m=-\infty}^{n-1} \dfrac{2}{3}\sqrt{1.25}\, x(m)(-2)^{-m}$$

Using Eq. 8.120

$$q_1(n) = \dfrac{1}{\sqrt{2}} r_1(n) + \dfrac{1}{\sqrt{1.25}} r_2(n)$$

$$q_2(n) = -\dfrac{1}{\sqrt{2}} r_1(n) - \dfrac{0.5}{\sqrt{1.25}} r_2(n)$$

or

$$q_1(n) = -\frac{1}{3}(-1)^{n-1}\sum_{m=-\infty}^{n-1} x(m)(-1)^{-m} + \frac{2}{3}(-2)^{n-1}\sum_{m=-\infty}^{n-1} x(m)(-2)^{-m}$$

$$q_2(n) = \frac{1}{3}(-1)^{n-1}\sum_{m=-\infty}^{n-1} x(m)(-1)^{-m} - \frac{1}{3}(-2)^{n-1}\sum_{m=-\infty}^{n-1} x(m)(-2)^{-m}$$

or

$$q_1(n) = -\frac{1}{3}\sum_{m=-\infty}^{n-1} x(m)[(-1)^{n-m-1} - 2(-2)^{n-m-1}]$$

$$q_2(n) = \frac{1}{3}\sum_{m=-\infty}^{n-1} x(m)[(-1)^{n-m-1} - (-2)^{n-m-1}]$$

These are the state variables for the system. The (one) output is given by Eq. 8.98. This results in

$$y(n) = \frac{1}{3}\sum_{m=-\infty}^{n-1} x(m)[(-1)^{n-m-1} - 4(-2)^{n-m-1}] + \frac{1}{3}x(n)$$

When the input is $x(n) = 18 \cdot 2^n u(n)$, we should properly call the output the zero-state response:

$$y_{zs}(n) = 6\sum_{m=0}^{n-1} 2^m[(-1)^{n-m-1} - 4(-2)^{n-m-1}] + 6 \cdot 2^n u(n)$$

$$y_{zs}(n) = 6\sum_{m=0}^{n-1} 2^m[(-1)^{n-m+1} - (-2)^{n-m+1}] + 6 \cdot 2^n u(n)$$

or

$$y_{zs}(n) = 6\sum_{m=0}^{n+1} 2^m[(-1)^{n-m+1} - (-2)^{n-m+1}] + 6 \cdot 2^n u(n)$$

$$- 6\sum_{m=n}^{n+1} 2^m[(-1)^{n-m+1} - (-2)^{n-m+1}]$$

The second and third terms cancel, so

$$y_{zs}(n) = 6\sum_{m=0}^{n+1} 2^m[(-1)^{n-m+1} - (-2)^{n-m+1}]$$

which is identical to the result for $y_{zs}(n)$ in Example 5.

It should be clear to the reader that the last two (matrix) methods that we investigated are lengthy and complicated, even for a second-order system. If the system is a high-order one, then any method will be lengthy and the

386 DISCRETE-TIME SYSTEMS

organization of the state-variable method is a distinct advantage. In addition, the state-variable method gives us explicit information about how the internal states of the system are changing. Finally, it should be remembered that even a second-order system can become troublesome with multiple inputs and outputs if we use the unit-function approach, whereas the state-variable approach is made to order for this type problem.

Example 25

Find the state-variable and output equations in matrix form for the system shown in Fig. 8.30. We have by inspection

$$q_1(n + 1) = 2q_1(n) + \tfrac{1}{2}q_2(n) + x_1(n) - 2x_2(n)$$
$$q_2(n + 1) = 3q_2(n) + x_2(n)$$

or

$$\mathbf{q}(n + 1) = \begin{bmatrix} 2 & \tfrac{1}{2} \\ 0 & 3 \end{bmatrix} \begin{bmatrix} q_1(n) \\ q_2(n) \end{bmatrix} + \begin{bmatrix} 1 & -2 \\ 0 & -1 \end{bmatrix} \begin{bmatrix} x_1(n) \\ x_2(n) \end{bmatrix}$$

The outputs are

$$y_1(n) = q_1(n) + \tfrac{1}{2}x_2(n)$$
$$y_2(n) = -q_1(n) + q_2(n)$$

or

$$\mathbf{y}(n) = \begin{bmatrix} 1 & 0 \\ -1 & 0 \end{bmatrix} \begin{bmatrix} q_1(n) \\ q_2(n) \end{bmatrix} + \begin{bmatrix} 0 & \tfrac{1}{2} \\ 0 & 0 \end{bmatrix} \begin{bmatrix} x_1(n) \\ x_2(n) \end{bmatrix}$$

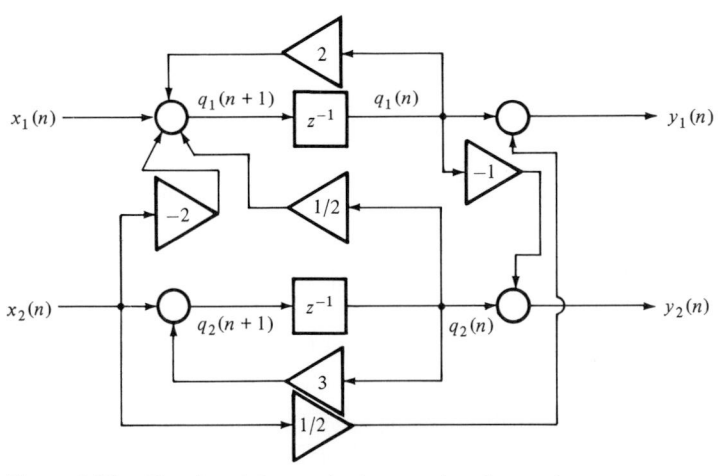

Figure 8.30. Two input–two output second-order system.

8.9 SIMULATION

As we have already seen, digital systems are used to solve recursion equations. We have also seen (Secs. 8.1 and 8.6) examples of digitalizing an analog filter and an example of designing a low-pass digital filter. Digital systems also can be used as an aid in designing analog systems. Consider an analog system H_a with its input $x(t)$ and output $y(t)$. That is, $y(t) = H_a\{x(t)\}$. Also, consider a digital system whose input sequence is the discrete-time signal $x(n) = x(nT)$. That is, the input is given (exactly) by the samples of the continuous-time signal $x(t)$. The discrete system H_d is an *exact digital simulator* of the analog system H_a if $y_d(n) = H_d\{x(n)\} = y(nT)$: that is, if the output sequence of the digital system *exactly* represents the samples of the output of the analog system. We know from the sampling theorem that we cannot achieve an *exact* digital simulator in a practical case (Why not?), but we should be able to obtain very useful *approximate* results if we sample at a high rate. Intuitively, we expect good results if the input $x(t)$ and the unit-impulse response $h(t)$ are nearly constant over a sampling period T. Thus, the simulation problem is that of determining H_d (that is, determining what elements make up the digital system) when H_a is specified. This problem is illustrated in Fig. 8.31.

Notice, first of all, that the solution of this problem (that is, the determination of H_d) reduces the problem of finding the solution of a *differential* equation for H_a to that of finding the solution of a *difference* equation for H_d. This can be done numerically using only delay elements and multipliers. Secondly, it means that the digital system can actually *replace* the analog system in applications such as signal processing, filtering, and so on. This is of the utmost importance. In order to do these things adequately, we only need to ensure that T is small enough.

Example 26

Solve the first-order differential equation

$$a_1 \frac{dy}{dt} + a_0 y = x(t) \qquad y(0) = y_0 \qquad t > 0 \qquad (8.127)$$

numerically. We can approximate the derivative in several ways. In Example 13,

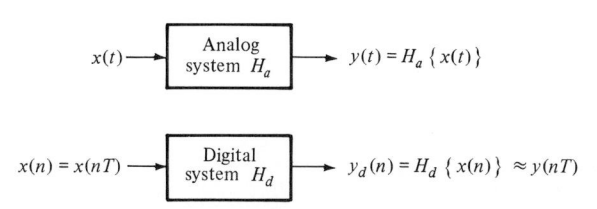

$x(t) \longrightarrow \boxed{\text{Analog system } H_a} \longrightarrow y(t) = H_a\{x(t)\}$

$x(n) = x(nT) \longrightarrow \boxed{\text{Digital system } H_d} \longrightarrow y_d(n) = H_d\{x(n)\} \approx y(nT)$

Figure 8.31. Digital simulation problem.

Chap. 3, we used the form

$$\frac{dy}{dt} \approx \frac{y(t+T) - y(t)}{T}$$

Here, as in Sec. 8.1, we choose to use

$$\frac{dy}{dt} \approx \frac{y(t) - y(t-T)}{T}$$

The difference between the two forms is insignificant for small T, but if we consider the simulator of a differentiator (whose output simulates the derivative of the input), then the former form leads to an anticipatory system and the latter does not. Thus, an approximation for Eq. 8.127 is

$$\frac{a_1}{T}[y(t) - y(t-T)] + a_0 y(t) \approx x(t)$$

With $t = nT$

$$(a_0 + a_1/T)y(nT) - (a_1/T)y(nT - T) \approx x(nT)$$

or

$$(a_0 T + a_1)y(nT) - a_1 y(nT - T) \approx Tx(nT)$$

The difference equation is therefore

$$(a_0 T + a_1)y_d(n) - a_1 y_d(n-1) = Tx(n) \qquad (8.128)$$

The digital simulation of the differential equation is shown in Fig. 8.32. Choosing $a_1 = 1, a_0 = 2, T = 0.5, y(0) = 0$, and $x(t) = u(t)$, we obtain

$$\frac{dy}{dt} + 2y = u(t) \qquad y(0) = 0$$

and

$$2y_d(n) - y_d(n-1) = 0.5u(n) \qquad y_d(0) = 0 \quad (T = 0.5)$$

for the analog and digital systems, respectively. If we choose $T = 0.25$ for comparison, then we have

$$1.5y_d(n) - y_d(n-1) = 0.25u(n) \qquad y_d(0) = 0 \quad (T = 0.25)$$

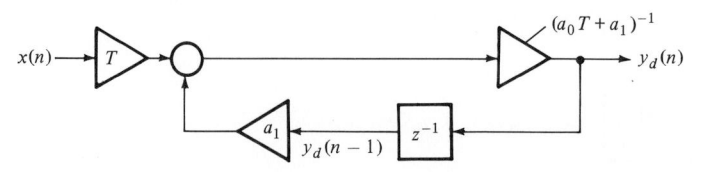

Figure 8.32. Digital simulator for a first-order differential equation.

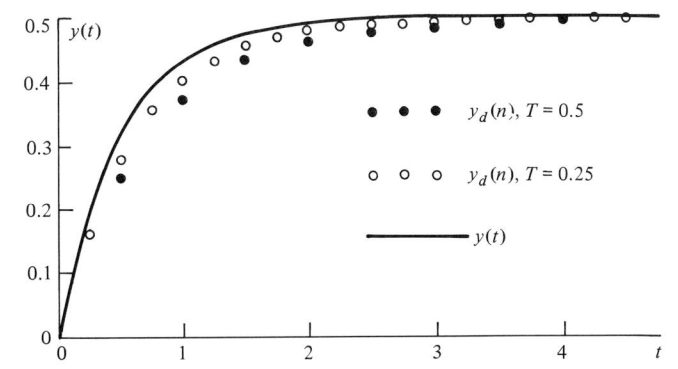

Figure 8.33. Results for the numerical solution (difference equation) of a first-order differential equation.

Results for these three cases are shown in Fig. 8.33. They are as expected: the smaller T gives the better approximation. Notice that, as in Example 13, Chap. 3, halving T essentially halves the error.

Now consider a series RL circuit driven by an ideal voltage source $v(t)$ with the resultant current $i(t)$ as the output. The differential equation is

$$L\frac{di}{dt} + Ri = v(t) \qquad i(0) = i_0, \quad t > 0$$

Comparing this equation to Eqs. 8.127 and 8.128 leads to the conclusion that the analog system can be replaced with the digital system of Fig. 8.34

$$y_d(n) - \frac{L}{L + RT} y_d(n-1) = \frac{T}{L + RT} x(n) \tag{8.129}$$

and the resulting discrete output approximately equals the samples of the current in the analog system for small T. Notice that the second form requires one less multiplier.

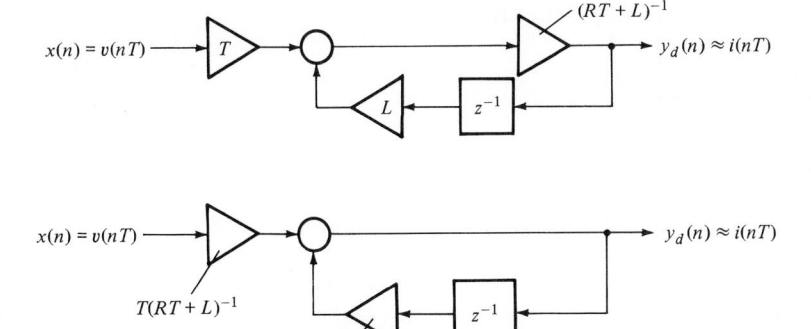

Figure 8.34. Digital simulators for the series RL circuit.

We can generalize the results that we have obtained up to this point.[4] Consider the single-input, single-output, causal, time-invariant analog system. The zero-state response is given by Eq. 2.60 with $h(t, \tau) = h(t - \tau)$:

$$y_{zs}(t) = \int_{-\infty}^{t} x(\tau) h(t - \tau) \, d\tau$$

This simple change of variable $t - \tau = \eta$ gives

$$y_{zs}(t) = \int_{0}^{\infty} x(t - \eta) h(\eta) \, d\eta$$

Using side-by-side rectangular pulses of width T as an approximation (that becomes exact for $T \to 0$) for representing an integral by a finite sum

$$y_{zs}(t) \approx T[x(t)h(0) + x(t - T)h(T) + \cdots + x(t - mT)h(mT) + \cdots]$$

or

$$y_{zs}(t) \approx T \sum_{m=0}^{\infty} x(t - mT) h(mT)$$

The samples of this continuous-time function at $t = nT$ are

$$y_{zs}(nT) \approx T \sum_{m=0}^{\infty} x(nT - mT) h(mT) \qquad (8.130)$$

Now, the single-input, single-output, causal, shift-invariant digital system has a zero-state response given by Eq. 8.15 with $h(l) = 0$ for $l < 0$:

$$y_{dzs}(n) = \sum_{l=0}^{\infty} x(n - l) h(l)$$

or with $l = m$

$$y_{dzs}(n) = \sum_{m=0}^{\infty} x(n - m) h(m) \qquad (8.131)$$

A comparison of Eqs. 8.130 and 8.131 reveals that if the unit-function response of the digital system equals T multiplied by the samples of the unit-impulse response of the analog system: that is, if[†]

$$\boxed{h_d(n) = T h_a(nT)} \qquad (8.132)$$

then

$$y_{dzs}(n) \approx y_{zs}(nT)$$

for small T, and we have a *time-sampling* digital simulator. Again, it is only necessary for $x(t)$ and $h(t)$ to be nearly constant over the interval of time T in order to obtain acceptable results. Thus, better results simply require more computation time.

[†] The subscripts d and a in Eq. 8.132 refer to digital and analog, respectively, to avoid any confusion in the notation.

Returning to the RL series circuit that was just simulated, we have

$$H_a(s) = \frac{1}{L}\frac{1}{s + R/L}$$

$$h_a(t) = \frac{1}{L}e^{-Rt/L}u(t)$$

and using Eq. 8.132

$$h_d(n) = \frac{T}{L}e^{-RnT/L}u(n) \qquad (8.133)$$

This result gives us the unit-function response of the digital system that simulates the RL series circuit. Notice that Eq. 8.133 does not (by itself) allow us to specify the elements of the simulator. If we take the z transform of Eq. 8.133, we can easily specify these elements. We will do this in the next chapter. For now we will deduce the difference equation in order to specify the digital system. Since

$$h_d(n-1) = \frac{T}{L}e^{-RnT/L}e^{RT/L}u(n-1)$$

we have

$$h_d(n) - e^{-RT/L}h_d(n-1) = \frac{T}{L}e^{-RnT/L}[u(n) - u(n-1)] = \frac{T}{L}\delta(n)$$

Consequently, the difference equation is

$$y(n) - e^{-RT/L}y(n-1) = \frac{T}{L}x(n) \qquad (8.134)$$

We can now easily draw the block diagram of the simulator. Notice that it will *not* be the same as Fig. 8.34, nor is Eq. 8.134 the same as Eq. 8.129. Why is this? The answer is that since T must be much less than L/R [$h(t)$ must be almost constant over the interval T], we have ($RT \ll L$)

$$e^{-RT/L} = \frac{1}{e^{RT/L}} \approx \frac{1}{1 + RT/L} = \frac{L}{L + RT}$$

and

$$\frac{T}{L + RT} \approx \frac{T}{L}$$

Thus, we merely have *two* approximations for the same thing.

Returning briefly to the design of the ideal low-pass filter (Example 18), we had

$$H(\omega) = u(\omega + \omega_0) - u(\omega - \omega_0)$$

so

$$h(t) = \frac{\omega_0}{\pi} \text{Sa}(\omega_0 t)$$

Therefore, with Eq. 8.132

$$h_d(n) = \frac{\omega_0 T}{\pi} \text{Sa}(\omega_0 n T)$$

and with $\omega_0 T = \pi/2$, as specified

$$h_d(n) = \tfrac{1}{2} \text{Sa}(n\pi/2)$$

This agrees with Eq. 8.79.

In this section we have found explicit methods for both solving constant-coefficient differential equations using difference equations and determining time-sampling digital simulators for analog systems.

8.10 STABILITY

Several times in this chapter there were questions asked about stability. The questions were not answered in the hope that the reader would do this for himself or herself. It should be clear from the preceding sections that *a discrete system will be stable if and only if the eigenvalues of the state matrix (or roots of the characteristic equation) have a magnitude less than unity*. Put another way, the unit-function response must go to 0 as n goes to ∞ if the system is to be stable, and this requires that the roots of the characteristic equation lie *inside* the unit circle in the complex plane. If one or more roots lie *on* the unit circle, then it is always possible to find a bounded input sequence that causes the output sequence to be unbounded as n increases (marginal stability). This last case is described by the second-order system of Sec. 8.6 for $-1 \leq a \leq 1$ (Eq. 8.50). See Fig. 8.19.

Example 27

Find the range of values of K that makes the second-order system

$$y(n) - y(n-1) + Ky(n-2) = x(n)$$

stable. The characteristic equation is

$$\lambda^2 - \lambda + K = 0$$

Thus, the eigenvalues are

$$\lambda_1 = \tfrac{1}{2} - \tfrac{1}{2}\sqrt{1 - 4K}$$
$$\lambda_2 = \tfrac{1}{2} + \tfrac{1}{2}\sqrt{1 - 4K}$$

If we consider the cases $1 - 4K > 0$ and $1 - 4K < 0$ separately, then it can be shown that $0 < K < 1$ gives $|\lambda_1| < 1$ and $|\lambda_2| < 1$. It is left as an exercise for the reader to prove this.

Further comments on the stability of discrete systems will be made in the next chapter.

8.11 CONCLUDING REMARKS

The material in this chapter concerning discrete systems closely paralleled that in Chaps. 2, 3, and 4 for continuous systems. We found that, for the most part, the discrete system is simpler than its continuous counterpart, and most of our problems were usually due to unfamiliarity with these systems rather than any difficulty with the techniques of analysis.

We only considered linear, shift-invariant, discrete-time systems (except for Example 11). Initially, we considered the input-output characterizations of the system through the difference equation approach and the unit-function response approach. These approaches care not what elements make up the block diagram of the system but, given the input, simply answer the question: What is the output? Superposition applies, and so multiple input-multiple output systems are relatively easy to handle also, unless the order of the system and (or) the number of inputs and outputs becomes large. In these cases the aforementioned methods become unwieldy, lacking organization.

It would be wise for us to summarize the method developed in this chapter for treating the frequently encountered *causal* and *shift-invariant discrete linear system*. The same precedure was followed at the end of Chap. 2 for the continuous-time case. The difference equation description is

$$a_N y(n + N) + \cdots + a_2 y(n + 2) + a_1 y(n + 1) + a_0 y(n)$$
$$= b_0 x(n) + b_1 x(n + 1) + \cdots + b_M x(n + M) = x_f(n) \quad (8.18)$$

The solution is (no repeated roots)

$$y(n) = \sum_{m=-\infty}^{n} x_f(m) h_f(n - m) + \sum_{m=1}^{N} E_m (z_m)^n \quad (8.135)$$

where

$$h_f(n) = [D_1 z_1^{n-1} + D_2 z_2^{n-1} + \cdots + D_N z_N^{n-1}] u(n - 1) = y_{zi}(n - 1) u(n - 1) \quad (8.136)$$

subject to the *special* initial conditions

$$y_{zi}(0) = 0, \, y_{zi}(1) = 0, \ldots, y_{zi}(N - 1) = \frac{1}{a_N} \quad (8.137)$$

and where $z_1, z_2, \ldots, z_N$ are the N solutions to the Nth order polynomial (characteristic equation)

$$a_N z^N + a_{N-1} z^{N-1} + \cdots + a_1 z + a_0 = 0 \quad (8.138)$$

The N constants $E_1, E_2, \ldots, E_N$ are determined from N initial conditions to form the zero-input response.

The state-variable approach provides organization through the use of matrix algebra. This formulation inherently treats the multiple input-multiple output case. In addition, it provides information about the internal workings of the discrete system where the other approaches do not. Thus, it would appear that we can obtain more information about the system using the state-variable formulation, and this is true. We pay for this increased information, however, because the formulation is more complex and the number of computations is increased.

Digital systems can be used to design analog systems, and in this regard they are simulators. We learned how to specify digital simulators. Additionally, digital systems are often used to process analog signals, and many of the same simulation principles apply in this application. We will investigate this again in the next chapter when the z transform is introduced.

PROBLEMS

1. State whether the following signals are discrete or continuous.
   (a) Elevation contours on a map.
   (b) Temperature in a refrigerator.
   (c) Digital clock display.
   (d) The score of a basketball game.
   (e) The price of gasoline.
   (f) The output of a loudspeaker.
2. Sketch the following sequences:
   (a) $f(n) = (\frac{1}{3})^n u(n)$ (b) $f(n) = (\frac{1}{3})^n u(n-3)$
   (c) $f(n) = u(n) - u(n-2)$ (d) $f(n) = \sum_{n=0}^{5} \delta(n)$
3. Express the sequence (for $0 \le n \le 5$) $\{2, 1, 3, 0, 2, 4\}$ as a superposition of unit functions.
4. Express the sequences shown in Fig. 8.35 in terms of $u(n)$.

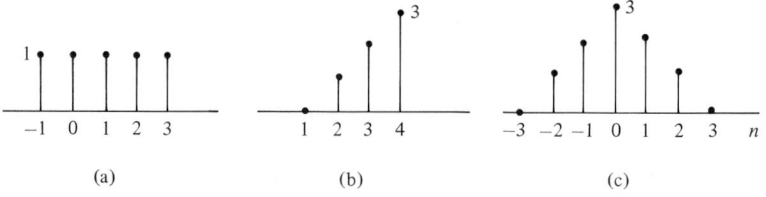

(a)   (b)   (c)

Figure 8.35. Sequences for Prob. 4.

5. Sketch the sequence $f(n)$ that is generated when the continuous function
$$f(t) = 2(1 - |t|)[u(t+1) - u(t-1)]$$
is uniformly sampled with
   (a) $T = \frac{1}{4}$ s   (b) $T = \frac{1}{2}$ s   (c) $T = 1$ s

6. Repeat Example 10 (savings account system) if $i = 0.05$, $k = 2$, and $x(0) = \$1000$, $x(1) = 0$, $x(2) = \$500$, $x(3) = -\$300$, $x(4) = \$250$, $x(5) = 0$, $x(6) = 0$, $x(7) = \$450$, $x(8) = \$500$, $x(9) = 0$, and $x(10) = 0$.

7. As an example of a *nonlinear* discrete system, consider the system described by

$$y(n) = \frac{1}{2}\left[y(n-1) + \frac{1}{y(n-1)}au(n)\right] \qquad n = 0, 1, \ldots$$

Show that this difference equation generates a number that approaches $\sqrt{a}$ ($a > 0$) for large n by simply taking $a = 3$ as an example. Use $y(-1) = 1.5$ as an initial trial, and find $y(6)$ iteratively.

8. A person deposits m dollars per conversion period in a savings account for which interest is compounded k times per year at a rate of i per year. Find a *closed-form* expression for the amount in the account after n conversion periods.

9. The block diagram representation of a loan payment scheme (amortization) is shown in Fig. 8.36. The interest rate is i, the remaining principal after n payments is $y(n)$, and the amount of the nth payment is $x(n)$.
   (a) Show that

$$y(n) = (1 + i)y(n-1) - x(n)$$

   (b) If the number of periods is selected to be N, and if the periodic payments are equal in amount, then $x(n) = p$, $n = 1, 2, \ldots, N$. Start the iteration at $n = 1$, and show that

$$p = \frac{i(1+i)^N}{(1+1)^N - 1}y(0)$$

   where $y(0)$ is the initial debt.

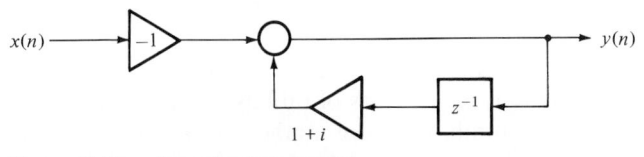

Figure 8.36. Amortization model.

10. (a) Find the unit-function response for the system in Prob. 9 using only the difference equation.
    (b) Find the zero-state response using convolution-summation, Eq. 8.17.

11. A discrete linear system is described by the linear difference equation

$$y(n) - \tfrac{5}{6}y(n-1) + \tfrac{1}{6}y(n-2) = x(n)$$

(a) Find the zero-input response.
(b) Find the unit-function response for $x_f(n) = x(n+2)$ using Eq. 8.32.
(c) Find the unit-function response for $x(n)$.
(d) Write the formal complete solution for $x(n) = nu(n)$.

12. Find a *particular* solution for the system in Prob. 11. Hint: try $y_p(n) = C_1 + C_2 n$. Is this a stable system?
13. Find the unit-function response $h(n)$ for the system shown in Fig. 8.37. Check the result.

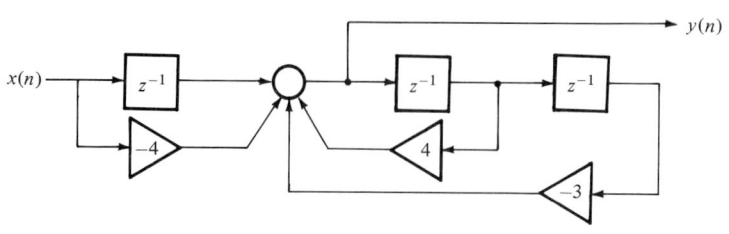

Figure 8.37. Linear system.

14. Find the zero-state response for the discrete system shown in Fig. 8.38.

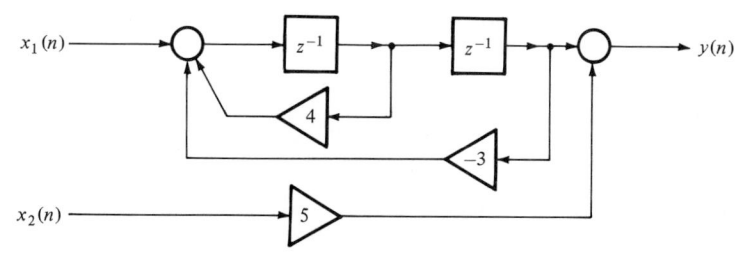

Figure 8.38. Linear system with two inputs.

15. Assume that[2]:
    1. A *pair* of rabbits (one male, one female) is born to each pair of adult rabbits once a month ($T = 1$ month).
    2. It requires two months for a newborn pair to produce offspring.
    3. Pairs remain paired indefinitely.
    4. There exists one pair of rabbits in an enclosed area initially.

    If we let $y(n)$ represent the number of pairs of rabbits within the enclosed area at the end of the nth month, then the mathematical model for the rabbit population is

    $$y(n) = y(n-1) + y(n-2)$$

    since $y(n-1)$ is the number of pairs that are one or more months old at the end of the nth month, and $y(n-2)$ is the number of newborn rabbits at the end of the nth month.
    (a) If $y(0) = 1$ and $y(-1) = 0$, find $y(n)$. This sequence is called the *Fibonacci sequence*.
    (b) Is this a stable system?
    (c) Rewriting the difference equation (above) as

    $$y(n+2) = y(n) + y(n+1)$$

show that
$$\sum_{n=1}^{10} y(n) = 11y(7)$$
for *any* $y(1)$ and $y(2)$.

16. National income[1] (the response here) essentially consists of three components:
$$y(n) = c(n) + i(n) + x(n)$$
where
  $c(n) =$ consumer expenditure,
  $i(n) =$ induced private investment, and
  $x(n) =$ government expenditure (the input here).

According to Samuelson,[1] we can assume that
$$c(n) = \alpha y(n)$$
and
$$i(n) = \beta[c(n) - c(n-1)]$$
where $\alpha > 0$ and $\beta > 0$ are proportionality constants.
   (a) Draw the block diagram model.
   (b) Find the (zero-state) unit-sequence response for $\alpha = \frac{1}{8}$, $\beta = 1$ in (real) closed form.
   (c) Is this a stable system?

17. Find the *noncausal* unit-function response for the system
$$y(n+2) + 3y(n+1) + 2y(n) = x_f(n)$$

18. Find the zero-state response of the nonrecursive filter of Fig. 8.39 for (a) $x(n) = u(n)$ and (b) $x(n) = nu(n)$. Plot these and comment ($0 \le n \le 10$). Is this a stable system?

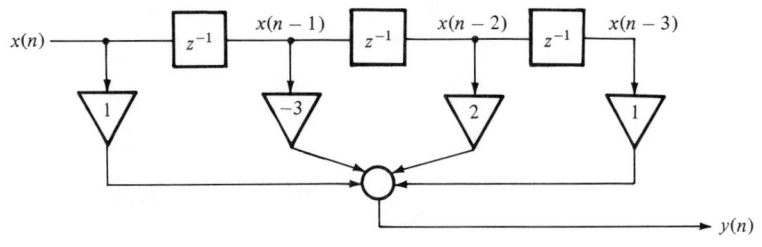

Figure 8.39. Nonrecursive filter.

19. Draw a block diagram of a system that halves the frequency of a unit sequence, $u(n)$.

20. (a) Show that the frequency response of the first-order system
$$y(n) - \alpha y(n-1) = \beta x(n)$$

is

$$H_d(\theta) = \frac{\beta e^{j\theta}}{-\alpha + e^{j\theta}}$$

(b) Show that a low-pass filter results if $\beta = 1 - \alpha$ and α is less than, but close to, unity. Plot $|H_d(\theta)|$ and $\underline{|H_d(\theta)|}$ versus θ. Use $\alpha = 0.5$ and $\alpha = 0.9$.

(c) Show that a high-pass filter results if $\beta = 1 + \alpha$ and α is greater than, but close to, minus one. Use $\alpha = -0.5$ and $\alpha = -0.9$.

21. Verify the results of Prob. 20 by
   (a) finding and plotting the zero-state response of the low-pass filter to $x(n) = u(n)$ and $x(n) = u(n)(-1)^n$ ($\alpha = 0.9$) and
   (b) finding and plotting the zero-state response of the high-pass filter to $x(n) = u(n)$ and $x(n) = u(n)(-1)^n$ ($\alpha = -0.9$).

22. Design an *ideal* high-pass filter with $\omega_s = 4\omega_0$, $T = 1$, and $\theta_0 = \pi/2$ as in Example 15. Use $H_d(\theta) = 1$ for $-\pi \leq \theta < -\pi/2$, $\pi/2 < \theta \leq \pi$.
   (a) Find $h(n)$.
   (b) Truncate $h(n)$ ($-5 \leq n \leq 5$) and shift $h(n)$ so that it is causal. Plot $|H_d(\theta)|$ versus θ.

23. The numerical integrator (Chap. 9) that uses the "rectangle rule" for integration is described by

$$y(n) - y(n-1) = Tx(n)$$

   (a) Find $h_f(n)$ and $h(n)$.
   (b) Is it stable?
   (c) Find a bounded input that produces an unbounded output.

24. Repeat Prob. 23 for the numerical integrator (Chap. 9) that uses the "trapezoidal rule":

$$y(n) - y(n-1) = \frac{T}{2}[x(n) + x(n-1)]$$

25. The numerical differentiator (Chap. 9) corresponding to the numerical integrator of Prob. 23 is described by

$$y(n) + y(n-1) = \frac{1}{T}x(n)$$

   (a) Find $h_f(n)$ and $h(n)$.
   (b) Is it stable?

26. Repeat Prob. 25 for the differentiator (Chap. 9) that corresponds to the integrator of Prob. 24:

$$y(n) + y(n-1) = \frac{2}{T}[x(n) - x(n-1)]$$

27. Using the contents of the unit delays as the state variables, put the description of the system shown in Fig. 8.38 (Problem 14) in the form of Eqs. 8.92 and 8.93.

28. Find $\mathbf{A}^n$ for

(a) $\mathbf{A} = \begin{bmatrix} 3 & 0 \\ 2 & 2 \end{bmatrix}$ (b) $\mathbf{A} = \begin{bmatrix} 4 & 2 \\ \frac{1}{2} & 4 \end{bmatrix}$

(c) $\mathbf{A} = \begin{bmatrix} 1 & 2 \\ 1 & 1 \end{bmatrix}$ (d) $\mathbf{A} = \begin{bmatrix} 1 & 0 \\ 1 & 1 \end{bmatrix}$

29. (a) Find $h(n)$ for the system shown in Fig. 8.37 using Eq. 8.107.
    (b) Find the zero-state response using Eq. 8.105.
30. Find the zero-state response for the system used in Prob. 29 using the method that employs the modal matrix $\mathbf{T}$.
31. Find the zero-state response of the system shown in Fig. 8.30 when $x_1(n) = u(n)$ and $x_2(n) = nu(n)$.
32. Find the difference equation for the *continuous* system described by

$$\frac{d^2y}{dt^2} + 6\frac{dy}{dt} + 5y = x(t)$$

using the numerical method of Example 26 with $T = 0.02$ s. Draw the digital block diagram.

33. Repeat Prob. 32 using the simulation scheme that incorporates Eq. 8.132.
34. Repeat Prob. 32 using $y'(t) \approx [y(t + T) - y(t)]/T$. Compare the results to those in Prob. 33.
35. A discrete linear system is described by

$$y(n - 1) - y(n - 3) = -x(n) + x(n - 1)$$

(a) Is this an anticipatory system?
(b) Can a block diagram that uses only multipliers, summers, and delay elements be drawn for this system?

36. The approximation of a derivative that was mentioned, but not used, in Example 26 leads to the simulator of a differentiator with output y (the approximation of the derivative) and input x:

$$y(t) = \frac{x(t + T) - x(t)}{T} \approx \frac{dx}{dt}$$

Show that it cannot be built.

References

1. Cadzow, J. A. *Discrete Time Systems.* Englewood Cliffs, N.J.: Prentice-Hall, 1973.
2. Gabel, R. A., and Roberts, R. A. *Signals and Linear Systems.* New York: John Wiley & Sons, 1980.
3. McGillem, C. D., and Cooper, G. R. *Continuous and Discrete Signal and System Analysis.* New York: Holt, Rinehart and Winston, 1974.
4. Papoulis, A. *Circuits and Systems.* Holt, Rinehart and Winston, 1980.

Chapter 9
The z Transform

We have seen the advantages of the Fourier and Laplace transform methods in Chaps. 5, 6, and 7 for analyzing continuous-time linear systems. In the simplest sense, perhaps, they convert differential equations into algebraic equations that are usually easy to manipulate. After the manipulations have been performed, we inverse transform back to the time domain. It is often easier to visualize system performance in the s (or $j\omega$) domain. For example, it is relatively easy to visualize system performance with regard to stability in terms of the variation of a parameter with the *root-locus technique* in the s plane. Convolution (integration) corresponds to multiplication (as in cascading blocks within an overall system) in the s (or $j\omega$) domain. We can think of the transform technique as being *auxiliary* in the process of linear-system analysis. That is, it is an *aid* rather than a necessity, although many times a problem solution that is easy to obtain with the transform approach may be extremely difficult to obtain with a strictly time-domain approach.

We should hope that the same general benefits are available for discrete-time linear systems. As we shall shortly see, they are, and the transformation that provides this is called the z transform. It will be found that many operations can be better understood or performed in the z domain than in the discrete-time (or sample) domain. In Chap. 8, for example, we found a way to simulate a continuous (analog) system with a discrete (digital) one, and we were even able to find its unit-function response without much difficulty. It was not so easy, however, to specify the multipliers and delay elements that made up the required digital system. Perhaps the z transform can help us here.

A word of caution: Be careful with the symbols that are used in this chapter. There is no easy way to avoid subscripted and (or) superscripted symbols at certain points, and this can be confusing unless the reader takes care. Generally, a subscript d means $d$igital, while a means $a$nalog.

9.1 THE z TRANSFORM

Rather than introducing the z transform as an abstract mathematical formula we will attempt to relate it to past experience. We used this approach when we introduced the Fourier transform as an extension of the Fourier series. Consider the three related signals[1,2] shown in Fig. 9.1. The first, $f(t)$, is a continuous-time signal and is the only one that is a *function* in the strict mathematical sense. The sequence of impulses can be obtained in theory with a balanced modulator (Chap. 5) that multiplies a periodic unit-impulse train with period T by $f(t)$:

$$f_s(t) = f(t) \sum_{n=0}^{\infty} \delta(t - nT) = \sum_{n=0}^{\infty} f(nT)\delta(t - nT) \qquad (9.1)$$

Notice that $f_s(t)$ is not unique with respect to $f(t)$. That is, two *different* $f(t)$ can lead to the same $f_s(t)$. Consider, for example, $f_1(t) = 1$, and $f_2(t) = \cos(2\pi t/T)$. They give the same $f_s(t)$. Notice here, as before, that in order to distinguish $f_1(t)$ and $f_2(t)$ we must sample at a higher rate. The representation in Fig. 9.1(c) is that of a sequence, $f(n)$, and nothing more.

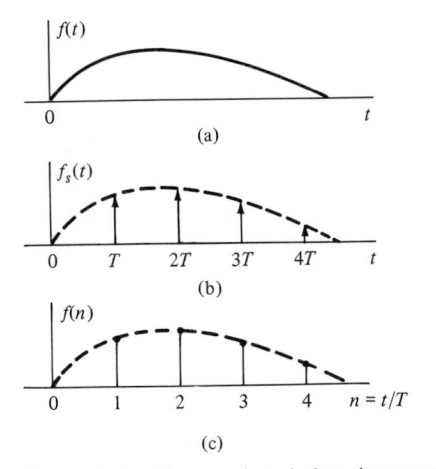

Figure 9.1. Three related signal representations.

The unilateral Laplace transform of $f_s(t)$ is $F_s(s)$:

$$F_s(s) = \sum_{n=0}^{\infty} f(nT)e^{-nTs} \qquad (9.2)$$

which is a *periodic function*. That is, $F_s(s) = F_s(s + j2n\pi)$, $n = 0, 1, 2, \ldots$. This periodicity complicates the process of performing the inverse Laplace transformation, taking away some of the attractive features that occur when aperiodic functions of s appear. On the other hand, if we let $s = (1/T)\ln z$, or $z = e^{sT}$, then we obtain a function of z called the *unilateral z transform of $f(nT)$*:

$$\boxed{F(z) = \sum_{n=0}^{\infty} f(nT)z^{-n}} \qquad \text{(unilateral z transform)} \qquad (9.3)$$

Since $|z| = |e^{sT}| = |e^{\sigma T}e^{j\omega T}| = |e^{\sigma T}|$, it is easy to see that when $\sigma < 0$ (that is, when in the left half of the s plane), then $|z| < 1$, and thus the left half of the s plane corresponds to the interior of the unit circle in the complex z plane. In the same way the right half of the s plane corresponds to the exterior of the unit circle in the z plane. The $j\omega$ axis in the s plane and the unit circle in the z plane correspond. These results should not be surprising since we already know that a stable continuous system has a characteristic equation with roots in the left half of the s plane, while a stable discrete system has a characteristic equation with roots inside the unit circle in the z plane. At this point we leave the Laplace transform and concern ourselves only with the z transform. Notice that taking the *inverse z transform* of a function of z will *always* give *only* a sequence of samples.

Thus

$$f(nT) \leftrightarrow F(z) \qquad (9.4)$$

or†

$$\mathscr{Z}\{f(nT)\} = F(z) \qquad (9.5)$$

and

$$\mathscr{Z}^{-1}\{F(z)\} = f(nT) \qquad (9.6)$$

Equation 9.3 gives the unilateral z transform and, therefore, as the equation shows, is only concerned with sequences for which $n \geq 0$. This is sufficient for most cases of a practical nature, and when finite or semi-infinite duration sequences with $n \leq 0$ for some samples must be considered, we can always shift up to 0, or redefine the origin, since we are dealing only with shift-invariant systems. When infinite-duration sequences must be considered, a bilateral z transform

$$\mathscr{Z}\{f(nT)\} = \sum_{n=-\infty}^{\infty} f(nT)z^{-n} \qquad \text{(bilateral z transform)} \qquad (9.7a)$$

† Warning! Note the similarity between the capital script "zee" ($\mathscr{Z}$) and capital script "ell" ($\mathscr{L}$).

is defined. This bilateral z transform is analogous to the bilateral Laplace transform (Chap. 6) and suffers from the same sort of difficulties. That is, two *different* sequences, one for $n < 0$ and one for $n \geq 0$ (for example), can have the *same* z transform. This requires that the *regions of convergence*, for which the sums converge absolutely, must be specified in order to uniquely specify the sequence $f(nT)$ that generates $F(z)$. The region of convergence is simply that region in the complex z plane for which

$$\sum_{n=-\infty}^{\infty} |f(nT)z^{-n}| \qquad (9.7b)$$

exists. Recall that this same problem existed for the bilateral Laplace transform. There, the regions of convergence were planes with edges parallel to the $j\omega$ axis in the s plane: that is, strips, semi-infinite planes, or the infinite plane. Here, as one might guess, the regions of convergence involve circles that are concentric with the origin in the z plane: that is, within a circle, outside a circle, inside an annular ring, outside an annular ring, etc.

Some of the more important properties of the bilateral z transform are listed in App. G. One that we need concern ourselves with here is the shifting property that holds for a delay or for an advance. This property is easy to establish by simply substituting into Eq. 9.7a. It is

$$f(n \pm m) \leftrightarrow z^{\pm m} F(z) \quad \text{(shifting property, bilateral } z \text{ transform)} \qquad (9.7c)$$

The perceptive reader may be able to anticipate that pair 9.7c will be extremely valuable to us because it gives the z transform of linear difference equations directly. This will, as we shall shortly see, enable us to easily solve for the *zero-state response* (only) for these systems. A zero-input response may be added if necessary. On the other hand, the *unilateral z* transform will allow us to find *both* the zero-state response and the zero-input response for sequences that start at $n = 0$. The other important property of the bilateral z transform that we mention here is that involving convolution-summation.

It is not necessary to specify a region of convergence to obtain a unique sequence $f(nT)$ with the unilateral z transform, but we will indicate the *radius of convergence* when it is appropriate to do so.

The formula for obtaining $f(nT)$ from $F(z)$, called the *inverse z transform*,[2] is

$$\boxed{f(nT) = \frac{1}{2\pi j} \oint_c F(z) z^{n-1} \, dz} \qquad (9.8)$$

where c is a circle in the complex plane that encircles all the *singularities* of $F(z)z^{n-1}$ (loci of points where $F(z)z^{n-1} \to \infty$). Recall that it requires an integration along a line parallel to the $j\omega$ axis *to the right*, a distance σ for inversion with the Laplace transform:

$$f(t) = \frac{1}{2\pi j} \int_{\sigma - j\omega}^{\sigma + j\omega} F(s) e^{ts} \, ds$$

We decided to avoid Eq. 6.5 at this level of exposure, and relied on partial fraction expansions (for example) to construct a table of pairs for use when inversion became necessary. Since we are probably no better off now than we were three chapters back with regard to integration in the complex plane, we will also avoid Eq. 9.8 for inversion. Since $F(z)$ is given in Eq. 9.3 as an infinite series in powers of z^{-1}, we can find $f(nT)$ if we can arrange $F(z)$ into such a series by long division. This will be done shortly. We intend to construct a table of pairs for our use, just like we did for the Fourier and Laplace transforms. We intend for the z transform to "stand on its own."

9.2 z-TRANSFORM PAIRS

In this section we will present some of the properties of the z transform that form pairs themselves and also enable us to deduce other pairs.

Linearity

The transform of a linear sum of functions is the sum of the transforms of the individual functions. If $af(nT) \leftrightarrow aF(z)$ and $bg(nT) \leftrightarrow bG(z)$, then[†]

$$af(nT) + bg(nT) \leftrightarrow aF(z) + bG(z) \qquad (9.9)$$

This is easy to prove with Eq. 9.3.

Shifting (Delay)

The transform of a delayed sequence[‡] $f(n - m)$ is found by substitution:

$$\mathscr{L}\{f(n - m)\} = \sum_{n=0}^{\infty} f(n - m)z^{-n}$$

With $l = n - m$

$$f(n - m) \leftrightarrow \sum_{l=-m}^{\infty} f(l)z^{-l-m} = z^{-m} \sum_{l=-m}^{\infty} f(l)z^{-l}$$

or

$$f(n - m) \leftrightarrow z^{-m} \sum_{l=0}^{\infty} f(l)z^{-l} + z^{-m} \sum_{l=-m}^{-1} f(l)z^{-l}$$

or

$$f(n - m) \leftrightarrow z^{-m}F(z) + z^{-m} \sum_{l=-m}^{-1} f(l)z^{-l} \qquad (9.10)$$

[†] We assume, of course, that $F(z)$ and $G(z)$ exit. The region or *radius* of convergence for pair 9.9 is the intersection (the common part) of that for $F(z)$ and $G(z)$ separately.
[‡] Notice that we have dropped T for notational simplicity.

That is, a delay of m samples leads to a transform that is z^{-m} multiplied by the transform of the undelayed samples plus terms due to $f(n)$ for $n < 0$. If $f(n) = 0$ for $n < 0$, the last term in pair 9.10 is 0, and

$$f(n-m) \leftrightarrow z^{-m}F(z) \qquad [f(n) = 0, n < 0] \qquad (9.11)$$

This last result can be stated in a better way. If $f(n) = 0$ for $n < 0$, we can replace $f(n)$ with $f(n)u(n)$, so

$$f(n-m)u(n-m) \leftrightarrow \sum_{n=0}^{\infty} f(n-m)u(n-m)z^{-n}$$

$$f(n-m)u(n-m) \leftrightarrow \sum_{n=m}^{\infty} f(n-m)z^{-n}$$

With $l = n - m$,

$$f(n-m)u(n-m) \leftrightarrow \sum_{l=0}^{\infty} f(l)z^{-l-m}$$

$$f(n-m)u(n-m) \leftrightarrow z^{-m}F(z) \qquad (9.12)$$

Shifting (Advance)

$$f(n+m) \leftrightarrow \sum_{n=0}^{\infty} f(n+m)z^{-n}$$

with $l = n + m$

$$f(n+m) \leftrightarrow \sum_{l=m}^{\infty} f(l)z^{m-l} = z^m \sum_{l=m}^{\infty} f(l)z^{-l}$$

$$f(n+m) \leftrightarrow z^m \sum_{l=0}^{\infty} f(l)z^{-l} - z^m \sum_{l=0}^{m-1} f(l)z^{-l}$$

$$f(n+m) \leftrightarrow z^m F(z) - z^m \sum_{l=0}^{m-1} f(l)z^{-l} \qquad (9.13)$$

In the same way

$$f(n+m)u(n+m) \leftrightarrow \sum_{n=0}^{\infty} f(n+m)u(n+m)z^{-n}$$

but since $m > 0$ for an advance shift, $u(n+m) = 1, n \geq 0$, and

$$f(n+m)u(n+m) \leftrightarrow \sum_{n=0}^{\infty} f(n+m)z^{-n}$$

so

$$f(n+m)u(n+m) \leftrightarrow z^m F(z) - z^m \sum_{l=0}^{m-1} f(l)z^{-l} \qquad (9.14)$$

as in pair 9.13. Examples are

$$f(n+1) \leftrightarrow z[F(z) - f(0)] \qquad (9.15)$$
$$f(n+2) \leftrightarrow z^2[F(z) - f(0)] - zf(1) \qquad (9.16)$$
$$f(n+3) \leftrightarrow z^3[F(z) - f(0)] - z^2 f(1) - zf(2) \qquad (9.17)$$
$$\vdots$$

Convolution-Summation

The zero-state response of a shift-invariant, causal, linear discrete system starting at $n = 0$ was found in Chap. 8 to be

$$y_{zs}(n) = \sum_{m=0}^{n} x(m)h(n-m) \qquad (8.17)$$

Taking the z transform gives

$$\mathscr{L}\{y_{zs}(n)\} = Y_{zs}(z) = \sum_{n=0}^{\infty} \left\{ \sum_{m=0}^{n} x(m)h(n-m) \right\} z^{-n}$$

but since n goes to infinity in the outer sum, the upper index in the inner sum can be replaced with infinity. Thus

$$Y_{zs}(z) = \sum_{n=0}^{\infty} \left\{ \sum_{m=0}^{\infty} x(m)h(n-m) \right\} z^{-n}$$

Interchanging the order of summation

$$Y_{zs}(z) = \sum_{m=0}^{\infty} x(m) \sum_{n=0}^{\infty} h(n-m)z^{-n}$$

With $l = n - m$ in the second sum

$$Y_{zs}(z) = \sum_{m=0}^{\infty} x(m) \sum_{l=-m}^{\infty} h(l)z^{-l-m}$$

$$Y_{zs}(z) = \sum_{m=0}^{\infty} x(m)z^{-m} \sum_{l=-m}^{\infty} h(l)z^{-l}$$

But $h(l) = 0$ for $l < 0$ since this is a causal system, so

$$Y_{zs}(z) = \sum_{m=0}^{\infty} x(m)z^{-m} \sum_{l=0}^{\infty} h(l)z^{-l}$$

or

$$Y_{zs}(z) = X(z)H_d(z) \tag{9.18}$$

This result is not unexpected. It simply states that the z transform of the zero-state response of a discrete system (causal, shift-invariant, and starting at $t = 0$) is given by the product of the transform of the input and a *discrete transfer function*,[†] $H_d(z)$, that is the z transform of the unit-function response. Thus

$$x(n) * h(n) \leftrightarrow X(z)H_d(z) \tag{9.19}$$

or, in general

$$f_1(n) * f_2(n) \leftrightarrow F_1(z)F_2(z) \tag{9.20}$$

Equation 9.18 can also be written as

$$H_d(z) = \frac{Y_{zs}(z)}{X(z)} \tag{9.21}$$

with obvious implications in analyzing systems, designing filters, and so on.

Multiplication by n

$$nf(n) \leftrightarrow \sum_{n=0}^{\infty} nf(n)z^{-n} = z \sum_{n=0}^{\infty} f(n)nz^{-n-1}$$

$$= z \sum_{n=0}^{\infty} f(n)\left[-\frac{dz^{-n}}{dz}\right]$$

$$= -z\frac{d}{dz}\sum_{n=0}^{\infty} f(n)z^{-n}$$

or

$$nf(n) \leftrightarrow -z\frac{dF(z)}{dz} \tag{9.22}$$

Multiplication by a^n

$$a^n f(n) \leftrightarrow \sum_{n=0}^{\infty} a^n f(n)z^{-n} = \sum_{n=0}^{\infty} f(n)(z/a)^{-n}$$

$$a^n f(n) \leftrightarrow F(z/a) \tag{9.23}$$

[†] Sometimes called the *system function*. (See Eq. 8.68.) $Y_{zs}(z)$ converges in the common region of convergence of $X(z)$ and $H_d(z)$.

Division by $n + a$

$$\frac{f(n)}{n+a} \leftrightarrow \sum_{n=0}^{\infty} \frac{f(n)}{n+a} z^{-n} = \sum_{n=0}^{\infty} f(n) z^a \left[-\int_0^z \eta^{-n-a-1} \, d\eta \right]$$

$$\frac{f(n)}{n+a} \leftrightarrow -z^a \int_0^z \eta^{-a-1} \left[\sum_{n=0}^{\infty} f(n) \eta^{-n} \right] d\eta$$

$$\frac{f(n)}{n+a} \leftrightarrow -z^a \int_0^z \eta^{-a-1} F(\eta) \, d\eta \tag{9.24}$$

Initial Value

Using Eq. 9.3

$$F(z) = f(0) + f(1)z^{-1} + f(2)z^{-2} + \cdots$$

Thus

$$f(0) = \lim_{z \to \infty} F(z) \quad \text{(initial value)} \tag{9.25}$$

Also, if $f(0) \equiv 0$

$$f(1) = \lim_{z \to \infty} zF(z) \tag{9.26}$$

Final Value

Using pair 9.12 with $m = 1$

$$f(n)u(n) - f(n-1)u(n-1) \leftrightarrow F(z) - z^{-1}F(z)$$

or

$$F(z) - z^{-1}F(z) = \sum_{n=0}^{\infty} [f(n) - f(n-1)u(n-1)] z^{-n}$$

$$(1 - 1/z)F(z) = \lim_{N \to \infty} \sum_{n=0}^{N} [f(n) - f(n-1)u(n-1)] z^{-n}$$

Taking the limit as $z \to 1$ on both sides

$$\lim_{z \to 1} (1 - 1/z)F(z) = \lim_{z \to 1} \left\{ \lim_{N \to \infty} \sum_{n=0}^{N} [f(n) - f(n-1)u(n-1)] z^{-n} \right\}$$

$$= \lim_{N \to \infty} \sum_{n=0}^{N} [f(n) - f(n-1)u(n-1)]$$

$$= \lim_{N \to \infty} f(N)$$

or

$$\lim_{n \to \infty} f(n) = \lim_{z \to 1} (1 - 1/z)F(z) \quad \text{(final value)} \tag{9.27}$$

9.2 z-TRANSFORM PAIRS

We note that in order for Eq. 9.27 to be valid $(z - 1)F(z)$ must be analytic for $|z| \geq 1$. This simply means that $(z - 1)F(z)$ cannot have poles on or outside the unit circle in the complex plane.

Unit Function

$$\delta(n) \leftrightarrow \sum_{n=0}^{\infty} \delta(n) z^{-n}$$

$$\delta(n) \leftrightarrow 1 \tag{9.28}$$

$$\delta(n - m) \leftrightarrow \sum_{n=0}^{\infty} \delta(n - m) z^{-n}$$

$$\delta(n - m) \leftrightarrow z^{-m} \quad \text{(radius of convergence, } |z| > 0\text{)}$$

Unit Sequence

$$u(n) \leftrightarrow \sum_{n=0}^{\infty} u(n) z^{-n} = \sum_{n=0}^{\infty} z^{-n} = \frac{1}{1 - z^{-1}}$$

$$u(n) \leftrightarrow \frac{1}{1 - z^{-1}} = \frac{z}{z - 1} \quad (|z| > 1) \tag{9.29}$$

Long division can be used to verify this result. The delayed unit sequence can be transformed with the aid of pairs 9.12 and 9.29:

$$u(n - m) \leftrightarrow \frac{z^{-m}}{1 - z^{-1}} = \frac{z}{z - 1} z^{-m} \tag{9.30}$$

Ramp Sequence

Using pair 9.22

$$nu(n) \leftrightarrow -z \frac{d}{dz} \left[\frac{z}{z - 1} \right]$$

$$nu(n) \leftrightarrow \frac{z}{(z - 1)^2} \quad (|z| > 1) \tag{9.31}$$

Parabolic Sequence

Using pair 9.22 again

$$n^2 u(n) \leftrightarrow -z \frac{d}{dz} \left[\frac{z}{(z - 1)^2} \right]$$

$$n^2 u(n) \leftrightarrow \frac{z^2 + z}{(z - 1)^3} \quad (|z| > 1) \tag{9.32}$$

It is left for the reader to show that

$$na^{n-1} \leftrightarrow \frac{z}{(z-a)^2} \qquad (|z| > |a|) \qquad (9.33)$$

and

$$\frac{n(n-1)\cdots(n-m+1)}{m!} a^{n-m} \leftrightarrow \frac{z}{(z-a)^{m+1}} \qquad (|z| > |a|) \qquad (9.34)$$

Exponential

Since $e^{nb} = (e^b)^n$, we can use pair 9.23 with $f(n) = u(n)$ and $a = e^b$ and pair 9.29 to obtain

$$e^{nb} \leftrightarrow \frac{ze^{-b}}{ze^{-b}-1} = \frac{z}{z-e^b} \qquad (|z| > |e^b|) \qquad (9.35)$$

If $b = j\omega$

$$e^{jn\omega} \leftrightarrow \frac{z}{z-e^{j\omega}} \qquad (|z| > 1) \qquad (9.36)$$

It is now easy to show that

$$\cos(n\omega) \leftrightarrow \frac{z[z - \cos(\omega)]}{z^2 - 2z\cos(\omega) + 1} \qquad (|z| > 1) \qquad (9.37)$$

$$\sin(n\omega) \leftrightarrow \frac{z\sin(\omega)}{z^2 - 2z\cos(\omega) + 1} \qquad (|z| > 1) \qquad (9.38)$$

$$\cosh(nb) \leftrightarrow \frac{z[z - \cosh(b)]}{z^2 - 2z\cosh(b) + 1} \qquad (|z| > \max\{|e^b|, |e^{-b}|\}) \qquad (9.39)$$

$$\sinh(nb) \leftrightarrow \frac{z\sinh(b)}{z^2 - 2z\cosh(b) + 1} \qquad (|z| > \max\{|e^b|, |e^{-b}|\}) \qquad (9.40)$$

Pair 9.35 with $a = e^b$ gives

$$a^n \leftrightarrow \frac{z}{z-a} \qquad (|z| > |a|) \qquad (9.41)$$

Exponentially Damped Function

Using pair 9.23 with $a = e^{-b}$ gives

$$f(n)e^{-nb} \leftrightarrow F(ze^b) \qquad (9.42)$$

The results that we have obtained in this section for the unilateral z transform (plus a few from the references) are listed in Table 9.1. Notice particularly that it has been assumed that $f(n) = 0$ for $n < 0$.

9.2 z-TRANSFORM PAIRS

Table 9.1 UNILATERAL z TRANSFORM PROPERTIES AND PAIRS

NO.		$f(n), n \geq 0, T = 1$	$F(z)$
		$f(n) = \dfrac{1}{2\pi j} \oint_c F(z) z^{n-1} \, dz$	$F(z) = \displaystyle\sum_{n=0}^{\infty} f(n) z^{-n}$
1	Linearity	$\displaystyle\sum_{n=1}^{N} a_n f_n(n)$	$\displaystyle\sum_{n=1}^{N} a_n F_n(z)$
2	Shifting† (Delay)	$f(n-m)u(n-m)$	$z^{-m} F(z) \quad m \geq 0$ (see pair 9.10)
3	Shifting† (Advance)	$f(n+m)u(n)$	$z^m F(z) - z^m \displaystyle\sum_{l=0}^{m-1} f(l) z^{-l} \quad m \geq 0$
4	Convolution-Summation	$\displaystyle\sum_{m=0}^{n} f_1(m) f_2(n-m)$	$F_1(z) F_2(z)$
5	Sequence Multiplication	$f_1(n) f_2(n)$	$\dfrac{1}{2\pi j} \oint_c \dfrac{F_1(\eta) F_2(z/\eta) \, d\eta}{\eta}$
6	Multiplication by n	$n f(n) u(n)$	$-z \dfrac{dF(z)}{dz}$
7	Multiplication by a^n	$a^n f(n) u(n)$	$F(z/a)$
8	Division by $n + a$	$\dfrac{f(n)}{n+a} u(n)$	$-z^a \displaystyle\int_0^z \eta^{-a-1} F(\eta) \, d\eta$
9	Initial Value	$\displaystyle\lim_{n \to 0} f(n) \;=$	$\displaystyle\lim_{z \to \infty} F(z)$
10	Final Value	$\displaystyle\lim_{n \to \infty} f(n) \;=$	$\displaystyle\lim_{z \to 1} (1 - 1/z) F(z)$
11		$\delta(n)$	1
12		$\delta(n-m)$	z^{-m}
13		$u(n)$	$\dfrac{z}{z-1}$
14		$u(n-m)$	$\dfrac{z}{z-1} z^{-m}$
15		$n u(n)$	$\dfrac{z}{(z-1)^2}$
16		$n^2 u(n)$	$\dfrac{z^2 + z}{(z-1)^3}$
17		$a^n u(n)$	$\dfrac{z}{z-a}$
18		$n a^{n-1} u(n)$	$\dfrac{z}{(z-a)^2}$
19		$\dfrac{n(n-1) \cdots (n-m+1)}{m!} a^{n-m} u(n)$	$\dfrac{z}{(z-a)^{m+1}}$
20		$e^{nb} u(n)$	$\dfrac{z}{z - e^b}$

Table 9.1 (continued)

NO.	$f(n), n \geq 0, T = 1$	$F(z)$
	$f(n) = \dfrac{1}{2\pi j} \oint_c F(z) z^{n-1}\, dz$	$F(z) = \sum\limits_{n=0}^{\infty} f(n) z^{-n}$
21	$\cos(n\omega) u(n)$	$\dfrac{z[z - \cos(\omega)]}{z^2 - 2z\cos(\omega) + 1}$
22	$\sin(n\omega) u(n)$	$\dfrac{z \sin(\omega)}{z^2 - 2z\cos(\omega) + 1}$
23	$\cosh(nb) u(n)$	$\dfrac{z[z - \cosh(b)]}{z^2 - 2z\cosh(b) + 1}$
24	$\sinh(nb) u(n)$	$\dfrac{z \sinh(b)}{z^2 - 2z\cosh(b) + 1}$
25	$f(n) e^{-nb} u(n)$	$F(ze^b)$

† For completely general sequences that are not necessarily 0 for $n < 0$, the bilateral z transform (Eq. 9.7a) gives the pair 9.7c:

$$f(n \pm m) \leftrightarrow z^{\pm m} F(z) \tag{9.7c}$$

instead of pairs 2 and 3 in Table 9.1 that are obtained using the unilateral transform. We will need to use this (bilateral) pair later on. See Examples 8 and 12.

9.3 EXAMPLES

We will examine some relatively simple examples in this section in using the z transform.

Example 1

Find the inverse transform of

$$F(z) = \dfrac{z}{z^2 - 6z + 8}$$

A glance at pair 17, Table 9.1, suggests that we try an expansion in forms like $z/(z - a)$. To this end we write

$$\dfrac{F(z)}{z} = \dfrac{1}{z^2 - 6z + 8} = \dfrac{1}{(z-2)(z-4)} = \dfrac{A}{z - 2} + \dfrac{B}{z - 4}$$

Thus

$$A = (z - 2) F(z)/z \bigg|_{z=2} = \dfrac{1}{z - 4} \bigg|_{z=2} = -\dfrac{1}{2}$$

$$B = (z - 4) F(z)/z \bigg|_{z=4} = \dfrac{1}{z - 2} \bigg|_{z=4} = \dfrac{1}{2}$$

Therefore
$$F(z) = -\frac{1}{2}\left(\frac{z}{z-2} - \frac{z}{z-4}\right)$$
and with pair 17 (Table 9.1)
$$f(n) = -\tfrac{1}{2}(2^n - 4^n)u(n)$$

Example 2

Use the final-value theorem to find $f(\infty)$ if
$$F(z) = \frac{z}{z-a}$$
We take note of the fact that the final-value theorem is applicable only if the pole at $z = a$ is *inside* the unit circle. Hence we require that $|a| < 1$. Thus
$$\lim_{n\to\infty} f(n) = \lim_{z\to 1} (1 - 1/z)z/(z-a)$$
$$= \lim_{z\to 1} \frac{z-1}{z-a} = 0 \qquad |a| < 1$$
This is certainly an expected result since
$$a^n u(n) \leftrightarrow \frac{z}{z-a}$$
and
$$\lim_{n\to\infty} a^n = 0$$
if $|a| < 1$.

Notice that the final-value theorem does not apply for pair 21 or 22 in Table 9.1 since there are poles *on* the unit circle in both cases. This corresponds to the behavior of $\cos(n\omega)$ and $\sin(n\omega)$.

Example 3

Find the inverse z transform of $(z+a)^{-1}$. Pair 17 (Table 9.1) is
$$a^n u(n) \leftrightarrow \frac{z}{z-a}$$
or, replacing a with $-a$
$$(-a)^n u(n) \leftrightarrow \frac{z}{z+a}$$

414 THE z TRANSFORM

Using pair 2 with $m = 1$

$$(-a)^{n-1}u(n-1) \leftrightarrow \frac{1}{z+a}$$

That is

$$\mathcal{L}^{-1}\{(z+a)^{-1}\} = (-a)^{n-1}u(n-1)$$

This result can be obtained in a different form:

$$\frac{1}{z+a} = \frac{1}{a}\left(1 - \frac{z}{z+a}\right)$$

Therefore

$$\frac{1}{a}[\delta(n) - (-a)^n u(n)] \leftrightarrow \frac{1}{z+a}$$

This result is the same as the other result.

Example 4

Verify pair 16 (Table 9.1) using convolution-summation

$$n^2 u(n) \leftrightarrow \frac{z(z+1)}{(z-1)^3} = \frac{z}{(z-1)^2}\frac{z+1}{z-1} = \frac{z}{(z-1)^2}\left[\frac{z}{z-1} + \frac{1}{z-1}\right]$$

Let

$$F_1(z) = \frac{z}{(z-1)^2} \leftrightarrow nu(n) \quad \text{(pair 15)}$$

$$F_2(z) = \frac{z}{z-1} + \frac{1}{z-1} \leftrightarrow u(n) + u(n-1) \quad \text{(pair 14, } m = 1\text{)}$$

With pair 4

$$\sum_{m=0}^{n} m[u(n-m) + u(n-m-1)] \leftrightarrow F_1(z)F_2(z)$$

Checking the first few values of n shows that the sum is n^2 for $n \geq 0$; therefore

$$n^2 u(n) \leftrightarrow \frac{z(z+1)}{(z-1)^3}$$

Example 5

Find the inverse transform of $F(z)$ that has repeated roots in the denominator:

$$F(z) = \frac{3z^2 - z}{(z-1)(z-2)^2}$$

Following the expansion in Example 1 (see Example 10, Chap. 6), we try

$$\frac{F(z)}{z} = \frac{3z-1}{(z-1)(z-2)^2} = \frac{A}{z-1} + \frac{B}{(z-2)^2} + \frac{C}{z-2}$$

$$A = \left.\frac{3z-1}{(z-2)^2}\right|_{z=1} = 2$$

$$B = \left.\frac{3z-1}{z-1}\right|_{z=2} = 5$$

Thus

$$\frac{3z-1}{(z-1)(z-2)^2} = \frac{2}{z-1} + \frac{5}{(z-2)^2} + \frac{C}{z-2}$$

and setting $z = 0$ (for example) gives

$$\frac{1}{4} = -2 + \frac{5}{4} - \frac{C}{2}$$

Thus, $C = -2$. Therefore

$$F(z) = 2\frac{z}{z-1} + 5\frac{z}{(z-2)^2} - 2\frac{z}{z-2}$$

and using Table 9.1

$$f(n) = 2u(n) + 5n(2)^{n-1}u(n) - 2^{n+1}u(n)$$
$$f(n) = [2 - 2^{n+1} + 5n(2)^{n-1}]u(n)$$

What happens if we try

$$F(z) = \frac{3z^2 - z}{(z-1)(z-2)^2} = \frac{D}{z-1} + \frac{E}{(z-2)^2} + \frac{F}{z-2}$$

$$D = \left.\frac{3z^2 - z}{(z-2)^2}\right|_{z=1} = 2$$

$$E = \left.\frac{3z^2 - z}{(z-1)}\right|_{z=2} = 10$$

Substituting $D = 2$, $E = 10$, and then using $z = -1$ (for example) gives $F = 1$, so that

$$F(z) = 2\frac{1}{z-1} + 10\frac{1}{(z-2)^2} + \frac{1}{z-2}$$

Example 3 shows that

$$(-a)^{n-1}u(n-1) \leftrightarrow \frac{1}{z+a} \qquad (9.43)$$

Pair 18 with pair 2 ($m = 1$) gives

$$(n - 1)(-a)^{n-2}u(n - 1) \leftrightarrow \frac{1}{(z + a)^2} \qquad (9.44)$$

These give

$$f(n) = 2u(n - 1) + 10(n - 1)(2)^{n-2}u(n - 1) + (2)^{n-1}u(n - 1)$$
$$f(n) = [2 + (2)^{n-1} + 10(n - 1)^{n-2}]u(n - 1)$$

This result agrees with the other.

It is worth pointing out at this time, even though it should be obvious, that the inverse of $F(z)$ is giving *closed forms* for $f(n)$. This does not usually occur when working exclusively in the (discrete) time domain with superposition-summation.

Example 6

Verify the results of Example 5 using long division. We have

$$\begin{array}{r}
3/z + 14/z^2 + 46/z^3 + \cdots \\
z^3 - 5z^2 + 8z - 4 \overline{\smash{\big)}\, 3z^2 - z } \\
\underline{3z^2 - 15z + 24 - 12/z} \\
14z - 24 + 12/z \\
\underline{14z - 70 + 112/z - 56/z^2} \\
46 - 100/z + 56/z^2 \\
46 \cdots
\end{array}$$

Thus

$$F(z) = \frac{3}{z} + \frac{14}{z^2} + \frac{46}{z^3} + \cdots$$

or

$$f(n) = 3\delta(n - 1) + 14\delta(n - 2) + 46\delta(n - 3) + \cdots$$

Comparing this result with the others in Example 5 for $n = 0, 1, 2, 3$, shows that they are the same: that is $f(0) = 0$, $f(1) = 3$, $f(2) = 14$, $f(3) = 46$, and so on.

Example 7

Find the unit-function response of the discrete feedback system shown in the z domain in Fig. 9.2. Notice that $X(z) = R(z)$, $Y(z) = C(z)$, and the transfer function will be $T_d(z)$. This is done to conform to the conventional symbols in a feedback system. We have $[C(z) = C_{zs}(z)]$

$$C(z) = H_1(z)H_2(z)E(z) = H_1(z)H_2(z)[R(z) + H_3(z)C(z)]$$

Solving for $C(z)$

$$C(z) = \frac{H_1(z)H_2(z)}{1 - H_1(z)H_2(z)H_3(z)} R(z)$$

Consequently, the *overall* transfer function is

$$T_d(z) = \frac{H_1(z)H_2(z)}{1 - H_1(z)H_2(z)H_3(z)} = \frac{C(z)}{R(z)}$$

Substituting the given functions of z and simplifying

$$T_d(z) = \frac{6(z-1)}{z(z-4)(z+3)} = \frac{6}{z}\left(\frac{A}{z-4} + \frac{B}{z+3}\right)$$

$$T_d(z) = \frac{18}{7}\frac{1}{z(z-4)} + \frac{24}{7}\frac{1}{z(z+3)}$$

Using pairs 2 and 17 (Table 9.1) we obtain the unit-function response

$$t(n) = \left[\tfrac{18}{7}(4)^{n-2} + \tfrac{24}{7}(-3)^{n-2}\right]u(n-2) = \mathscr{L}^{-1}\{T_d(z)\}$$

Example 8

Draw the discrete-time block diagram of the system shown in Fig. 9.2, and verify the result of Example 7 using the time-domain technique of Chap. 8. Using the transfer function of Example 7 we have

$$z(z-4)(z+3)C(z) = 6(z-1)R(z)$$
$$(z^2 - z - 12)C(z) = (6 - 6/z)R(z)$$

There are several ways to obtain the difference equation of the system since we are only interested in the zero-state response. The simplest is to use the *bilateral* pair 9.7c:

$$f(n \pm m) \leftrightarrow z^{\pm m}F(z) \quad \text{(bilateral } z \text{ transform)} \tag{9.7c}$$

This easily gives

$$c(n+2) - c(n+1) - 12c(n) = 6r(n) - 6r(n-1) \equiv r_f(n)$$

On the other hand the transformed equation can be multiplied by z^{-2}, giving

$$(1 - z^{-1} - 12z^{-2})C(z) = (6z^{-2} - 6z^{-3})R(z)$$

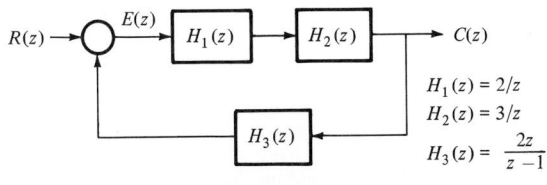

$H_1(z) = 2/z$
$H_2(z) = 3/z$
$H_3(z) = \dfrac{2z}{z-1}$

Figure 9.2. Discrete feedback system.

418 THE z TRANSFORM

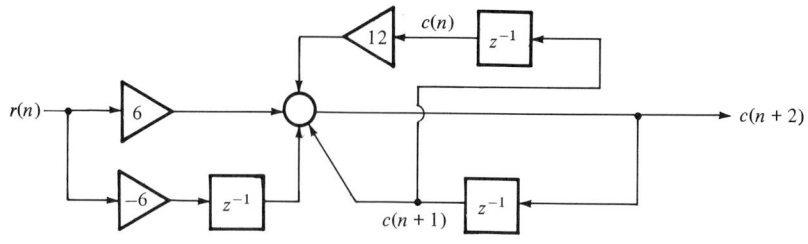

Figure 9.3. Discrete-time form of Fig. 9.2.

and this is a form such that pair 2 (Table 9.1) can be used. This gives

$$c(n)u(n) - c(n-1)u(n-1) - 12c(n-2)u(n-2)$$
$$= 6r(n-2)u(n-2) - 6r(n-3)u(n-3)$$

It is not obvious that these two forms of the difference equation give the same results for *the zero-state response* (the unit-function response, for example), but they do, and it is left to the reader to show this. The discrete-time block diagram obtained from the former form (that was obtained from the bilateral pair) is shown in Fig. 9.3.†

The use of the bilateral pair in this case has been very helpful. In this regard it is worth noting that we often do the same thing in continuous-time systems. Suppose that we wanted the differential equation represented by

$$(s^2 + 4s + 3)Y(s) = sX(s)$$

We would use the bilateral Laplace transform or Fourier transform ($s = j\omega$) for inversion, but in either case the result is

$$\frac{d^2y}{dt^2} + 4\frac{dy}{dt} + 3y = \frac{dx}{dt}$$

Taking the (unilateral) Laplace transform of this differential equation and dropping out all initial conditions for the zero-state response gives us back the original equation in s. The same kind of thing occurs with the unilateral z transform.

The characteristic equation in the present example is

$$z^2 - z - 12 = 0$$

hence

$$z_1 = 4 \qquad z_2 = -3$$

---

† Notice that the overall transfer function

$$\frac{C(z)}{R(z)} = \frac{G(z)}{1 - G(z)H(z)} = T_d(z)$$

has the same form (and can be drawn in canonical form) as that for the Laplace transform with continuous systems.

We now have
$$y_{zi}(n) = D_1(4)^n + D_2(-3)^n$$
$$\left. \begin{array}{l} 0 = D_1 + D_2 \\ 1 = 4D_1 - 3D_2 \end{array} \right\} \begin{array}{l} D_1 = \tfrac{1}{7} \\ D_2 = -\tfrac{1}{7} \end{array}$$

Therefore, the unit-function response for $r_f(n)$ is
$$t_f(n) = y_{zi}(n-1)u(n-1) = \tfrac{1}{7}[(4)^{n-1} - (-3)^{n-1}]u(n-1)$$
and the unit-function response for $r(n)$ is
$$t(n) = \tfrac{6}{7}[(4)^{n-1} - (-3)^{n-1}]u(n-1) - \tfrac{6}{7}[(4)^{n-2} - (-3)^{n-2}]u(n-2)$$

The first bracketed term on the right is 0 for $n = 1$, so
$$t(n) = \tfrac{6}{7}[4(4)^{n-2} + 3(-3)^{n-2} - (4)^{n-2} + (-3)^{n-2}]u(n-2)$$
$$t(n) = [\tfrac{18}{7}(4)^{n-2} + \tfrac{24}{7}(-3)^{n-2}]u(n-2)$$
which is identical to the result in Example 7.

We considered an RL series circuit driven by an ideal voltage source in Sec. 8.9. Based on Eq. 8.132, we found that a digital simulator must have the unit-function response
$$h_d(n) = \frac{T}{L} e^{-RnT/L} u(n) \tag{8.133}$$
and it was stated that the elements of the system could easily be specified if we could take the z transform of Eq. 8.133. Using pair 20 (Table 9.1) we have
$$H_d(z) = \frac{T}{L} \frac{z}{z - e^{RT/L}}$$
so
$$Y(z) = \frac{T}{L} \frac{z}{z - e^{-RT/L}} X(z)$$
or
$$Y(z) - e^{-RT/L} z^{-1} Y(z) = \frac{T}{L} X(z) \qquad Y(z) = Y_{dzs}(z)$$

This z-domain equation gives the digital system shown in Fig. 9.4. Notice that this simulator is essentially the same as those in Fig. 8.34 since we must require that $RT \ll L$ if we are to obtain useful results. Thus
$$e^{-RT/L} \approx L(L + RT)^{-1}$$
and
$$T/L \approx T(RT + L)^{-1}$$

```
x(n) = v(nT) ──▷ T/L ──→ ○ ─────────────────→ y_d(n) ≈ i(nT)
                        ↑                    │
                        │    ◁──── z⁻¹ ◀─────┤
                        e^(-RT/L)
```

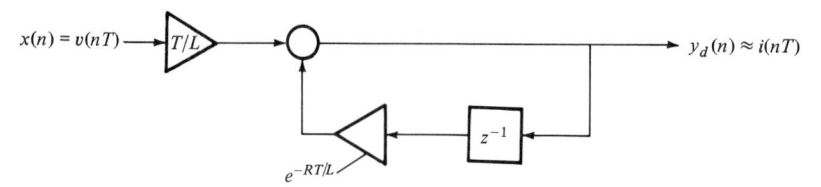

Figure 9.4. Digital simulator for the RL series circuit ($RT \ll L$).

Other examples, particularly those of a practical nature, will be considered in the material that follows.

9.4 LINEAR DIFFERENCE EQUATIONS (AGAIN)

Equation 8.18 gives a general form for the difference equation model for shift-invariant, linear, discrete systems:

$$\sum_{m=0}^{N} a_m y(n+m) = x_f(n) = \sum_{m=0}^{M} b_m x(n+m) \tag{8.18}$$

Using pairs 1 and 3 (Table 9.1) we apply the z transform to Eq. 8.18, obtaining

$$\sum_{m=0}^{N} a_m z^m \left[Y(z) - \sum_{l=0}^{m-1} y(l) z^{-l} \right] = \sum_{m=0}^{M} b_m z^m \left[X(z) - \sum_{l=0}^{m-1} x(l) z^{-l} \right] \tag{9.45}$$

Notice that, as in the case of the Laplace transform, *the initial conditions appear explicitly*. Solving *algebraically* for $Y(z)$, we obtain

$$Y(z) = \left[\frac{\sum_{m=0}^{M} b_m z^m}{\sum_{m=0}^{N} a_m z^m} X(z) - \frac{\sum_{m=0}^{M} b_m z^m \sum_{l=0}^{m-1} x(l) z^{-l}}{\sum_{m=0}^{N} a_m z^m} \right]$$

$$+ \left[\frac{\sum_{m=0}^{N} a_m z^m \sum_{l=0}^{m-1} y(l) z^{-l}}{\sum_{m=0}^{N} a_m z^m} \right] \tag{9.46}$$

$$Y(z) = Y_{zs}(z) + Y_{zi}(z)$$

As indicated, the first term is the transform of the zero-state response, and the second term is the transform of the zero-input response. Compare Eq. 9.46 to Eq. 7.1 for the Laplace transform of the general differential equation form for the time-invariant (continuous-time) linear system. The resemblance between the two equations is striking. The zero-state response is due entirely to the forcing function [that is, due to $x(n)$, $x(n+1)$, ..., $x(n+M)$], while the zero-input response is due entirely to the N initial conditions [that is, due to $y(0)$, $y(1), y(2), \ldots, y(N-1)$].

The transfer function (or *system function*) is given in Eq. 9.18:

$$Y_{zs}(z) = H_d(z)X(z) \tag{9.18}$$

and is easily identified in Eq. 9.46 as the term multiplying $X(z)$ (that is, when *all* initial conditions are 0). Thus

$$Y_{zs}(z) = \frac{\sum_{m=0}^{M} b_m z^m}{\sum_{m=0}^{N} a_m z^m} X(z) \tag{9.47}$$

or

$$H_d(z) = \frac{\sum_{m=0}^{M} b_m z^m}{\sum_{m=0}^{N} a_m z^m} \tag{9.48}$$

It is a ratio of polynomials in z, and its inverse z transform is the unit-function response of the system

$$h(n) \leftrightarrow H_d(z) \tag{9.49}$$

The characteristic equation is obtained by setting the denominator of Eq. 9.46 equal to 0. It is

$$\sum_{m=0}^{N} a_m z^m = 0 \tag{9.50}$$

as we saw in Chap. 8. The roots of this equation give the eigenvalues that are used to form the fundamental set.

Based on past experience with the Laplace transform, it is easy to formulate a procedure for finding the complete response of the discrete system using the z transform.

1. Given the difference equation (Eq. 8.18), take the z transform of both sides of the equation.
2. Transform the input sequence to find $X(z)$, then insert $X(z)$ (in Eq. 9.46).
3. Apply the initial conditions $y(0), y(1), \ldots, y(N-1)$ and $x(0), x(1), \ldots, x(M-1)$ to Eq. 9.46. Notice that in some cases it may be necessary to *calculate* the *required* initial conditions from the *given* conditions, but this can always be accomplished using the difference equation itself. This will be shown in the examples that follow.
4. Arrange $Y(z)$ into a form such that its inverse transform can be found:

   $$y(n) \leftrightarrow Y(z)$$

   We normally want a closed form for $y(n)$.

Example 9

Find the closed-form unit-function response for the system that is modeled by the difference equation

$$3y(n+2) + 9y(n+1) + 6y(n) = x(n+2)$$

by first finding the z transform of the difference equation and then identifying $H_d(z)$. (See Eq. 8.41.) Applying the z transform gives

$$3z^2 Y(z) - 3z^2 y(0) - 3zy(1) + 9zY(z) - 9zy(0) + 6Y(z)$$
$$= z^2 X(z) - z^2 x(0) - zx(1) \qquad (9.51)$$

Rearranging this equation into the form of Eq. 9.46

$$Y(z) = \left[\frac{z^2}{3z^2 + 9z + 6} X(z) - \frac{z^2 x(0) + zx(1)}{3z^2 + 9z + 6}\right] + \left[\frac{3z^2 y(0) + 9zy(0) + 3zy(1)}{3z^2 + 9z + 6}\right]$$

Therefore

$$H_d(z) = \frac{z^2}{3z^2 + 9z + 6} = \frac{1}{3}\frac{z^2}{z^2 + 3z + 2} \qquad (9.52)$$

Since the degree of the numerator polynomial is the same as that of the denominator polynomial, a unit function $\delta(n)$ is present in the unit-function response. That is, $H_d(z)$ may be written

$$H_d(z) = \frac{1}{3}\frac{z^2 + 3z + 2}{z^2 + 3z + 2} - \frac{1}{3}\frac{3z + 2}{z^2 + 3z + 2} = \frac{1}{3} - \frac{1}{3}\frac{3z + 2}{z^2 + 3z + 2}$$

$$H_d(z) = \frac{1}{3} - \frac{1}{3}\left(\frac{A}{z+1} + \frac{B}{z+2}\right) = \frac{1}{3} - \frac{1}{3}\frac{1}{z+1} - \frac{4}{3}\frac{1}{z+2}$$

Consequently

$$h(n) = \tfrac{1}{3}\delta(n) + \tfrac{1}{3}[(-1)^{n-1} - 4(-2)^{n-1}]u(n-1)$$

This sequence is the same as that given by Eq. 8.41:

$$h(n) = \tfrac{1}{3}[(-1)^{n+1} - (-2)^{n+1}]u(n+1)$$

Returning to Eq. 9.52

$$H_d(z) = \frac{z}{3}\left[\frac{z}{(z+1)(z+2)}\right] = \frac{z}{3}\left[\frac{C}{z+1} + \frac{D}{z+2}\right]$$

$$H_d(z) = \frac{1}{3}\left(-\frac{z}{z+1} + 2\frac{z}{z+2}\right)$$

Thus, from Table 9.1

$$h(n) = \tfrac{1}{3}[-(-1)^n + 2(-2)^n]u(n)$$

or

$$h(n) = \tfrac{1}{3}[(-1)^{n+1} - (-2)^{n+1}]u(n+1)$$

9.4 LINEAR DIFFERENCE EQUATIONS (AGAIN)

since the bracketed term in the last equation is 0 for $n = -1$. That is, it makes no difference whether we multiply by $u(n)$ or $u(n+1)$. This result is identical to Eq. 8.41 and was obtained with less effort. It should be pointed out as a reminder that if we are only interested in the unit-function response, all the initial conditions can be ignored at the outset (as in Example 8). Is this a stable system?

Example 10

Find the *complete* response (see Example 5, Chap. 8) of the system described in Example 9 if $x(n) = 18(2)^n u(n)$ and $y(-1) = 1$, $y(-2) = 0$. Notice that the two initial conditions would normally be given in terms of $y(0)$ and $y(1)$ as demanded in Eq. 9.51. Here we merely demonstrate that $y(0)$ and $y(1)$ can be found, as mentioned earlier, from the difference equation. With $n = -2$, the difference equation gives

$$3y(0) + 9y(-1) + 6y(-2) = x(0)$$

or

$$3y(0) + 9 + 0 = 18$$
$$y(0) = 3$$

For $n = -1$

$$3y(1) + 9y(0) + 6y(-1) = x(1)$$
$$3y(1) + 27 + 6 = 36$$
$$y(1) = 1$$

Also

$$x(n) = 18(2)^n u(n) \leftrightarrow 18\frac{z}{z-2} = X(z)$$

Substituting these results into the equation following Eq. 9.51

$$Y(z) = 6\frac{z^3}{(z-2)(z+1)(z+2)} - \frac{6z^2 + 12z}{(z+1)(z+2)} + \frac{3z^2 + 9z + z}{(z+1)(z+2)}$$

$$Y(z) = \frac{6z^3}{(z-2)(z+1)(z+2)} - \frac{3z^2 + 2z}{(z+1)(z+2)}$$

$$Y(z) = z\frac{3z^2 + 4z + 4}{(z-2)(z+1)(z+2)}$$

$$Y(z) = z\left[\frac{A}{z-2} + \frac{B}{z+1} + \frac{C}{z+3}\right]$$

$$Y(z) = z\left[\frac{2}{z-2} - \frac{1}{z+1} + \frac{2}{z+2}\right] = 2\frac{z}{z-2} - \frac{z}{z+1} + 2\frac{z}{z+2}$$

With the aid of Table 9.1

$$y(n) = 2(2)^n u(n) - (-1)^n u(n) + 2(-2)^n u(n)$$
$$y(n) = [(-1)^{n+1} + (2)^{n+1} - (-2)^{n+1}]u(n)$$

This result can be put in the form

$$y(n) = \begin{cases} 1 & n \text{ odd} \\ 2^{n+2} - 1 & n \text{ even} \end{cases}$$

which agrees exactly with Example 5, Chap. 8. The effort expended to obtain the final result was perhaps greater than using convolution-summation (Example 5, Chap. 8), but this is offset by the fact that the answer appears in closed form.

Example 11

Repeat Example 10 when the difference equation is written

$$3y(n) + 9y(n-1) + 6y(n-2) = x(n) = 18(2)^n u(n)$$

Taking the z transform using pairs 9.10 and 9.23

$$3Y(z) + 9z^{-1}Y(z) + 9y(-1) + 6z^{-2}Y(z) + 6z^{-1}y(-1) + 6y(-2) = 18\frac{z}{z-2}$$

$$(3 + 9z^{-1} + 6z^{-2})Y(z) + 9 + 6z^{-1} = 18\frac{z}{z-2}$$

Solving for $Y(z)$

$$Y(z) = 18\frac{z}{(z-2)(3 + 9z^{-1} + 6z^{-2})} - \frac{9 + 6z^{-1}}{3 + 9z^{-1} + 6z^{-2}}$$

$$Y(z) = 6\frac{z^3}{(z-2)(z^2 + 3z + 2)} - \frac{3z^2 + 2z}{z^2 + 3z + 2}$$

This $Y(z)$ is the same as that found in Example 10, so we need not proceed further.

Of particular interest in simulation is the digital simulation of a *differentiator*. In this regard it is wise for the reader to briefly review Sec. 8.9.

Example 12

Draw the block diagram and find the unit-function response of the digital simulator of a differentiator starting with the simulator of an integrator using the "rectangle rule" for integration:

$$y(n) = Tx(n) + y(n-1)$$

Notice that the "area" is $Tx(n)$, and the "constant of integration" is $y(n-1)$. Taking the z transform using pair 9.10, we obtain

$$Y(z) = TX(z) + z^{-1}Y(z) + y(-1)$$

or, setting $y(-1) = 0$, since we intend to first find the transfer function (which is based on the zero-state response) of the integrator:

$$Y(z) = TX(z) + z^{-1}Y(z)$$
$$Y(z)(1 - z^{-1}) = TX(z)$$

or

$$Y(z) = \frac{T}{1 - z^{-1}} X(z)$$

or

$$H_d^i(z) = \frac{T}{1 - z^{-1}} = \frac{Tz}{z - 1} \qquad (9.53)$$

This is the transfer function of a numerical integrator, and a numerical differentiator must have a transfer function that is the reciprocal of that of the numerical integrator:

$$H_d^d(z) = \frac{1}{H_d^i(z)} = \frac{1}{T} \frac{z - 1}{z} \qquad (9.54)$$

Thus, for the numerical differentiator

$$Y(z) = \frac{1}{T} \frac{z - 1}{z} X(z) \qquad (9.55)$$

The unit-function response is given by

$$h(n) = \mathscr{L}^{-1}\left\{\frac{1}{T} \frac{z-1}{z}\right\} = \frac{1}{T} \mathscr{L}^{-1}\{1 - z^{-1}\}$$

or

$$h(n) = \frac{1}{T}[\delta(n) - \delta(n-1)] \qquad (9.56)$$

which is shown in Fig. 9.5. Notice that for a continuous-time differentiator

$$y(t) = \frac{dx}{dt}$$

$$Y(s) = sX(s) - x(0^-)$$

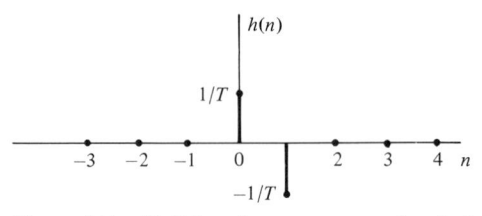

Figure 9.5. Unit-function response of a digital simulator of a differentiator.

and with $x(0^-) = 0$ for finding the transfer function

$$H(s) = s$$

Thus

$$h(t) = \frac{d\delta(t)}{dt} = \delta'(t)$$

This unit-impulse response *is not smooth*, so the time-sampling method of simulation (Chap. 8) should not work well.

The difference equation for the differentiator can be obtained in several ways. We can use convolution-summation with the unit-function response, for example

$$y(n) = \sum_{m=-\infty}^{\infty} x(m) \frac{1}{T} [\delta(n-m) - \delta(n-m-1)]$$

$$y(n) = \frac{1}{T}[x(n) - x(n-1)] \qquad (9.57)$$

We can inverse transform Eq. 9.55:

$$Y(z) = \frac{1}{T} X(z) - \frac{1}{T} z^{-1} X(z)$$

giving

$$y(n) = \frac{1}{T} x(n) u(n) - \frac{1}{T} x(n-1) u(n-1) \qquad (9.58)$$

Notice that $u(n)$ and $u(n-1)$ are present in Eq. 9.58 as a result of the fact that this form was obtained from the *unilateral* z transform[†] $[x(n) = 0, n < 0]$, whereas Eq. 9.57 was obtained using convolution-summation *without* assuming that $x(n) = 0$ for $n < 0$. The block diagram for Eq. 9.57 is shown in Fig. 9.6. This approximation for the derivative is, of course, well known.

[†] Notice that the pertinent *bilateral* z-transform pair (9.7c) is $y(n \pm m) \leftrightarrow z^{\pm m} Y(z)$, and this would give a result in agreement with Eq. 9.57.

9.4 LINEAR DIFFERENCE EQUATIONS (AGAIN)

Example 13

The derivative of $t^2/2$ is t. What does the simulator of Fig. 9.6 produce if $x(t) = t^2/2$ is sampled at $t = nT$, for $T = 1$ s and $T = 0.2$ s. We have

$$y(n) = \frac{1}{T}\left[\frac{n^2 T^2}{2} - \frac{(n-1)^2 T^2}{2}\right] = nT - T/2$$

$$y(n) = n - 0.5 \quad (T = 1 \text{ s})$$
$$y(n) = 0.2n - 0.1 \quad (T = 0.2 \text{ s})$$

Results are shown in Fig. 9.7.

The derivative of $t^2 u(t)/2$ is $tu(t)$ for which Eq. 9.58 gives

$$y(n) = \frac{1}{T}\left[\frac{n^2 T^2}{2} u(n) - \frac{(n-1)^2}{2} T^2 u(n-1)\right]$$

or

$$y(n) = \begin{cases} 0 & n \leq 0 \\ nT - T/2 & n > 0 \end{cases}$$

This will give results that agree with those shown in Fig. 9.7 for $n > 0$.

Numerical integration is often implemented using the *trapezoidal rule*

$$y(n) = \frac{T}{2}[x(n) + x(n-1)] + y(n-1) \tag{9.59}$$

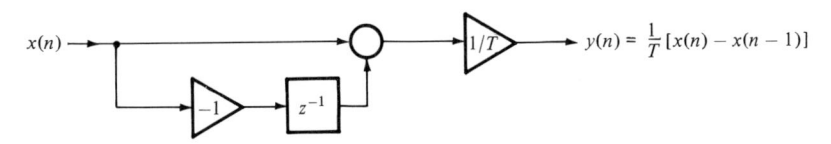

Figure 9.6. Digital simulator of a differentiator.

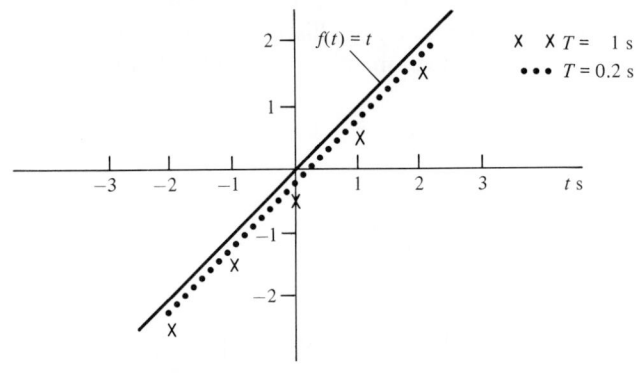

Figure 9.7. Discrete simulation of the derivative of $x(t) = t^2/2$ using the simulator of Fig. 9.6, $T = 1$ s and $T = 0.2$ s.

Example 14

Draw the block diagram and find the unit-function response for the numerical differentiator obtained from Eq. 9.59. The z transform of Eq. 9.59 is

$$Y(z) = \frac{T}{2}[X(z) + z^{-1}X(z) + x(-1)] + z^{-1}Y(z) + y(-1)$$

Since $x(-1) = y(-1) \equiv 0$ for finding the transfer function

$$Y(z)(1 - z^{-1}) = \frac{T}{2}(1 + z^{-1})X(z)$$

$$Y(z) = \frac{T}{2}\left(\frac{z+1}{z-1}\right)X(z)$$

or

$$H_d^i(z) = \frac{T}{2}\frac{z+1}{z-1}$$

This is the transfer function of another numerical integrator, so the transfer function of the numerical differentiator corresponding to it is its reciprocal:

$$H_d^d(z) = \frac{2}{T}\frac{z-1}{z+1} \tag{9.60}$$

or

$$H_d^d(z) = \frac{2}{T}[1 - 2z^{-1} + 2z^{-2} - 2z^{-3} + \cdots]$$

The unit-function response is[†]

$$h(n) = \frac{2}{T}[\delta(n) - 2\delta(n-1) + 2\delta(n-2) - \cdots] \tag{9.61}$$

In order to find the difference equation represented by this numerical differentiator, we will use convolution-summation. This avoids the difficulty encountered with the unilateral z transform as mentioned in Example 12. We have

$$y(n) = \sum_{m=-\infty}^{\infty} x(m)\frac{2}{T}[\delta(n-m) - 2\delta(n-m-1) + 2\delta(n-m-2) - \cdots]$$

$$y(n) = \frac{2}{T}[x(n) - 2x(n-1) + 2x(n-2) - \cdots]$$

Also

$$y(n-1) = \frac{2}{T}[x(n-1) - 2x(n-2) + 2x(n-3) - \cdots]$$

[†] See Prob. 19 at the end of the chapter.

Adding the last two equations:

$$y(n) + y(n-1) = \frac{2}{T}[x(n) - x(n-1)]$$

or

$$y(n) = \frac{2}{T}[x(n) - x(n-1)] - y(n-1) \quad (9.62)$$

Here, we will use Eq. 9.60 instead of 9.62 to obtain the block diagram. It is shown in Fig. 9.8. The output of the first summer involves a feedback loop and is $X(z)/(1+z^{-1})$ as determined from Example 7. Therefore the output of the unit delay is $-z^{-1}X(z)/(1+z^{-1})$. The output of the second summer is the sum of these two terms:

$$\frac{X(z)}{1+z^{-1}} - \frac{z^{-1}X(z)}{1+z^{-1}}$$

Then $Y(z)$ is given by $2/T$ multiplied by this term:

$$Y(z) = \frac{2}{T}\frac{1-z^{-1}}{1+z^{-1}}X(z) = \frac{2}{T}\frac{z-1}{z+1}X(z)$$

in agreement with Eq. 9.60.

Example 15

The derivative of

$$x(t) = e^{-100t}\sin(1000t)$$

is

$$x'(t) = e^{-100t}[1000\cos(1000t) - 100\sin(1000t)]$$

What is the output of the digital simulator of Fig. 9.8 if samples of $x(t)$ are taken at $t = nT$? Use $T = 10^{-3}$ s and $T = 10^{-4}$ s. Notice that in both cases $\omega_s = 2\pi/T$ is greater than $\omega = 10^3$ rad/s. That is, we are *sampling* at a rate considerably greater than 10^3 rad/s. This is mentioned even though it was

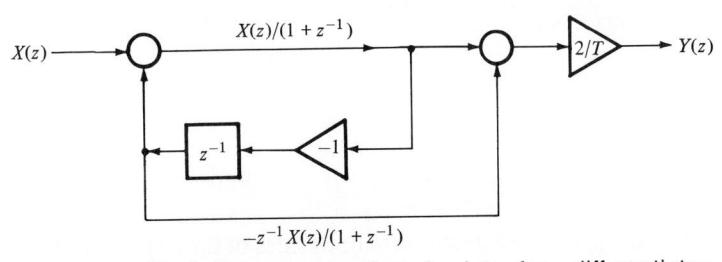

Figure 9.8. Block diagram of another simulator for a differentiator.

stated earlier that the time-sampling method of simulation does not work well (in principle). Equation 9.62 gives

$$y(n) = \frac{2}{T}[e^{-100nT}\sin(10^3 nT) - e^{-100(n-1)T}\sin(10^3 nT - 10^3 T)] - y(n-1)$$

or

$$y(n) = 2 \times 10^3 e^{-0.1n}[\sin(n) - e^{0.1}\sin(n-1)] - y(n-1) \quad (T = 10^{-3})$$

and

$$y(n) = 2 \times 10^4 e^{-0.01n}[\sin(0.1n) - e^{0.01}\sin(0.1n - 0.1)] - y(n-1)$$
$$(T = 10^{-4})$$

If we start at $n = 0$, for example, what number do we use for $y(-1)$? This question arose quite naturally in this example. In a more general case where we might be approximating an Nth-order differential equation with an Nth-order difference equation we would be given (or could find)

$$y(t)|_{t=0}, \frac{dy}{dt}\bigg|_{t=0}, \ldots, \frac{d^N y}{dt^N}\bigg|_{t=0}$$

and would need to find the *discrete initial conditions*

$$y(0), y(-1), \ldots, y(-N+1)$$

How is this accomplished? We have shown that Eq. 9.54 represents the transfer function of a numerical differentiator. This means that

$$\mathscr{L}\left\{\frac{dy}{dt}\bigg|_{t=nT}\right\} \approx \frac{1}{T}\frac{z-1}{z}Y(z)$$

$$\mathscr{L}\left\{\frac{d^2 y}{dt^2}\bigg|_{t=nT}\right\} \approx \left[\frac{1}{T}\frac{z-1}{z}\right]^2 Y(z)$$

The last approximation is represented by *two* such numerical differentiators (or simulators) in cascade. Thus, in general

$$\mathscr{L}\left\{\frac{d^m y}{dt^m}\bigg|_{t=nT}\right\} \approx \left[\frac{1}{T}\frac{z-1}{z}\right]^m Y(z) = \frac{1}{T^m}(1 - z^{-1})^m Y(z)$$

Using a binomial expansion

$$\mathscr{L}\left\{\frac{d^m y}{dt^m}\bigg|_{t=nT}\right\}$$

$$\approx \frac{1}{T^m}\left[1 + m(-z^{-1}) + \frac{m(m-1)}{2!}(-z^{-1})^2 + \cdots + (-z^{-1})^m\right]Y(z)$$

We now need to take the inverse z transform of both sides of the approximation, but on the right side we face the same difficulty as before. We can find the

9.4 LINEAR DIFFERENCE EQUATIONS (AGAIN)

inverse transform (as before) by using convolution-summation, or by using the bilateral transform (which we will not do), or by using the *unilateral* transform and simply dropping $u(n), u(n-1), \ldots, u(n-m)$. This latter possibility could have been exploited in Examples 13 and 14, but we chose convolution-summation instead. In any case the result is

$$\frac{d^m y}{dt^m}\bigg|_{t=nT} \approx \frac{1}{T^m}\left[y(n) - my(n-1) + \frac{m(m-1)}{2!}y(n-2) - \cdots (-1)^m y(n-m)\right]$$

For $n = 0$

$$y(0) - my(-1) + \frac{m(m-1)}{2}y(-2) - \cdots (-1)^m y(-m) \approx T^m \frac{d^m y}{dt^m}\bigg|_{t=0}$$

(9.63)

Equation 9.63 can be used iteratively to generate the initial conditions

$$y(-1), y(-2), \ldots, y(-N+1)$$

that we need. For $m = 1$

$$y(0) - y(-1) \approx T\frac{dy}{dt}\bigg|_{t=0}$$

or

$$y(-1) = y(0) - T\frac{dy}{dt}\bigg|_{t=0} \quad (9.64)$$

In the same way

$$y(-2) = y(0) - 2T\frac{dy}{dt}\bigg|_{t=0} + T^2\frac{d^2 y}{dt^2}\bigg|_{t=0} \quad (9.65)$$

and so forth.

Returning to the present example, from which we have strayed considerably, we still need $y(-1)$. We cannot find any of the terms in Eq. 9.64 (without cheating), but if T is small enough, we have the further approximation:[†]

$$y(-1) \approx y(0)$$

This gives us a way to start. For $n = 0$ our two equations for $y(n)$ become

$$y(0) = 2 \times 10^3[-e^{0.1}\sin(-1)] - y(0) \quad (T = 10^{-3})$$

and

$$y(0) = 2 \times 10^4[-e^{0.01}\sin(-0.1)] - y(0) \quad (T = 10^{-4})$$

[†] This is rather obvious in any case.

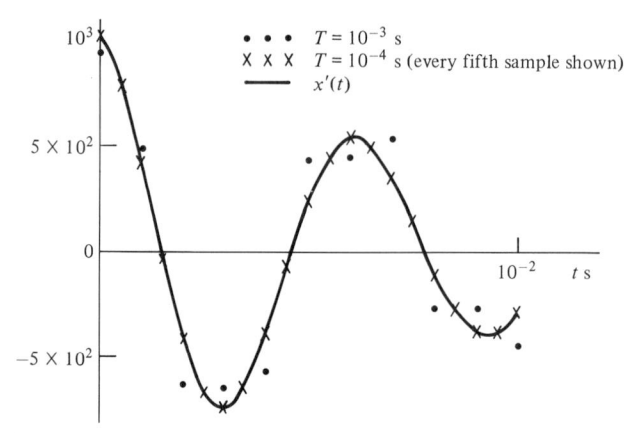

Figure 9.9. Numerical evaluation of $x'(t)$ using the simulator of Fig. 9.8 for $T = 10^{-3}$ s and $T = 10^{-4}$ s.

or

$$y(0) = 10^3[e^{0.1}\sin(1)] = 930 \qquad (T = 10^{-3})$$

and

$$y(0) = 10^4[e^{0.01}\sin(0.1)] = 1008 \qquad (T = 10^{-4})$$

Results are shown in Fig. 9.9. It is obvious that the smaller T gives very accurate results.

This digital simulator of an analog differentiator will be mentioned again and used later on.

As another example of using the z transform to solve a difference equation, consider Example 10, Chap. 8, for the resistance ladder network. The difference equation was ($a = 1$) given by

$$v(n+2) - 3v(n+1) + v(n) = 0$$

Using pair 3 (Table 9.1) we obtain

$$z^2 V(z) - z^2 v(0) - zv(1) - 3zV(z) + 3zv(0) + V(z) = 0$$

Solving for $V(z)$

$$V(z) = \frac{(z^2 - 3z)v(0) + zv(1)}{z^2 - 3z + 1} = \frac{z(z-3)v(0) + zv(1)}{(z - z_1)(z - z_2)}$$

where

$$z_1 = \frac{3 - \sqrt{5}}{2} \qquad z_2 = \frac{3 + \sqrt{5}}{2}$$

so

$$V(z) = v(0)z\frac{z-3}{(z-z_1)(z-z_2)} + v(1)z\frac{1}{(z-z_1)(z-z_2)}$$

$$V(z) = v(0)\left[\frac{z_1-3}{z_1-z_2}\frac{z}{z-z_1} + \frac{z_2-3}{z_2-z_1}\frac{z}{z-z_2}\right]$$

$$+ v(1)\left[\frac{1}{z_1-z_2}\frac{z}{z-z_1} + \frac{1}{z_2-z_1}\frac{z}{z-z_2}\right]$$

Therefore

$$v(n) = v(0)\left[\frac{z_1-3}{z_1-z_2}(z_1)^n u(n) + \frac{z_2-3}{z_2-z_1}(z_2)^n u(n)\right]$$

$$+ v(1)\left[\frac{1}{z_1-z_2}(z_1)^n u(n) + \frac{1}{z_2-z_1}(z_2)^n u(n)\right]$$

Now, we know that $v(0) = V_0$, but we must find $v(1)$ from the last equation. That is, with $n = N + 1$, $v(N + 1) \equiv 0$

$$0 = V_0\left[\frac{z_1-3}{z_1-z_2}(z_1)^{N+1} + \frac{z_2-3}{z_2-z_1}(z_2)^{N+1}\right]$$

$$+ v(1)\left[\frac{1}{z_1-z_2}(z_1)^{N+1} + \frac{1}{z_2-z_1}(z_2)^{N+1}\right]$$

or

$$v(1) = V_0\frac{(z_1-3)(z_1)^{N+1} - (z_2-3)(z_2)^{N+1}}{(z_1)^{N+1} - (z_2)^{N+1}}$$

Substituting this into the equation (above) for $v(n)$ and reducing will ultimately give Eq. 8.49 for $v(n)$. The z-transform method is somewhat indirect in this example.

Sometimes we are required to analyze a system which is continuous in one variable and discrete in another. Consider the ladder network that was just discussed with position index as a discrete variable. Suppose that the lumped electrical parameters are combinations of resistors, inductors, and capacitors. The series elements might be inductors with small resistance, and the shunt elements might be capacitors paralleled by large resistance. In this case we would be modeling an artificial transmission line or delay line, and the resulting equations would be integro-differential difference equations. Such a situation is shown in Fig. 9.10. Notice that in the interest of simplicity it has been assumed that there is zero initial energy storage and, as indicated in Fig. 9.10, the Laplace transform with respect to the continuous variable t (time) has already been taken. Thus, at the node where the voltage is labeled $V(n + 1, s)$ we have

$$\frac{V(n+1, s) - V(n, s)}{Z(s)} + V(n+1, s)Y(s) + \frac{V(n+1, s) - V(n+2, s)}{Z(s)} = 0$$

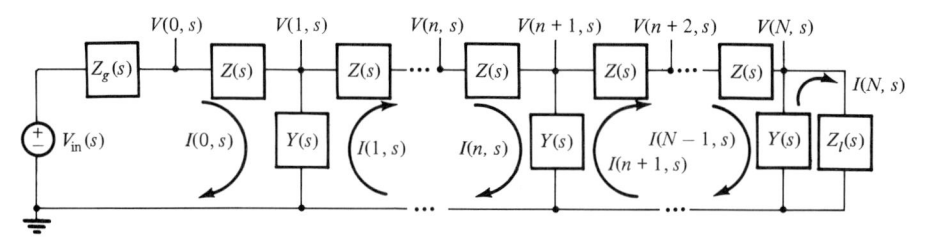

Figure 9.10. Artificial transmission line.

or

$$V(n+2, s) - [Z(s)Y(s) + 2]V(n+1, s) + V(n, s) = 0 \qquad (9.66)$$

Notice carefully that, in general, $Z(s)$ and $Y(s)$ *are not reciprocal quantities.* For the delay line they would be $R + sL$ and $G + sC$, respectively. With the definition

$$b(s) \equiv Z(s)Y(s) + 2$$

Eq. 9.66 becomes

$$V(n+2, s) - b(s)V(n+1, s) + V(n, s) = 0$$

and can be z transformed[†]:

$$z^2 V_d(z, s) - z^2 V(0, s) - zV(1, s) - b(s)zV_d(z, s) + b(s)zV(0, s) + V_d(z, s) = 0$$

Solving for $V_d(z, s)$

$$V_d(z, s) = \frac{(z^2 - bz)V(0, s) + zV(1, s)}{z^2 - bz + 1}$$

or

$$V_d(z, s) = 1 + V(1, s)\frac{z}{z^2 - bz + 1} - V(0, s)\frac{z}{z^2 - bz + 1}$$

$$V_d(z, s) = 1 + \frac{V(1, s)}{z_1 - z_2}\left[\frac{z}{z - z_1} - \frac{z}{z - z_2}\right] - \frac{V(0, s)}{z_1 - z_2}\left[\frac{1}{z - z_1} - \frac{1}{z - z_2}\right]$$

where

$$z_1 = \frac{b}{2} - \frac{1}{2}\sqrt{b^2 - 4} = z_1(s)$$

$$z_2 = \frac{b}{2} + \frac{1}{2}\sqrt{b^2 - 4} = z_2(s)$$

[†] Notice that a subscript d has been added for the z-transformed quantity to distinguish it from the untransformed (with z) quantity.

9.4 LINEAR DIFFERENCE EQUATIONS (AGAIN)

Taking the inverse z transform

$$V(n, s) = \delta(n) + \frac{V(1, s)}{z_1 - z_2}[(z_1)^n u(n) - (z_2)^n u(n)]$$

$$- \frac{V(0, s)}{z_1 - z_2}[(z_1)^{n-1} u(n-1) - (z_2)^{n-1} u(n-1)] \quad (9.67)$$

Also, from Fig. 9.10

$$I(n, s) = \frac{1}{Z(s)}[V(n, s) - V(n+1, s)] \quad (9.68)$$

$V(0, s)$ and $V(1, s)$ can be determined from boundary conditions at the input and load ends. At the input end

$$V_{in}(s) = V(0, s) + Z_g(s)I(0, s) \quad (9.69)$$

while at the load end

$$V(N, s) = Z_l(s)I(N, s) \quad (9.70)$$

Equation 9.68 (with $n = 0$) is substituted into Eq. 9.69, and Eqs. 9.67 and 9.68 (with $n = N$) are substituted into Eq. 9.70. This leaves us with two equations for the two unknowns; $V(0, s)$ and $V(1, s)$. These equations can be written

$$V_{in}(s) = A_{11}(s)V(0, s) + A_{12}(s)V(1, s)$$
$$0 = A_{21}(s)V(0, s) + A_{22}(s)V(1, s)$$

so that

$$V(0, s) = \frac{A_{22} V_{in}(s)}{A_{11}A_{22} - A_{12}A_{21}} \quad (9.71)$$

and

$$V(1, s) = \frac{-A_{21} V_{in}(s)}{A_{11}A_{22} - A_{12}A_{21}} \quad (9.72)$$

When Eqs. 9.71 and 9.72 are substituted into Eqs. 9.67 and 9.68, we have the s-domain solution. The time-domain solutions $v(n, t)$ and $i(n, t)$ are found by taking the inverse Laplace transforms. It is obvious that this will be a difficult task for the general case that we have treated. We will pursue this no further, but will mention, as a reminder, that if the excitation is sinusoidal (steady state) or is periodic and expressible as a sum of sinusoids, then we should have less difficulty because $(s \to j\omega)$ phasor methods are applicable.

Thus, we see that *both* the z transform and the Laplace transform can be employed as analysis techniques in the *same* problem if this is necessary. What form of solution would we obtain for the delay line of Fig. 9.10 if we use convolution-summation for solving Eq. 9.66? See Eq. 8.50. Would this be of any help in performing the inverse Laplace transform?

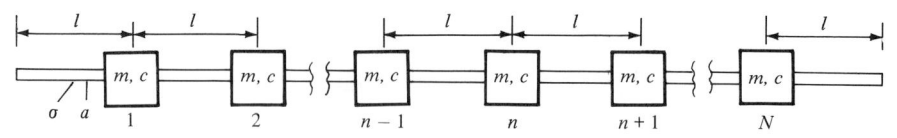

Figure 9.11. Heat transfer problem.

The same technique can be applied to the heat transfer problem[3] shown in Fig. 9.11. A number (N) of concentrated mass (m) points with specific heat c are connected by equal length (l) rods with cross-sectional area a and thermal conductivity σ. The net rate of flow of heat into the $n + 1$ mass point with time, $Q(n, t)$, is related to the temperature at the $n + 1$ mass point and the thermal "resistance" $l/(\sigma a)$ according to

$$Q(n+1, t) = \frac{\sigma a}{l}\{[T(n+2, t) - T(n+1, t)] - [T(n+1, t) - T(n, t)]\}$$

It is also given in terms of the specific heat, mass, and temperature of the nth mass point as

$$Q(n+1, t) = mc\frac{d}{dt}[T(n+1, t)]$$

Combining these equations

$$\frac{mcl}{\sigma a}\frac{d}{dt}[T(n+1, t)] = T(n+2, t) - 2T(n+1, t) + T(n, t)$$

Taking the Laplace transform and making the definition

$$k \equiv \frac{mcl}{\sigma a}$$

we have

$$ksT(n+1, s) - kT(n+1, 0^-) = T(n+2, s) - 2T(n+1, s) + T(n, s)$$

or

$$-kT(n+1, 0^-) = T(n+2, s) - (ks+2)T(n+1, s) + T(n, s)$$

Taking the z transform

$$-kz[T_d(z, 0^-) - T(0, 0^-)] = z^2[T_d(z, s) - T(0, s)] - zT(1, s)$$
$$\qquad - (ks+2)z[T_d(z, s) - T(0, 2)] + T_d(z, s)$$

Solving for $T_d(z, s)$

$$T_d(z, s) = \frac{[z^2 + (ks+2)z]T(0, s) + zT(1, s) - kz[T_d(z, 0^-) - T(0, 0^-)]}{z^2 - (ks+2)z + 1}$$

(9.73)

Now, given two boundary (end) conditions, we can find $T(0, s)$ and $T(1, s)$. Also, given the initial (in time) temperature distribution with position index,

9.4 LINEAR DIFFERENCE EQUATIONS (AGAIN)

$T(n + 1, 0^-)$, we can find its transform, which we symbolized by $z[T_d(z, 0^-) - T(0, 0^-)]$. Thus, the numerator is a function of z that we can find, and, therefore, we can take the inverse z transform of Eq. 9.73 to find $T(n, s)$. The last step is that of inverse Laplace transforming to find $T(n, t)$, and this may be a difficult task.

Difference equations occur in many other areas of study. Consider economics,[3] for example (see also Examples 10, 11, and 12 in Chap. 8). It is usually assumed to be true, even by those of us who know very little about economics, that "price is proportional to demand." Thus, during the interval of time n to $n + 1$ we may say that

$$p(n + 1) = k_1 d(n + 1) \tag{9.74}$$

where $p(n + 1)$ is the price of a commodity, $d(n + 1)$ is the demand, and k_1 is the constant of proportionality. Human nature (being what it is) is a factor in the following way. If prices increase, we expect them to increase more and so rush out to buy now before they do so, and this increases the demand. On the other hand, decreasing prices leads to the expectation of further decreases, so we hold off buying, and decrease the demand. This behavior can be modeled mathematically by assuming that the demand is equal to an initial or "rest" value $d(0)$ plus a term that is proportional to the *increase* in price over the previous interval. That is

$$d(n + 1) = d(0) + k_2[p(n + 1) - p(n)] \tag{9.75}$$

Combining Eqs. 9.74 and 9.75 gives a difference equation for the price:

$$p(n + 1) = k_1 d(0) + k_1 k_2 p(n + 1) - k_1 k_2 p(n)$$

or

$$(1 - k_1 k_2) p(n + 1) + k_1 k_2 p(n) = k_1 d(0) \tag{9.76}$$

Example 16

Solve Eq. 9.76 using the z-transform method assuming that it is valid for $n \geq 0$. Applying the z transform, we have

$$(1 - k_1 k_2) z P(z) - (1 - k_1 k_2) z p(0) + k_1 k_2 P(z) = k_1 d(0) \frac{z}{z - 1}$$

Solving for $P(z)$

$$P(z) = \frac{k_1 d(0)}{1 - k_1 k_2} \frac{z}{z - 1} \frac{1}{z + \frac{k_1 k_2}{1 - k_1 k_2}} + p(0) \frac{z}{z + \frac{k_1 k_2}{1 - k_1 k_2}}$$

Using a partial fraction expansion on the first term

$$P(z) = k_1 d(0) \frac{z}{z - 1} + [p(0) - k_1 d(0)] \frac{z}{z - \frac{k_1 k_2}{k_1 k_2 - 1}}$$

Inverse transforming

$$p(n) = k_1 d(0) u(n) + [p(0) - k_1 d(0)] \left(\frac{k_1 k_2}{k_1 k_2 - 1} \right)^n u(n) \qquad (9.77)$$

The first term is the steady-state (constant) price, while the second term is the transient price that can either go up or down with n depending on the numerical value of $k_1 k_2/(k_1 k_2 - 1)$. It is easy to see that this model predicts an unstable system if $k_1 k_2 - 1 > 0$ (or $k_1 k_2 > 1$) since in this case $[k_1 k_2/(k_1 k_2 - 1)] > 1$. This is also obvious from the characteristic equation whose single root is $z_1 = k_1 k_2/(k_1 k_2 - 1)$, and lies *outside* the unit circle for $k_1 k_2 > 1$. The price approaches $+\infty$ for $p(0) > k_1 d(0)$, and approaches $-\infty$ for $p(0) < k_1 d(0)$. For the case $|k_1 k_2/(k_1 k_2 - 1)| < 1$, $p(n)$ approaches $k_1 d(0)$ as n approaches ∞ (by the final-value theorem).

9.5 FREQUENCY RESPONSE

The frequency response of a discrete-time system was obtained in Sec. 8.6 using time-domain techniques. It is given as that part of the zero-state output that is multiplied by $e^{j\omega n T} = e^{jn\theta}$ when the input is $e^{jn\theta}$. Equation 8.68 gives the frequency response as

$$H_d(\theta) = \sum_{m=-\infty}^{\infty} h(m) e^{-jm\theta} \qquad [x(n) = e^{jn\theta}] \qquad (8.68)$$

In terms of the linear difference equation for shift-invariant systems

$$\sum_{m=0}^{N} a_m y(n+m) = \sum_{m=0}^{M} b_m x(n+m) \qquad (8.18)$$

we found that for the same input

$$H_d(\theta) = \frac{\sum_{m=0}^{M} b_m e^{jm\theta}}{\sum_{m=0}^{N} a_m e^{jm\theta}} \qquad (8.71)$$

In terms of the z transform we conclude that

$$H_d(\theta) \equiv H_d(e^{j\theta}) = H_d(z)\big|_{z=e^{j\omega T} = e^{j\theta}}$$

That is, $H_d(\theta)$ is simply that value of the *transfer function* obtained for $z = e^{j\theta}$. Thus, the value of the transfer function on the unit circle, $H_d(e^{j\theta})$, is the coefficient of the zero-state response to the input $e^{jn\theta}$:

$$y_{zs}(n) = H_d(e^{j\theta}) e^{jn\theta} = H_d(\theta) e^{jn\theta} \qquad (9.78)$$

The real and imaginary parts of Eq. 9.78 equal the responses to $\cos(n\theta)$ and $\sin(n\theta)$, respectively. The use of the z transform gives us (usually) a *closed form* for the frequency response, as opposed to the series of Eqs. 8.68 and 8.71.

Example 17

What is the frequency response of the simulator of a differentiator that was investigated in Examples 12 and 13 (Fig. 9.6)? Equation 9.54 was

$$H_d^d(z) = \frac{1}{T}(1 - z^{-1})$$

so

$$H_d^d(\theta) = \frac{1}{T}(1 - e^{-j\theta}) = \frac{1}{T}(1 - e^{-j\omega T})$$

$$H_d^d(\theta) = \frac{2j}{T}\sin(\omega T/2)e^{-j\omega T/2} \tag{9.79}$$

Notice that for small ω, $\sin(\omega T/2) \approx \omega T/2$ and $e^{-j\omega T/2} \approx 1$

$$H_d^d(\theta) \approx j\omega \tag{9.80}$$

The transfer function or frequency response of a continuous-time differentiator (via the Fourier transform or by simply using phasors) is $H(\omega) = j\omega$. It is intuitively obvious (and proved in the next section) that for the digital system to be an *exact* simulator of a differentiator we must have $H_d(\theta) \equiv j\omega$. The digital system (Fig. 9.6) only approximates this condition at low frequencies (9.80). We also know (Eq. 8.68) that $H_d(\theta)$ is always periodic, but the frequency response of the analog differentiator is not. Thus, we cannot build an *exact* digital simulator for a differentiator that will work for *all* signals. Results for Example 17 are shown in Fig. 9.12. We conclude that this simulator will work well only for signals with predominantly low-frequency content.

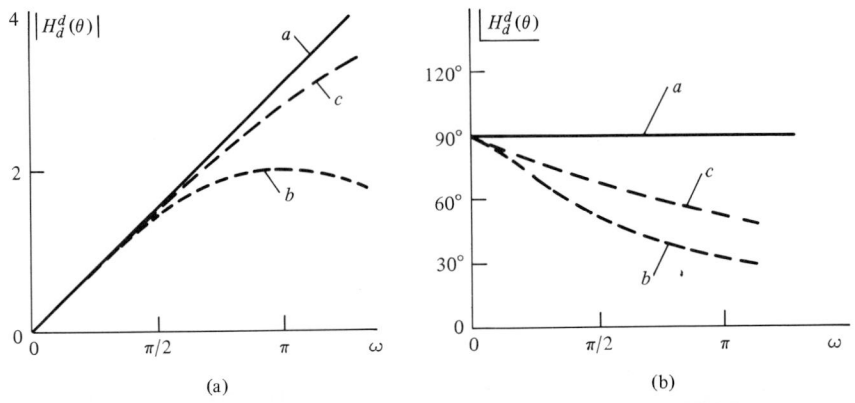

Figure 9.12. Frequency response of the digital simulator of a differentiator (Fig. 9.6). Analog differentiator: (a) Digital simulator, $T = 1$. (b) Digital simulator, $T = 0.5$. (c) $H_d^d(z) = (1/T)(z - 1)/z$.

Example 18

Repeat Example 17 for the numerical differentiator of Fig. 9.8. Equation 9.60 was

$$H_d^d(z) = \frac{2}{T} \frac{z-1}{z+1}$$

so

$$H_d^d(\theta) = \frac{2}{T} \frac{e^{j\theta}-1}{e^{j\theta}+1} = \frac{2j}{T} \tan \frac{\theta}{2} = \frac{2j}{T} \tan \frac{\omega T}{2} \tag{9.81}$$

and, once again, for small ω

$$H_d^d(\theta) \approx j\omega$$

Notice that the phase in Eq. 9.81 is exact. Results for this simulator are shown in Fig. 9.13 ($T = 1$ and $T = 0.5$). Generally speaking, this simulator is a better numerical differentiator than the other (Fig. 9.12), although it is somewhat more difficult to use numerically. That is, Eq. 9.57 can be used directly to calculate $y(n)$ for any n, whereas Eq. 9.62 must be used recursively and requires an initialization process.

We discussed three simple digital low-pass filters in Sec. 8.6. The respective frequency responses were found in Eqs. 8.84, 8.88, and 8.90 for the rectangular, triangular, and Hamming windows, respectively. We did not specify the elements of the digital system at the time, but now, with the availability of the z transform, it is easy to do so.

Example 19

Specify the elements making up the digital filter of Example 18, Chap. 8. As we have seen, the frequency response $H_d(\theta)$ is just $H_d(z)$ with z replaced by $e^{j\omega T}$,

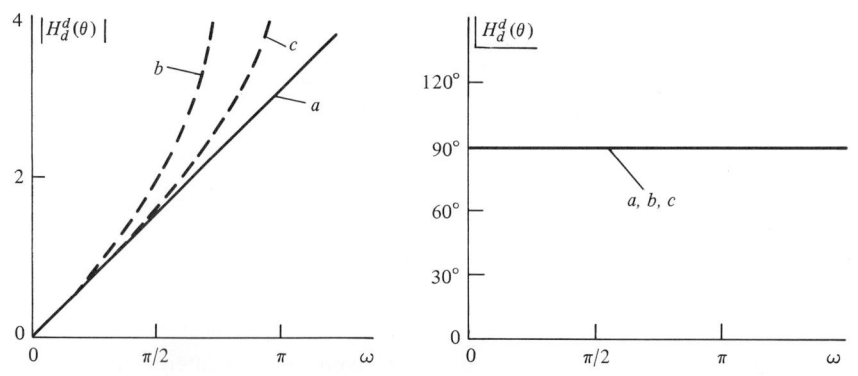

Figure 9.13. Frequency response of the digital simulator of a differentiator (Fig. 9.8). Analog differentiator: (a) Digital simulator, $T = 1$. (b) Digital simulator, $T = 0.5$. (c) $H_d^d(z) = (2/T)(z-1)/(z+1)$.

9.5 FREQUENCY RESPONSE

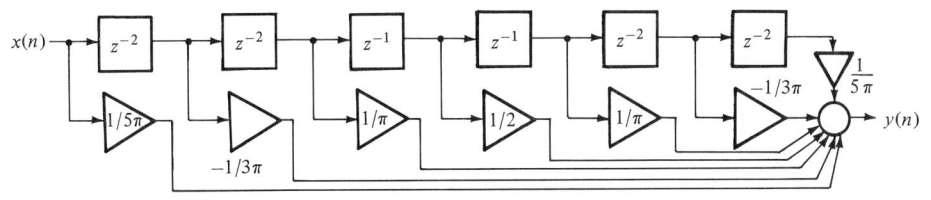

Figure 9.14. Low-pass digital filter (Eq. 8.84).

$\theta = \omega T$. Then, in order to obtain $H_d(z)$, we merely replace $e^{j\theta}$ by z in Eq. 8.84:

$$H_d(\theta) = e^{-j5\theta}\left[\frac{1}{2} + \frac{2}{\pi}\cos(\theta) - \frac{2}{3\pi}\cos(3\theta) + \frac{2}{5\pi}\cos(5\theta)\right] \quad (8.84)$$

$$H_d(\theta) = e^{-j5\theta}\left[\frac{1}{2} + \frac{1}{\pi}e^{j\theta} + \frac{1}{\pi}e^{-j\theta} - \frac{1}{3\pi}e^{j3\theta} - \frac{1}{3\pi}e^{-j3\theta} + \frac{1}{5\pi}e^{j5\theta} + \frac{1}{5\pi}e^{-j5\theta}\right]$$

giving

$$H_d(z) = z^{-5}\left[\frac{1}{2} + \frac{z}{\pi} + \frac{z^{-1}}{\pi} - \frac{z^3}{3\pi} - \frac{z^{-3}}{3\pi} + \frac{z^5}{5\pi} + \frac{z^{-5}}{5\pi}\right]$$

$$H_d(z) = \frac{1}{5\pi} - \frac{z^{-2}}{3\pi} + \frac{z^{-4}}{\pi} + \frac{z^{-5}}{2} + \frac{z^{-6}}{\pi} - \frac{z^{-8}}{3\pi} + \frac{z^{-10}}{5\pi} \quad (9.82)$$

Equation 9.82 easily leads to Fig. 9.14 which is the digital low-pass filter whose normalized frequency response is given by Eq. 8.84 and shown in Fig. 8.26.

The digital low-pass filter in the preceding example was designed with a cutoff frequency of one-fourth the sampling frequency. That is, $\omega_s = 4\omega_0$, or $2\pi/T = 4\omega_0$. Suppose that the analog signal $x(t) = \cos(\omega t)$ is sampled at $t = nT$, so that $x(nT) = \cos(\omega nT)$, and then applied to the digital filter. Does it actually behave as a low-pass filter? In order to answer this question we calculate $y(n)$ for four values of ω. First, we try $\omega = \omega_0/2$ (or $\omega T = \omega_0 T/2 = \pi/4$, or $\omega nT = n\pi/4$), which is well within the passband of $H_d(\theta)$ (Fig. 8.26). Notice that we have not specified T, but, rather, have specified $\omega T = \pi/4$. If, for example, $\omega = 10^3$ radians per second, then $T = (\pi/4) \times 10^{-3}$ seconds. The response is easy to obtain with Eq. 9.82. According to this equation, $y(n)$ will merely consist of seven terms, each of which is delayed and attenuated version of $\cos(n\omega T) = \cos(n\pi/4)$. Thus

$$y(n) = \frac{\cos(n\pi/4)}{5\pi}u(n) - \frac{\cos(n-2)\pi/4}{3\pi}u(n-2)$$

$$+ \frac{\cos(n-4)\pi/4}{\pi}u(n-4) + \frac{\cos(n-5)\pi/4}{2}u(n-5)$$

$$+ \frac{\cos(n-6)\pi/4}{\pi}u(n-6) - \frac{\cos(n-8)\pi/4}{3\pi}u(n-8)$$

$$+ \frac{\cos(n-10)\pi/4}{5\pi}u(n-10) \quad (9.83)$$

This discrete signal is shown in Fig. 9.15(a). It is essentially what we would expect: samples of a delayed cosine signal with very little attenuation. Notice that $y(n)$ is periodic for $n \geq 10$. There is a transient buildup for $0 \leq n < 10$.

Next, we try $\omega = \omega_0$, or $\omega nT = n\pi/4$, which leads to the result shown in Fig. 9.15(b). It too is as expected: a periodic, half-amplitude signal for $n \geq 10$. Its frequency is twice that of the signal in Fig. 9.14(a). Notice that there are four samples in a period, and if we continue to increase the frequency of the input signal, we should expect difficulties (according to the sampling theorem, Sec. 5.8).

Suppose we try $\omega = 2\omega_0$ or $\omega nT = n\pi$, which is in the stop band. This gives the result shown in Fig. 9.15(c), and is what we expect: large attenuation (25.7 dB) in the stop band. Notice that there are now only two samples per period. This is the minimum number, and if ω is increased, we will definitely have aliasing effects to contend with. If, for example, $\omega = 4\omega_0$, or $\omega nT = 2n\pi$, then

$$y(n) = \frac{\cos(n2\pi)}{5\pi} u(n) - \frac{\cos(n-2)2\pi}{3\pi} u(n-2)$$

$$+ \frac{\cos(n-4)2\pi}{\pi} u(n-4) + \frac{\cos(n-5)2\pi}{2} u(n-5)$$

$$+ \frac{\cos(n-6)2\pi}{\pi} u(n-6) - \frac{\cos(n-8)2\pi}{3\pi} u(n-8)$$

$$+ \frac{\cos(n-10)2\pi}{5\pi} u(n-10)$$

$$y(n) = \frac{1}{5\pi} u(n) - \frac{1}{3\pi} u(n-2) + \frac{1}{\pi} u(n-4) + \frac{1}{2} u(n-5)$$

$$+ \frac{1}{\pi} u(n-6) - \frac{1}{3\pi} u(n-8) + \frac{1}{5\pi} u(n-10) \qquad (9.84)$$

This response is shown in Fig. 9.15(d). Equation 9.84 shows that it is *exactly the same* as the zero frequency or step response ($\omega = 0$). This aliasing effect is something we must always be aware of. Notice that in Fig. 8.26 (if we imagine it to be extended to $\theta = 2\pi$), we are now (with $\omega = 4\omega_0$) precisely in the middle of the passband where the transfer function, because of its periodicity, has the same value as it has at $\theta = 0$. That is, $H_d(0) = H_d(2\pi)$. Another way to state the situation is to say that we now have only *one* sample per period for the input signal, and it occurs precisely at the *peak* of the input (cosine) signal. Thus, the digital filter cannot discern the difference between the input cosine signal and a step of the same amplitude. It "sees" the same thing. As we have seen before, an obvious cure, if we want the filter to attenuate the samples of $x(t) = \cos(4\omega_0 t)$, is to increase the sampling frequency $\omega_s = 2\pi/T$. We used $\omega_s = 4\omega_0$ in the present filter (Example 18, Chap. 8), so increasing ω_s with respect to ω_0 means redesigning the filter.

9.5 FREQUENCY RESPONSE 443

$\omega nT = n\pi/4$

(a)

$\omega nT = n\pi/2$

(b)

$\omega nT = n\pi$

(c)

$\omega nT = 2n\pi$

(d)

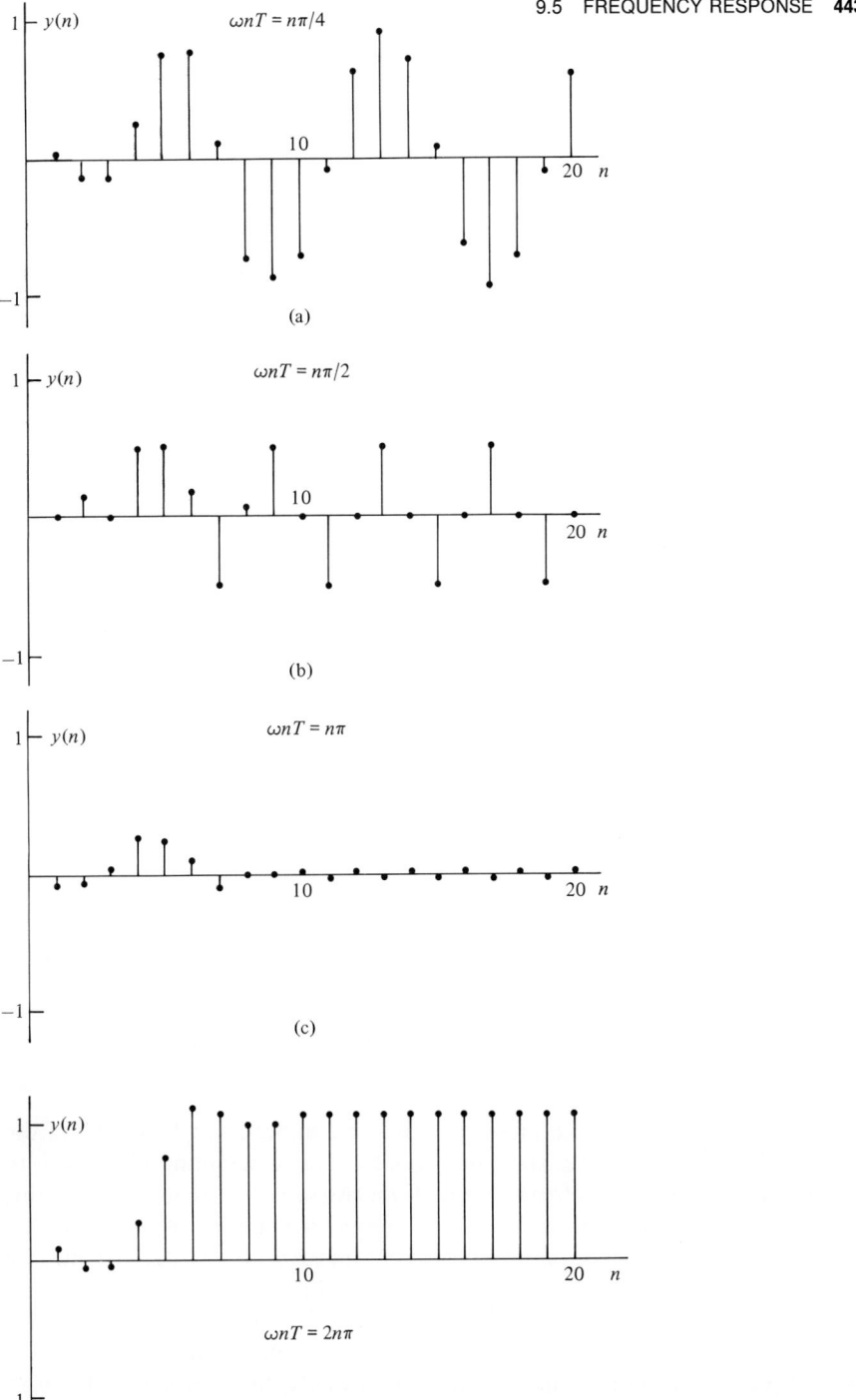

Figure 9.15. Discrete output of the digital low-pass filter of Fig. 9.13: (a) $\omega = \omega_0/2$. (b) $\omega = \omega_0$. (c) $\omega = 2\omega_0$. (d) $\omega = 4\omega_0$.

Notice that the use of windows, as in Examples 18 and 19, Chap. 8, does not alleviate the aliasing problem. We could easily analyze these filters for various inputs in exactly the same manner as above. After all, the only difference between the filters is in the coefficients (in the multipliers of Fig. 9.14).

Figure 9.14 and Eq. 9.82 indicate the need for a *tapped delay line*. In fact, this device could be used in any of the discrete-time systems that we have encountered since we have never used anything but (unit or otherwise) delay elements and multipliers. Of course, a tapped transmission line could be employed, but the available time delays are much too small for low-frequency applications, and, furthermore, the delays are not easily altered. The *artificial transmission line* (Fig. 9.10) can be designed with the appropriate delays, but the delays are also not easily altered. These deficiencies have been removed with the advent of integrated circuit technology that has produced charge-coupled devices that yield useful time delays over a very broad range of frequencies.[4] These devices are called electronic *tapped delay lines* since they are equivalent to tapped transmission lines. They can be used to design a system that is equivalent to a digital system without the need for an A/D (analog-to-digital) converter. The output of an A/D converter is a digital signal in the sense that the values that it can take on are multiples of the *smallest* available digit.

9.6 SIMULATION

The *time-sampling* method of digital simulation was investigated in Sec. 8.9 (see Fig. 8.31). We discovered that if the unit-function response of the digital system equals T, the sampling interval, multiplied by the samples of the unit-impulse response of the analog system:

$$h_d(n) = Th_a(nT) \tag{8.132}$$

then the zero-state output of the digital system will be approximately equal to the samples of the zero-state output of the analog system:

$$y_{dzs}(n) \approx y_{zs}(nT) \tag{9.85}$$

It is only necessary that the input $x(t)$ and unit-impulse response $h(t)$ of the analog system be nearly constant (or smooth) over a sampling interval T to obtain useful results. We already know from the sampling theorem what this implies. It means that we must be able to sample $x(t)$ at a rate which is *at least* twice the highest frequency ω_h in $x(t)$, and this, in turn, means that $x(t)$ must be band-limited. That is, $x(t)$ contains no frequencies higher than ω_h. Furthermore, the sampling theorem tells us that we can recover $y_{zs}(t)$ from its samples $y_{zs}(nT)$ that are available at the output of the digital simulator. Thus, if $x(t)$ is truly band-limited, we can in principle recover $y(t)$ exactly. In other words, Eq. 9.85 becomes an equality rather than an approximation. It has already been pointed out in Chap. 5, where the sampling theorem was introduced, that this can never happen in practice because if $x(t)$ is truly band-limited, it must be of *infinite* extent in the time domain, thus requiring an *infinite number* of samples, regardless of the sampling interval. Nevertheless, approximate equality in 9.85 is sufficient for nearly all engineering applications.

We now need to establish criteria for simulation based on frequency-domain ideas (as promised in Sec. 9.5). The problem is, of course, the same as before. If[†] $x_d(n) = x_a(nT)$, then what digital system produces $y_d(n) = y_a(nT)$? It is sufficient to consider sinusoidal inputs since an arbitrary input signal can be expressed as a superposition of sinusoids by way of the inverse Fourier transform (Chap. 5). Thus, if $x_a(t) = e^{j\omega t}$, then $x_d(n) = x_a(nT) = e^{jn\omega T}$. The zero-state response of the analog system was given in Sec. 8.7, Fig. 8.21(a), by

$$y_{zs}(t) = H(\omega)e^{j\omega t} \tag{9.86}$$

and the zero-state response for the digital system was given in Fig. 8.21(b):

$$y_d(n) = y_{zs}(n) = H_d(\theta)e^{jn\theta} = H_d(e^{j\omega T})e^{jn\omega T} \tag{9.87}$$

Therefore, if $y_d(n) = y_a(nT)$, we must have

$$H(\omega)e^{jn\omega T} = H_d(e^{j\omega T})e^{jn\omega T} \tag{9.88}$$

or

$$\boxed{H_d(e^{j\omega T}) = H(\omega)} \tag{9.89}$$

Thus, we reach the general conclusion that the frequency-domain condition for exact simulation is that the discrete transfer function must equal the continuous transfer function. This was pointed out and used earlier, but not proved. It was also pointed out earlier that Eq. 9.89 cannot hold for *every* ω because $H_d(\theta)$ is periodic, while $H(\omega)$ is not. The period of $H_d(\theta)$ is $\theta = \omega T = 2\pi$, or $\omega = 2\pi/T$. It is possible, at least in theory, to satisfy Eq. 9.89 for every ω that is less than π/T. This is shown in Fig. 9.16. Notice that this (once again) requires that all inputs to be used must be band-limited with $\omega_h \leq \pi/T$, or $T \leq \pi/\omega_h$, or $2\pi/\omega_s \leq \pi/\omega_h$, or $\omega_s \geq 2\omega_h$. Thus, we reach a conclusion that we were already aware of: The sampling frequency must be at least twice the highest frequency in the input signal. If this is the case, then Eq. 9.78 holds, and the digital system is a simulator for the analog system. We have already pointed out that perfect results will never be obtained in practice. We can now give

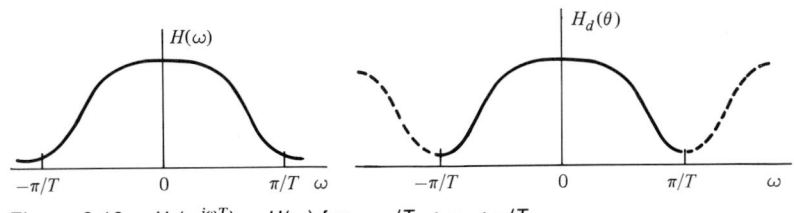

Figure 9.16. $H_d(e^{j\omega T}) = H(\omega)$ for $-\pi/T < \omega < \pi/T$.

[†] Subscripts *d* and *a* refer to digital and analog, respectively, as before.

another (frequency-domain) reason why this is so.[4] Any *real* system is limited to the implementation of a *finite* number of elements, and $H_d(e^{j\omega T})$ is a rational function of $e^{j\omega T}$ (as we have seen) whereas $H(\omega)$ is a rational function of just ω. Thus, Eq. 9.89 cannot hold exactly in an interval in a *real* system.

Consider the implementation of the analog system described by its s-domain transfer function[4] (Eq. 7.2):

$$H(s) = \frac{\sum_{n=0}^{M} b_n s^n}{\sum_{n=0}^{N} a_n s^n} \tag{7.2}$$

In order to simplify matters, let $M = N = 2$, so that we are dealing with a second-order system whose forcing function consists of a single input and its first and second derivatives:

$$H(s) = \frac{b_0 + b_1 s + b_2 s^2}{a_0 + a_1 s + a_2 s^2} = \frac{Y_{zs}(s)}{X(s)} \tag{9.90}$$

or

$$a_2 \frac{d^2 y}{dt^2} + a_1 \frac{dy}{dt} + a_0 y = b_2 \frac{d^2 x}{dt^2} + b_1 \frac{dx}{dt} + b_0 x \tag{9.91}$$

The analog system for this differential equation is shown in Fig. 9.17. It consists of differentiators and multipliers. Notice that Eq. 9.90 can be written

$$\frac{Y_{zs}(s)}{X(s)} = \frac{b_0 + b_1 s + b_2 s^2}{a_0 + a_1 s + a_2 s^2} \frac{I(s)}{I(s)}$$

or

$$I(s) = \frac{1}{a_0 + a_1 s + a_2 s^2} X(s) \tag{9.92}$$

Since it is also true that

$$Y_{zs}(s) = (b_0 + b_1 s + b_2 s^2) I(s) \tag{9.93}$$

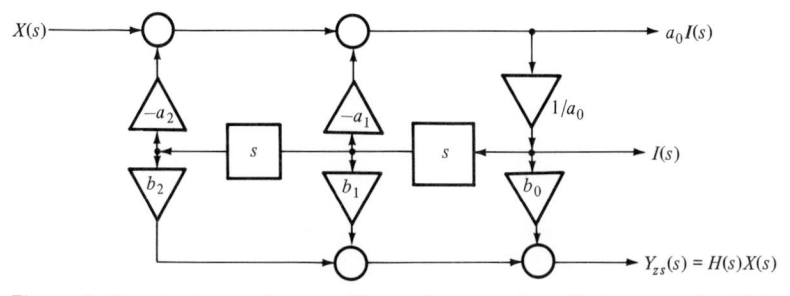

Figure 9.17. Analog system of differentiators and multipliers for Eq. 9.91.

9.6 SIMULATION 447

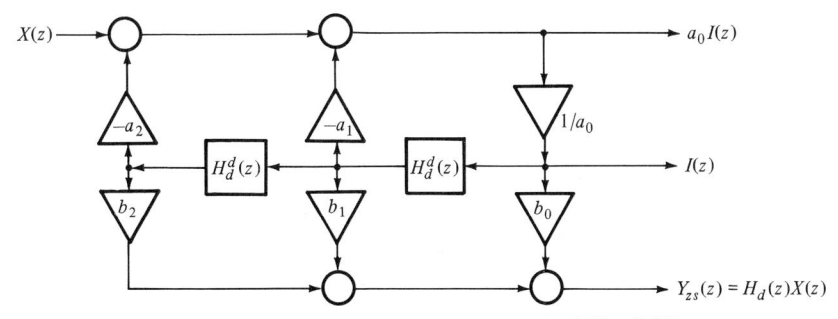

Figure 9.18. Digital simulator of the analog system of Fig. 9.17.

it follows that Eq. 9.91 is represented by Fig. 9.17. It is easy to recognize that the system shown in Fig. 9.17 can be extended to include the general case (Eq. 7.2) for any M and N.

If we replace the analog differentiators of Fig. 9.17 with digital simulators of a differentiator whose transfer function we found earlier, we obtain the digital system of Fig. 9.18. This is the desired simulator of the analog system of Fig. 9.17. In the same way that we verified $H(s)$ for Fig. 9.17, we can easily show that

$$H_d(z) = \frac{b_0 + b_1 H_d^d(z) + b_2[H_d^d(z)]^2}{a_0 + a_1 H_d^d(z) + a_2[H_d^d(z)]^2} \tag{9.94}$$

It also follows that the digital system of Fig. 9.18 can be extended to include the general case:

$$H_d(z) = \frac{\sum_{n=0}^{M} b_n [H_d^d(z)]^n}{\sum_{n=0}^{N} a_n [H_d^d(z)]^n} \tag{9.95}$$

We have already found two digital simulators of a differentiator in Sec. 9.5. We choose the second (Fig. 9.8) for use here. Its transfer function was given by Eq. 9.60:

$$H_d^d(z) = \frac{2}{T} \frac{z-1}{z+1} \tag{9.60}$$

Example 20

We considered an RC low-pass filter and its digital version in Example 16, Chap. 8 (Fig. 8.22). We obtained the digital version by using the time-sampling method (without explicitly calling it that). The analog transfer function for the capacitor voltage was given by

$$H(\omega) = \frac{\omega_2}{j\omega + \omega_2} \tag{8.72}$$

The frequency response of this digital filter is shown in Fig. 8.21, for three values of ω_2, and for the larger value of ω_s/ω_2 we see that $H_d(\theta) \approx H(\omega)$ for small ω, as expected. We would now like to realize the digital version of the RC filter using the method just developed and compare it to the one shown in Fig. 8.22. In terms of s

$$H(s) = \frac{\omega_2}{s + \omega_2} \tag{9.96}$$

where $\omega_2 = 1/(RC)$. Equation 9.94 with $b_1 = b_2 = a_2 = 0$, $b_0 = a_0 = \omega_2$, and $a_1 = 1$ gives

$$H_d(z) = \frac{\omega_2}{H_d^d(z) + \omega_2} = \frac{1}{1 + \dfrac{1}{\omega_2} H_d^d(z)} \tag{9.97}$$

which is shown in Fig. 9.19. Substituting Eq. 9.60 into 9.97 gives

$$H_d(z) = \frac{1}{1 + \dfrac{2}{\omega_2 T}\dfrac{1 - z^{-1}}{1 + z^{-1}}} = \frac{1 + z^{-1}}{1 + z^{-1} + \dfrac{2}{\omega_2 T} - \dfrac{2}{\omega_2 T} z^{-1}}$$

or

$$H_d(z) = \frac{1 + z^{-1}}{\left(1 + \dfrac{2}{\omega_2 T}\right) + \left(1 - \dfrac{2}{\omega_2 T}\right) z^{-1}} = \frac{1}{1 + \dfrac{2}{\omega_2 T}} \frac{1 + z^{-1}}{1 - \dfrac{2 - \omega_2 T}{2 + \omega_2 T} z^{-1}} \tag{9.98}$$

The realization of Eq. 9.98 is shown in Fig. 9.20 along with the frequency responses for the case that corresponds to Fig. 8.23(c): $\omega_s/\omega_2 = 20\pi$, $2\pi/(\omega_2 T) = 20\pi$, or $\omega_2 T = 0.1$. This last value of $\omega_2 T$ gives

$$H_d(z) = \frac{1}{21} \frac{1 + z^{-1}}{1 - (1.9/2.1) z^{-1}}$$

The frequency responses of Fig. 9.20(c) can be obtained by any of several methods, but it is worthwhile to use a general approach at this point. The fre-

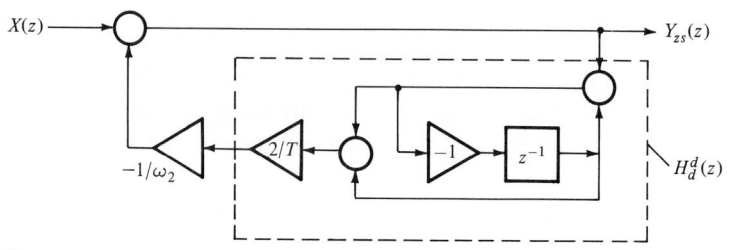

Figure 9.19. Block diagram of the realization of Eq. 9.97.

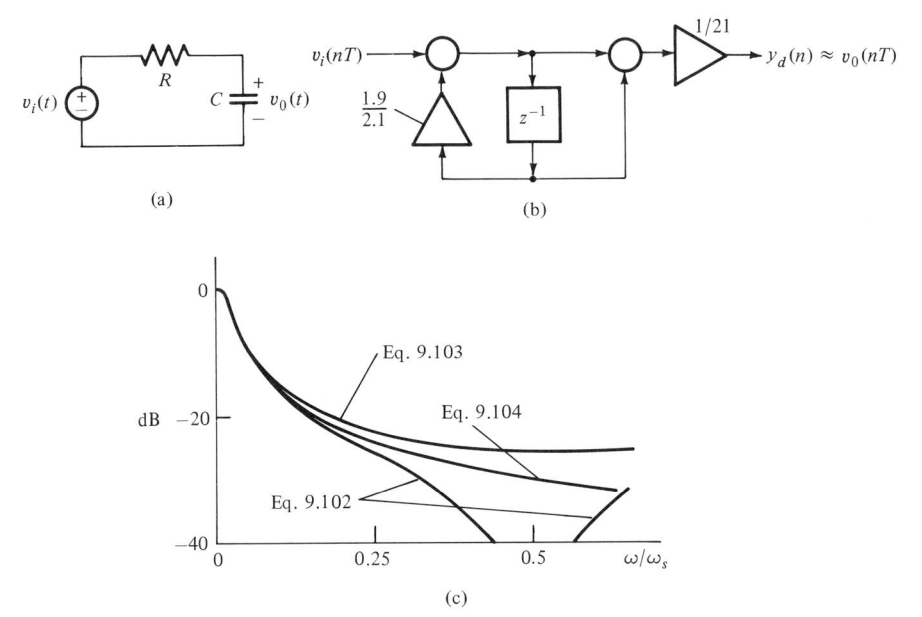

Figure 9.20. (a) Analog RC filter. (b) Digital version based on Eqs. 7.2 and 9.95. (c) Frequency responses for the analog and digital systems.

quency response of the analog system is given by Eq. 7.2 with $s = j\omega$:

$$H(\omega) = \frac{\sum_{n=0}^{M} b_n (j\omega)^n}{\sum_{n=0}^{N} a_n (j\omega)^n} \qquad (9.99)$$

while the frequency response of the digital system is given by Eqs. 9.95, 9.60, and 9.81 ($z = e^{j\omega T}$):

$$H_d(\theta) = H_d(e^{j\omega T}) = \frac{\sum_{n=0}^{M} b_n \left(\frac{2j}{T} \tan \frac{\omega T}{2}\right)}{\sum_{n=0}^{N} a_n \left(\frac{2j}{T} \tan \frac{\omega T}{2}\right)} \qquad (9.100)$$

Comparing Eqs. 9.99 and 9.100, we see that

$$\boxed{H_d(\theta) = H\left(j\frac{2}{T} \tan \frac{\omega T}{2}\right) = H(H_d^d(\omega))} \qquad (9.101)$$

Because of Eq. 9.101 the simulation technique we have been discussing is called the *frequency transformation* method of simulation.

Returning briefly to Example 18, we see that

$$|H_d(\theta)| = |H(H_d^d(\omega))|$$

Since

$$H(\omega) = \frac{1}{1 + j\omega/\omega_2}$$

it follows that

$$H(H_d^a(\omega)) = \frac{1}{1 + j\dfrac{2}{\omega_2 T}\tan(\omega T/2)}$$

Therefore

$$|H_d(\theta)| = \frac{1}{\left[1 + \left[\dfrac{2}{\omega_2 T}\tan(\omega T/2)\right]^2\right]^{1/2}}$$

or

$$|H_d(\theta)| = \frac{1}{[1 + 400\tan^2(0.05\omega/\omega_2)]^{1/2}}$$

since $\omega/\omega_2 = 20\pi\omega/\omega_s$

$$|H_d(\theta)| = \frac{1}{[1 + 400\tan^2(\pi\omega/\omega_s)]^{1/2}} \tag{9.102}$$

For Fig. 8.20(b) we have ($\omega_s/\omega_2 = 20\pi$)

$$|H_d(\theta)| = \frac{0.1}{[1 + e^{-0.2} - 2e^{-0.1}\cos(2\pi\omega/\omega_s)]^{1/2}} \tag{9.103}$$

while for the analog system we have

$$|H(\omega)| = \frac{1}{[1 + (20\pi\omega/\omega_s)^2]^{1/2}} \tag{9.104}$$

These are the three responses shown in Fig. 9.20(c). They compare favorably at low frequencies. The digital system of Fig. 8.22(b) is not the same as that shown in Fig. 9.20(b).

Let us now look at the frequency-domain approach to simulation in a slightly different manner. The simulator that results from the frequency transformation method will provide an approximate numerical solution to a differential equation. The method that will be outlined here is based on the same principle as the frequency transformation method for simulation, and is called the *Tustin substitution method*.[5] Again, in order to simplify matters, consider a second-order analog system described by Eqs. 9.90 and 9.91 and Fig. 9.17. At the sampling instant $t = nT$ Eq. 9.91 becomes

$$a_2 \left.\frac{d^2 y}{dt^2}\right|_{t=nT} + a_1 \left.\frac{dy}{dt}\right|_{t=nT} + a_0 y|_{t=nT} = x_f|_{t=nT} \tag{9.105}$$

Notice that we have used

$$x_f(t) = \sum_{n=0}^{M} b_n \frac{d^n x}{dt^n}$$

as in Eq. 2.4. Notice also that Eq. 9.105 is a linear combination of four sequences. Taking the z transform of both sides of Eq. 9.105 gives

$$a_2 \mathscr{L}\left\{\frac{d^2 y}{dt^2}\bigg|_{t=nT}\right\} + a_1 \mathscr{L}\left\{\frac{dy}{dt}\bigg|_{t=nT}\right\} + a_0 \mathscr{L}\{y|_{t=nT}\} = \mathscr{L}\{x_f|_{t=nT}\} \qquad (9.106)$$

But, using $H_d^d(z)$ given by Eq. 9.60

$$\mathscr{L}\left\{\frac{dy}{dt}\bigg|_{t=nT}\right\} \approx \frac{2}{T} \frac{1-z^{-1}}{1+z^{-1}} Y(z) \qquad (9.107)$$

and

$$\mathscr{L}\left\{\frac{d^2 y}{dt^2}\bigg|_{t=nT}\right\} \approx \left(\frac{2}{T} \frac{1-z^{-1}}{1+z^{-1}}\right)^2 Y(z) \qquad (9.108)$$

so *substituting* Eqs. 9.107 and 9.108 into 9.106, we obtain the *approximate* result

$$\left[a_2 \left(\frac{2}{T} \frac{1-z^{-1}}{1+z^{-1}}\right)^2 + a_1 \left(\frac{2}{T} \frac{1-z^{-1}}{1+z^{-1}}\right) + a_0\right] Y(z) = X_f(z) \qquad (9.109)$$

or

$$[4a_2(1-z^{-1})^2 + 2a_1 T(1-z^{-1})(1+z^{-1}) + a_0 T^2(1+z^{-1})^2] Y(z)$$
$$= T^2(1+z^{-1})^2 X_f(z)$$

Collecting like powers of z^{-1}, we have

$$[(4a_2 + 2a_1 T + a_0 T^2) + (-8a_2 + 2a_0 T^2)z^{-1} + (4a_2 - 2a_1 T + a_0 T^2)z^{-2}] Y(z)$$
$$= [(T^2) + (2T^2)z^{-1} + (T^2)z^{-2}] X_f(z) \qquad (9.110)$$

which has the form

$$(\alpha_0 + \alpha_1 z^{-1} + \alpha_2 z^{-2}) Y(z) = (\beta_0 + \beta_1 z^{-1} + \beta_2 z^{-2}) X_f(z) \qquad (9.111)$$

We can now inverse transform[†] Eq. 9.111 to find

$$\alpha_0 y(n) + \alpha_1 y(n-1) + \alpha_2 y(n-2) = \beta_0 x_f(n) + \beta_1 x_f(n-1) + \beta_2 x_f(n-2)$$

or

$$y(n) = \frac{\beta_0}{\alpha_0} x_f(n) + \frac{\beta_1}{\alpha_0} x_f(n-1) + \frac{\beta_2}{\alpha_0} x_f(n-2) - \frac{\alpha_1}{\alpha_0} y(n-1) - \frac{\alpha_2}{\alpha_0} y(n-2)$$

$$(9.112)$$

The solution of this difference equation gives an approximate numerical solution to the original differential equation. Notice that it is second order,

---

[†] The same remarks concerning the use of the unilateral z transform that were made for Example 12 apply here. Refer to the footnote below Eq. 9.58.

recursive, and requires the calculation of two initial conditions. We have already found a method for accomplishing this in Eq. 9.63 that gives Eq. 9.64 for the present second-order case. Tustin's substitution method can be generalized to an Nth-order system with very little difficulty.

We mention in passing that we could also use Eq. 9.54:

$$\mathscr{L}\left\{\frac{dy}{dt}\bigg|_{t=nT}\right\} = \frac{1}{T}(1-z^{-1})Y(z) \tag{9.113}$$

and

$$\mathscr{L}\left\{\frac{d^2y}{dt^2}\bigg|_{t=nT}\right\} = \frac{1}{T^2}(1-z^{-1})^2 Y(z) \tag{9.114}$$

[that is, $H_d^d(z) = (z-1)/(Tz) = (1-z^{-1})/T$ as in Example 13] instead of Eqs. 9.107 and 9.108. Following the same procedure as outlined above leads to a slightly different form for the difference equation. This is called the *first-difference method* because it is based on

$$H_d^d(z) = \frac{1}{T}\frac{z-1}{z} \tag{9.53}$$

and

$$y(n) = \frac{1}{T}[x(n) - x(n-1)] \tag{9.57}$$

which is a first difference. Normally, the Tustin method gives a more accurate solution for the same T than the first-difference method.

Example 21

Find the unit-step response for the capacitor voltage in the overdamped RLC series circuit of Fig. 2.12(a) [redrawn in Fig. 9.21(a)], Example 5, Chap. 2,

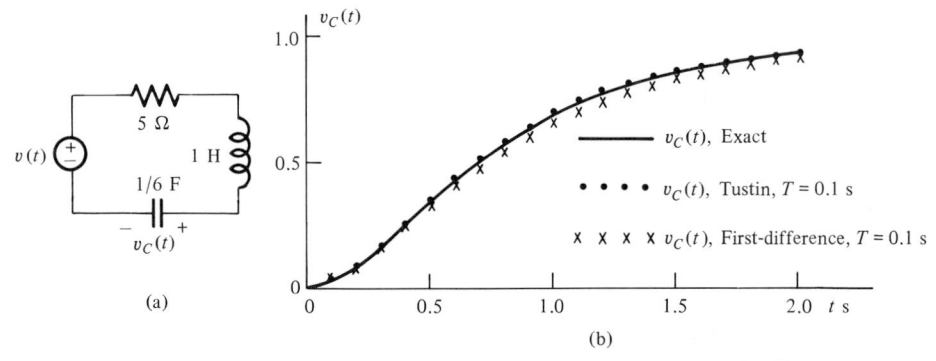

Figure 9.21. (a) Series RLC circuit. (b) $v_C(t)$, analytically and numerically ($T = 0.1$ s) solved (Tustin and first difference).

using Tustin's method and the first-difference method with $T = 0.1$ s. The differential equation for the capacitor voltage is easily found:

$$\frac{d^2 v_C}{dt^2} + 5\frac{dv_C}{dt} + 6v_C = 6u(t) \qquad i(0) = v_C(0) = 0$$

and its solution was found to be

$$v_C(t) = [1 + 2e^{-3t} - 3e^{-2t}]u(t)$$

Tustin's substitution method gives $(nT \to n)$

$$(4 + 10T + 6T^2)v_C(n) = 6T^2 u(n) + 12T^2 u(n-1) + 6T^2 u(n-2)$$
$$+ (8 - 12T^2)v_C(n-1)$$
$$+ (-4 + 10T - 6T^2)v_C(n-2)$$

or, with $T = 0.1$

$$5.06 v_C(n) = 0.06 u(n) + 0.12 u(n-1) + 0.06 u(n-2)$$
$$+ 7.88 v_C(n-1) - 3.06 v_C(n-2) \qquad (9.115)$$

The first-difference method gives

$$(1 + 5T + 6T^2)v_C(n) = T^2 u(n) + (2 + 5T)v_C(n-1) - v_C(n-2)$$

or

$$1.56 v_C(n) = 0.06 u(n) + 2.5 v_C(n-1) - v_C(n-2)$$

The initial conditions are $i(0) = v_C(0) = 0$, so that

$$\left.\frac{dv_C}{dt}\right|_{t=0} = 0$$

(all in the analog case). Equation 9.64 gives $v_C(-1) = 0$ for $v_C(0) = 0$ in the numerical case. Results are given in Fig. 9.21(b), and are good in both cases.

The differential equation for the current is

$$\frac{d^2 i}{dt^2} + 5\frac{di}{dt} + 6i = \frac{dv}{dt}$$

and if we attempt to find its unit-step response, dv/dt (the forcing function) is $\delta(t)$. The unit impulse is difficult to treat numerically!

Example 22

Find the digital simulator of the system of Example 21. In terms of the input $x(t)$, that was $u(t)$ in Example 21, we have

$$H(s) = \frac{1}{(1/6)s^2 + (5/6)s + 1} = \frac{6}{s^2 + 5s + 6}$$

454 THE z TRANSFORM

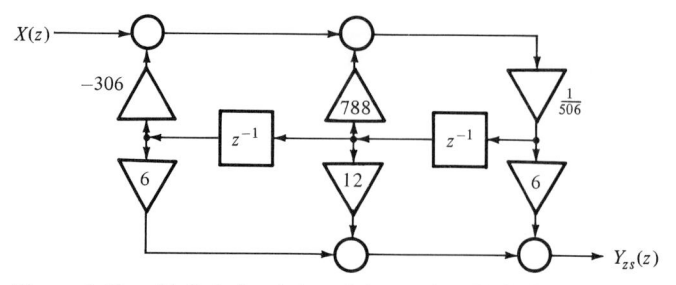

Figure 9.22. Digital simulator of the series *RLC* circuit of Fig. 9.21(a).

Therefore

$$H_d(z) = \frac{6}{\left(\dfrac{2}{T}\dfrac{1-z^{-1}}{1+z^{-1}}\right)^2 + 5\left(\dfrac{2}{T}\dfrac{1-z^{-1}}{1+z^{-1}}\right) + 6}$$

Simplifying, with $T = 0.1$ s, we obtain familiar numbers (Eq. 9.115):

$$H_d(z) = \frac{6 + 12z^{-1} + 6z^{-2}}{506 - 788z^{-1} + 306z^{-2}} \qquad (9.116)$$

The simulator that is described by this transfer function is shown in Fig. 9.22. It is helpful to compare Eq. 9.116 and Fig. 9.22 to Eq. 9.94 and Fig. 9.18.

In order to conclude this section on simulation and relate it more directly to frequency response and to the time-sampling method of simulation, consider once again Eq. 8.68:

$$H_d(\theta) = \sum_{n=-\infty}^{\infty} h(n)e^{-jn\theta} \qquad (8.68)$$

or more explicitly

$$H_d(e^{j\omega T}) = \sum_{n=-\infty}^{\infty} h(n)e^{-jn\omega T} \qquad (9.117)$$

According to the time-sampling criterion for simulation (Eq. 8.132)

$$h_d(n) = Th_a(nT) \qquad (8.132)$$

where $h_d(n) = h(n)$, the unit-function response. Thus, Eq. 8.68 can be written

$$H_d(e^{j\omega T}) = \sum_{n=-\infty}^{\infty} Th_a(nT)e^{-jn\omega T} \qquad (9.118)$$

Now, consider the periodic train of unit impulses that represent a periodic function of ω with period ω_s:

$$g_p(\omega) = \sum_{n=-\infty}^{\infty} \delta(\omega + n\omega_s) \qquad \omega_s = 2\pi/T$$

9.6 SIMULATION

Since this is a periodic function of ω, it has a complex Fourier series whose coefficients are given by the Euler formula (Eq. 4.24 with T replaced with ω_s and t replaced with ω):

$$C_n = \frac{1}{\omega_s} \int_{-\omega_s/2}^{\omega_s/2} \sum_{n=-\infty}^{\infty} \delta(\omega + n\omega_s) e^{jn2\pi\omega/\omega_s} d\omega$$

or

$$C_n = \frac{1}{\omega_s} \sum_{n=-\infty}^{\infty} \int_{-\omega_s/2}^{\omega_s/2} \delta(\omega + n\omega_s) e^{jn2\pi\omega/\omega_s} d\omega$$

The sampling property of the impulse reduces this to

$$C_n = 1/\omega_s$$

Therefore, the train of impulses can be expressed as

$$g_p(\omega) = \sum_{n=-\infty}^{\infty} \delta(\omega + n\omega_s) = \frac{1}{\omega_s} \sum_{n=-\infty}^{\infty} e^{-jn2\pi\omega/\omega_s} \quad (9.119)$$

The right side of Eq. 9.119 is the complex Fourier series. Next, convolve both sides of Eq. 9.119 with $F(\omega)$, where $F(\omega)$ is the Fourier transform of $f(t)$: that is, $f(t) \leftrightarrow F(\omega)$. For the left side

$$g_p(\omega) * F(\omega) = \frac{1}{2\pi} \int_{-\infty}^{\infty} \sum_{n=-\infty}^{\infty} \delta(u + n\omega_s) F(\omega - u) du$$

$$= \frac{1}{2\pi} \sum_{n=-\infty}^{\infty} \int_{-\infty}^{\infty} \delta(u + n\omega_s) F(\omega - u) du$$

$$= \frac{1}{2\pi} \sum_{n=-\infty}^{\infty} F(\omega + n\omega_s) \quad (9.120)$$

For the right side

$$g_p(\omega) * F(\omega) = \frac{1}{2\pi} \int_{-\infty}^{\infty} \frac{1}{\omega_s} \sum_{n=-\infty}^{\infty} e^{-jn2\pi(\omega-u)/\omega_s} F(u) du$$

$$= \frac{T}{4\pi^2} \sum_{n=-\infty}^{\infty} e^{-jn2\pi\omega/\omega_s} \int_{-\infty}^{\infty} F(u) e^{jn2\pi u/\omega_s} du$$

The integral on the right is $2\pi f(n2\pi/\omega_s) = 2\pi f(nT)$ since $f(t) \leftrightarrow F(\omega)$. Therefore

$$g_p(\omega) * F(\omega) = \frac{T}{2\pi} \sum_{n=-\infty}^{\infty} f(nT) e^{-jn\omega T} \quad (9.121)$$

and equating Eqs. 9.120 and 9.121 gives

$$T \sum_{n=-\infty}^{\infty} f(nT) e^{-jn\omega T} = \sum_{n=-\infty}^{\infty} F(\omega + n\omega_s) \quad (9.122)$$

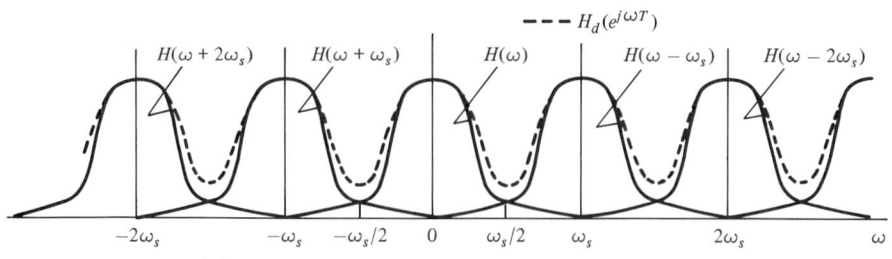

Figure 9.23. $H_d(e^{j\omega T})$ equals the sum of the displaced replications of $H(\omega)$.

This paragraph has been devoted to the derivation of Eq. 9.122 which is known as *Poisson's sum formula*.[4]

Using Eq. 9.122 in 9.118 we have the important result

$$H_d(e^{j\omega T}) = \sum_{n=-\infty}^{\infty} H(\omega + n\omega_s) \qquad \omega_s = 2\pi/T, \; h_a(t) \leftrightarrow H(\omega) \qquad (9.123)$$

It states that in order to find $H_d(e^{j\omega T})$ we merely add $H(\omega)$ to all of its displacements (or replications that are displaced by $n\omega_s$). This shown in Fig. 9.23 for an arbitrary $H(\omega)$.

Equation 9.89

$$H_d(e^{j\omega T}) = H(\omega) \qquad (9.89)$$

stated that for proper simulation $H_d(e^{j\omega T})$ must be equal to $H(\omega)$ for $-\omega_s/2 < \omega < \omega_s/2$, and we see in Fig. 9.23 that this is essentially true if $H(\omega)$ is negligible for $\omega > \omega_s/2$. In fact, we know from the sampling theorem that if $H(\omega) = 0$ for $\omega > \omega_s/2$, we can (in principle) recover $h_a(t)$ exactly. Thus, we have found the connection between time-sampling simulation, frequency-domain simulation, and the sampling theorem.

9.7 STATE-SPACE ANALYSIS

The normal-form difference equations were given in Chap. 8:

$$\mathbf{q}(n+1) = \mathbf{A}\mathbf{q}(n) + \mathbf{B}\mathbf{x}(n) \qquad (8.92)$$

with the output vector given by

$$\mathbf{y}(n) = \mathbf{C}\mathbf{q}(n) + \mathbf{D}\mathbf{x}(n) \qquad (8.93)$$

In choosing the state variables we made the *natural* choice

$$q_i(n) = y(n-i) \qquad i = 1, 2, \ldots, N \qquad (9.124)$$

for an Nth-order system. Given the N initial conditions $y(-1), y(-2), \ldots, y(-N)$ and $x(n)$, we have the minimum information required to find the system

9.7 STATE-SPACE ANALYSIS

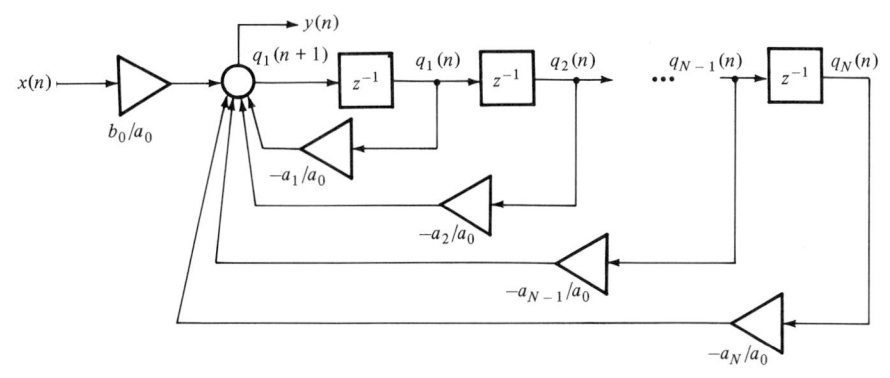

Figure 9.24. Block diagram for the system described by Eqs. 9.124 and 9.125.

response *if the system is described by*

$$a_0 y(n) + a_1 y(n-1) + a_2 y(n-2) + \cdots + a_N y(n-N) = b_0 x(n) \quad (9.125)$$

The block diagram representation for the system described by Eqs. 9.124 and 9.125 is shown in Fig. 9.24. This, of course is a special case, and the state variables specified by Eq. 9.124 are called the *natural* state variables.

If the right side of Eq. 9.125 is changed to the more general form $b_0 x(n) + b_1 x(n-1) + \cdots + b_M x(n-M)$, then it is easy to show that the selection of the natural state variables will lead to forms that cannot be represented by Eqs. 8.92 and 8.93. There are alternate state-variable representations[5] that overcome these difficulties. They are best described with the aid of the z transform, which is the reason why they were not presented in Chap. 8.

Consider the more general form indicated in the previous paragraph with $M = N$:

$$a_0 y(n) + a_1 y(n-1) + \cdots + a_N y(n-N) = b_0 x(n) + \cdots + b_N x(n-N) \quad (9.126)$$

whose transfer function is given by

$$H_d(z) = \frac{b_0 + b_1 z^{-1} + \cdots + b_N z^{-N}}{a_0 + a_1 z^{-1} + \cdots + a_N z^{-N}} \quad (9.127)$$

Multiply the numerator and denominator of Eq. 9.127 by z^N/a_0, and define new coefficients ($a_i/a_0 \equiv \alpha_i$ and $b_i/a_0 \equiv \beta_i$) so that

$$H_d(z) = \frac{\beta_0 z^N + \beta_1 z^{N-1} + \cdots + \beta_N}{z^N + \alpha_1 z^{N-1} + \cdots + \alpha_N} \quad (9.128)$$

Now, factor the denominator:

$$H_d(z) = \frac{\beta_0 z^N + \beta_1 z^{N-1} + \cdots + \beta_N}{(z - p_1)(z - p_2)(z - p_N)} \quad (9.129)$$

458 THE z TRANSFORM

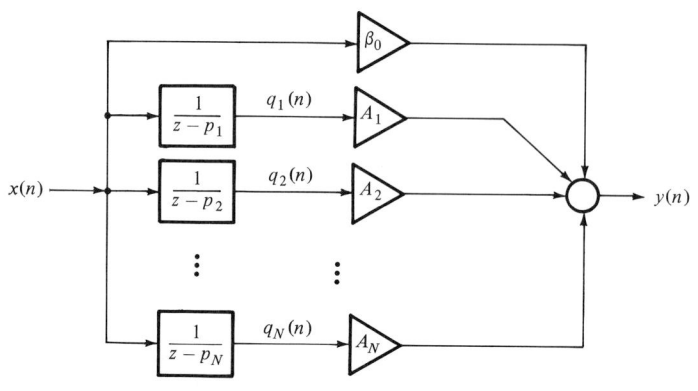

Figure 9.25. State-variable block diagram for Eq. 9.131.

where p_i are the poles of $H_d(z)$, or roots of the characteristic equation. Assuming that there are no repeated roots, we next obtain a particular partial fraction expansion:

$$H_d(z) = A_0 + \frac{A_1}{z - p_1} + \frac{A_2}{z - p_2} + \cdots + \frac{A_N}{z - p_N} \qquad (9.130)$$

where

$$A_0 = \lim_{|z| \to \infty} H_d(z) = \beta_0 \qquad A_i = (z - p_i)H_d(z)|_{z = p_i}$$

Using Eq. 9.130 we have

$$Y(z) = A_0 X(z) + \frac{A_1}{z - p_1} X(z) + \cdots + \frac{A_N}{z - p_N} X(z) \qquad (9.131)$$

This gives the block diagram representation of Fig. 9.25. This is called the *parallel programming method* for obvious reasons. It selects the state variables so that [with $q_i(n) \leftrightarrow Q_i(z)$]

$$Q_i(z) = \frac{1}{z - p_i} X(z) \qquad (9.132)$$

Equation 9.132 gives

$$zQ_i(z) = p_i Q_i(z) + X(z)$$

and inverse transforming,[†] we obtain the difference equation for the ith state variable:

$$q_i(n + 1) = p_i q_i(n) + x(n) \qquad (9.133)$$

The output is given in Fig. 9.25 by

$$y(n) = A_1 q_1(n) + A_2 q_2(n) + \cdots + \beta_0 x(n) \qquad (9.134)$$

[†] Refer to the footnote below Eq. 9.111.

The normal-form equation now becomes

$$\begin{bmatrix} q_1(n+1) \\ q_2(n+1) \\ \vdots \\ q_N(n+1) \end{bmatrix} = \begin{bmatrix} p_1 & 0 & 0 & \cdots & 0 \\ 0 & p_2 & 0 & \cdots & 0 \\ \vdots & \vdots & \vdots & & \vdots \\ 0 & 0 & 0 & & p_N \end{bmatrix} \begin{bmatrix} q_1(n) \\ q_2(n) \\ \vdots \\ q_N(n) \end{bmatrix} + \begin{bmatrix} 1 \\ 1 \\ \vdots \\ 1 \end{bmatrix} x(n) \quad (9.135)$$

and the output $y(n)$ is

$$y(n) = [A_1 \, A_2 \, \cdots \, A_N] \begin{bmatrix} q_1(n) \\ q_2(n) \\ \vdots \\ q_N(n) \end{bmatrix} + \beta_0 x(n) \quad (9.136)$$

Notice that the system matrix **A** is *diagonal* in this formulation. This is highly advantageous in matrix manipulations (like those of Chap. 8) for analyzing a high-order system.

Example 23

Find the parallel programming state variables for the discrete system of Fig. 8.29. The difference equation was 8.40:

$$3y(n) + 9y(n-1) + 6y(n-2) = 2x(n) + x(n-1) \quad (8.40)$$

so

$$H_d(z) = \frac{2 + z^{-1}}{3 + 9z^{-1} + 6z^{-2}}$$

or, in the form of Eq. 9.129

$$H_d(z) = \frac{(2/3)z^2 + (1/3)z}{(z+1)(z+2)}$$

Notice that $\beta_0 = \tfrac{2}{3}$ and $\beta_N = \beta_2 = 0$. The coefficients in the partial fraction expansion are easy to find:

$$A_1 = \left.\frac{(2/3)z^2 + (1/3)z}{z+2}\right|_{z=-1} = \frac{1}{3}$$

$$A_2 = \left.\frac{(2/3)z^2 + (1/3)z}{z+1}\right|_{z=-2} = -2$$

Therefore

$$\begin{bmatrix} q_1(n+1) \\ q_2(n+1) \end{bmatrix} = \begin{bmatrix} -1 & 0 \\ 0 & -2 \end{bmatrix} \begin{bmatrix} q_1(n) \\ q_2(n) \end{bmatrix} + \begin{bmatrix} 1 \\ 1 \end{bmatrix} x(n)$$

460 THE z TRANSFORM

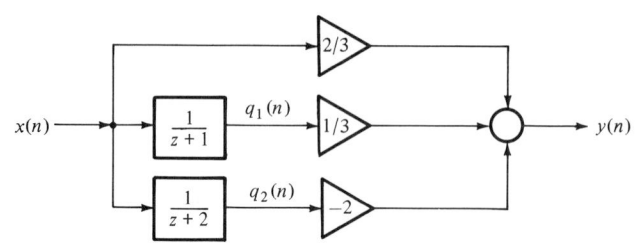

Figure 9.26. Block diagram representation for the parallel programming method.

and

$$y(n) = \begin{bmatrix} \frac{2}{3} & -2 \end{bmatrix} \begin{bmatrix} q_1(n) \\ q_2(n) \end{bmatrix} + \frac{1}{3}x(n)$$

The block diagram representation is shown in Figure 9.26.

When there are repeated roots, there will be nonzero terms off the main diagonal of the system matrix **A**. We demonstrate this with an example.

Example 24

Given

$$H_d(z) = \frac{1}{(z+2)(z-1)^2}$$

find the state-variable representation using the parallel programming method. First of all, $A_0 = 0$, and the partial fraction expansion takes the form

$$H_d(z) = \frac{A_1}{z+2} + \frac{A_2}{z-1} + \frac{A_3}{(z-1)^2}$$

and the coefficients are determined as before (see Example 5):

$$A_1 = \frac{1}{(z-1)^2}\bigg|_{z=-2} = \frac{1}{9}$$

$$A_3 = \frac{1}{z+2}\bigg|_{z=1} = \frac{1}{3}$$

$$\frac{1}{(z+2)(z-1)^2} = \frac{1/9}{z+2} + \frac{A_2}{z-1} + \frac{1/3}{(z-1)^2}$$

Setting $z = 0$ and solving for A_2 in the last equation gives

$$A_2 = -\tfrac{1}{9}$$

Therefore

$$H_d(z) = \frac{1/9}{z+2} - \frac{1/9}{z-1} + \frac{1/3}{(z-1)^2} \qquad (9.137)$$

9.7 STATE-SPACE ANALYSIS

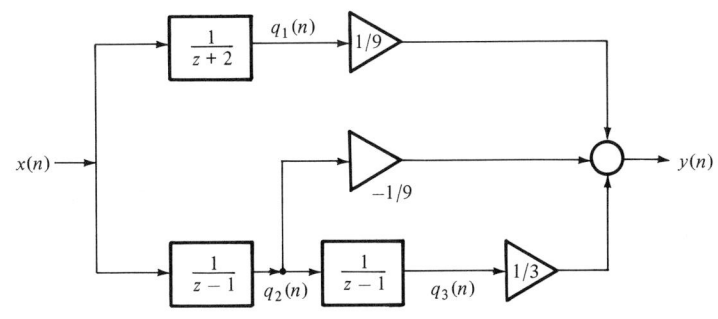

Figure 9.27. Block diagram for Eq. 9.137.

and the block diagram for this system is shown in Fig. 9.27. The normal-form equations are

$$\begin{bmatrix} q_1(n+1) \\ q_2(n+1) \\ q_3(n+1) \end{bmatrix} = \begin{bmatrix} -2 & 0 & 0 \\ 0 & 1 & 0 \\ 0 & 1 & 1 \end{bmatrix} \begin{bmatrix} q_1(n) \\ q_2(n) \\ q_3(n) \end{bmatrix} + \begin{bmatrix} 1 \\ 1 \\ 0 \end{bmatrix} x(n) \quad (9.138)$$

and the output is

$$y(n) = [\tfrac{1}{9} \;\; -\tfrac{1}{9} \;\; \tfrac{1}{3}] \begin{bmatrix} q_1(n) \\ q_2(n) \\ q_3(n) \end{bmatrix} + 0 \cdot x(n) \quad (9.139)$$

Equation 9.138 can easily be verified:

$$Q_1(z) = \frac{1}{z+2} X(z)$$

$$Q_2(z) = \frac{1}{z-1} X(z)$$

$$Q_3(z) = \frac{1}{(z-1)^2} X(z) \quad \text{or} \quad Q_3(z) = \frac{1}{z-1} Q_2(z)$$

or

$$zQ_1(z) = -2Q_1(z) + X(z)$$
$$zQ_2(z) = Q_2(z) + X(z)$$
$$zQ_3(z) = Q_3(z) + Q_2(z)$$

or

$$\left. \begin{array}{l} q_1(n+1) = -2q_1(n) + x(n) \\ q_2(n+1) = q_2(n) + x(n) \\ q_3(n+1) = q_2(n) + q_3(n) \end{array} \right\} \quad (9.140)$$

Repeated roots of higher multiplicity are treated the same way as in the preceding example.

Next, return to the general form for the transfer function, Eq. 9.43 ($m \to n$):

$$H_d(z) = \frac{\sum_{n=0}^{M} b_n z^n}{\sum_{n=0}^{N} a_n z^n}$$

$$= \frac{b_0 + b_1 z + b_2 z^2 + \cdots + b_M z^M}{a_0 + a_1 z + a_2 z^2 + \cdots + a_N z^N}$$

$$= \frac{b_M}{a_N} \frac{z^M + \frac{b_{M-1}}{b_M} z^{M-1} + \cdots + \frac{b_0}{b_M}}{z^N + \frac{a_{N-1}}{a_N} z^{N-1} + \cdots + \frac{a_0}{a_N}}$$

Factoring both numerator and denominator

$$H_d(z) = b \frac{(z - z_1)(z - z_2) \cdots (z - z_M)}{(z - p_1)(z - p_2) \cdots (z - p_N)} \qquad (9.141)$$

where $b = b_M/a_N$ and z_i represents the zeros of the transfer function. First assume that $M < N$, so that

$$H_d(z) = b H_1(z) H_2(z) \cdots H_i(z) \cdots H_N(z) \qquad (9.142)$$

where

$$H_i(z) = \frac{z - z_i}{z - p_i} \qquad i = 1, 2, \ldots, M$$

$$H_i(z) = \frac{1}{z - p_i} \qquad i = M+1, M+2, \ldots, N$$

Equation 9.142 indicates a cascaded arrangement shown in Figure 9.28. We make the rather obvious choice of state variables that is indicated in this figure. Thus, we have by inspection

$$\left. \begin{array}{l} Q_N(z) = \dfrac{b}{z - p_N} X(z) \\[6pt] Q_{N-1}(z) = \dfrac{1}{z - p_{N-1}} Q_N(z) \\[6pt] \vdots \\[6pt] Q_1(z) = \dfrac{z - z_1}{z - p_1} Q_2(z) = Y(z) \end{array} \right\} \qquad (9.143)$$

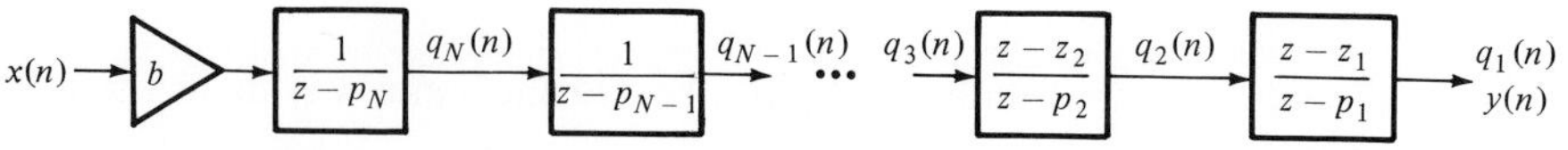

Figure 9.28. Block diagram for Eq. 9.141.

9.7 STATE-SPACE ANALYSIS

from which the first-order difference equations are obtained. This is the *iterative programming method* for state-variable selection.

Example 25

Find the state-variable representation using the iterative programming method for the system of Example 23:

$$H_d(z) = \frac{(2/3)z^2 + (1/3)z}{z^2 + 3z + 2}$$

Since M is not less than N, $H_d(z)$ must be modified:

$$H_d(z) = \frac{2}{3}\frac{z^2 + (1/2)z}{z^2 + 3z + 2} = \frac{2}{3}\frac{z^2 + 3z + 2 - 2.5z - 2}{z^2 + 3z + 2} \qquad (9.144)$$

$$H_d(z) = \frac{2}{3} - \frac{2}{3}\frac{2.5z + 2}{z^2 + 3z + 2} = \frac{2}{3} - \frac{5}{3}\frac{z + 0.8}{(z+1)(z+2)}$$

$$H_d(z) = H_a(z) + H_b(z)$$

$H_a(z)$ and $H_b(z)$ are parallel, and $H_a(z) = \frac{2}{3}$ is simply a multiplier ($\frac{2}{3}$). The block diagram is shown in Fig. 9.29. Thus, we have

$$Q_2(z) = -\frac{5}{3}\frac{1}{z+1}X(z)$$

$$Q_1(z) = \frac{z + 0.8}{z + 2}Q_2(z)$$

or

$$zQ_2(z) = -Q_2(z) - \tfrac{5}{3}X(z)$$
$$zQ_1(z) = -2Q_1(z) + zQ_2(z) + 0.8Q_2(z)$$

or

$$q_2(n + 1) = -q_2(n) - \tfrac{5}{3}x(n) \qquad (9.145)$$
$$q_1(n + 1) = -2q_1(n) + q_2(n + 1) + 0.8q_2(n) \qquad (9.146)$$

Substituting Eq. 9.145 into 9.146 gives the state-space equations in normal form:

$$q_1(n + 1) = -2q_1(n) - 0.2q_2(n) - \tfrac{5}{3}x(n)$$
$$q_2(n + 1) = -q_2(n) - \tfrac{5}{3}x(n)$$

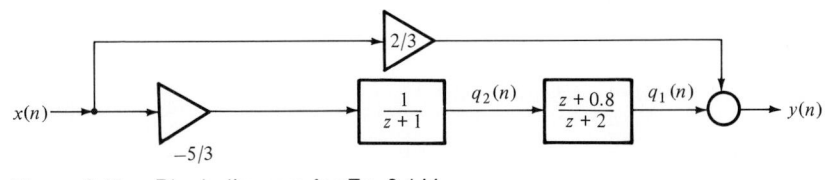

Figure 9.29. Block diagram for Eq. 9.144.

The output is obviously

$$y(n) = q_1(n) + \tfrac{2}{3}x(n)$$

In matrix form

$$\begin{bmatrix} q_1(n+1) \\ q_2(n+1) \end{bmatrix} = \begin{bmatrix} -2 & -0.2 \\ 0 & -1 \end{bmatrix} \begin{bmatrix} q_1(n) \\ q_2(n) \end{bmatrix} + \begin{bmatrix} -\tfrac{5}{3} \\ -\tfrac{5}{3} \end{bmatrix} x(n)$$

$$y(n) = \begin{bmatrix} 1 & 0 \end{bmatrix} \begin{bmatrix} q_1(n) \\ q_2(n) \end{bmatrix} + \tfrac{2}{3}x(n)$$

Both of the preceding methods for obtaining the state-space equations require factoring, and this certainly represents an undesirable feature, particularly for high-order systems. The *direct programming method* avoids this problem. Assume that $H_d(z)$ is given by Eq. 9.128 ($M = N$):

$$H_d(z) = \frac{\beta_0 + \beta_1 z^{-1} + \cdots + \beta_N z^{-N}}{1 + \alpha_1 z^{-1} + \cdots + \alpha_N z^{-N}} = \frac{Y_{zs}(z)}{X(z)} \qquad (9.147)$$

Therefore

$$\frac{Y_{zs}(z)}{\beta_0 + \beta_1 z^{-1} + \cdots + \beta_N z^{-N}} = \frac{X(z)}{1 + \alpha_1 z^{-1} + \cdots + \alpha_N z^{-N}} \equiv R(z)$$

We can safely drop the subscript on $Y_{zs}(z)$ at this point. We have

$$R(z)(1 + \alpha_1 z^{-1} + \alpha_2 z^{-2} + \cdots + \alpha_N z^{-N}) = X(z)$$

so

$$R(z) = -\alpha_1 z^{-1} R(z) - \alpha_2 z^{-2} R(z) - \cdots - \alpha_N z^{-N} R(z) + X(z) \qquad (9.148)$$

Also

$$Y(z) = \beta_0 R(z) + \beta_1 z^{-1} R(z) + \beta_2 z^{-2} R(z) + \cdots + \beta_N z^{-N} R(z) \qquad (9.149)$$

For reasons that are not apparent, except that the terms do appear in the last two equations, we choose the state variables as

$$\left. \begin{aligned} Q_1(z) &= z^{-1} R(z) \\ Q_2(z) &= z^{-2} R(z) \\ &\vdots \\ Q_N(z) &= z^{-N} R(z) \end{aligned} \right\} \qquad (9.150)$$

This choice does give simple forms:

$$\left. \begin{aligned} zQ_2(z) &= z^{-1} R(z) = Q_1(z) \\ zQ_3(z) &= z^{-2} R(z) = z^{-1} Q_1(z) = Q_2(z) \\ &\vdots \\ zQ_N(z) &= Q_{N-1}(z) \end{aligned} \right\} \qquad (9.151)$$

9.7 STATE-SPACE ANALYSIS

or

$$q_2(n+1) = q_1(n)$$
$$q_3(n+1) = q_2(n)$$
$$\vdots$$
$$q_N(n+1) = q_{N-1}(n)$$

(9.152)

From the first equation in the set 9.150 we have $R(z) = zQ_1(z)$. This is substituted into Eq. 9.148 giving

$$zQ_1(z) = -\alpha_1 z^{-1} R(z) - \alpha_2 z^{-2} R(z) - \cdots - \alpha_N z^{-N} R(z) + X(z)$$

Using 9.150

$$zQ_1(z) = -\alpha_1 Q_1(z) - \alpha_2 Q_2(z) - \cdots - \alpha_N Q_N(z) + X(z)$$

or

$$q_1(n+1) = -\alpha_1 q_1(n) - \alpha_2 q_2(n) - \cdots - \alpha_N q_N(n) + x(n) \qquad (9.153)$$

The output is found from Eqs. 9.149 and 9.150:

$$y(n) = \beta_0 q_1(n+1) + \beta_1 q_1(n) + \beta_2 q_2(n) + \cdots + \beta_N q_N(n)$$

Using Eq. 9.153

$$y(n) = \beta_0 [-\alpha_1 q_1(n) - \alpha_2 q_2(n) - \cdots - \alpha_N q_N(n) + x(n)]$$
$$+ \beta_1 q_1(n) + \beta_2 q_2(n) + \cdots + \beta_N q_N(n)$$

or

$$y(n) = (\beta_1 - \alpha_1 \beta_0) q_1(n) + (\beta_2 - \alpha_2 \beta_0) q_2(n) + \cdots + (\beta_N - \alpha_N \beta_0) + \beta_0 x(n)$$

(9.154)

Using the set 9.152 and Eqs. 9.153 and 9.154 we obtain the matrix forms

$$\begin{bmatrix} q_1(n+1) \\ q_2(n+1) \\ q_3(n+1) \\ \vdots \\ q_N(n+1) \end{bmatrix} = \begin{bmatrix} -\alpha_1 & -\alpha_2 & \cdots & -\alpha_{N-1} & -\alpha_N \\ 1 & 0 & \cdots & 0 & 0 \\ \vdots & & \ddots & & \\ 0 & 0 & \cdots & 1 & 0 \end{bmatrix} \begin{bmatrix} q_1(n) \\ q_2(n) \\ q_3(n) \\ \vdots \\ q_N(n) \end{bmatrix} + \begin{bmatrix} 1 \\ 0 \\ 0 \\ \vdots \\ 0 \end{bmatrix} x(n)$$

(9.155)

$$y(n) = [\gamma_1 \gamma_2 \gamma_3 \cdots \gamma_N] \begin{bmatrix} q_1(n) \\ q_2(n) \\ q_3(n) \\ \vdots \\ q_N(n) \end{bmatrix} + \beta_0 x(n) \qquad (9.156)$$

where

$$\gamma_i = \beta_i - \alpha_i \beta_0$$

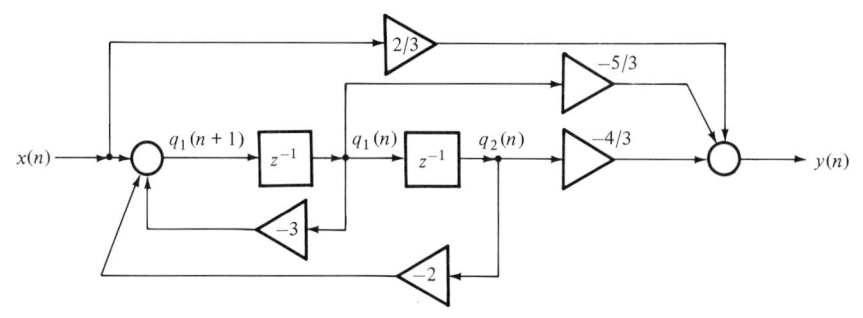

Figure 9.30. Block diagram of the system of Example 24.

Example 26

Repeat Example 25 (and 23) using the direct programming approach. We had

$$H_d(z) = \frac{2 + z^{-1}}{3 + 9z^{-1} + 6z^{-2}} = \frac{(2/3) + (1/3)z^{-1}}{1 + 3z^{-1} + 2z^{-2}}$$

Here, we take the easy way and simply substitute $\beta_0 = \frac{2}{3}$, $\beta_1 = \frac{1}{3}$, $\beta_2 = 0$, $\alpha_1 = 3$, and $\alpha_2 = 2$ into Eqs. 9.155 and 9.156:

$$\begin{bmatrix} q_1(n+1) \\ q_2(n+1) \end{bmatrix} = \begin{bmatrix} -3 & -2 \\ 1 & 0 \end{bmatrix} \begin{bmatrix} q_1(n) \\ q_2(n) \end{bmatrix} + \begin{bmatrix} 1 \\ 0 \end{bmatrix} x(n) \qquad (9.157)$$

$$y(n) = \begin{bmatrix} -\frac{5}{3} & -\frac{4}{3} \end{bmatrix} \begin{bmatrix} q_1(n) \\ q_2(n) \end{bmatrix} + \frac{2}{3} x(n) \qquad (9.158)$$

The block diagram representation of this state-space description is shown in Fig. 9.30. It is left as an exercise to show that it is the correct representation of Eqs. 9.157 and 9.158.

Comparing Figures 9.26, 9.29, and 9.30 for the same system, we see three different representations. The third is obviously the most complicated looking, and yet it is probably the most useful for higher-order systems whose transfer functions can be put in the form of Eq. 9.147. This includes *all* linear, shift-invariant, and causal systems, as the reader can verify. The z transform has been a tremendous aid in developing these three systems.

9.8 CONCLUDING REMARKS

The z transform for discrete linear systems was introduced in this concluding chapter. We discovered that although it is not absolutely necessary for purposes of analysis, it does offer some shortcuts in the description and analysis of linear systems. In this regard it should be thought of as an aid and should be used in conjunction with the discrete time-domain methods of Chap. 8.

As was the case for the Laplace transform with respect to continuous-time systems, we found both a unilateral and bilateral z transform. We discovered that the same sort of advantages and disadvantages exist when the unilateral and bilateral z transforms are compared that existed when the unilateral and bilateral Laplace transforms were compared. The unilateral transforms, ideal for inputs starting at $t = 0$ (or $n = 0$), *explicitly* give terms for the initial conditions, so initial conditions are easily handled. On the other hand this feature, which is so desirable in many cases, also means that transients are created in many cases where we are interested not in transient behavior but in the steady state. In the case of the Laplace transform we mentioned that we could circumvent the difficulty by using the bilateral Laplace transform (which we did not do) or by using the Fourier transform (which we did do). Remember that the Fourier transform can treat any input or excitation that can be generated in practice. In the case of the z transform we mentioned that (see Example 13) we could use convolution-summation as an alternate method, or we could use the bilateral z transform. We, in fact, did use one bilateral pair:

$$f(n \pm m) \leftrightarrow z^{\pm m} F(z)$$

in order to circumvent the transient problem mentioned above. Why *not* use the bilateral z (or Laplace) transform? The answer is that, first of all, they are rarely needed (the pair above being an exception), and, secondly, as pointed out below Eq. 9.7, two different sequences, one for $n < 0$ and one for $n > 0$, for example, can have the *same* bilateral z transform. This requires careful specification of the regions of convergence for the transforms, and we chose to avoid that course in this material. Thus, for the majority of situations the unilateral or one-sided Laplace and z transforms suffice. We also avoided integration in the complex plane for determining inverse transforms and instead chose to rely on a table of pairs.

The z transform facilitates the solution of difference equations that arise when discrete systems are analyzed. Quite often, knowledge of the frequency response of a discrete system is helpful in the analysis problem, or in cases where one is interested in simulating a continuous system with a discrete one (or simply processing discrete data). The z transform approach again is very helpful. Finally, the internal states of a discrete system can be described in many ways, and the block diagrams for the resulting systems are easy to obtain if we work in the z domain. This is primarily due to the transfer function concept that arises because the z transform converts a linear difference equation (in n) into an algebraic equation (in z).

PROBLEMS

1. Find the z transform of
   (a) $n^3 u(n)$
   (b) $(a^n/n!)u(n)$
   (c) $na^n u(n)$
   (d) $n^2 a^n u(n)$
   (e) $e^{n\theta} u(n)$
   (f) $ne^{(n-1)\theta} u(n)$
   (g) $u(n) - u(n-2)$
   (h) $nf(n-1)u(n)$
   (i) $\sum_{m=0}^{n} \delta(m)$

2. Show that for the *bilateral* z transform
$$f(n \pm m) \leftrightarrow z^{\pm m}F(z)$$
Refer to the footnote of Table 9.1.
3. Show that for a *periodic* sequence $f(n)u(n) = f(n + N)u(n)$
$$f(n)u(n) \leftrightarrow \frac{z^N}{z^N - 1} F_1(z)$$
where
$$F_1(z) = \sum_{n=0}^{N-1} f(n)z^{-n}$$
is the z transform of the first cycle of $f(n)u(n)$.
4. Use the results of Prob. 3 to find the z transform of
 (a) $u(n)$ (b) $(-1)^n u(n)$ (c) $\cos(n\pi/2)u(n)$
5. (a) Using the bilateral z transform (Prob. 2), transform the difference equation
$$a_2 y(n+2) + a_1 y(n+1) + a_0 y(n) = x_f(n)$$
 (b) Using the unilateral z transform, repeat (a).
6. Show that if $f_1(t) = u(t)$ and $f_2(t) = u[\cos(2\pi t)]$ are sampled with $T = 1$, the resulting sequences have the same z transform. Do $f_1(t)$ and $f_2(t)$ have the same Laplace transform?
7. Find $f(n)$ for the following $F(z)$:

 (a) $\dfrac{1}{z-a}$ (b) $\dfrac{z^2}{(z-a)^2}$ (c) $\dfrac{1}{(z-a)^2}$ (d) $\dfrac{z}{(z-a)^2}$

 (e) $\dfrac{z}{(z-b)(z-b)}$ (f) $\dfrac{dF(z)}{dz}$ (g) $\dfrac{1}{(z-1)^2(z-2)}$ (h) $\dfrac{1}{(z-a)^m}$

8. Find the (unilateral) z transform of the sequences shown in Fig. 9.31.

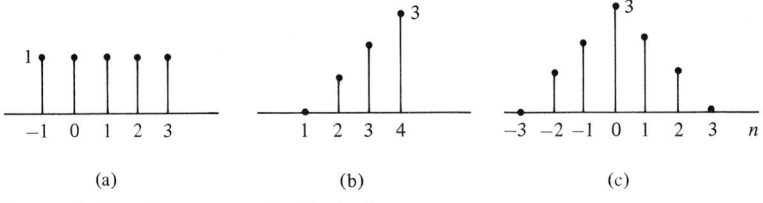

Figure 9.31. Sequences for Prob. 8.

9. Find $f(0)$ for the following $F(z)$:

 (a) $\dfrac{z}{z-1}$ (b) $\dfrac{z\sin(\omega)}{z^2 - 2z\cos(\omega) + 1}$ (c) $\dfrac{z}{z-a}$

 (d) $\dfrac{z}{(z-a)(z-b)}$ (e) $z^{-m}F(z)$ (f) $\dfrac{1}{(z-a)^m}$

10. Use the final-value theorem (if applicable) to find $f(\infty)$ for

   (a) 1
   (b) $\dfrac{z}{z-1}$
   (c) $\dfrac{z^2 - z\cos(\omega)}{z^2 - 2z\cos(\omega) + 1}$
   (d) $\dfrac{1}{(z-a)^m}$
   (e) $\dfrac{z+2}{(z-2)^2}$
   (f) $\dfrac{z}{z-a}$

11. Use the z transform to solve Prob. 14, Chap. 8. Find $y_{zs}(n)$ in Figure 9.32.

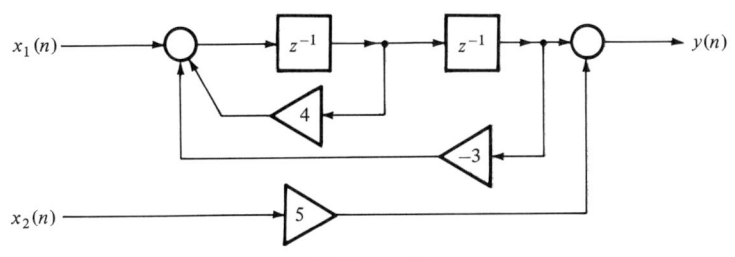

Figure 9.32. Linear system for Prob. 11.

12. Find the zero-state response of the system shown in Figure 9.33 using the z transform. See Prob. 29, Chap. 8.

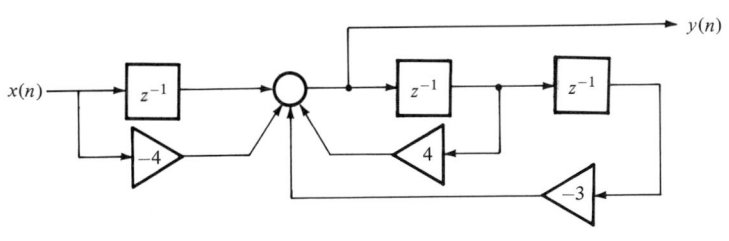

Figure 9.33. Linear system for Prob. 12.

13. Find the response of the compound savings account system $[y(-1) = 0]$, Example 10, Chap. 8, to the inputs
   (a) $x(n) = \delta(n)$
   (b) $x(n) = u(n)$
   (c) $x(n) = nu(n)$
   The difference equation is
   $$y(n) = (1 + i/k)y(n-1) + x(n)$$

14. Find the transfer function for
   (a) Prob. 9, Chap. 8
   (b) Prob. 16, Chap. 8

15. Two numerical integrators (Eq. 9.59) are cascaded. What is the overall transfer function? What is the response to $x(n) = u(n)$?

16. (a) Solve the *differential-difference* equation
   $$\frac{dy}{dt} + y(t-1)u(t-1) = tu(t) \qquad y(0) = 0$$
   (b) Find $y(2)$.
   (c) Find $y(2.3)$.

470 THE z TRANSFORM

17. A certain linear system is shown in Fig. 9.34.
  (a) Is this system anticipatory?
  (b) Find the inverse system (difference equation).
  (c) Is this system anticipatory?

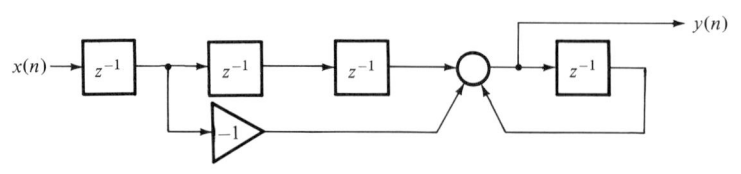

Figure 9.34. Linear system for Prob. 17.

18. (a) What is the transfer function of the low-pass filter of Fig. 9.35 and Eq. 8.84?
  (b) What is the transfer function of the system that consists of two such filters in cascade?

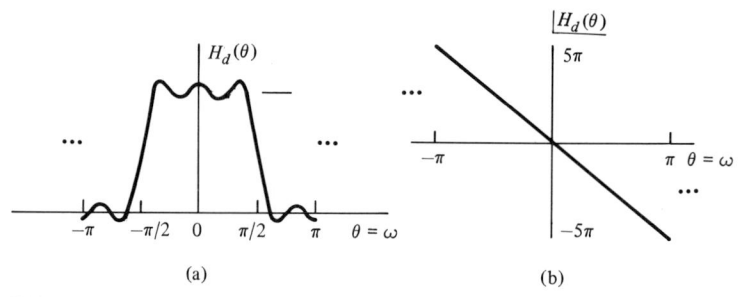

Figure 9.35. Low-pass filter.

19. What is the closed-form unit-function response for the numerical differentiator of Example 14?

20. Find the transfer function and difference equation for two numerical differentiators of Eq. 9.60 in cascade.

21. (a) Repeat Prob. 20 for the differentiator of Eq. 9.54. Notice that the numerical calculation of the second derivative can be made in a *direct manner* (nonrecursive system) in this case, as opposed to that of Prob. 20.
  (b) Calculate the second derivative of $x(t) = \sin(10t)$ at $t = 0.1$ with $T = 0.01$ s, and compare with the exact value.
  (c) Repeat (b) for $T = 0.001$ s.

22. In Example 15, $x'(0) = 10^3$ (exactly), whereas the numerical derivative for $T = 10^{-4}$ is 1008. If the differentiator of Eq. 9.54 is used (instead of Eq. 9.60), what is $x'(0)$ ($T = 10^{-4}$)? What is $x'(60)$ ($n = 60$)?

23. (a) Plot the frequency response for an analog integrator, $H(\omega) = 1/(j\omega)$.
  (b) Repeat for the digital integrator, Eq. 9.59, with $T = 1$.
  (c) Repeat (b) for $T = 0.5$.

24. The symmetrical square wave of Fig. 9.36 is sampled at $t = nT$, where $T = 0.15T_p$, and the samples are applied to the low-pass digital filter of Fig. 9.14 and Eq. 9.82. Find and plot $y(n)$ versus n for $0 \le n \le 20$ as in Fig. 9.15.

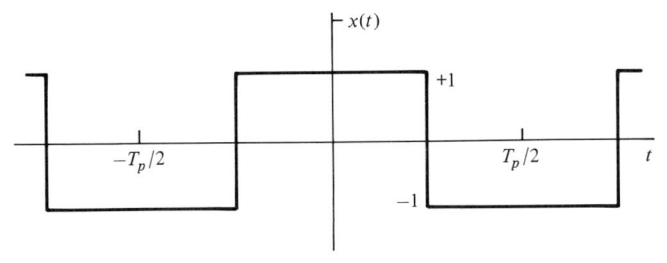

Figure 9.36. Symmetrical square wave.

25. Verify Fig. 9.15 for the case $\omega nT = n\pi/4$ using convolution-summation. Notice that $x(nT) = \cos(n\pi/4)u(n)$.
26. Following the procedure of Example 18, Chap. 8, design a low-pass filter using 11 unit delays and $\omega_s = 16\omega_0$.
   (a) Find the causal $h(n)$, $0 \le n \le 10$.
   (b) Plot $|H_d(\theta)|$ for $0 \le \theta \le \pi$.
27. (a) Find $y(t)$ (Fig. 9.37) analytically if $y(0) = y'(0) = 0$.
   (b) Find $y(n)$ numerically using Tustin's substitution method ($T = 0.1$). Compare (a) and (b) for $t = 1$ s.
   (c) Draw the block diagram for the difference equation obtained in (b).

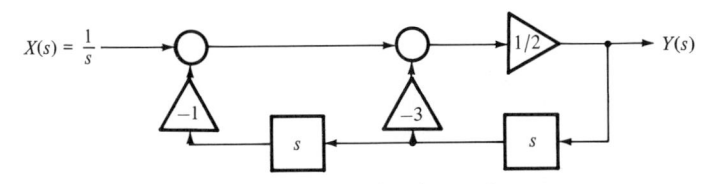

$X(s) = \dfrac{1}{s}$

Figure 9.37. Overdamped second-order system.

28. Repeat Prob. 27 [(b) and (c)] using the *first-difference method*.
29. The sampled signal shown in Fig. 9.38 is applied to the digital system found in Prob. 28. Find $y(n)$ for $n = 5$ if $y(0) = y(-1) = 0$. What is $y(n)$ for $n \to \infty$?

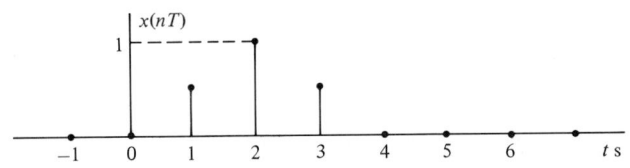

Figure 9.38. Sampled input signal.

472 THE z TRANSFORM

30. The unit-impulse response of a linear system is given by

$$h(t) = [e^{-3t} - e^{-4t}]u(t)$$

(a) Using the time-sampling method of simulation, find the difference equation that describes the system for $T = 10^{-2}$.
(b) Repeat (a) using the frequency transformation method of simulation.
(c) Repeat (b) using Eq. 9.54 as the representation of the simulator of a differentiator.

31. Consider a series RL circuit with the current as the output and voltage source as the input. Then

$$h(t) = \frac{1}{L} e^{-\alpha t} u(t) \qquad \alpha = R/L$$

For small T

$$h(t) \approx T \sum_{n=0}^{\infty} h(nT)\delta(t - nT) \leftrightarrow T \sum_{n=0}^{\infty} h(nT) e^{-nTs}$$

(a) Draw a block diagram of the sampled analog system with one analog delay element (e^{-sT}).
(b) Truncate the series above at $n = N$ and draw the *tapped delay line* that is *nonrecursive* and representative of the system.
(c) Replacing the analog delay element in (a) with a digital delay element (z^{-1}), draw the block diagram of the digital simulator of the system.
(d) Compare the result in (c) to that in Fig. 9.4.

32. Simpson's rule for integration is implemented by the system shown in Fig. 9.39.
(a) Show that

$$y(n) = T/3\{x(0) + 4[x(0) + x(3) + \cdots + x(n-1)] + 2[x(2) + x(4) + \cdots + (n-2)] + x(n)\}$$

if n is even and $y(0) = 0$.
(b) Compare $|H_d(0)| = |H_d(e^{j\omega T})|$ to $|1/(j\omega)|$ graphically.
(c) Find

$$\int_0^4 t \, dt = 8$$

numerically using $T = 1$.

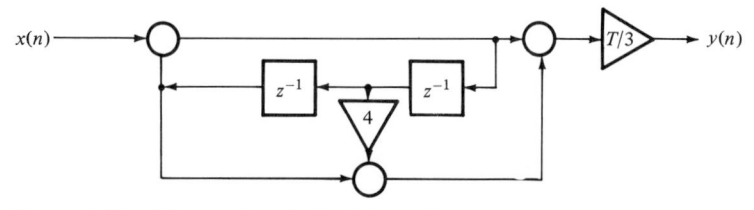

Figure 9.39. Simpson's rule for integration.

33. Find the digital simulator of the RLC circuit of Fig. 9.40 using the frequency transformation method with $T = 1$ μs. Sketch $|H_d(\theta)|$ versus $\theta = \omega T$.

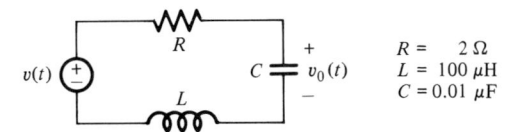

```
        R
   ────WWW──────┬──────
                │      +
v(t) (+)      C ═══   v_0(t)
                │      -
        L       │
   ────mmm─────┴──────
```

$R = 2\,\Omega$
$L = 100\,\mu H$
$C = 0.01\,\mu F$

Figure 9.40. Series RLC circuit.

34. A digital low-pass filter is characterized by

$$y(n) - 0.9y(n - 1) = 0.1x(n)$$

If two such filters are cascaded is the combination low pass?

35. Two digital systems

$$y(n) - 0.3y(n - 1) = x(n) - 0.6x(n - 1)$$

and

$$y(n) + 0.6y(n - 1) = 0.9x(n)$$

are cascaded. Find the difference equation that characterizes the overall system.

36. A digital system is described by

$$y(n) - 2y(n - 1) = 3x(n - 1)$$

(a) Is it stable?
(b) A feedback element, $H_d(z)$, whose input-output is characterized by

$$y(n) - \alpha y(n - 1) = \beta x(n - 1)$$

is added to stabilize the system as in Fig. 9.41. Find α and β so that the new system has a double pole at $z = 0.5$. Is it stable?
(c) Repeat (b) for a double pole at $z = -0.25$.

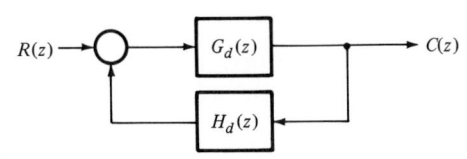

```
R(z) ──▶○──────▶│ G_d(z) │──────▶ C(z)
        ▲                │
        │                │
        └───│ H_d(z) │◀──┘
```

Figure 9.41. Feedback system for Prob. 36.

37. A discrete system is described by

$$H_d(z) = \frac{1}{1 + z^{-1} + \alpha z^{-2}}$$

For what range of values of α is it stable?

38. A discrete system is described by

$$H_d(z) = \frac{z^{-1}}{1 - (1/4)z^{-1} - (3/8)z^{-2}}$$

Decompose its response to $x(n) = nu(n)$ into the transient and steady-state parts.

39. A linear system is characterized by

$$y(n) = 2x(n) + x(n-1) - \tfrac{1}{2}x(n-2) + \tfrac{13}{12}y(n-1) - \tfrac{3}{8}y(n-2) + \tfrac{1}{24}y(n-3)$$

Is it stable?

40. Obtain the state-space representation of the system of Prob. 39 using the *natural* state variables. Draw the block diagram.

41. Repeat Prob. 40 using the *parallel programming method*.

42. Repeat Prob. 40 using the *iterative programming method*.

43. Repeat Prob. 40 using the *direct programming method*.

References

1. Cheng, D. K. *Analysis of Linear Systems.* Reading, Mass.: Addison-Wesley, 1959.
2. McGillem, C. D., and Cooper, G. R. *Continuous and Discrete Signal and System Analysis.* New York: Holt, Rinehart and Winston, 1974.
3. Gabel, R. A., and Roberts, R. A. *Signals and Linear Systems.* New York: John Wiley & Sons, 1980.
4. Papoulis, A. *Circuits and Systems.* New York: Holt, Rinehart and Winston, 1980.
5. Cadzow, J. A. *Discrete Time Systems.* Englewood Cliffs, N. J.: Prentice-Hall, 1973.

Appendices

A. MATRICES

The system of equations

$$\left.\begin{aligned} a_{11}x_1 + a_{12}x_2 + \cdots + a_{1n}x_n &= b_1 \\ a_{21}x_1 + a_{22}x_2 + \cdots + a_{2n}x_n &= b_2 \\ &\vdots \\ a_{m1}x_1 + a_{m2}x_2 + \cdots + a_{mn}x_n &= b_m \end{aligned}\right\} \quad \text{A.1}$$

can be written as

$$\sum_{j=1}^{n} a_{ij} x_j = b_i \qquad i = 1, 2, \ldots, m \qquad \text{A.2}$$

It can also be written in *matrix* notation as

$$\mathbf{A}\mathbf{x} = \mathbf{b} \qquad \text{A.3}$$

where $\mathbf{A}$ is the ordered array or matrix with m rows and n columns ($m \times n$):

$$\mathbf{A} = \begin{bmatrix} a_{11} & a_{12} & \cdots & a_{1n} \\ a_{21} & a_{22} & \cdots & a_{2n} \\ \vdots & \vdots & & \vdots \\ a_{m1} & a_{m2} & \cdots & a_{mn} \end{bmatrix} \qquad \text{A.4}$$

x is the matrix or *column vector* containing 1 column and n rows:

$$\mathbf{x} = \begin{bmatrix} x_1 \\ x_2 \\ \vdots \\ x_n \end{bmatrix} \qquad \text{A.5}$$

b is the column vector containing 1 column and m rows:

$$\mathbf{b} = \begin{bmatrix} b_1 \\ b_2 \\ \vdots \\ b_m \end{bmatrix} \qquad \text{A.6}$$

In order that matrix multiplication be defined, the two matrices involved must be such that the first has the same number of columns as the second has rows. The matrices are said to be *conformable*. **A** in Eq. A.4, for example, has n columns, and **x** in Eq. A.5 has n rows. As another example, consider

$$\mathbf{CD} = \begin{bmatrix} 2 & 1 & 0 \\ -1 & 3 & 2 \end{bmatrix} \begin{bmatrix} 1 & 2 \\ 3 & 0 \\ 1 & -2 \end{bmatrix} = \mathbf{E}$$

C has 3 columns, and **D** has 3 rows, so matrix multiplication is defined. The multiplication of an $r \times m$ matrix with an $m \times n$ matrix produces an $r \times n$ matrix. Thus, **CD** is a 2×3 matrix times a 3×2 matrix, producing a 2×2 matrix:

$$\mathbf{E} = \begin{bmatrix} e_{11} & e_{12} \\ e_{21} & e_{22} \end{bmatrix}$$

The elements of **E** are

$$e_{11} = 2 \cdot 1 + 1 \cdot 3 + 0 \cdot 1 = 5$$

That is, the elements of the first row of **C** are multiplied by the elements of the first column of **D**, and added. In the same way

$$e_{12} = 2 \cdot 2 + 1 \cdot 0 + 0 \cdot (-2) = 4$$

That is, the elements of the first row of **C** are multiplied by the elements of the second column of **D**, and added. Likewise

$$e_{21} = -1 \cdot 1 + 3 \cdot 3 + 2 \cdot 1 = 10$$

and

$$e_{22} = -1 \cdot 2 + 3 \cdot 0 + 2 \cdot (-2) = -6$$

Therefore

$$\mathbf{CD} = \mathbf{E} = \begin{bmatrix} 5 & 4 \\ 10 & -6 \end{bmatrix}$$

On the other hand

$$\mathbf{DC} = \begin{bmatrix} 1 & 2 \\ 3 & 0 \\ 1 & -2 \end{bmatrix} \begin{bmatrix} 2 & 1 & 0 \\ -1 & 3 & 2 \end{bmatrix} = \begin{bmatrix} 0 & 7 & 4 \\ 6 & 3 & 0 \\ 4 & -5 & -4 \end{bmatrix}$$

Thus, in general

$\mathbf{CD} \neq \mathbf{DC}$

Since two matrices are added by merely adding their elements, it is necessary that the two matrices have the same *dimensions*. For example

$$\mathbf{F} = \begin{bmatrix} 2 & 1 & 0 \\ -1 & 3 & 2 \end{bmatrix} \qquad \mathbf{G} = \begin{bmatrix} 1 & 2 & 3 \\ 2 & 0 & -1 \end{bmatrix}$$

$$\mathbf{F} + \mathbf{G} = \begin{bmatrix} 2+1 & 1+2 & 0+3 \\ -1+2 & 3+0 & 2-1 \end{bmatrix} = \begin{bmatrix} 3 & 3 & 3 \\ 1 & 3 & 1 \end{bmatrix}$$

$$= \mathbf{G} + \mathbf{F}$$

The *identity* matrix is a (*diagonal*) *square* ($m \times m$) matrix with all elements on the main *diagonal* equal to unity and all others zero:

$$\mathbf{I} = \begin{bmatrix} 1 & 0 & 0 & \cdots & 0 \\ 0 & 1 & 0 & \cdots & 0 \\ 0 & 0 & 1 & \cdots & 0 \\ & & & \ddots & \\ 0 & 0 & 0 & \cdots & 1 \end{bmatrix} \qquad \text{A.7}$$

If we follow the rules for matrix multiplication that were outlined above, it becomes apparent that

$\mathbf{IB} = \mathbf{B}$ \hfill A.8

if $\mathbf{I}$ is $m \times m$ and $\mathbf{B}$ is $m \times n$, or

$\mathbf{BI} = \mathbf{B}$ \hfill A.9

if $\mathbf{I}$ is $n \times n$.

The *transpose* of a matrix is written $\mathbf{A}^t$ (where $a^t_{ij} = a_{ji}$), and is the new matrix obtained when the elements of $\mathbf{A}$ are transposed. For example

$$\mathbf{D} = \begin{bmatrix} 1 & 2 \\ 3 & 0 \\ 1 & -2 \end{bmatrix} \quad \text{and} \quad \mathbf{D}^t = \begin{bmatrix} 1 & 3 & 1 \\ 2 & 0 & -2 \end{bmatrix}$$

A *symmetric* matrix is one for which $a_{ij} = a_{ji}$.

The *left inverse* of a matrix $\mathbf{A}$ is a matrix $\mathbf{B}$ such that

$\mathbf{BA} = \mathbf{I}$ \hfill A.10

where $\mathbf{B}$ is $n \times m$, and $\mathbf{A}$ is $m \times n$, if $\mathbf{I}$ is $n \times n$. The right inverse of $\mathbf{A}$ is $\mathbf{C}$ such that

$$\mathbf{AC} = \mathbf{I} \qquad \text{A.11}$$

$\mathbf{I}$ is $m \times m$ if $\mathbf{A}$ is $m \times n$, and $\mathbf{C}$ is $n \times m$. For a square matrix, $\mathbf{B} = \mathbf{C}$, and we simply say that $\mathbf{B}$ (or $\mathbf{C}$) is the *inverse* of $\mathbf{A}$:

$$\mathbf{B} = \mathbf{A}^{-1} \qquad \text{A.12}$$

Notice that $\mathbf{I}^{-1} = \mathbf{I}$ since

$$\mathbf{I}^{-1}\mathbf{I} = \mathbf{I}$$

Also, the inverse of a diagonal matrix

$$\mathbf{A} = \begin{bmatrix} a_1 & 0 & \cdots & 0 \\ 0 & a_2 & & \\ & & \ddots & \\ 0 & 0 & \cdots & a_n \end{bmatrix} \quad (n \times n)$$

is

$$\mathbf{A}^{-1} = \begin{bmatrix} 1/a_1 & 0 & \cdots & 0 \\ 0 & 1/a_2 & & 0 \\ & & \ddots & \\ 0 & 0 & \cdots & 1/a_n \end{bmatrix} \quad (n \times n)$$

since

$$\mathbf{A}^{-1}\mathbf{A} = \mathbf{I} \qquad \text{A.13}$$

Not all matrices have inverses.

The set of Eq. A.3 (or A.1, or A.2) can be solved for $\mathbf{x}$ if $\mathbf{A}^{-1}$ can be found. To show this we *premultiply* $\mathbf{A}$ by $\mathbf{A}^{-1}$ in Eq. A.3:

$$\mathbf{A}^{-1}(\mathbf{A}\mathbf{x}) = \mathbf{A}^{-1}\mathbf{b} = (\mathbf{A}^{-1}\mathbf{A})\mathbf{x}$$

Using Eq. A.13

$$\mathbf{I}\mathbf{x} = \mathbf{A}^{-1}\mathbf{b}$$

and using Eq. A.8

$$\mathbf{x} = \mathbf{A}^{-1}\mathbf{b} \qquad \text{A.14}$$

the desired solution.

The *adjoint* matrix for an $n \times n$ (square) matrix is

$$\text{ad } \mathbf{A} = \begin{bmatrix} \Delta_{11} & \Delta_{21} & \cdots & \Delta_{n1} \\ \Delta_{12} & \Delta_{22} & \cdots & \Delta_{n2} \\ & & \ddots & \\ \Delta_{1n} & \Delta_{2n} & \cdots & \Delta_{nn} \end{bmatrix} \qquad \text{A.15}$$

That is, to find the adjoint matrix for $\mathbf{A}$, we take the matrix $\mathbf{A}$, replace each element by its *cofactor* (which is the determinant of the matrix that remains when both the row and column containing the element are removed) and then transpose the resultant matrix. For example

$$A = \begin{bmatrix} 2 & 1 \\ -1 & 3 \end{bmatrix}$$

$\Delta_{11} = (-1)^2(3) = 3 \qquad \Delta_{12} = (-1)^3(-1) = 1$
$\Delta_{21} = (-1)^3(1) = -1 \qquad \Delta_{22} = (-1)^4(2) = 2$

$$\text{ad } \mathbf{A} = \begin{bmatrix} 3 & -1 \\ 1 & 2 \end{bmatrix}$$

The adjoint matrix is such that

$$\mathbf{A}(\text{ad } \mathbf{A}) = \Delta \mathbf{I}$$

where Δ is the determinant of $\mathbf{A}$. Thus

$$\mathbf{A}^{-1}\mathbf{A}(\text{ad } \mathbf{A}) = \mathbf{A}^{-1}\Delta \mathbf{I} = \Delta \mathbf{A}^{-1}$$
$$\text{ad } \mathbf{A} = \Delta \mathbf{A}^{-1}$$

or

$$\boxed{\mathbf{A}^{-1} = \frac{1}{\Delta} \text{ad } \mathbf{A}} \qquad \text{A.16}$$

This is the desired formula for finding the inverse of a square matrix. If $\Delta \neq 0$, then the rows of $\mathbf{A}$ are linearly independent, and $\mathbf{A}^{-1}$ exists.

Equations A.14 and A.16 give

$$\begin{bmatrix} x_1 \\ x_2 \\ \vdots \\ x_n \end{bmatrix} = \frac{1}{\Delta} \begin{bmatrix} \Delta_{11} & \Delta_{21} & \cdots & \Delta_{n1} \\ \Delta_{12} & \Delta_{22} & \cdots & \Delta_{n2} \\ & & \ddots & \\ \Delta_{1n} & \Delta_{2n} & \cdots & \Delta_{nn} \end{bmatrix} \begin{bmatrix} b_1 \\ b_2 \\ \vdots \\ b_n \end{bmatrix} \qquad \text{A.17}$$

and so the ith unknown is given by

$$x_i = \frac{1}{\Delta}(b_1\Delta_{1i} + b_2\Delta_{2i} + \cdots + b_n\Delta_{ni}) \qquad \text{A.18}$$

or

$$x_i = \frac{\Delta_{bi}}{\Delta} \qquad \text{A.19}$$

where Δ_{bi} is the determinant of the matrix obtained when the elements of the ith column of $\mathbf{A}$ are replaced with the elements of the column vector $\mathbf{b}$. This result is called *Cramer's rule*.

Consider, as a simple example, the set of equations

$$x_1 + 2x_2 = 3$$
$$-2x_1 + 3x_2 = 2$$

or

$$\begin{bmatrix} 1 & 2 \\ -2 & 3 \end{bmatrix} \begin{bmatrix} x_1 \\ x_2 \end{bmatrix} = \begin{bmatrix} 3 \\ 2 \end{bmatrix}$$

$$\mathbf{A} = \begin{bmatrix} 1 & 2 \\ -2 & 3 \end{bmatrix} \quad \text{ad } \mathbf{A} = \begin{bmatrix} 3 & -2 \\ 2 & 1 \end{bmatrix} \quad \Delta = 3 + 4 = 7$$

$$\mathbf{A}^{-1} = \tfrac{1}{7} \begin{bmatrix} 3 & -2 \\ 2 & 1 \end{bmatrix}$$

$$\begin{bmatrix} x_1 \\ x_2 \end{bmatrix} = \tfrac{1}{7} \begin{bmatrix} 3 & -2 \\ 2 & 1 \end{bmatrix} \begin{bmatrix} 3 \\ 2 \end{bmatrix} = \tfrac{1}{7} \begin{bmatrix} 5 \\ 8 \end{bmatrix} = \begin{bmatrix} \tfrac{5}{7} \\ \tfrac{8}{7} \end{bmatrix}$$

$$x_1 = \tfrac{5}{7}$$
$$x_2 = \tfrac{8}{7}$$

Although the method presented does give a formal solution for the x_i in a set of n linearly independent equations, it is not as (computationally) efficient as other methods. We briefly consider here one of these called *Gaussian elimination*. Its object is to first put the set of equations in triangular form by substitution. A simple example will suffice. Consider the set

$$x_1 + 2x_2 + x_3 = 1 \quad \text{(a)}$$
$$x_1 + x_2 + 2x_3 = 2 \quad \text{(b)}$$
$$x_1 - x_2 + x_3 = 0 \quad \text{(c)}$$

We first eliminate x_1 from all equations except the first. To do this, first multiply (a) by -1, add the new equation to (b), and replace (b) with the result [called Eq. (d)]:

$$\begin{array}{r} -x_1 - 2x_2 - x_3 = -1 \\ \underline{x_1 + x_2 + 2x_3 = 2} \\ -x_2 + x_3 = 1 \end{array} \quad \text{(d)}$$

$$x_1 + 2x_2 + x_3 = 1 \quad \text{(a)}$$
$$-x_2 + x_3 = 1 \quad \text{(d)}$$
$$x_1 - x_2 + x_3 = 0 \quad \text{(c)}$$

Now multiply (a) by -1, add this to (c) to eliminate x_1, and replace (c) with the result:

$$\begin{array}{r} -x_1 - 2x_2 - x_3 = -1 \\ \underline{x_1 - x_2 + x_3 = 0} \\ -3x_2 = -1 \end{array} \quad \text{(e)}$$

$$x_1 + 2x_2 + x_3 = 1 \quad \text{(a)}$$
$$-x_2 + x_3 = 1 \quad \text{(d)}$$
$$-3x_2 = -1 \quad \text{(e)}$$

Finally, use (d) to eliminate x_2 from (e):

$$x_1 + 2x_2 + x_3 = 1 \qquad (a)$$
$$-x_2 + x_3 = 1 \qquad (d)$$
$$-3x_3 = -4 \qquad (f)$$

The equations are in triangular form, and we find the unknowns by *back substitution*:

$$x_3 = \tfrac{4}{3}$$

Substitute this (back) into (d):

$$x_2 = \tfrac{1}{3}$$

Substitute these (back) into (a):

$$x_1 = -1$$

B. DERIVATION OF THE UNIT-IMPULSE RESPONSE FOR CONTINUOUS-TIME SYSTEMS

The general form of the Nth order linear differential equation is

$$\sum_{n=0}^{N} a_n(t) \frac{d^n y}{dt^n} = x_f(t) = \sum_{n=0}^{M} b_n(t) \frac{d^n x}{dt^n} \qquad \text{B.1}$$

where $x_f(t)$ is the forcing function and $x(t)$ is the input function. The zero-input response is $y_{zi}(t, \tau)$, and it satisfies the homogeneous equation

$$\sum_{n=0}^{\infty} a_n(t) \frac{d^n}{dt^n} [y_{zi}(t, \tau)] = 0 \qquad \text{B.2}$$

The unit-impulse response for the forcing function $x_f(t)$ is called $h_f(t, \tau)$, and it must satisfy

$$\sum_{n=0}^{\infty} a_n(t) \frac{d^n}{dt^n} [h_f(t, \tau)] = \delta(t - \tau) \qquad \text{B.3}$$

We want to show that

$$h_f(t, \tau) = \frac{A}{A+B} h^a(t, \tau) + \frac{B}{A+B} h^b(t, \tau) \qquad \text{B.4}$$

where

$$h^a(t, \tau) = y_{zi}(t, \tau) u(t - \tau) \qquad \text{B.5}$$

and

$$h^b(t, \tau) = -y_{zi}(t, \tau) u(\tau - t) \qquad \text{B.6}$$

subject to the particular initial conditions

$$\left.\begin{array}{r}y_{zi}\big|_{t=\tau} = 0 \\ \dfrac{dy_{zi}}{dt}\bigg|_{t=\tau} = 0 \\ \vdots \\ \dfrac{d^{N-1}y_{zi}}{dt^{N-1}}\bigg|_{t=\tau} = \dfrac{1}{a_N(\tau)}\end{array}\right\} \quad \text{B.7}$$

Consider $h^a(t, \tau)$ (alone) first and substitute Eq. B.5 into B.3. After the differentiation is performed we obtain

$$\begin{aligned} & a_0(t)y_{zi}(t,\tau)u(t-\tau) \\ & + a_1(\tau)y_{zi}(\tau,\tau)\delta(t-\tau) + a_1(t)y'_{zi}(t,\tau)u(t-\tau) \\ + a_2(\tau)y_{zi}(\tau,\tau)\delta'(t-\tau) & + a_2(\tau)y'_{zi}(\tau,\tau)\delta(t-\tau) + a_2(t)y''_{zi}(t,\tau)u(t-\tau) \\ + \cdots\cdots\cdots\cdots\cdots\cdots & \cdots\cdots\cdots\cdots\cdots\cdots\cdots\cdots\cdots\cdots\cdots\cdots\cdots\cdots \\ a_N(\tau)y_{zi}(\tau,\tau)\delta^{N-1}(t-\tau) + \cdots & + \boxed{a_N(\tau)y_{zi}^{N-1}(\tau,\tau)\delta(t-\tau)} \\ & + a_N(t)y_{zi}^N(t,\tau)u(t-\tau) = \delta(t-\tau) \end{aligned}$$

B.8

The last column on the left side of Eq. B.8 is 0 because of Eq. B.2, while all the other terms on the left, except the one enclosed, are 0 because of Eq. B.7. Thus, Eq. B.8 reduces to

$$a_N(\tau)y_{zi}^{N-1}(\tau,\tau)\delta(t-\tau) = \delta(t-\tau)$$

or

$$y_{zi}^{N-1}(\tau,\tau) = \dfrac{1}{a_n(\tau)}$$

and the proof is complete for $h^a(t-\tau)$. Following the same procedure for $h^b(t, \tau)$, we arrive at

$$a_N(\tau)y_{zi}^{N-1}(\tau,\tau)\delta(\tau-t) = \delta(t-\tau)$$

or since $\delta(\tau - t) = \delta(t - \tau)$

$$y_{zi}^{N-1}(\tau,\tau) = \dfrac{1}{a_n(\tau)}$$

and the proof is complete for $h^b(t, \tau)$.

This is a linear system so superposition applies, and insofar as Eq. B.3 is concerned, $[A/(A+B)]h^a(t,\tau)$, when substituted into the left side, gives $[A/(A+B)]\delta(t-\tau)$. Likewise $[B/(A+B)]h^b(t,\tau)$, when substituted into the left side, gives $[B/(A+B)]\delta(t-\tau)$. Thus, by superposition

$$\frac{A}{A+B}h^a(t,\tau)+\frac{B}{A+B}h^b(t,\tau) \text{ gives } \frac{A}{A+B}\delta(t-\tau)+\frac{B}{A+B}\delta(t-\tau)=\delta(t-\tau)$$

Thus, Eq. B.3 is satisfied by Eq. B.4, and it is indeed true that if we can find the zero-input response, $y_{zi}(t)$, satisfying Eq. B.2, then we immediately have the unit-impulse response, Eq. B.4 [with B.5, B.6, and B.7]. The unit-impulse response can then be used in the superposition integral to find the response for any *excitation*.

C. USEFUL INTEGRALS AND SERIES

Indefinite Integrals

$$\int x \cos ax\, dx = \frac{\cos ax}{a^2} + \frac{x \sin ax}{a} \quad \text{C.1}$$

$$\int x^2 \cos ax\, dx = \frac{2x}{a^2}\cos ax + \left(\frac{x^2}{a}-\frac{2}{a^3}\right)\sin ax \quad \text{C.2}$$

$$\int x^3 \cos ax\, dx = \left(\frac{3x^2}{a^2}-\frac{6}{a^4}\right)\cos ax + \left(\frac{x^3}{a}-\frac{6x}{a^3}\right)\sin ax \quad \text{C.3}$$

$$\int \cos ax \cos px\, dx = \frac{\sin(a-p)x}{2(a-p)} + \frac{\sin(a+p)x}{2(a+p)} \quad \text{C.4}$$

$$\int x \sin ax\, dx = \frac{\sin ax}{a^2} - \frac{x \cos ax}{a} \quad \text{C.5}$$

$$\int x^2 \sin ax\, dx = \frac{2x}{a^2}\sin ax + \left(\frac{2}{a^3}-\frac{x^2}{a}\right)\cos ax \quad \text{C.6}$$

$$\int x^3 \sin ax\, dx = \left(\frac{3x^2}{a^2}-\frac{6}{a^4}\right)\sin ax + \left(\frac{6x}{a^3}-\frac{x^3}{a}\right)\cos ax \quad \text{C.7}$$

$$\int \sin px \sin qx\, dx = \frac{\sin(p-q)x}{2(p-q)} - \frac{\sin(p+q)x}{2(p+q)} \quad \text{C.8}$$

$$\int \sin px \cos qx\, dx = -\frac{\cos(p-q)x}{2(p-q)} - \frac{\cos(p+q)x}{2(p+q)} \quad \text{C.9}$$

$$\int xe^{ax}\, dx = \frac{e^{ax}}{a}\left(x-\frac{1}{a}\right) \quad \text{C.10}$$

$$\int x^2 e^{ax}\, dx = \frac{e^{ax}}{a}\left(x^2-\frac{2x}{a}+\frac{2}{a^2}\right) \quad \text{C.11}$$

484 APPENDICES

$$\int e^{ax}\sin bx\, dx = \frac{e^{ax}(a\sin bx - b\cos bx)}{a^2 + b^2} \qquad \text{C.12}$$

$$\int e^{ax}\cos bx\, dx = \frac{e^{ax}(a\cos bx + b\sin bx)}{a^2 + b^2} \qquad \text{C.13}$$

$$\int xe^{ax}\sin bx\, dx = \frac{xe^{ax}(a\sin bx - b\cos bx)}{a^2 + b^2}$$
$$\qquad - \frac{e^{ax}\{(a^2 - b^2)\sin bx - 2ab\cos bx\}}{(a^2 + b^2)^2} \qquad \text{C.14}$$

$$\int xe^{ax}\cos bx\, dx = \frac{xe^{ax}(a\cos bx + b\sin bx)}{a^2 + b^2}$$
$$\qquad - \frac{e^{ax}\{(a^2 - b^2)\cos bx + 2ab\sin bx\}}{(a^2 + b^2)^2} \qquad \text{C.15}$$

Definite Integrals

$$\int_0^\infty e^{-a^2x^2}\, dx = \sqrt{\pi}/2a \qquad 0 < a \qquad \text{C.16}$$

$$\int_0^\infty x^2 e^{-x^2}\, dx = \sqrt{\pi}/4 \qquad \text{C.17}$$

$$\int_0^\infty \text{Sa}(x)\, dx = \int_0^\infty \frac{\sin(x)}{x}\, dx = \frac{\pi}{2} \qquad \text{C.18}$$

$$\int_0^\infty \text{Sa}^2(x)\, dx = \pi/2 \qquad \text{C.19}$$

Finite Series

$$\sum_{n=1}^N n = \frac{N(N+1)}{2} \qquad \text{C.20}$$

$$\sum_{n=1}^N n^2 = \frac{N(N+1)(2N+1)}{6} \qquad \text{C.21}$$

$$\sum_{n=1}^N n^3 = \frac{N^2(N+1)^2}{4} \qquad \text{C.22}$$

$$\sum_{n=0}^N x^n = \frac{x^{N+1} - 1}{x - 1} \qquad \text{(geometric series)} \qquad \text{C.23}$$

$$\sum_{n=0}^N nx^n = \frac{x}{(1-x)^2}[1 - (N+1)x^N + Nx^{N+1}] \qquad \begin{array}{l}\text{(arithmetic-}\\ \text{geometric series)}\end{array} \qquad \text{C.24}$$

$$\sum_{n=0}^N n^2 x^n = \frac{x}{(1-x)^3}[(1+x) - (N+1)^2 x^n + (2N^2 + 2N - 1)x^{n+1} - N^2 x^{N+2}] \qquad \text{C.25}$$

Taylor Series

$$f(x) = f(a) + f'(a)(x-a) + \frac{f''(a)(x-a)^2}{2!} + \cdots + \frac{f^{(n-1)}(a)(x-a)^{n-1}}{(n-1)!} + R_n$$ C.26

$$R_n = \frac{f^{(n)}(\xi)(x-a)^n}{n!} \qquad a < \xi < x$$

$$R_n = \frac{f^{(n)}(\xi)(x-\xi)^{n-1}(x-a)}{(n-1)!}$$

Binomial Series

$$(a+x)^n = a^n + na^{n-1}x + \frac{n(n-1)}{2!}a^{n-2}x^2 + \frac{n(n-1)(n-2)}{3!}a^{n-3}x^3 + \cdots$$

$$= a^n + \binom{n}{1}a^{n-1}x + \binom{n}{2}a^{n-2}x^2 + \binom{n}{3}a^{n-3}x^3 + \cdots \qquad \text{C.27}$$

D. ROUTH'S TEST

The transfer function of a continuous-time system is given by Eq. 7.2:

$$H(s) = \frac{\sum_{n=0}^{M} b_n s^n}{\sum_{n=0}^{N} a_n s^n} \qquad 7.2$$

and the characteristic equation is obtained by setting the denominator equal to 0:

$$\sum_{n=0}^{N} a_n s^n = 0$$

or

$$a_0 + a_1 s + a_2 s^2 + \cdots + a_N s^N = 0 \qquad \text{D.1}$$

If all of the roots of this equation lie in the left half of the s plane, then the system described by Eq. 7.2 is stable. In order for this to be the case, it is necessary, but not sufficient, that the coefficients of Eq. D.1 have the same sign. This can, of course, be determined by inspection. Another requirement on Eq. D.1 that can be determined visually is that all coefficients (with the possible exception of a_0) must be nonzero.[†] These two tests are easily applied.

In order to determine how many roots of the characteristic equation lie in the right half of the s plane, the *Routh test* can be applied. We first form the Routh table (or array). Its first row consists of alternate coefficients beginning

---

[†] If the polynomial is an *even* or an *odd* function of s, all of the roots of the characteristic equation could lie on the imaginary axis.

with a_N, while the second row consists of the remaining coefficients beginning with a_{N-1}:

$$
\begin{array}{ccc}
a_N & a_{N-2} & a_{N-4} \\
a_{N-1} & a_{N-3} & a_{N-5}
\end{array}
$$

The third row is obtained from the first two rows using the following multiplications:

$$\frac{(a_{N-1})(a_{N-2}) - (a_{N-3})(a_N)}{a_{N-1}} \quad \frac{(a_{N-1})(a_{N-4}) - (a_{N-5})(a_N)}{a_{N-1}} \ldots$$

The fourth row is obtained using the same multiplication rule applied to the second and third rows. This process is repeated until there is no remaining term. We can multiply (or divide) any row by a positive constant without altering the final result. This may help simplify the numerical work. When the table is completed, the number of roots of the characteristic equation lying in the right half of the s plane is equal to the number of *sign changes* in the first column of the Routh table.

Consider, as a first example, a system that is obviously (why?) unstable:

$$s^3 + 2s^2 - 5s - 6 = 0$$

The Routh table is

s^3	1	-5		s^3	1	-5
s^2	2	-6		s^2	2	-6
s^1	$\frac{(2)(-5)-(-6)(1)}{2}$	0		s^1	-2	
s^0	-6			s^0	-6	

There is one *sign change* in the first column, thus there is one root of the characteristic equation in the right half of the s plane, and the system is unstable. In fact, it is not too difficult (not too much fun, either) to factor the (cubic) characteristic equation for this example:

$$(s+1)(s-2)(s+3) = 0$$

and the one root that is in the right half of the s plane is located at $s = 2$.

Next, consider

$$s^4 + 3s^3 + 3s^2 + 3s + 2 = 0$$

s^4	1	3	2
s^3	3	3	
s^2	2	2	
s^1	0		

The presence of 0 in the s^1 (fourth) row is puzzling. Does this, or does it not, represent a sign change? The answer can be found by asking another question. What sign changes (if any) would occur in the first column of the Routh table

D. ROUTH'S TEST

if the system possessed conjugate imaginary roots *on* the imaginary axis ($j\omega$) in the s plane? These roots are neither *in* the left half or right half of the s plane. The row above that which has the 0 in the first column (that is, the third, or s^2, row) gives us an *auxiliary equation*:

$$2s^2 + 2 = 0$$
$$s^2 + 1 = 0 \qquad \text{D.2}$$
$$(s + j)(s - j) = 0$$

Thus, we have conjugate roots, $s = -j$ and $s = +j$, on the imaginary axis, as suspected. The table may be completed by differentiating Eq. D.2:

$$4s = 0$$

and using the coefficient 4 in the table:

s^4	1	3	2
s^3	3	3	
s^2	2	2	
s^1	4		
s^0	2		

Thus, we have no roots *in* the right half of the s plane.

On the other hand since we know that $s^2 + 1$ is a factor of the characteristic equation, the other roots can be determined by long division:

$$
\begin{array}{r}
s^2 + 3s + 2 \\
s^2 + 1 \overline{\smash{\big)}\, s^4 + 3s^3 + 3s^2 + 3s + 2} \\
\underline{s^4 + s^2 } \\
3s^3 + 2s^2 + 3s \\
\underline{3s^3 + 3s } \\
2s^2 + 2 \\
\underline{2s^2 + 2} \\
\end{array}
$$

$$s^2 + 3s + 2 = (s + 1)(s + 2)$$

Thus

$$(s + 1)(s + 2)(s + j)(s - j) = s^4 + 3s^3 + 3s^2 + 3s + 2$$

and this is a *marginally stable* system.

Consider a system with a parameter K (perhaps the gain of an amplifier):

$$s^3 + 3s^2 + Ks + 1 = 0$$

s^3	1	K
s^2	3	1
s^1	$\dfrac{3K - 1}{3}$	
s^0	1	

It is immediately obvious that $3K - 1 > 0$, or $K > \frac{1}{3}$, for a stable system. For $K = \frac{1}{3}$ the auxiliary equation (from the s^2 row) is

$$3s^2 + 1 = 0$$
$$s^2 + \tfrac{1}{3} = 0$$
$$(s + j/\sqrt{3})(s - j/\sqrt{3}) = 0$$

and the third root is $s = -3$, as is easily verified.

Consider the unstable (a_3 is missing) system:

$$s^4 + 3s^2 + s + 2 = 0$$

How many roots are in the right half of the s plane? The Routh table is

s^4	1	3	2
s^3	0	1	

One method of overcoming the difficulty is that of forming a new polynomial by substituting $P = 1/s$:

$$1/P^4 + 3/P^2 + 1/P + 2 = 0$$

and then multiplying by P^4:

$$P^4 + P^3 + 3P^2 + 1 = 0$$

This places the missing coefficient in a different position, and we can now proceed to form a new Routh table from the polynomial in P:

P^4	1	3	1
P^3	1	0	
P^2	3	1	
P^1	$-\frac{1}{3}$		
P^0	1		

There are two sign changes and two roots of the characteristic equation in the right half of the s plane.

E. ANALOGOUS SYSTEMS

Systems that are governed by the same kinds of equations are called *analogous systems*, and if we know how to analyze one of them, then we know how to analyze all of them. Some of the systems that an electrical engineer may encounter that usually have an electrical analog are the mechanical translational, mechanical rotational, acoustical, fluid, and thermal systems. If the electrical analog is known [that is, if an electrical circuit can be drawn (or built)], the electrical engineer can analyze it (or measure its performance), and thus the performance of the original system under all kinds of conditions can easily be found.

E. ANALOGOUS SYSTEMS

In addition, it is becoming more important for the electrical engineer to have some knowledge of how to *find* the equations that govern these other systems since he, or she, often finds them *combined* with an electrical system, forming a *hybrid*. This is especially true in a *control system*, for example. It certainly is not only beyond the scope of the material in this book, but also impossible, to consider all systems that have electrical analogs (or all hybrid systems). We will consider a few examples.

Consider the simple three-element mechanical translational system shown in Fig. E.1(a). There is one forcing function. It is the applied force, so we have a system with an *ideal force source* that is assumed to be independent of what is connected to it. We mention that it is also possible to have an ideal (independent) *velocity* source in mechanical systems. There are three forces *resisting* the applied force. According to Newton's second law the *inertia* force is given by *mass* times acceleration, or

$$f_M = M\frac{du}{dt} = M\frac{d^2z}{dt^2} \qquad u = \frac{dz}{dt} \qquad \text{E.1}$$

where M is the mass (in kilograms). The damping force is proportional to velocity in linear systems if the friction is *viscous*. The force is only approximately proportional to velocity for *dry* friction. If we assume linear damping, then

$$f_D = Du = D(dz/dt) \qquad \text{E.2}$$

where D is the *damping coefficient* (newton-seconds per meter). The linear spring force is given by Hooke's law:

$$f_K = \frac{1}{K}\int u\,dt = \frac{1}{K}z \qquad \text{E.3}$$

where K is the *compliance* (that is, the reciprocal of the *stiffness*) of the spring (meters per newton).

These forces are combined to form an equation with the aid of D'Alembert's principle. It states that:

The algebraic sum of the externally applied forces and the resisting forces in any direction that are acting on a body is 0. Focusing attention on the *one* mass in Fig. E.1(a), we have

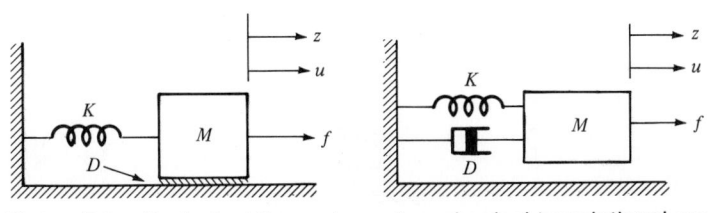

$$f - f_M - f_D - f_K = 0 \qquad \text{E.4}$$

Figure E.1. Equivalent three-element mechanical translational systems.

or

$$f = M\frac{du}{dt} + Du + \frac{1}{K}\int u\,dt = M\frac{d^2z}{dt^2} + D\frac{dz}{dt} + \frac{1}{K}z \qquad \text{E.5}$$

According to Kirchhoff's voltage law, the equation describing the series RLC circuit driven by an ideal voltage source (Fig. E.2) is

$$v(t) = L\frac{di}{dt} + Ri + \frac{1}{C}\int i\,dt = L\frac{d^2q}{dt^2} + R\frac{dq}{dt} + \frac{1}{C}q \qquad \text{E.6}$$

Comparing Eqs. E.5 and E.6, we see that they are *dual*. They are the same, except for symbols, and analogous quantities are

$$f \longleftrightarrow v$$
$$u \longleftrightarrow i$$
$$z \longleftrightarrow q$$
$$M \longleftrightarrow L$$
$$D \longleftrightarrow R$$
$$K \longleftrightarrow C$$

The series *RLC* circuit is an analog for the mechanical translational system, and we can determine the complete behavior of the mechanical system by investigating the electrical system. This is obviously a *force-voltage* (*f-v*) analog.

Again, if we focus attention on the *one* mass in Fig. E.1(b), and recognize that for force-voltage analogs *each junction (all points on a rigid mass) in the mechanical system corresponds to a single closed loop in the electrical analog*, we see that the equation governing this system is the same as E.5, and Fig. E.2 is also its analog. Notice that the damping in Fig. E.1(b) is provided by a dashpot (shock absorber) with a piston in a cylinder.

A more natural analogy exists in the *force-current* case since *force through* is analogous to *current through* and *velocity across* is analogous to *voltage across*. Furthermore, a junction in the mechanical system becomes a junction (node) in the electrical circuit. Also, velocity and displacement in a mechanical system can be measured with negligible loading effect just as can voltage in an electrical system. This is not true for force and current, respectively. It is helpful to recognize that all points on a rigid mass represent a junction, and one terminal (analogous to a mass) is always connected to the reference node

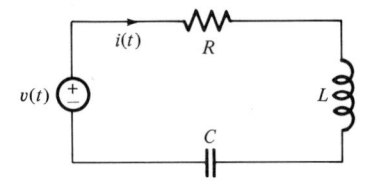

Figure E.2. Series RLC circuit.

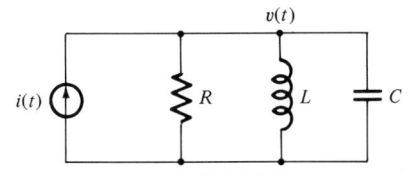

Figure E.3. Parallel *RLC* circuit.

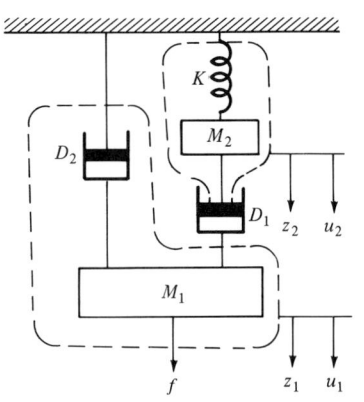

Figure E.4. More complex (two-coordinate) mechanical translational system.

(ground). The analogous quantities are

$$f \longrightarrow i$$
$$u \longrightarrow v$$
$$z \longrightarrow \Psi, \text{ flux linkage}$$
$$M \longrightarrow C$$
$$D \longrightarrow 1/R = G$$
$$K \longrightarrow L$$

We can easily draw the analogous force-current analog for Fig. E.1. It is shown in Fig. E.3, and Kirchhoff's current law gives

$$i(t) = C\frac{dv}{dt} + \frac{1}{R}v + \frac{1}{L}\int v\, dt = C\frac{d^2\Psi}{dt^2} + \frac{1}{R}\frac{d\Psi}{dt} + \frac{1}{L}\Psi \qquad \text{E.7}$$

where $v(t) = d\Psi/dt$. The reader can easily see that Eqs. E.5, E.6, and E.7 are duals.

Now consider Fig. E.4.[†] Using D'Alembert's principle, we can deduce an equation for the forces on M_1 (for that portion of the system enclosed by the dashed line, including M_1):

$$M_1\frac{du_1}{dt} + D_1(u_1 - u_2) + D_2 u_1 = f$$

[†] See Cheng or Dorf in the list of references in Chap. 7.

or

$$M_1 \frac{du_1}{dt} + (D_1 + D_2)u_1 - D_1 u_2 = f \qquad \text{E.8}$$

In the same way (for that portion of the system enclosed by the other dashed line, including M_2)

$$M_2 \frac{du_2}{dt} + D_1(u_2 - u_1) + \frac{1}{K}\int u_2\, dt = 0 \qquad \text{E.9}$$

The electrical analog, using the force-voltage analogy, is relatively easy to obtain from Fig. E.4 (or, alternatively, from Eqs. E.8 and E.9). There are two masses and two coordinates, z_1 and z_2; therefore, there are two loops in the electrical analog. Using the element analogs listed earlier, we see that one loop contains the voltage source v (f), an inductance L_1 (M_1) and two resistances R_1 (D_1) and R_2 (D_2). The second loop also contains R_1 (D_1) (the element common to both loops with current $i_1 - i_2$). Other elements in the second loop are an inductance L_2 (M_2) and a capacitance C (K). This circuit is shown in Fig. E.5. Kirchhoff's law gives

$$L(di_1/dt) + (R_1 + R_2)i_1 - R_1 i_2 = v \qquad \text{E.10}$$

and

$$L_2(di_2/dt) + R_1(i_2 - i_1) + (1/C)\int i_2\, dt = 0 \qquad \text{E.11}$$

Equations E.10 and E.11 are obviously dual to E.8 and E.9, respectively.

We follow a similar procedure for the force-current analog. The elements enclosed by the dashed lines in Fig. E.4 are connected to a node with R_1 (D_1) as the common element once again. The electric circuit analog is shown in Fig. E.6. It is left as an exercise for the reader to write the equations given by

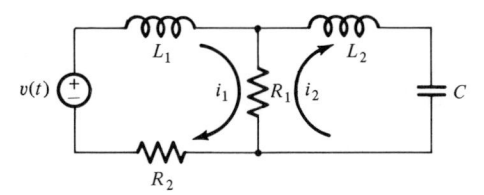

Figure E.5. Electrical analog for the mechanical system of Fig. E.4. (force-voltage).

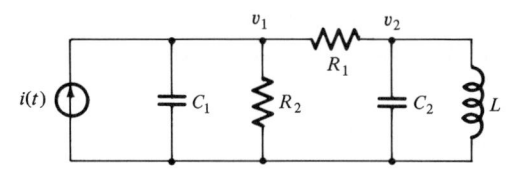

Figure E.6. Electrical analog for the mechanical system of Fig. E.4. (force-current).

Kirchhoff's current law for the circuit of Fig. E.6. They are dual to Eqs. E.10 and E.11.

The algebraic sum of the applied torques in a rotational system and those resisting rotation about an axis is 0 in a manner analogous to the forces in a translational system. The torques that are resisting rotation in Fig. E.7 are the inertia torque given by the product of moment of inertia I and angular acceleration

$$\tau_I = I\frac{d\omega}{dt} = I\frac{d^2\theta}{dt^2} \qquad \omega = \frac{d\theta}{dt} \qquad \text{E.12}$$

the damping torque (assumed to be linear) given by the damping coefficient times the angular velocity

$$\tau_D = D\omega = D\frac{d\theta}{dt} \qquad \text{E.13}$$

and the spring torque given by

$$\tau_K = \frac{1}{K}\int \omega \, dt = \frac{1}{K}\theta \qquad \text{E.14}$$

Notice that in Fig. E.7 the torsional compliance is $2K$ *on each side* of the flywheel and, also, the shaft *is clamped* at each end. The applied (independent) torque is τ. Thus, we obtain

$$\tau - \tau_I - \tau_D - \tau_K = 0 \qquad \text{E.15}$$

or

$$\tau = I\frac{d\omega}{dt} + D\omega + \frac{1}{2K}\int \omega \, dt + \frac{1}{2K}\int \omega \, dt$$

$$\tau = I\frac{d\omega}{dt} + D\omega + \frac{1}{K}\int \omega \, dt = I\frac{d^2\theta}{dt^2} + D\frac{d\theta}{dt} + \frac{1}{K}\theta \qquad \text{E.16}$$

Equation E.16 is dual to E.5, so the rotational system of Fig. E.7 is analogous to the translational system of Fig. E.1. The reader can derive the units for the rotational constants I, D, and K.

Mechanical coupling devices such as gears, nonslipping friction wheels, and levers also have electrical analogs, and it is rather obvious that under

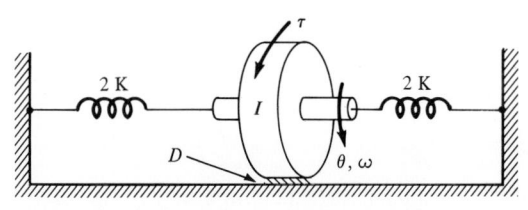

Figure E.7. Mechanical rotational system.

ideal conditions, the analog is an *ideal* transformer with infinite mutual and self inductances. Its behavior is governed only by the turns ratio between the primary and secondary windings, N_1/N_2. Figure E.8 shows a pair of meshed gears for which the reader can easily show that

$$\frac{\omega_1}{\omega_2} = \frac{r_2}{r_1} = \frac{\tau_2}{\tau_1} \qquad \text{E.17}$$

Thus, if we wish torque (τ) to be analogous to voltage and angular velocity (ω) to be analogous to current, then

$$\frac{v_1}{v_2} = \frac{N_1}{N_2} = \frac{r_1}{r_2} = \frac{\tau_1}{\tau_2} = \frac{i_2}{i_1} \qquad \text{E.18}$$

and this gives a force-voltage analog as shown in Fig. E.9. Notice the polarity reversal for the voltages with respect to the current arrows. This accounts for the reversal of both τ and ω for meshed gears.

If we want a force-current analog

$$\frac{v_1}{v_2} = \frac{N_1}{N_2} = \frac{r_2}{r_1} = \frac{\tau_2}{\tau_1} = \frac{i_2}{i_1} \qquad \text{E.19}$$

and Fig. E.10 applies.

A simple mechanical rotational system with meshed gears and its electrical analog using the force-voltage analog are shown in Fig. E.11.

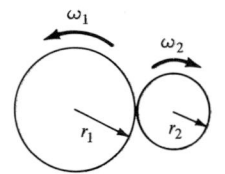

Figure E.8. Meshed gears.

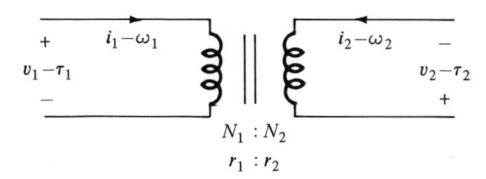

Figure E.9. Ideal transformer as the force-voltage analog of meshed gears.

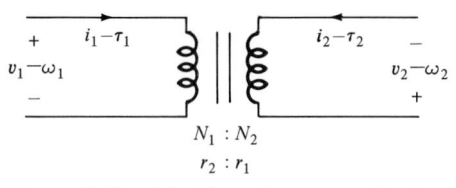

Figure E.10. Ideal transformer as the force-current analog of meshed gears.

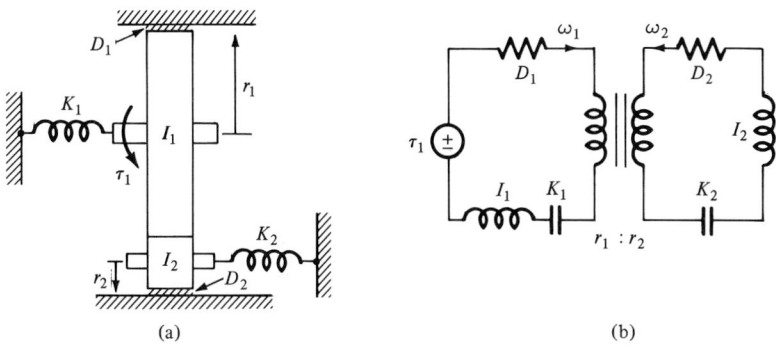

Figure E.11. (a) Meshed gear mechanical rotational system. (b) Electrical analog for (a).

Motors and generators are governed by very simple basic equations even though they are best described as vector equations. The same equations are used to describe other electromechanical systems that convert electrical energy into mechanical energy or vice versa. The first of these equations is (in differential vector form)[†]

$$d\mathbf{F} = I\, d\mathbf{l} \times \mathbf{B} \qquad \text{E.20}$$

and gives the differential force on a differential length of filamentary current in a magnetic field. The symbol **B** is the uniform magnetic flux density. If we are dealing with a straight conductor of length l that is perpendicular to **B**, the three vectors **F**, **l**, and **B** are mutually perpendicular, and we can dispense with the vector notation and simply use

$$f = Bil \qquad i = f/(Bl) \qquad \text{E.21}$$

with Fig. E.12 giving the proper directions.

The second equation is (also in differential form)[‡]

$$dv = \mathbf{u} \times \mathbf{B} \cdot d\mathbf{l} \qquad \text{E.22}$$

and gives the differential emf (voltage) induced in a differential length conductor moving with velocity **u** in a magnetic field. Again, for a straight conductor of

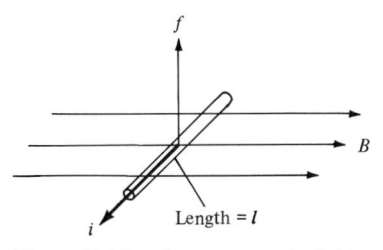

Figure E.12. Force on a straight current-bearing conductor in a magnetic field.

[†] See Neff in the list of references in Chap. 7.
[‡] Sometimes called the *motional emf equation*.

length l perpendicular to **B** moving with velocity u such that **u**, **B**, and l are perpendicular, we have

$$v = (Bl)u \qquad \text{E.23}$$

as in Fig. E.13. Notice the voltage polarity.

Equations E.21 and E.23 can be combined since Bl is a common factor. As a matter of fact, they can be combined so that we effectively have an ideal transformer with electrical quantities v and i in the primary, turns ratio $Bl/1$, and mechanical quantities f and u in the secondary. This is shown in Fig. E.14. Since f is analogous to i (secondary) and u is analogous to v, we have a force-current analog.

Figure E.15 shows an electromechanical system that could represent a loudspeaker, microphone, or, perhaps, a solenoid-type relay. The device is

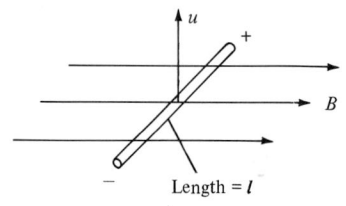

Figure E.13. Voltage induced in a straight conductor moving in a magnetic field.

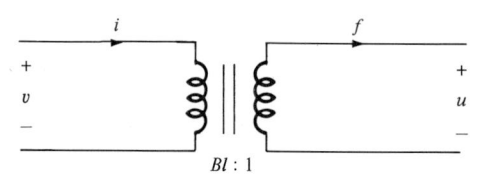

Figure E.14. Force-current analog for motor (E.21) and generator (E.23) equations combined.

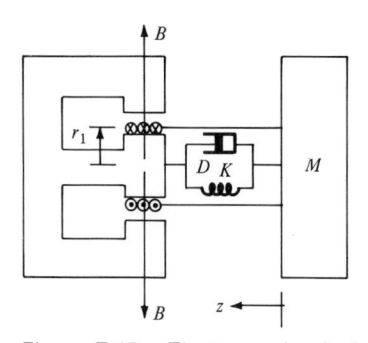

Figure E.15. Electromechanical system (loudspeaker).

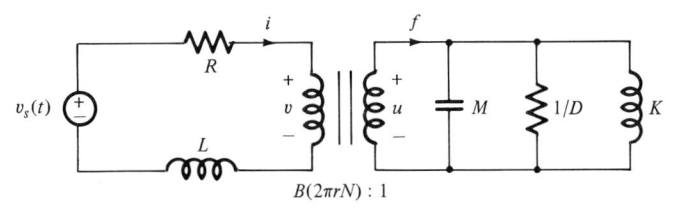

Figure E.16. Circuit for the loudspeaker of Fig. E.15.

circular in cross-section, and the *radial* magnetic field is produced by a permanent magnet. The N-turn coil has a radius r and is rigidly connected to the mass M. Here, we choose to let the system represent a loudspeaker, so the coil is driven by a source voltage $v_s(t)$. The coil has inductance L and resistance R. Since the conductors making up the coil are always perpendicular to B, Fig. E.14 applies with $Bl = B(2\pi rN) = 2\pi rBN$ and with v_s, R, and L in series in the primary circuit. Thus, for the primary, Kirchhoff's voltage law gives

$$L(di/dt) + Ri + v = v_s(t) \qquad \text{E.24}$$

The secondary (mechanical) circuit has K (inductance), M (capacitance), and $1/D$ (resistance) in parallel, so Kirchhoff's current law (nodal analysis) gives

$$M\frac{du}{dt} + Du + \frac{1}{K}\int u\,dt = f \qquad \text{E.25}$$

The circuit is shown in Fig. E.16.

Since $v = (2\pi rBN)u$, and $f = (2\pi rBN)i$, v and f can be eliminated from the last two equations giving

$$L\frac{di}{dt} + Ri + (2\pi rBN)u = v_s(t) \qquad \text{E.26}$$

$$M\frac{du}{dt} + Du + \frac{1}{K}\int u\,dt - (2\pi rBN)i = 0 \qquad \text{E.27}$$

These contain the same two unknowns (i and u), and can be solved by any of the methods that are discussed in the text.

F. BILATERAL LAPLACE TRANSFORM PROPERTIES

$$f(t) \leftrightarrow F(s) = \int_{-\infty}^{\infty} f(t)e^{-st}\,dt \qquad g(t) \leftrightarrow G(s)$$

1. Linearity, $\displaystyle\sum_{n=1}^{N} a_n f_n(t) \leftrightarrow \sum_{n=1}^{N} a_n F_n(s)$

2. Time Shift, $f(t - t_0) \leftrightarrow F(s)e^{-st_0}$

3. Frequency Shift, $e^{s_0 t} \leftrightarrow F(s - s_0)$

4. Scaling, $f(\alpha t) \leftrightarrow \dfrac{1}{|\alpha|} F(s/\alpha)$

5. Time Differentiation, $\dfrac{df}{dt} \leftrightarrow sF(s)$

$\dfrac{d^n f}{dt^n} \leftrightarrow s^n F(s)$

6. Frequency Differentiation, $(-t)f(t) \leftrightarrow \dfrac{dF}{ds}$

$(-t)^n f(t) \to \dfrac{d^n F}{ds^n}$

7. Time Integration, $\displaystyle\int_{-\infty}^{t} f(\tau)\,d\tau \leftrightarrow F(s)/s$

8. Time Convolution, $f(t) * g(t) = \displaystyle\int_{-\infty}^{\infty} f(\tau)g(t-\tau)\,d\tau \leftrightarrow F(s)G(s)$

9. Frequency Convolution,

$$f(t)g(t) \leftrightarrow \dfrac{1}{2\pi j}\int_{c-j\infty}^{c+j\infty} F(u)G(s-u)\,du = F(s) * G(s)$$

A comparison of these properties with those for the Fourier transform (Table 5.1) reveals that with the exception of 7 and 9 they are the same when $s = j\omega$ wherever s appears explicitly.

Also, 1, 3, 6, and 9 are the same (in form) for the unilateral Laplace transform.

G. BILATERAL z-TRANSFORM PROPERTIES

$f(n) \leftrightarrow F(z) = \displaystyle\sum_{n=-\infty}^{\infty} f(n)z^{-n} \qquad g(n) \leftrightarrow G(z)$

1. Linearity, $af(n) + bg(n) \leftrightarrow aF(z) + bG(z)$
2. Shifting, $f(n \pm m) \leftrightarrow z^{\pm m}F(z)$
3. Convolution, $f(n) * g(n) = \displaystyle\sum_{m=-\infty}^{\infty} f(m)g(n-m) \leftrightarrow F(z)G(z)$
4. Multiplication by n, $nf(n) \leftrightarrow -z\dfrac{d}{dz}F(z)$

$n^m f(n) \leftrightarrow \left(-z\dfrac{d}{dz}\right)^m F(z)$

G. BILATERAL z-TRANSFORM PROPERTIES

5. Division by $n + a$, $\quad \dfrac{f(n)}{n + a} \leftrightarrow -z^a \int \eta^{-a-1} F(\eta)\, d\eta$

6. Multiplication by a^n, $\quad a^n f(n) \leftrightarrow F(z/a)$

7. $\displaystyle\sum_{m=-\infty}^{n} f(m) \leftrightarrow \dfrac{F(z)}{1 - 1/z}$

8. $\displaystyle\sum_{m=-\infty}^{\infty} f(m) \leftrightarrow F(1)$

Identical forms exist for the unilateral z transform for 1, 4, 5, and 6.

Answers to Odd-Numbered Problems

CHAPTER 2

1. (a) $A = \dfrac{1}{\pi\tau}$; (b) $A = \dfrac{1}{\pi\tau}$; (c) $A = \dfrac{1}{2\tau}$; (d) $A = \dfrac{1}{\tau}$

3. An even, symmetrical, square wave

5. (a) $\delta[f(t)]\dfrac{df}{dt}$; (b) $f(0)\delta(r) + \dfrac{df}{dr}u(r)$; (c) $\cos(t)u(t)$; (d) $\delta(t) - \sin(t)u(t)$;

(e) 0; (f) $2u(t)$

7. $y(t) = e^t u(-t) + (2 - e^{-t})u(t)$

9. (a) time-invariant, $k = $ constant
(b) time-variable
(c) time-invariant, $\dfrac{dy}{dt} + y = 0$
(d) time-variable
(e) $\cos(t) \approx 1$ for $0 \le t \le 0.01$
$\dfrac{dy}{dt} + y = 0$, time-invariant for $0 \le t \le 0.01$

ANSWERS TO ODD-NUMBERED PROBLEMS

11. $\dfrac{d^2y}{dt^2} + 4y = x(t),\ 0 < y < 1$

$\dfrac{d^2y}{dt^2} + 2y = x(t) - 2,\ 1 < y < 2$

$\dfrac{d^2y}{dt^2} + 3y = x(t) - 1,\ 2 < y$

13. $\dfrac{d^2y}{dt^2} + 4y = x$, valid for small variations about $y = x = 0$

15. Proof
17. Proof

19. $\dfrac{d^3y}{dt^3} + 9\dfrac{d^2y}{dt^2} + 26\dfrac{dy}{dt} + 24y = x(t)$

21. $e^{-(1+j)t},\ e^{-(1-j)t},\ e^{-2t}$

23. $y_{zs}(t) = 7(e^{-2t} - e^{-3t})u(t) - 4te^{-2t}u(t)$

25. Proof

27. $\alpha = 1;\ \omega_n = 2;\ \xi = 1/2;\ \omega_d = \sqrt{3};\ t = \pi/\sqrt{3}$ (gives max. y, 2.103)

29. (a) linear; (b) linear; (c) nonlinear, y^2; (e) nonlinear, $\sin(y)\dfrac{dy}{dt}$;

(f) nonlinear, $y\dfrac{dy}{dt},\ y^2$; (g) nonlinear, superposition invalid

31. (a) $v(t) = 289(e^{-0.536 \times 10^3 t} - e^{-7.464 \times 10^3 t})u(t)$
(b) $v(t) = 250 \sin(2 \times 10^3 t)$
The response in (a) is entirely transient.
The response in (b) is entirely steady-state.

33. $y_1 = \cos(t) - 6\sin(3t)$
$y_2 = 1 - \sin(t) - 18\cos(t)$

35. $-2y'' - y' + \dfrac{1}{3}y = x$

37. $-2y'' - y' + \dfrac{1}{3}y = x_1 + 2x_2$

39. Proof

41. $LM\dfrac{d^3u}{dt^3} + (LD + MR)\dfrac{d^2u}{dt^2} + (L/K + RD + A^2)\dfrac{du}{dt} + (R/K)u = A\dfrac{dv_s}{dt}$

CHAPTER 3

1. (e) $y_1 = \displaystyle\int_{-\infty}^{t} [z_1 + 2z_2 + z'_1]\sin(t - \tau)\,d\tau$

$y_2 = \displaystyle\int_{-\infty}^{t} [z_1 + z_2 - z'_2]\sin(t - \tau)\,d\tau$

ANSWERS TO ODD-NUMBERED PROBLEMS 503

3. (e) $y_1 = \dfrac{2}{\sqrt{3}} \displaystyle\int_{-\infty}^{t} \left[z_2 + \dfrac{1}{2}z_1' + \dfrac{1}{2}z_2' \right] e^{-3(t-\tau)/2} \sin[\sqrt{3}(t-\tau)/2] \, d\tau$

$y_2 = \dfrac{2}{\sqrt{3}} \displaystyle\int_{-\infty}^{t} \left[\dfrac{3}{2}z_1 - \dfrac{1}{2}z_2 + \dfrac{1}{2}z_1' - \dfrac{1}{2}z_2' \right] e^{-3(t-\tau)/2} \sin[\sqrt{3}(t-\tau)/2] \, d\tau$

5. (e) $y_1 = \dfrac{1}{8} \displaystyle\int_{-\infty}^{t} [3z_1 - 3z_3 - 3z_2' + z_1''][-2 + e^{-2(t-\tau)} + e^{2(t-\tau)}] \, d\tau$

$y_2 = \dfrac{1}{8} \displaystyle\int_{-\infty}^{t} [-4z_2 + z_1' - z_3' - z_2''][-2 + e^{-2(t-\tau)} + e^{2(t-\tau)}] \, d\tau$

$y_3 = \dfrac{1}{8} \displaystyle\int_{-\infty}^{t} [z_3 - z_1 + z_2' + z_3''][-2 + e^{-2(t-\tau)} + e^{2(t-\tau)}] \, d\tau$

7. $y_1(t) = 5\sin(t) + 3\cos(t) - 2,\ t \geq 0$
   $y_2(t) = 4\sin(t) - \cos(t) + 1,\ t \geq 0$

9. $y_1(t) = \dfrac{1}{3}\cos(2t + 216.87°)$

   $y_2(t) = 0.637\cos(2t + 173.99°)$

11. $y_1(t) = u(t)\left[-3t + \dfrac{1}{2}e^{-2t} - \dfrac{1}{2}e^{2t} \right]$

    $y_2(t) = u(t)\left[-\dfrac{3}{4} + \dfrac{t^2}{2} + \dfrac{3}{8}e^{-2t} + \dfrac{3}{8}e^{2t} \right]$

    $y_3(t) = u(t)\left[t + \dfrac{1}{4}e^{-2t} - \dfrac{1}{4}e^{2t} \right]$

13. Proof

15. $i_L = \dfrac{1}{\lambda_1 - \lambda_2} \displaystyle\int_{-\infty}^{t} [8i_1' + 4i_1 + 4i_2][e^{\lambda_1(t-\tau)} - e^{\lambda_2(t-\tau)}] \, d\tau$

    $v_C = \dfrac{1}{\lambda_1 - \lambda_2} \displaystyle\int_{-\infty}^{t} [20i_2' + 20i_1' + 160i_2][e^{\lambda_1(t-\tau)} - e^{\lambda_2(t-\tau)}] \, d\tau$

    The last result agrees with equation 2.75. The state-variable method is faster.

17. $i_L(t) = 0.417 \times 10^{-3} - 1.042 \times 10^{-3} u(t) + 1.077 \times 10^{-3} e^{-10.42t} \cos(39.47t - 14.79°) u(t)$

19. Same result as in Problem 1

21. $y_1(t + h) = y_1(t) - 0.05 y_2(t)$
    $y_2(t + h) = -0.025 y_1(t) + 0.975 y_2(t) + 0.025 u(t)$

23. $y_1' = 0 + y_2 + 0$

    $y_2' = -\dfrac{K}{M} y_1 - \dfrac{D}{M} y_2 + \dfrac{1}{M} u(t),\ x_2 = \dfrac{1}{M} u(t)$

25. (a) $q_1(t) = e^t - 1$
        $q_2(t) = 1 + e^t - 2e^{t/2}$

    (b) $q_1(t) = \dfrac{e^t}{2}[\cos(t) + \sin(t)] - \dfrac{1}{2}$

        $q_2(t) = \dfrac{e^t}{2}[\sin(t) - \cos(t)] + \dfrac{1}{2}$

    (c) $q_1(t) = \dfrac{1}{3}(e^{3t} - 1)$

        $q_2(t) = 0$

504 ANSWERS TO ODD-NUMBERED PROBLEMS

(d) $q_1(t) = \dfrac{1}{2(1+\sqrt{6})}[e^{(1+\sqrt{6})t} - 1] + \dfrac{1}{2(1-\sqrt{6})}[e^{(1-\sqrt{6})t} - 1]$

$q_2(t) = \dfrac{1}{\sqrt{6}(1+\sqrt{6})}[e^{(1+\sqrt{6})t} - 1] - \dfrac{1}{\sqrt{6}(1-\sqrt{6})}[e^{(1-\sqrt{6})t} - 1]$

27. There is no value of K that makes the system stable.

CHAPTER 4

1. (a) $H(\omega) = 0.943 \underline{|-45°}$; (b) $v_C(t) = 94.273 \cos(10^3 t - 45°)$

3. (a) $H(\omega) = \dfrac{1}{\left[\dfrac{1}{R^2} + \left(\omega C - \dfrac{1}{\omega L}\right)^2\right]^{1/2}} \underline{\left|-\tan^{-1}\left(\dfrac{\omega C - \dfrac{1}{\omega L}}{1/R}\right)\right.}$

(b) $v(t) = 10^4 \cos(10^3 t)$
(c) Underdamped

5. (a) $f(t) \approx \dfrac{1}{2}\sin(2\pi t)$

(b) $f(t) \approx \sqrt{2}\sin(2\pi t) - 0.75\sin(4\pi t)$

(c) $f(t) \approx \dfrac{2}{3}\sin(2\pi t) - \dfrac{2}{3\sqrt{3}}\sin(4\pi t) + \dfrac{1}{6}\sin(6\pi t)$

7. $f(t) = \dfrac{1}{2} - \sum_{n=1}^{\infty} \dfrac{1}{n\pi}\sin(n\pi t)$

9. $f(t) = \sum_{\substack{n=1 \\ (odd)}}^{\infty} \dfrac{40}{n\pi}(-1)^{(n-1)/2}\cos\left(\dfrac{n\pi t}{20}\right)$

11. (a) $a_n = \dfrac{1}{\pi}\int_0^{2\pi} x^2\cos(nx)\,dx = 4/n^2,\ n \ne 0$

13. $\varepsilon_{RMS} = 0.010 V_m$

15. $v_0(t) = \dfrac{V_m}{\pi}\dfrac{R_2}{R_1 + R_2} - \dfrac{2V_m R_2}{\pi}\sum_{n=1}^{\infty}\dfrac{\cos(n\pi/2)}{n^2 - 1}\dfrac{\cos(n\omega_T t + \theta_n)}{\sqrt{(R_1 + R_2)^2 + (n\omega_T CR_1 R_2)^2}}$

$\theta_n = -\tan^{-1}\left(\dfrac{n\omega_T CR_1 R_2}{R_1 + R_2}\right)$

17. (a) $h + \dfrac{dh}{dt} = \delta(t),\ i_{zi}(t) = e^{-t},\ h(t) = e^{-t}u(t)$

(b) $i(t) = \displaystyle\int_{-\infty}^{t} u(\tau)e^{-(t-\tau)}\,d\tau = u(t)(1 - e^{-t})$

19. Graph

21. (b) $C = \dfrac{1}{\omega R} = 159.2\ \text{pF}$

23. Graph

25. 4.18, $f_{rms} = 1/\sqrt{3}$
4.19, $f_{rms} = 1/\sqrt{3}$
4.20, $v_{rms} = V_m/2$

ANSWERS TO ODD-NUMBERED PROBLEMS 505

4.21, $v_{rms} = V_m/\sqrt{2}$
4.22, $f_{rms} = 1/(2\sqrt{2})$
4.24, $v_{rms} = 1/\sqrt{2}$
27. $\langle P \rangle = 7.5(10) = 75w$
    $W = 360(75) = 0.27 \times 10^6 j$
29. Graph
31. The *average* error might be small due to the cancellation of large positive and negative errors. This cannot happen with rms error.

CHAPTER 5

1. Proof
3. Proof
5. Proof
7. Proof

9. (a) $u(t) - \dfrac{1}{2} = \dfrac{1}{2} sgn(t)$

   (b) $1 - u(t) = u(-t)$

   (c) $\dfrac{1}{\pi} \cos(t)$

   (d) $\dfrac{A}{\tau}(1 - |t|/\tau)[u(t + \tau) - u(t - \tau)]$

   (e) $\dfrac{1}{2}(e^{-t} - e^{-3t})u(t)$

11. (a) $v(t) = 288.7(e^{-at} - e^{-bt})u(t)$
    (b) $v(t) = 250 \sin(2000t)$
13. The transform technique is not helpful (generally speaking) in time-variable systems.
15. See Example 12, Chapter 2
17. See Problem 9, Chapter 3
19. The current due to the 2 v battery is $2/(4800) = 0.417$ mA. The current due to the $5u(t)$ v source is

    $i_L(t) = -1.042 \times 10^{-3} u(t) + 1.077 \times 10^{-3} e^{-10.42t} \cos(39.47t - 14.79°)u(t)$

    The total current is sum of the two.

21. $y(x) = \mathscr{F}^{-1}\left\{\dfrac{G(\eta)}{1 - H(\eta)}\right\}$

23. $F(\omega) = \pi A_m [\delta(\omega - \omega_0) + \delta(\omega + \omega_0)] + \dfrac{\pi m_a A_m}{2}[\delta(\omega - \omega_0 - \omega_m) + \delta(\omega + \omega_0 + \omega_m)]$
    $+ \dfrac{\pi m_a A_m}{2}[\delta(\omega - \omega_0 + \omega_m) + \delta(\omega + \omega_0 - \omega_m)]$

25. Proof

27. $E_\theta = \dfrac{j\omega\mu\, e^{-jkr}}{4\pi\, r} \sin(\theta) \mathscr{F}^{-1}\{I_z(z')\}|_{\eta = -k\cos(\theta)}$

    where η is the transform variable.

506 ANSWERS TO ODD-NUMBERED PROBLEMS

29. $f(0) = 3/2 \ (n = 0)$
$f(1/2) = -1/2 \ (n = 4)$

31. (a) $y(t) = t[u(t) - u(t-2)] - (t-4)[u(t-2) - u(t-4)]$

(b) $y(n) = \dfrac{1}{2} \sum_{m=0}^{7} x(n-m)x(m)$

(c) $y(n) = \dfrac{1}{4} \sum_{m=0}^{15} x(n-m)x(m)$

(d) Results are better for large N. If $N \to \infty$ with $NT = 4$, then exact agreement with the continuous case occurs.

33. The output is identical to the input except for a scale change K and a time delay τ.

35. $F(\omega) \approx \dfrac{1}{\omega^2 T} \{ [f(T) - f(0)][1 - e^{-j\omega T}] + \cdots$
$+ [f(NT) - f(NT - T)][e^{-j(N-1)\omega T} - e^{-jN\omega T}] \}$

37. Proof

CHAPTER 6

1. Proof

3. (a) $\dfrac{5}{s-3}$

(b) $\dfrac{5}{s^2} - \dfrac{3}{s}$

(c) $\dfrac{e^{-s}}{s^2} + 2\dfrac{e^{-s}}{s}$

(d) $\dfrac{1}{s^2+1} - \cos(1)\dfrac{e^{-s}}{s^2+1} - \sin(1)\dfrac{se^{-s}}{s^2+1}$

(e) $\dfrac{1}{(s+1)^2} + \dfrac{1}{s+1}$

(f) Does not exist.

(g) $\dfrac{2e^{-s}}{s^3} + \dfrac{2e^{-s}}{s^2} + \dfrac{2e^{-s}}{s}$

(h) $-\dfrac{1}{s}\dfrac{d}{ds}\left(\dfrac{1}{s^2+1}\right) = \dfrac{2}{(s^2+1)^2}$

5. For 3. (a) unstable (b) unstable (c) unstable
     (d) stable   (e) stable   (f) unstable
     (g) unstable (h) unstable

For 4. (a) stable   (b) unstable
     (c) unstable (d) marginally stable (unstable)
     (e) stable   (f) stable
     (g) stable   (h) unstable

7. (a) $\dfrac{y_{tr}(t)}{V_m/\pi} = -1.811e^{-300t}$

(b) 0

ANSWERS TO ODD-NUMBERED PROBLEMS   507

9. (a) $\tan^{-1}(1/s)$
   (b) $\tan^{-1}(a/s)$
   (c) Proof

11. (a) $\dfrac{1}{\sqrt{s^2+1}}$

   (b) $\dfrac{1}{\sqrt{s^2+a^2}}$

   (c) Proof

   (d) $\dfrac{1}{(s+\sqrt{s^2+1})\sqrt{s^2+1}}$

   (e) $\dfrac{1}{\sqrt{s^2-1}}$

13. (a) 1
    (b) 0
    (c) FVT does not apply.
    (d) 0

15. $\pi \dfrac{1+e^{-s}}{s^2+\pi^2}$

17. Proof
19. Proof

21. $y(t) = \mathcal{L}^{-1}\left\{\dfrac{F(s)}{(1-H(s))}\right\}$

23. (a) $F(s) \approx \dfrac{T}{2}[f(0^+) + f(NT)e^{-NsT}] + T\sum_{n=1}^{N-1} f(nT)e^{-nsT}$

    (b) Proof
25. Proof

CHAPTER 7

1. (a) $H(s) = \dfrac{s+1/(\alpha\tau)}{s+1/\tau}$;   (b) $h(t) = \delta(t) - \dfrac{1}{\tau}(1/\alpha - 1)e^{-t/\tau}u(t)$

3. $H(s) = \dfrac{[s+1/(\alpha_1\tau_1)][s+1/(\alpha_2\tau_2)]}{(s+1/\tau_1)(s+1/\tau_2)}$

5. (a) See Example 8, Chapter 2
   (b) See Example 9, Chapter 2
7. Proof
9. See Problem 6, Chapter 3

11. (b)

Q_2	0	0.25	$\sqrt{2}-1$	0.5	0.75	1.0	
τ	0.1099	0.0768	0.0739	0.0705	0.0593	0.0470	μsec

   (c)

Q_2	0	0.25	$\sqrt{2}-1$	0.5	0.75	1.0	
$\tau\beta$	0.3497	0.3456	0.4049	0.4034	0.3485	0.2719	β in Hz

13. (a) $H(s) = \dfrac{Y_0(s)}{Y_1(s)} = \dfrac{-Ms^2}{Ms^2+Ds+K} = -\dfrac{s^2}{s^2+(D/M)s+K/M}$

15. $v(t)_{min} = -25,770$ V at $10^5 t = \tan^{-1} 10 = 1.4711$
The inductor does not survive.

17. See Example 12

19. (a) $\theta = \tan^{-1}\left(-\dfrac{1}{\omega RC}\right)$

(b) $\dfrac{V_0}{RC} - \dfrac{\omega V_m \sin(\theta)}{RC\left(\omega^2 + \dfrac{1}{R^2C^2}\right)} - \dfrac{V_m \cos(\theta)}{R^2C^2\left(\omega^2 + \dfrac{1}{R^2C^2}\right)} = 0$

that must be solved for θ.

21. (a) $H(s) = \dfrac{MR_2}{R_1L_2 + R_2L_1} \dfrac{s}{s + \dfrac{R_1R_2}{R_1L_2 + R_2L_1}}$

(b) $v_2(t) = \dfrac{10Aa^2}{a^2 + \omega^2}\left[e^{-at} - \dfrac{\omega}{a^2}\sqrt{a^2 + \omega^2}\sin(\omega t - \theta)\right]$, $\theta = \tan^{-1} b/a$

(c) $v_2(t) = 10A \displaystyle\int_0^t \cos(\omega \tau)[\delta(t-\tau) - ae^{-a(t-\tau)}]\, d\tau$

This gives the same answer.

23. $v_0(t) = V_0[e^{-10^6 t/2}u(t) - e^{-10^6(t-\tau)/2}u(t-\tau)$
$+ e^{-10^6(t-T)/2}u(t-T) - e^{-10^6(t-\tau-T)/2}u(t-\tau-T) + \cdots]$
The circuit differentiates.

25. $v_s(t) = 7.5u(-t) + [7.5 + 22.5(1 - e^{-1})e^{-4(t-1)}]u(t-1)$
$v_s(t)$ becomes an impulse at $t = 1$ if $R \to \infty$.

27. (a) $V_2(s) = \dfrac{1/LC^2}{s^3 + 2(G/C)s^2 + (G/C + 2/LC)s + 2G/LC^2} I_s(s)$

(b) $H(s) = \dfrac{1}{LC^2 s^3 + 2GLC s^2 + (2C + G^2 L)s + 2G}$

29. (a) $i(t) = 0.289[e^{-t} + 1.0494 \sin(\pi t - 72.34°)]u(t) + 0.289[e^{-(t-1)} - 1.0494 \sin(\pi t - 72.34°)]u(t-1)$

(b) $i(t) = \dfrac{1}{4}\displaystyle\sum_{n=1}^{3} \sin(n\pi/4)e^{-(t-n/4)}u(t-n/4)$

31. Stable systems are (b) and (c).

33. $y(t) = t^2 u(t) + \dfrac{t^4}{12}u(t)$

35. Same as Problem 41, Chapter 2, with p replaced by s.
37. Block diagrams
39. Plots
41. Plots

43. (b) $\dfrac{d^2 c}{dt^2} + (\alpha + \beta)\dfrac{dc}{dt} + (\alpha\beta + 1)c = \dfrac{dr}{dt} + r\beta$

45. Block Diagram

CHAPTER 8

1. (a) discrete; (b) continuous; (c) neither, quantized; (d) neither, quantized; (e) neither, quantized; (f) continuous

3. $f(n) = 2\delta(n) + \delta(n-1) + 3\delta(n-2) + 2\delta(n-4) + 4\delta(n-5)$

ANSWERS TO ODD-NUMBERED PROBLEMS 509

5. Plots
7. $y(6) = 1.732051 \quad \sqrt{3} = 1.732051$
9. Proof
11. (a) $y_{zi}(n) = D_1(1/3)^n + D_2(1/2)^n$
    (b) $h_f(n) = 6[(1/2)^{n-1} - (1/3)^{n-1}]u(n-1)$
    (c) $h(n) = h_f(n+2) = 6[(1/2)^{n+1} - (1/3)^{n+1}]u(n+1)$
    (d) $y(n) = 6 \sum_{m=0}^{n+1} m[(1/2)^{n-m+1} - (1/3)^{n-m+1}] + C_1(1/2)^n + C_2(1/3)^n$

13. $h(n) = 2(1 - 3^{n+1})u(n+1) - \frac{1}{2}(1 - 3^n)u(n)$

15. (a) $y(n) = \frac{1}{\sqrt{5}}\left[\left(\frac{1+\sqrt{5}}{2}\right)^{n+1} - \left(\frac{1-\sqrt{5}}{2}\right)^{n+1}\right] = y_{zi}(n)$

    (b) Not stable: $y_{zi} \to \infty$ as $n \to \infty$.
    (c) Repeated substitution gives the result.
17. $h_f(n) = -[(-1)^{n-1} - (-2)^{n-1}]u(-n)$
19. Block diagram
21. Plots
23. (a) $h_f(n) = u(n-1)$
        $h(n) = Tu(n)$
    (b) Not stable

25. (a) $h_f(n) = (-1)^{n-1}u(n-1), \; h(n) = \frac{1}{T}(-1)^n u(n)$

    (b) Not stable.

27. $\begin{bmatrix} q_1(n+1) \\ q_2(n+1) \end{bmatrix} = \begin{bmatrix} 4 & -3 \\ 1 & 0 \end{bmatrix}\begin{bmatrix} q_1(n) \\ q_2(n) \end{bmatrix} + \begin{bmatrix} 1 & 0 \\ 0 & 0 \end{bmatrix}\begin{bmatrix} x_1(n) \\ x_2(n) \end{bmatrix}$

    $[y(n)] = [0 \;\; 0]\begin{bmatrix} q_1(n) \\ q_2(n) \end{bmatrix} + [0 \;\; 5]\begin{bmatrix} x_1(n) \\ x_2(n) \end{bmatrix}$

29. (a) $h(n) = \delta(n) + \left[\frac{1}{2}(3)^{n+1} - \frac{1}{2}\right]u(n-1)$ for $x_1(n)$

    (b) $y_{zs}(n) = x_1(n) + \frac{1}{2}\sum_{m=0}^{n-1} x_1(m)[(3)^{n-m+1} - 1]$ for $x_1(n)$

31. $y_{zs}(n) = \sum_{m=0}^{n-1} \begin{bmatrix} (1-2m)2^{n-m-1} & -\frac{m}{2}(3^{n-m-1} - 2^{n-m-1}) \\ -(1-2m)2^{n-m-1} & +\frac{m}{2}(3^{n-m-1} - 2^{n-m-1}) \end{bmatrix} + \begin{bmatrix} \frac{1}{2}x_2(n) \\ 0 \end{bmatrix}$

33. $y(n+1) = \frac{1}{2654}[x(n) + 5003y(n) - 2534y(n-1)]$

35. (a) $y(n) = y(n-2) - x(n+1) + x(n)$
        This system is anticipatory since the output depends on a future value of the input (i.e. $x(nT + T)$)
    (b) No.

CHAPTER 9

1. (a) $\dfrac{z(z^2 + 4z + 1)}{(z-1)^4}$

(b) $e^{a/z}$

(c) $\dfrac{z/a}{(z/a - 1)^2}$

(d) $\dfrac{(z/a)^2 + z/a}{(z/a - 1)^2}$

(e) $\dfrac{z}{z - e^{\theta}}$

(f) $\dfrac{z}{(z - e^{\theta})^2}$

(g) $1 + 1/z$

(h) $\dfrac{F(z)}{z} - \dfrac{dF}{dz}$

(i) 1

3. Proof

5. (a) $(a_2 z^2 + a_1 z + a_0) Y(z) = X_f(z)$
   (b) $(a_2 z^2 + a_1 z + a_0) Y(z) = y(0)(a_2 z^2 + a_1 z) + y(1) a_2 z + X_f(z)$

7. (a) $a^{n-1} u(n - 1)$
   (b) $(n + 1) a^n u(n)$
   (c) $(n - 1) a^{n-2} u(n - 1)$
   (d) $n a^{n-1} u(n)$
   (e) $n b^{n-1} u(n)$
   (f) $-(n - 1) f(n - 1) u(n - 1)$
   (g) $(2^{n-1} - n) u(n - 1)$
   (h) $\dfrac{n(n - 1) \cdots (n - m + 1)}{m!} a^{n-m} \left[u(n) - \dfrac{n - m}{n} u(n - 1) \right]$

9. (a) $f(0) = 1$; (b) $f(0) = 0$; (c) $f(0) = 1$; (d) $f(0) = 0$;
   (e) $f(0) = 0, m > 0; f(0) = \lim_{z \to \infty} F(z), m = 0$; (f) $f(0) = 0, m > 0; f(0) = 1, m = 0$

11. $Y_{zs}(z) = \dfrac{5(z^2 - 4z + 3) X_2(z) + X_1(z)}{z^2 - 4z + 3} - \dfrac{5z[z x_2(0) + 4 x_2(0) + x_2(1)]}{z^2 - 4z + 3}$

13. (a) $(1.05)^n u(n)$
    (b) $\delta(n) + 20[(1.05)^{n+1} - 1] u(n - 1)$
    (c) $20[20(1.05)^{n+1} - n + 21] u(n - 1)$

15. $H_d(z) = \dfrac{T^2}{4} \left(\dfrac{z + 1}{z - 1} \right)^2$

    $y_{zs}(n) = \dfrac{T^2}{4} [\delta(n) + 5n^2 u(n) + (4n^2 - 19n + 15) u(n - 1)]$

17. (a) $y(n) = -x(n - 1) + x(n - 3) + y(n - 1)$, not anticipatory.
    (b) $y(n) = -x(n + 1) + x(n) + y(n - 2)$

    The inverse system is anticipatory.

19. $h(n) = \dfrac{2}{T} (-1)^n [u(n) - u(n - 1)]$

21. (a) $H_d^d(z) = \dfrac{1}{T^2} (1 - z^{-1})^2$

    $y(n) = \dfrac{1}{T^2} [x(n) - 2x(n - 1) + x(n - 2)]$

(b) $y(10) = -78.27$, $y'' = -100 \sin(1) = -84.15$
(c) $y(100) = -83.6$, $y'' = -84.15$
23. Plots
25. Proof
27. (a) $y(t) = \left(\dfrac{1}{2} - e^{-t} + \dfrac{1}{2}e^{-2t}\right)u(t)$

(b) $y(n) = \dfrac{1}{462}[u(n) + 2u(n-1) + u(n-2) + 796y(n-1) - 342y(n-2)]$

$y(10) = 0.2055$ ($t = 1$ sec.)
In (a) $y(1) = 0.1998$ ($t = 1$ sec.)

29. $y(n) \to 0$, $n \to \infty$
31. Plots
33. $H_d(z) = \dfrac{1 + 2z^{-1} + z^{-2}}{5.04 - 6z^{-1} + 4.96z^{-2}}$

35. $y(n) + 0.3y(n-1) - 0.18y(n-2) = 0.9x(n) - 0.54x(n-1)$
37. Stable for $|\alpha| < 1$, $0 < \alpha < 1$.
39. Yes poles at $z = 1/2, 1/3, 1/4$

41. $\begin{bmatrix} q_1(n+1) \\ q_2(n+1) \\ q_3(n+1) \end{bmatrix} = \begin{bmatrix} 1/2 & 0 & 0 \\ 0 & 1/3 & 0 \\ 0 & 0 & 1/4 \end{bmatrix} \begin{bmatrix} q_1(n) \\ q_2(n) \\ q_3(n) \end{bmatrix} + \begin{bmatrix} 1 \\ 1 \\ 1 \end{bmatrix} x(n)$

$y(n) = \begin{bmatrix} 6 & -4/3 & -3 \end{bmatrix} \begin{bmatrix} q_1(n) \\ q_2(n) \\ q_3(n) \end{bmatrix} + 2x(n)$

43. $\begin{bmatrix} q_1(n+1) \\ q_2(n+1) \\ q_3(n+1) \end{bmatrix} = \begin{bmatrix} 13/12 & -3/8 & 1/24 \\ 1 & 0 & 0 \\ 0 & 1 & 0 \end{bmatrix} \begin{bmatrix} q_1(n) \\ q_2(n) \\ q_3(n) \end{bmatrix} + \begin{bmatrix} 1 \\ 0 \\ 0 \end{bmatrix} x(n)$

$y(n) = \begin{bmatrix} 19/6 & -5/4 & 1/12 \end{bmatrix} \begin{bmatrix} q_1(n) \\ q_2(n) \\ q_3(n) \end{bmatrix} + 2x(n)$

Index

Accelerometer, 324
Additivity, 4
Aliased coefficients (DFS), 157
Aliasing error, 154, 157, 209, 211, 216, 369, 442
Amortization, 359, 395
Analog, 9
Analagous systems, 488
Analog-to-digital converter, 333
Angular acceleration, 493
Anticipatory system, 12, 13
Arithmetic-geometric series, 354, 484
Artificial transmission line, 434, 444
Associative property of convolution, 67
Asymptotes, 193
Augmented zeros, 216, 220, 228
Auto-correlation
    function, 231
    integral, 174, 231
Average
    error, 165
    power, 145, 148, 229
    value of signal, 132, 175

Back substitution, 481
Balanced modulator, 5, 174, 202, 401
Bandwidth, 165, 196, 199, 204
Bessel function, 23, 131, 269, 272, 273
    modified, 269, 272, 273
Bessel's equation, 16, 45
Binary digit (bit), 213
Binomial series, 191, 192, 485
Block diagram reduction, 318
Bode diagram, 192
Boundary conditions, 361
Boundedness, 71, 134, 308, 354
Break frequency, 193
Butterfly, 227

Cadzow, J. A., 399, 474
Campbell and Foster tables, 269

Canononical form, feedback system, 309, 418
Capacitor
current, 24
voltage, 23, 84
Carslaw, H. S., and Jaeger, J. C., 278
Causal system, 12
Cayley-Hamilton theorem, 104
Characteristic
equation, 25, 40, 45, 89, 186, 307, 311, 346, 379, 421
impedance, 301
Cheng, D. K., 278, 331
Coefficient of reflection, 301
Combinations, 358
Commutative property of convolution, 67
Compensator
lag, 322
lead, 321
lead-lag, 322
Complementary
error function, 269, 272
function, 19, 342
Complete response, continuous-time, 44, 47, 123
summary, 48, 73
Complete response, discrete-time, 346
summary, 346, 393
Complex
conjugate, 147, 174, 265
Fourier series, 136
Compliance, 489
Conformable, 86, 476
Conservation
of charge, 295, 296
of energy, 42, 294, 300
Continuous spectrum, 160, 168
Continuous-time system, 8, 9
Convergence, 132
factor, 244
strip, 244
Convolution
circular, 220, 375
discrete, 333, 339
integral, 46, 66, 172
Cooley-Tukey formulation (FFT), 225, 241
Corner frequency, 193

Correlation integral, 174
auto, 174
cross, 174
Cramer's rule, 64, 479
Critically damped system, 59, 272
Current source, 23
Cutoff frequency, 12

D'Alembert's principle, 489
Damped natural frequency, 58
Damping
coefficient, 58, 489, 493
ratio, 58
torque, 493
Decade, 193
Decibel (dB), 193
Del Toro, V., 80
Difference equations, 12, 333, 342, 420
Differential-difference equation, 469
Differential equations, 11
linear, 14, 15, 16, 17, 185
nonlinear, 15, 16
partial, 14
Differentiator
analog, 327
digital, 424, 427, 428, 429, 439, 470
Diffusivity, 305
Digital
computer, 9
filter, 368
Digitalization, 368
Digital-to-analog converter, 333
Dirac delta function, 26
Direct programming method, 464
Dirichlet conditions, 134, 170
Discrete
Fourier series (DFS), 151, 153
Fourier transform (DFT), 213
spectrum, 160, 168
state-transition equation, 377
transfer function, 221
transition matrix, 377
Discrete-time system, 8, 9, 220, 332
Dispersion, 304
Di Stefano, J. J., III, Stubberud, A. R., and Williams, I. J., 241
Distortionless system, 240, 304
Distributed system, 301

Distribution, 26
Dorf, R. C., 80, 331
Double-sideband, suppressed carrier (DSB) signal, 202
Dual
 equations, 57, 490
 networks, 57
Duality, 171, 214
Dynamic system, 2

Economics, 437
Effective value, 149
Eigenvalues (characteristic values), 45, 89, 379
Eigenvector, 88, 379
Elements (of fundamental set), 40
Energy
 signal, 174, 229
 spectral density, 229, 233
 state, 82
 stored, 18
Error function, 269, 272
Extension of the Fourier series, 167
Euler formulas, 132
Even function, 128, 138

Fast Fourier transform (FFT), 224
Feedback, 307
Fibonacci sequence, 396
Final value theorem, 258, 408
Finite duration impulse response (FIR) filter, 355
First difference method, 452
Fixed system, 10
Flux linkage, 491
Force
 current analog, 490, 492, 494
 voltage analog, 490, 492, 494
Forced response, 19
Forcing function, 17, 342
 vector, 86
Fourier
 integral, 169
 inverse transform, 169
 transform, 169, 171–182
 transform pairs, 171
 transform pair table, 178
 trigonometric series, 132, 135, 167
Free response, 19

Frequency
 corner, 193
 doubler, 145, 163
 fundamental, 145
 response, 193, 365, 438
 shifting, 171
Fundamental set, 40, 87

Gabel, R. A., and Roberts, R. A., 121, 399, 474
Gamma function, 270, 275
Gaussian elimination, 480
Gears, 494
Geometric series, 360, 484
Goldman, S., 331

Half-power, 165, 198
 frequencies, 198
Hamming window, 374, 440
Harmonics, 144, 193
Harrington, R. F., 165
Hayt, W. H., Jr., and Kimmerly, J. E., 80, 121, 241, 331
Heat conduction equation, 239, 305
Heat transfer, 436
Heaviside expansion theorem, 263
Helmholtz integral, 46
High-pass filter, 204, 238, 398
Homogeneity, 4
Homogeneous equation, 19, 20
Hooke's law, 58, 498

Identity matrix, 89, 477
Impedance, 126
Improper fraction, 262
Impulse invariant (IIR) filter, 368
Inductor
 current, 21, 84
 voltage, 21
Inertia
 force, 489
 torque, 493
Inhomogeneous equation, 19
Initial condition, 6, 22, 280, 420, 430
Initial value theorem, 256, 408
Instantaneous system, 2
Integral equation, 239, 277, 329
Integrals, table of, 483

Integrator
  analog, 326
  digital, 424, 427, 469, 470
Interest, 357, 359
Interpolation
  formulas, 211
  function, 211
Inverse
  discrete Fourier transform, 214
  Fourier transform, 169
  Laplace transform, 244, 248
  z transform, 403
Iterative programming method, 463

Jackson, J. D., 80
James, M. L., Smith, G. M., and Wolford, J. C., 121

Kirchhoff's law, 2, 7, 18, 24, 294, 490, 491
Kronecker delta (unit function), 222, 336

Laplace transform, 200, 242
  bilateral, 244, 497
  inverse (bilateral), 244
  inverse (unilateral), 248
  properties, 250–253, 270
  table of pairs, 270
Legendre function, 131
Linear
  difference equation, 12, 333, 342
  differential equation, 11, 14, 17
  system, 2, 3, 8, 15, 185, 221
  term, 15
Linearity, 2–9
Logarithmic scale, 192
Long division, 416
Loudspeaker, 80, 496
Low-pass filter, 12, 141, 193, 202, 205, 370, 371, 398, 440, 447

McGillem, C. D., and Cooper, G. R., 80, 241, 331, 399, 474
Magnetic coupling, 298
Magnetic flux density, 495–496
Mason's gain formula, 311
Matching points, 129
Matrix, 63, 82, 347, 475
  adjoint, 478
  cofactor, 478
  diagonal, 108, 110, 132, 383
  dimension, 85
  identity, 89, 477
  inversion, 154, 477
  modal, 108, 382
  symmetric, 477
  system, 37
  transpose, 477
Mechanical system
  rotational, 493
  translational, 489
Modes (fundamental set), 40
Moment, 28
Moment of inertia, 493
Motional emf, 495
Motor (dc), 315, 495
  armature controlled, 315, 317, 322
  field controlled, 330
Multiple input-output, 4, 18, 62

Natural response, 19
Natural state-variables, 457
Neff, H. P., Jr., 13, 80, 331, 495
Newton's law, 58, 489
Noise, 317
Noncausal system, 12, 13, 201
Nonlinear system, 16, 395
Nonrecursive (transversal) system, 352, 355
Normal form, 82, 85, 375
  block diagrams, 96
Normalizaton, 109, 384
Numerical methods, 82, 113, 224
Nyquist interval, 209

Octave, 193
Odd function, 134, 138
Ohm's law, 3, 15
Operational amplifier, 324
Operator
  p, 52
  system, 3
Orthogonality, 131, 151
Overdamped system, 42, 59, 272

p operator, 52
Papoulis, A., 165, 241, 399, 474

INDEX 517

Parallel programming method, 458
Parseval's theorem, 174
Partial fraction expansion, 263
Particular
  integral, 19
  solution, 19, 342, 351
Peak
  response, 61
  time, 61
Peebles, P. Z., Jr., 80
Percent overshoot, 61
Period, 129, 133
Periodic
  excitation, 122, 127, 189
  extension, 134
  function, 127, 132, 177
  sequence, 153, 215, 468
Phase-lock loop, 325
Phase shifter, 200
Phasor, 55, 122, 124, 186
Piecewise linear system, 75
Point matching, 129, 155
Poisson's
  equation, 15, 16
  sum formula, 456
Polar form (phasor), 126
Poles, 245
Position index, 342, 360
Power series, 103, 485
Power signal, 229
Power spectral density, 230
Power supply, 141
Proper fraction, 263

Quality factor (Q), 145, 198
Quantized system, 9, 213

Radius of convergence, 403, 404
Random signals, 229
Reciprocity theorem, 65
Rectangular window, 373
Rectified cosine signal, 122, 141, 162
Recursive relation, 344
Repeated roots, 45, 70, 112, 266, 352, 363, 381
Resonance, 144
Response to periodic excitation, 141
Reversed bit order, 226
Ripple, 122, 143

Rise time, 61, 194
  bandwidth product, 194
Root loci, 70, 308, 309, 311
Root-mean-square (rms), 124
  error, 136, 137, 139, 151
Roots of the characteristic equation, 70, 346
Rotational mechanical system, 493
Routh criterion, 309, 485
Runge-Fox, 116, 120

s plane, 71
Samples, 10, 152
Sampling
  frequency, 154, 209, 369
  impulse, 210
  nonuniform, 10
  property of the unit function, 337
  property of the unit impulse, 28
  theorem, 154, 207, 332, 387
  uniform, 10
Samuelson, P. A., 397
Savings account system, 357, 395, 469
Scaling, 4, 20, 171, 250
Second-order
  differential equation, 24
  system, 24, 57, 194, 362
Sensitivity, 313
Series, 484
Settling time, 61
Shift-invariant discrete system, 12, 220, 339
Shift register, 8
Shunt peaking network, 164, 323, 324
Signal, 9
Signum function, 175, 176
Similarity transformation, 108
Simpson's rule for integration, 472
Simulation, 387, 472, 473
  of differential equations, 388
  digital, 387, 390, 424, 444
  frequency transformation, 449, 472
  time sampling, 390, 472
Simultaneous equations, 63, 290
Sine integral, 185, 268
Single sideband
  signal (SSB), 203
  system, filter method, 203, 204
  system, phase method, 205

518 INDEX

Smoothness, 154
Source
  function, 17
  function vector, 86
Spectral analysis, 230
Spectrum
  continuous, 160, 168
  discrete, 160, 168
Spiegel, M. R., 80, 331
Spring torque, 493
Square wave, 129
Stability
  continuous-time, 70, 307, 312, 438
  discrete-time, 392, 473
  marginal, 71
Stable system
  continuous, 70, 258, 312
  discrete, 392
State
  space analysis, 81, 319, 375, 456
  transition matrix, 103
  variable, 81, 319, 375, 456
  vector, 85
Stationary system, 10
Steady-state solution, 19, 26
Stiffness, 489
Stremler, F. G., 241
Superposition
  of impulses, 28
  integral, 15, 46
  principle, 4
  of sinusoids, 127
Symmetry, 138, 171
Sync signal, 326, 327
System
  anticipatory, 13
  causal, 12
  concept, 1
  continuous-time, 9
  discrete-time, 9
  discretized, 10
  distortionless, 240
  fixed, 10
  general, 1
  linear, 4, 15
  linearization, 4, 14
  matrix, 84
  noncausal, 12, 13, 201
  nonlinear, 16, 395
  quantized, 10
  recursive, 344
  shift-invariant, 12
  stationary, 10
  time-invariant, 10, 11

Tachometer feedback, 315
Tapped delay line, 444, 472
Temperature, 305
Time-invariant system, 10, 11
Time shifting, 171
Total excitation function, 281
Transfer function, 124, 140, 186, 280
  closed-loop, 310, 418
  discrete, 221, 407, 421
Transformer, 298
  ideal, 494, 496
Transient solution, 19, 26
Transmission line
  distortionless, 305
  with loss, 304, 305
  lossless, 239, 301
Trapezoidal rule, 240, 277, 427
Traveling waves, 302
Triangular waveshape, 138
Triangular window, 373
Trigonometric series, 122, 128
Tustin substitution method, 450
Twin Tee (notch) filter, 164

Undamped natural frequency, 58
Underdamped system, 60, 272
Undersampling, 154
Undetermined coefficients, 263
Unit
  circle, 362, 402
  function, 222, 336, 409
  function response, 175, 333, 337, 343–344, 346, 421
  impulse function, 26, 175, 293
  impulse response, 30, 32, 70, 87, 163, 307, 481
  impulse train, 164
  ramp response, 77
  sequence response, 354
  step function, 29–30, 176, 252
  step response, 56, 163
  step sequence, 336–337, 409

Variation of parameters (feedback), 313
Vector, 4, 63, 85, 476

Walsh functions, 131
Wave equation, 235, 239
Weight, 28
Weighting factor, 211
Window sequence, 373
    Hamming, 374
    rectangular, 373
    triangular, 373

z plane, 363, 402
z transform, 342, 400
    bilateral, 402, 498
    inverse, 403
    pairs, table of (unilateral), 411
    properties, 404–408
    unilateral, 402

Zero-input response, 8, 19, 31, 70, 104, 280, 307, 333, 346, 378
Zero memory system, 2
Zero-order hold (ZOH), 79, 333, 334
Zero-state response, 8, 19, 30, 31, 46, 47, 56, 57, 91, 104, 125, 141, 172, 173, 186, 280, 333, 343
    causal system (continuous-time), 36, 47
    causal system (discrete-time), 338
    general case (continuous-time), 31, 46
    general case (discrete-time), 338, 378
    shift-invariant system, 339
    step response, 56
    time-invariant system, 46
Zero vector, 89